Hybrid Materials

Edited by
Guido Kickelbick

1807–2007 Knowledge for Generations

Each generation has its unique needs and aspirations. When Charles Wiley first opened his small printing shop in lower Manhattan in 1807, it was a generation of boundless potential searching for an identity. And we were there, helping to define a new American literary tradition. Over half a century later, in the midst of the Second Industrial Revolution, it was a generation focused on building the future. Once again, we were there, supplying the critical scientific, technical, and engineering knowledge that helped frame the world. Throughout the 20th Century, and into the new millennium, nations began to reach out beyond their own borders and a new international community was born. Wiley was there, expanding its operations around the world to enable a global exchange of ideas, opinions, and know-how.

For 200 years, Wiley has been an integral part of each generation's journey, enabling the flow of information and understanding necessary to meet their needs and fulfill their aspirations. Today, bold new technologies are changing the way we live and learn. Wiley will be there, providing you the must-have knowledge you need to imagine new worlds, new possibilities, and new opportunities.

Generations come and go, but you can always count on Wiley to provide you the knowledge you need, when and where you need it!

William J. Pesce
President and Chief Executive Officer

Peter Booth Wiley
Chairman of the Board

Hybrid Materials

Synthesis, Characterization, and Applications

Edited by
Guido Kickelbick

WILEY-VCH Verlag GmbH & Co. KGaA

The Editor

Prof. Dr. Guido Kickelbick
Technische Universität Wien
Institut für Materialchemie
Getreidemarkt 9/165
1060 Wien
Austria

■ All books published by Wiley-VCH are carefully produced. Nevertheless, authors, editors, and publisher do not warrant the information contained in these books, including this book, to be free of errors. Readers are advised to keep in mind that statements, data, illustrations, procedural details or other items may inadvertently be inaccurate.

Library of Congress Card No.:
applied for

British Library Cataloguing-in-Publication Data
A catalogue record for this book is available from the British Library.

Bibliographic information published by the Deutsche Nationalbibliothek
The Deutsche Nationalbibliothek lists this publication in the Deutsche Nationalbibliografie; detailed bibliographic data are available in the Internet at http://dnb.d-nb.de.

© 2007 WILEY-VCH Verlag GmbH & Co. KGaA, Weinheim

All rights reserved (including those of translation into other languages). No part of this book may be reproduced in any form – by photoprinting, microfilm, or any other means – nor transmitted or translated into a machine language without written permission from the publishers. Registered names, trademarks, etc. used in this book, even when not specifically marked as such, are not to be considered unprotected by law.

Composition SNP Best-set Typesetter Ltd., Hong Kong

Printing Betz-Druck GmbH, Darmstadt

Bookbinding Litges & Dopf Buchbinderei GmbH, Heppenheim

Printed in the Federal Republic of Germany
Printed on acid-free paper

ISBN 978-3-527-31299-3

Contents

1 Introduction to Hybrid Materials *1*
Guido Kickelbick
1.1 Introduction *1*
1.1.1 Natural Origins *1*
1.1.2 The Development of Hybrid Materials *2*
1.1.3 Definition: Hybrid Materials and Nanocomposites *3*
1.1.4 Advantages of Combining Inorganic and Organic Species in One Material *7*
1.1.5 Interface-determined Materials *10*
1.1.6 The Role of the Interaction Mechanisms *11*
1.2 Synthetic Strategies towards Hybrid Materials *12*
1.2.1 In situ Formation of Inorganic Materials *13*
1.2.1.1 Sol–Gel Process *14*
1.2.1.2 Nonhydrolytic Sol–Gel Process *16*
1.2.1.3 Sol–Gel Reactions of Non-Silicates *16*
1.2.1.4 Hybrid Materials by the Sol–Gel Process *17*
1.2.1.5 Hybrid Materials Derived by Combining the Sol–Gel Approach and Organic Polymers *19*
1.2.2 Formation of Organic Polymers in Presence of Preformed Inorganic Materials *20*
1.2.3 Hybrid Materials by Simultaneous Formation of Both Components *22*
1.2.4 Building Block Approach *23*
1.2.4.1 Inorganic Building Blocks *24*
1.2.4.2 Organic Building Blocks *32*
1.3 Structural Engineering *35*
1.4 Properties and Applications *39*
1.5 Characterization of Materials *41*
1.6 Summary *46*

2 Nanocomposites of Polymers and Inorganic Particles *49*
Walter Caseri
2.1 Introduction *49*
2.2 Consequences of Very Small Particle Sizes *53*

2.3 Historical Reports on Inorganic Nanoparticles and Polymer Nanocomposites *63*
2.4 Preparation of Polymer Nanocomposites *65*
2.4.1 Mixing of Dispersed Particles with Polymers in Liquids *67*
2.4.2 Mixing of Particles with Monomers Followed by Polymerization *71*
2.4.3 Nanocomposite Formation by means of Molten or Solid Polymers *73*
2.4.4 Concomitant Formation of Particles and Polymers *74*
2.5 Properties and Applications of Polymer Nanocomposites *75*
2.5.1 Properties *75*
2.5.2 Applications *78*
2.5.2.1 Catalysts *78*
2.5.2.2 Gas Sensors *79*
2.5.2.3 Materials with Improved Flame Retardance *80*
2.5.2.4 Optical Filters *80*
2.5.2.5 Dichroic Materials *81*
2.5.2.6 High and Low Refractive Index Materials *81*
2.6 Summary *83*

3 Hybrid Organic/Inorganic Particles *87*
Elodie Bourgeat-Lami
3.1 Introduction *87*
3.2 Methods for creating Particles *92*
3.2.1 Polymer Particles *92*
3.2.1.1 Oil-in-water Suspension Polymerization *92*
3.2.1.2 Precipitation and Dispersion Polymerizations *93*
3.2.1.3 Oil-in-water Emulsion Polymerization *94*
3.2.1.4 Oil-in-water Miniemulsion Polymerization *95*
3.2.1.5 Oil-in-water Microemulsion Polymerization *95*
3.2.2 Vesicles, Assemblies and Dendrimers *95*
3.2.2.1 Vesicles *95*
3.2.2.2 Block Copolymer Assemblies *96*
3.2.2.3 Dendrimers *97*
3.2.3 Inorganic Particles *98*
3.2.3.1 Metal Oxide Particles *98*
3.2.3.2 Metallic Particles *99*
3.2.3.3 Semiconductor Nanoparticles *101*
3.2.3.4 Synthesis in Microemulsion *102*
3.3 Hybrid Nanoparticles Obtained Through Self-assembly Techniques *103*
3.3.1 Electrostatically Driven Self-assembly *103*
3.3.1.1 Heterocoagulation *103*
3.3.1.2 Layer-by-layer Assembly *107*
3.3.2 Molecular Recognition Assembly *109*
3.4 O/I Nanoparticles Obtained by in situ Polymerization Techniques *111*

3.4.1   Polymerizations Performed in the Presence of Preformed Mineral Particles   *111*
3.4.1.1   Surface Modification of Inorganic Particles   *112*
3.4.1.2   Polymerizations in Multiphase Systems   *113*
3.4.1.3   Surface-initiated Polymerizations   *124*
3.4.2   In situ Formation of Minerals in the Presence of Polymer Colloids   *130*
3.4.2.1   Polymer Particles Templating   *130*
3.4.2.2   Block Copolymers, Dendrimers and Microgels Templating   *134*
3.5   Hybrid Particles Obtained by Simultaneously Reacting Organic Monomers and Mineral Precursors   *137*
3.5.1   Poly(organosiloxane/vinylic) Copolymer Hybrids   *137*
3.5.2   Polyorganosiloxane Colloids   *140*
3.6   Conclusion   *142*

4   **Intercalation Compounds and Clay Nanocomposites**   *151*
Jin Zhu and Charles A. Wilkie
4.1   Introduction   *151*
4.2   Polymer Lamellar Material Nanocomposites   *153*
4.2.1   Types of Lamellar Nano-additives   *153*
4.2.2   Montmorillonite Layer Structure   *154*
4.2.3   Modification of Clay   *154*
4.3   Nanostructures and Characterization   *156*
4.3.1   X-ray Diffraction and Transmission Electron Microscopy to Probe Morphology   *156*
4.3.2   Other Techniques to Probe Morphology   *158*
4.4   Preparation of Polymer-clay Nanocomposites   *160*
4.4.1   Solution Mixing   *161*
4.4.2   Polymerization   *161*
4.4.3   Melt Compounding   *163*
4.5   Polymer-graphite and Polymer Layered Double Hydroxide Nanocomposites   *164*
4.5.1   Nanocomposites Based on Layered Double Hydroxides and Salts   *166*
4.6   Properties of Polymer Nanocomposites   *167*
4.7   Potential Applications   *168*
4.8   Conclusion and Prospects for the Future   *169*

5   **Porous Hybrid Materials**   *175*
Nicola Hüsing
5.1   General Introduction and Historical Development   *175*
5.1.1   Definition of Terms   *177*
5.1.2   Porous (Hybrid) Matrices   *179*
5.1.2.1   Microporous Materials: Zeolites   *180*
5.1.2.2   Mesoporous Materials: M41S and FSM Materials   *182*
5.1.2.3   Metal–Organic Frameworks (MOFs)   *184*

5.2   General Routes towards Hybrid Materials *185*
5.2.1 Post-synthesis Modification of the Final Dried Porous Product by Gaseous, Liquid or Dissolved Organic or Organometallic Species *185*
5.2.2 Liquid-phase Modification in the Wet Nanocomposite Stage or – for Mesostructured Materials and Zeolites – Prior to Removal of the Template *187*
5.2.3 Addition of Molecular, but Nonreactive Compounds to the Precursor Solution *188*
5.2.4 Co-condensation Reactions by the use of Organically-substituted Co-precursors *188*
5.2.5 The Organic Entity as an Integral Part of the Porous Framework *190*
5.3   Classification of Porous Hybrid Materials by the Type of Interaction *192*
5.3.1 Incorporation of Organic Functions Without Covalent Attachment to the Porous Host *192*
5.3.1.1 Doping with Small Molecules *192*
5.3.1.2 Doping with Polymeric Species *196*
5.3.1.3 Incorporation of Biomolecules *199*
5.3.2 Incorporation of Organic Functions with Covalent Attachment to the Porous Host *201*
5.3.2.1 Grafting Reactions *201*
5.3.2.2 Co-condensation Reactions *203*
5.3.3 The Organic Function as an Integral Part of the Porous Network Structure *209*
5.3.3.1 ZOL and PMO: Zeolites with Organic Groups as Lattice and Periodically Mesostructured Organosilicas *209*
5.3.3.2 Metal–Organic Frameworks *213*
5.4   Applications and Properties of Porous Hybrid Materials *219*

6   Sol–Gel Processing of Hybrid Organic–Inorganic Materials Based on Polysilsesquioxanes *225*
         Douglas A. Loy
6.1   Introduction *225*
6.1.1 Definition of Terms *226*
6.2   Forming Polysilsesquioxanes *228*
6.2.1 Hydrolysis and Condensation Chemistry *228*
6.2.2 Alternative Polymerization Chemistries *234*
6.2.3 Characterizing Silsesquioxane Sol–Gels with NMR *235*
6.2.4 Cyclization in Polysilsesquioxanes *237*
6.3   Type I Structures: Polyhedral Oligosilsesquioxanes (POSS) *240*
6.3.1 Homogenously Functionalized POSS *240*
6.3.2 Stability of Siloxane Bonds in Silsesquioxanes *242*
6.4   Type II Structures: Amorphous Oligo- and Polysilsesquioxanes *243*
6.4.1 Gelation of Polysilsesquioxanes *243*
6.4.2 Effects of pH on Gelation *245*

6.4.3 Polysilsesquioxane Gels *246*
6.4.4 Polysilsesquioxane–Silica Copolymers *247*
6.5 Type III: Bridged Polysilsesquioxanes *248*
6.5.1 Molecular Bridges *248*
6.5.2 Macromolecule-bridged Polysilsesquioxanes *252*
6.6 Summary *252*
6.6.1 Properties of Polysilsesquioxanes *253*
6.6.2 Existing and Potential Applications *253*

7 Natural and Artificial Hybrid Biomaterials *255*
Heather A. Currie, Siddharth V. Patwardhan, Carole C. Perry, Paul Roach, Neil J. Shirtcliffe

7.1 Introduction *255*
7.2 Building Blocks *256*
7.2.1 Inorganic Building Blocks *256*
7.2.1.1 Nucleation and Growth *259*
7.2.2 Organic Building Blocks *262*
7.2.2.1 Proteins and DNA *262*
7.2.2.2 Carbohydrates *264*
7.2.2.3 Lipids *266*
7.2.2.4 Collagen *266*
7.3 Biomineralization *269*
7.3.1 Introduction *269*
7.3.1.1 Biomineral Types and Occurrence *269*
7.3.1.2 Functions of Biominerals *270*
7.3.1.3 Properties of Biominerals *270*
7.3.2 Control Strategies in Biomineralization *272*
7.3.3 The Role of the Organic Phase in Biomineralization *275*
7.3.4 Mineral or Precursor – Organic Phase Interactions *276*
7.3.5 Examples of Non-bonded Interactions in Bioinspired Silicification *279*
7.3.5.1 Effect of Electrostatic Interactions *279*
7.3.5.2 Effect of Hydrogen Bonding Interactions *279*
7.3.5.3 Effect of the Hydrophobic Effect *280*
7.3.6 Roles of the Organic Phase in Biomineralization *280*
7.4 Bioinspired Hybrid Materials *281*
7.4.1 Natural Hybrid Materials *283*
7.4.1.1 Bone *283*
7.4.1.2 Dentin *285*
7.4.1.3 Nacre *287*
7.4.1.4 Wood *287*
7.4.2 Artificial Hybrid Biomaterials *289*
7.4.2.1 Ancient materials *289*
7.4.2.2 Structural Materials *290*
7.4.2.3 Non-structural Materials *290*
7.4.3 Construction of Artificial Hybrid Biomaterials *291*

7.4.3.1 Organic Templates to Dictate Shape and Form *291*
7.4.3.2 Integrated Nanoparticle–Biomolecule Hybrid Systems *292*
7.4.3.3 Routes to Bio-nano Hybrid Systems *292*
7.5 Responses *294*
7.5.1 Biological Performance *294*
7.5.2 Protein Adsorption *295*
7.5.3 Cell Adhesion *295*
7.5.4 Evaluation of Biomaterials *296*
7.6 Summary *298*

8 Medical Applications of Hybrid Materials *301*
Kanji Tsuru, Satoshi Hayakawa, and Akiyoshi Osaka
8.1 Introduction *301*
8.1.1 Composites, Solutions, and Hybrids *301*
8.1.2 Artificial Materials for Repairing Damaged Tissues and Organs *306*
8.1.3 Tissue–Material Interactions *310*
8.1.4 Material–Tissue Bonding; Bioactivity *313*
8.1.5 Blood-compatible Materials *318*
8.2 Bioactive Inorganic–Organic Hybrids *319*
8.2.1 Concepts of Designing Hybrids *319*
8.2.2 Concepts of Organic–Inorganic Hybrid Scaffolds and Membranes *321*
8.2.3 PDMS–Silica Hybrids *323*
8.2.4 Organoalkoxysilane Hybrids *324*
8.2.5 Gelatin–Silicate Hybrids *326*
8.2.6 Chitosan–Silicate Hybrids *327*
8.3 Surface Modifications for Biocompatible Materials *328*
8.3.1 Molecular Brush Structure Developed on Biocompatible Materials *328*
8.3.2 Alginic Acid Molecular Brush Layers on Metal Implants *329*
8.3.3 Organotitanium Molecular Layers with Blood Compatibility *330*
8.4 Porous Hybrids for Tissue Engineering Scaffolds and Bioreactors *331*
8.4.1 PDMS–Silica Porous Hybrids for Bioreactors *331*
8.4.2 Gelatin–Silicate Porous Hybrids *332*
8.4.3 Chitosan–Silicate Porous Hybrids for Scaffold Applications *333*
8.5 Chitosan-based Hybrids for Drug Delivery Systems *334*
8.6 Summary *335*

9 Hybrid Materials for Optical Applications *337*
Luís António Dias Carlos, R.A. Sá Ferreira and V. de Zea Bermudez
9.1 Introduction *337*
9.2 Synthesis Strategy for Optical Applications *339*
9.3 Hybrids for Coatings *343*
9.4 Hybrids for Light-emitting and Electro-optic Purposes *353*
9.4.1 Photoluminescence and Absorption *353*
9.4.2 Electroluminescence *359*
9.4.3 Quantifying Luminescence *365*

9.4.3.1	Color Coordinates, Hue, Dominant Wavelength and Purity	365
9.4.3.2	Emission Quantum Yield and Radiance	368
9.4.4	Recombination Mechanisms and Nature of the Emitting Centers	372
9.4.5	Lanthanide-doped Hybrids	374
9.4.6	Solid-state Dye-lasers	379
9.5	Hybrids for Photochromic and Photovoltaic Devices	381
9.6	Hybrids for Integrated and Nonlinear Optics	387
9.6.1	Planar Waveguides and Direct Writing	387
9.6.2	Nonlinear Optics	393
9.7	Summary	398

10	**Electronic and Electrochemical Applications of Hybrid Materials**	**401**
	Jason E. Ritchie	
10.1	Introduction	401
10.2	Historical Background	402
10.3	Fundamental Mechanisms of Conductivity in Hybrid Materials	403
10.3.1	Electrical Conductivity	403
10.3.2	Li^- Conductivity	407
10.3.3	H Conductivity	409
10.4	Explanation of the Different Materials	411
10.4.1	Sol–Gel Based Systems	411
10.4.2	Nanocomposites	412
10.4.3	Preparation of Electrochemically Active Films (and Chemically Modified Electrodes)	414
10.5	Special Analytical Techniques	415
10.5.1	Electrochemical Techniques	415
10.5.2	Pulsed Field Gradient NMR	418
10.6	Applications	419
10.6.1	Electrochemical Sensors	419
10.6.2	Optoelectronic Applications	421
10.6.3	H-conducting Electrolytes for Fuel Cell Applications	423
10.6.4	Li-conducting Electrolytes for Battery Applications	426
10.6.5	Other Ion Conducting Systems	429
10.7	Summary	430

11	**Inorganic/Organic Hybrid Coatings**	**433**
	Mark D. Soucek	
11.1	General Introduction to Commodity Organic Coatings	433
11.2	General Formation of Inorganic/Organic Hybrid Coatings	435
11.2.1	Acid and Base Catalysis within an Organic Matrix	436
11.2.2	Thermally Cured Inorganic/Organic Seed Oils Coatings	443
11.2.3	Drying Oil Auto-oxidation Mechanism	444
11.2.4	Metal Catalysts	445
11.3	Alkyds and Other Polyester Coatings	449
11.3.1	Inorganic/Organic Alkyd Coatings	450

11.4	Polyurethane and Polyurea Coatings	*451*
11.4.1	Polyurea Inorganic/Organic Hybrid Coatings	*452*
11.4.2	Polyurethane/Polysiloxane Inorganic/Organic Coating System	*455*
11.5	Radiation Curable Coatings	*459*
11.5.1	UV-curable Inorganic/Organic Hybrid Coatings	*461*
11.5.2	Models for Inorganic/Organic Hybrid Coatings	*465*
11.5.3	Film Morphology	*468*
11.6	Applications	*470*
11.7	Summary	*471*

Index *477*

Preface

Research and development of novel hybrid materials and nanocomposites with extraordinary properties has become one of the most expanding fields in materials chemistry in recent years. One reason for this trend is that this class of materials bridges various scientific disciplines and combines the best attributes of the different worlds in one system. Traditional materials, such as polymers or ceramics, can be combined with substances of a dissimilar type, such as biological molecules and diverse chemical functional groups to form novel functional materials using a building block approach. In a truly interdisciplinary manner, inorganic and organic chemistry, physical and biological sciences are united in the search for novel recipes to create unique materials. The compounds formed often possess exciting new properties for future functional materials and technological applications. The hype concerning nanotechnology in recent years has given an additional boost to this topic. Natural materials often act as a model for these systems and many examples for biomimetic approaches can be found in the development of hybrid materials. The requirements of future technologies act as a driving force for the research and development of these materials.

Many excellent reviews and books have already been published, that summarize research and developments in hybrid materials and nanocomposites. However there was still no broad and educational introductory text. This prompted my co-authors and myself to cooperate to fill this gap and you hold the outcome in your hands.

Our major goal was to give an introduction to the topic of hybrid materials and nanocomposites, written by experts in the field, but assuming that the reader does not have any previous experience of the topic. Therefore it provides graduate students, scientists entering the field, and also interested people from other branches of science, with the opportunity to learn about the basic synthetic and characterization approaches to this type of material and to get an overview of potential applications. Of course we could not cover all aspects of this broad topic but I would say that you find a good cross-section and we leave some space for the fantasy and creativity of the readers.

The book is divided into three sections: the first chapter is an introduction to the basic chemical principles and characterization of hybrid materials; the second chapter gives an overview of specific types of hybrid materials in several subchapters; and the third part shows some applications. The breadth of the topic means that not all topics can be covered; however the interested reader will find additional references at the end of each chapter. I would like to thank all contributors for their effort to write the chapters in a way that experts and non-experts alike are able to read them. I let the reader judge whether they succeeded. Enjoy reading this book and if you find mistakes or if you have suggestions for improvements please let me know.

Vienna, October 2006   Guido Kickelbick

List of Contributors

Elodie Bourgeat-Lami
UMR 140 CNRS-CPE
Laboratoire de Chimie et Procédés
  de Polymérisation
Bât. 308F
43, Bd. du 11 novembre 1918
69616 Villeurbanne
France

Luís António Dias Carlos
Departamento de Física and
  CICECO
Universidade de Aveiro
3810-193 Aveiro
Portugal

Walter Caseri
Department of Materials
Institute of Polymers
ETH Zürich
Wolfgang-Pauli Strasse 10
8093 Zürich
Switzerland

Heather A. Currie
Interdisciplinary Biomedical
  Research Centre
School of Biomedical and Natural
  Sciences
Nottingham Trent University
Clifton Lane
Nottingham, NG11 8NS
UK

V. de Zea Bermudez
Departamento de Química and
  CQ-VR
Universidade de Trás-os-Montes e
  Alto Douro
5001-911 Vila Real Codex
Portugal

Satoshi Hayakawa
Biomaterial Laboratory
Faculty of Engineering
Okayama University
Tsushima
Okayama, 700-8530
Japan

Nicola Hüsing
University of Ulm
Inorganic Chemistry I
89069 Ulm
Germany

Guido Kickelbick
Vienna University of Technology
Institute of Materials Chemistry
Getreidemarkt 9-165
1060 Wien
Austria

Hybrid Materials. Synthesis, Characterization, and Applications. Edited by Guido Kickelbick
Copyright © 2007 Wiley-VCH Verlag GmbH & Co. KGaA, Weinheim
ISBN: 978-3-527-31299-3

Douglas A. Loy
Polymers and Coatings Group
MS E549
Los Alamos National Laboratory
Los Alamos, NM 87545
USA
and Dept Materials Science & Engineering
University of Arizona
Tucson, AZ 85721
USA

Akiyoshi Osaka
Biomaterial Laboratory
Faculty of Engineering
Okayama University
Tsushima
Okayama, 700-8530
Japan

Siddharth V. Patwardhan
Interdisciplinary Biomedical Research Centre
School of Biomedical and Natural Sciences
Nottingham Trent University
Clifton Lane
Nottingham, NG11 8NS
UK

Carole C. Perry
Interdisciplinary Biomedical Research Centre
School of Biomedical and Natural Sciences
Nottingham Trent University
Clifton Lane
Nottingham, NG11 8NS
UK

Jason E. Ritchie
Department of Chemistry
The University of Mississippi
University, MS 38677
USA

Paul Roach
Interdisciplinary Biomedical Research Centre
School of Biomedical and Natural Sciences
Nottingham Trent University
Clifton Lane
Nottingham, NG11 8NS
UK

R.A. Sá Ferreira
Departamento de Física and CICECO
Universidade de Aveiro
3810-193 Aveiro
Portugal

Neil J. Shirtcliffe
Interdisciplinary Biomedical Research Centre
School of Biomedical and Natural Sciences
Nottingham Trent University
Clifton Lane
Nottingham, NG11 8NS
UK

Mark D. Soucek
Department of Polymer Engineering
The University of Akron
250 South Forge
Akron, OH 44325-0301
USA

Kanji Tsuru
Biomaterial Laboratory
Faculty of Engineering
Okayama University
Tsushima
Okayama, 700-8530
Japan

Charles A. Wilkie
Department of Chemistry
Marquette University
PO Box 1881
535 N 14th Street
Milwaukee, WI 53233
USA

Jin Zhu
OPTEM, Inc.
1030 West Smith Road
Medina, OH 44256
USA

1
Introduction to Hybrid Materials
Guido Kickelbick

1.1
Introduction

Recent technological breakthroughs and the desire for new functions generate an enormous demand for novel materials. Many of the well-established materials, such as metals, ceramics or plastics cannot fulfill all technological desires for the various new applications. Scientists and engineers realized early on that mixtures of materials can show superior properties compared with their pure counterparts. One of the most successful examples is the group of composites which are formed by the incorporation of a basic structural material into a second substance, the *matrix*. Usually the systems incorporated are in the form of particles, whiskers, fibers, lamellae, or a mesh. Most of the resulting materials show improved mechanical properties and a well-known example is inorganic fiber-reinforced polymers. Nowadays they are regularly used for lightweight materials with advanced mechanical properties, for example in the construction of vehicles of all types or sports equipment. The structural building blocks in these materials which are incorporated into the matrix are predominantly inorganic in nature and show a size range from the lower micrometer to the millimeter range and therefore their heterogeneous composition is quite often visible to the eye. Soon it became evident that decreasing the size of the inorganic units to the same level as the organic building blocks could lead to more homogeneous materials that allow a further fine tuning of materials' properties on the molecular and nanoscale level, generating novel materials that either show characteristics in between the two original phases or even new properties. Both classes of materials reveal similarities and differences and an attempt to define the two classes will follow below. However, we should first realize that the origin of hybrid materials did not take place in a chemical laboratory but in nature.

1.1.1
Natural Origins

Many natural materials consist of inorganic and organic building blocks distributed on the (macro)molecular or nanoscale. In most cases the inorganic part

Hybrid Materials. Synthesis, Characterization, and Applications. Edited by Guido Kickelbick
Copyright © 2007 Wiley-VCH Verlag GmbH & Co. KGaA, Weinheim
ISBN: 978-3-527-31299-3

provides mechanical strength and an overall structure to the natural objects while the organic part delivers bonding between the inorganic building blocks and/or the soft tissue. Typical examples of such materials are bone, or nacre.

The concepts of bonding and structure in such materials are intensively studied by many scientists to understand the fundamental processes of their formation and to transfer the ideas to artificial materials in a so-called biomimetic approach. The special circumstances under which biological hybrid inorganic–organic materials are formed, such as ambient temperatures, an aqueous environment, a neutral pH and the fascinating plethora of complex geometries produced under these conditions make the mimicking of such structures an ultimate goal for scientists. In particular the study of biomineralization and its shape control is an important target of many scientific studies. This primarily interface-controlled process still reveals many questions, in particular how such a remarkable level of morphological diversity with a multiplicity of functions can be produced by so few building blocks. In addition to questions concerning the composition of the materials, their unique structures motivate enquiry to get a deeper insight in their formation, often not only because of their beauty but also because of the various functions the structures perform. A complex hierarchical order of construction from the nanometer to the millimeter level is regularly found in nature, where every size level of the specific material has its function which benefits the whole performance of the material. Furthermore these different levels of complexity are reached by soft chemical self-assembly mechanisms over a large dimension, which is one of the major challenges of modern materials chemistry.

Chapter 7 describes the fundamental principles of biomineralization and hybrid inorganic–organic biomaterials and many applications to medical problems are shown in Chapter 8.

1.1.2
The Development of Hybrid Materials

Although we do not know the original birth of hybrid materials exactly it is clear that the mixing of organic and inorganic components was carried out in ancient world. At that time the production of bright and colorful paints was the driving force to consistently try novel mixtures of dyes or inorganic pigments and other inorganic and organic components to form paints that were used thousands of years ago. Therefore, hybrid materials or even nanotechnology is not an invention of the last decade but was developed a long time ago. However, it was only at the end of the 20th and the beginning of the 21st century that it was realized by scientists, in particular because of the availability of novel physico–chemical characterization methods, the field of nanoscience opened many perspectives for approaches to new materials. The combination of different analytical techniques gives rise to novel insights into hybrid materials and makes it clear that bottom-up strategies from the molecular level towards materials' design will lead to novel properties in this class of materials.

Apart from the use of inorganic materials as fillers for organic polymers, such as rubber, it was a long time before much scientific activity was devoted to mixtures of inorganic and organic materials. One process changed this situation: the sol–gel process. This process, which will be discussed in more detail later on, was developed in the 1930s using silicon alkoxides as precursors from which silica was produced. In fact this process is similar to an organic polymerization starting from molecular precursors resulting in a bulk material. Contrary to many other procedures used in the production of inorganic materials this is one of the first processes where ambient conditions were applied to produce ceramics. The control over the preparation of multicomponent systems by a mild reaction method also led to industrial interest in that process. In particular the silicon based sol–gel process was one of the major driving forces what has become the broad field of inorganic–organic hybrid materials. The reason for the special role of silicon was its good processability and the stability of the Si—C bond during the formation of a silica network which allowed the production of organic-modified inorganic networks in one step.

Inorganic–organic hybrids can be applied in many branches of materials chemistry because they are simple to process and are amenable to design on the molecular scale. Currently there are four major topics in the synthesis of inorganic–organic materials: (a) their molecular engineering, (b) their nanometer and micrometer-sized organization, (c) the transition from functional to multifunctional hybrids, and (d) their combination with bioactive components.

Some similarities to sol–gel chemistry are shown by the stable metal sols and colloids, such as gold colloids, developed hundreds of years ago. In fact sols prepared by the sol–gel process, i.e. the state of matter before gelation, and the gold colloids have in common that their building blocks are nanosized particles surrounded by a (solvent) matrix. Such metal colloids have been used for optical applications in nanocomposites for centuries. Glass, for example, was already colored with such colloids centuries ago. In particular many reports of the scientific examination of gold colloids, often prepared by reduction of gold salts, are known from the end of the 18th century. Probably the first nanocomposites were produced in the middle of the 19th century when gold salts were reduced in the presence of gum arabic. Currently many of the colloidal systems already known are being reinvestigated by modern instrumental techniques to get new insights into the origin of the specific chemistry and physics behind these materials.

1.1.3
Definition: Hybrid Materials and Nanocomposites

The term hybrid material is used for many different systems spanning a wide area of different materials, such as crystalline highly ordered coordination polymers, amorphous sol–gel compounds, materials with and without interactions between the inorganic and organic units. Before the discussion of synthesis and properties of such materials we try to delimit this broadly-used term by taking into account various concepts of composition and structure (Table 1.1). The most wide-ranging

Table 1.1 Different possibilities of composition and structure of hybrid materials.

Matrix:	crystalline ↔ amorphous
	organic ↔ inorganic
Building blocks:	molecules ↔ macromolecules ↔ particles ↔ fibers
Interactions between components:	strong ↔ weak

definition is the following: a hybrid material is a material that includes two moieties blended on the molecular scale. Commonly one of these compounds is inorganic and the other one organic in nature. A more detailed definition distinguishes between the possible interactions connecting the inorganic and organic species. *Class I* hybrid materials are those that show weak interactions between the two phases, such as van der Waals, hydrogen bonding or weak electrostatic interactions. *Class II* hybrid materials are those that show strong chemical interations between the components. Because of the gradual change in the strength of chemical interactions it becomes clear that there is a steady transition between weak and strong interactions (Fig. 1.1). For example there are

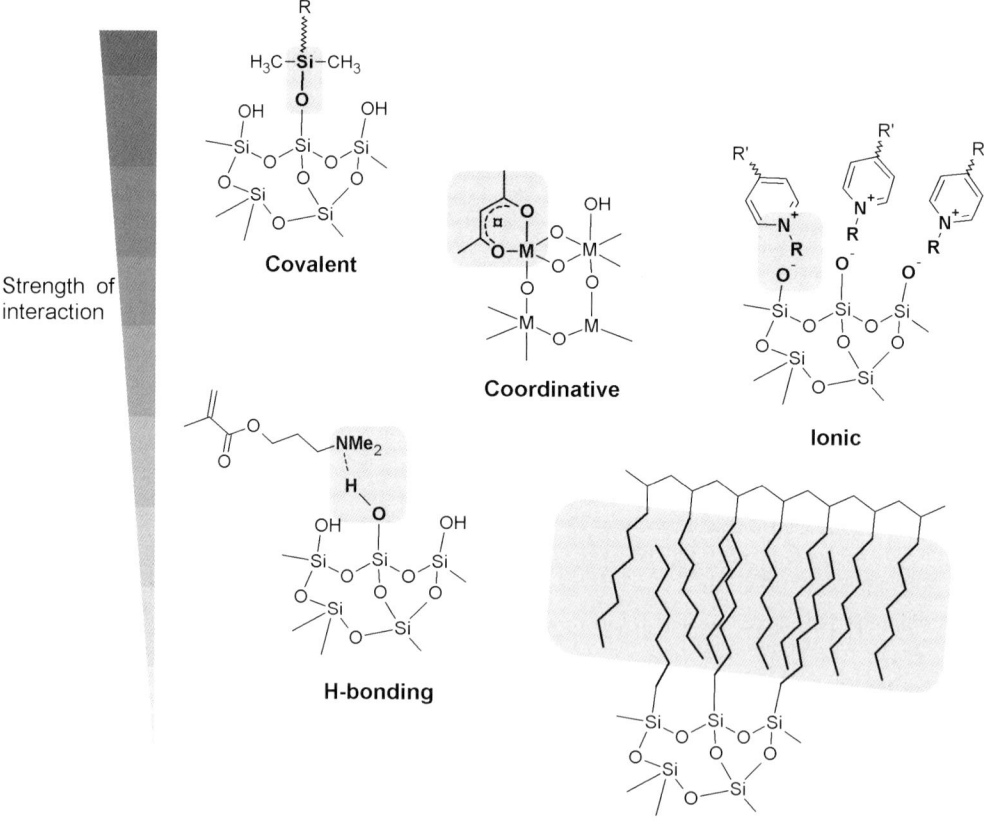

Fig. 1.1 Selected interactions typically applied in hybrid materials and their relative strength.

Table 1.2 Different chemical interactions and their respective strength.

Type of interaction	Strength [kJ mol^{-1}]	Range	Character
van der Waals	ca. 50	Short	nonselective, nondirectional
H-bonding	5–65	Short	selective, directional
Coordination bonding	50–200	Short	directional
Ionic	50–250[a]	Long	nonselective
Covalent	350	Short	predominantly irreversible

a Depending on solvent and ion solution; data are for organic media.

hydrogen bonds that are definitely stronger than for example weak coordinative bonds. Table 1.2 presents the energetic categorization of different chemical interactions depending on their binding energies.

In addition to the bonding characteristics structural properties can also be used to distinguish between various hybrid materials. An organic moiety containing a functional group that allows the attachment to an inorganic network, e.g. a trialkoxysilane group, can act as a network modifying compound because in the final structure the inorganic network is only modified by the organic group. Phenyltrialkoxysilanes are an example for such compounds; they modify the silica network in the sol–gel process via the reaction of the trialkoxysilane group (Scheme 1.1a) without supplying additional functional groups intended to undergo further chemical reactions to the material formed. If a reactive functional group is incorporated the system is called a network functionalizer (Scheme 1.1c). The situation is different if two or three of such anchor groups modify an organic segment; this leads to materials in which the inorganic group is afterwards an integral part of the hybrid network (Scheme 1.1b). The latter systems are described in more detail in Chapter 6.

Blends are formed if no strong chemical interactions exist between the inorganic and organic building blocks. One example for such a material is the combination of inorganic clusters or particles with organic polymers lacking a strong (e.g. covalent) interaction between the components (Scheme 1.2a). In this case a material is formed that consists for example of an organic polymer with entrapped discrete inorganic moieties in which, depending on the functionalities of the components, for example weak crosslinking occurs by the entrapped inorganic units through physical interactions or the inorganic components are entrapped in a crosslinked polymer matrix. If an inorganic and an organic network interpenetrate each other without strong chemical interactions, so called interpenetrating networks (IPNs) are formed (Scheme 1.2b), which is for example the case if a sol–gel material is formed in presence of an organic polymer or vice versa. Both materials described belong to class I hybrids. Class II hybrids are formed when the discrete inorganic building blocks, e.g. clusters, are covalently bonded to the

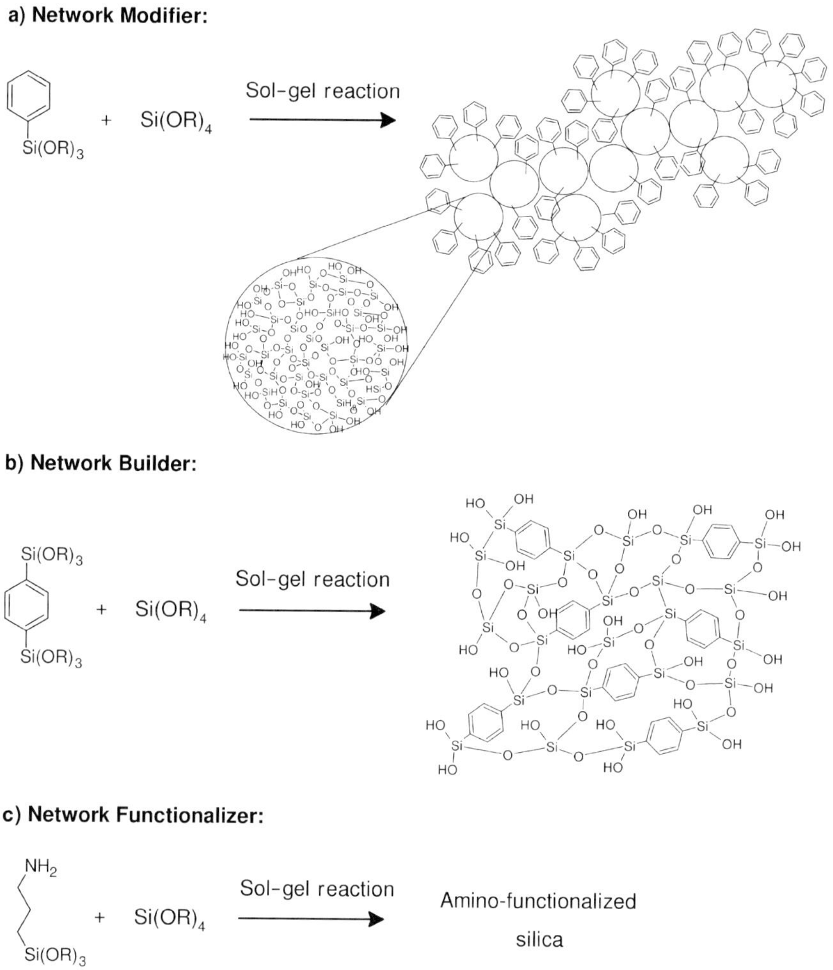
a) Network Modifier:
Si(OR)3
+ Si(OR)4
Sol-gel reaction
b) Network Builder:
Si(OR)3
Si(OR)3
+ Si(OR)4
Sol-gel reaction
c) Network Functionalizer:
NH2
Si(OR)3
+ Si(OR)4
Sol-gel reaction
Amino-functionalized
silica

Scheme 1.1 Role of organically functionalized trialkoxysilanes in the silicon-based sol–gel process.

organic polymers (Scheme 1.2c) or inorganic and organic polymers are covalently connected with each other (Scheme 1.2d).

Nanocomposites After having discussed the above examples one question arises: what is the difference between inorganic–organic hybrid materials and inorganic–organic nanocomposites? In fact there is no clear borderline between these materials. The term nanocomposite is used if one of the structural units, either the organic or the inorganic, is in a defined size range of 1–100 nm. Therefore there is a gradual transition between hybrid materials and nanocomposites,

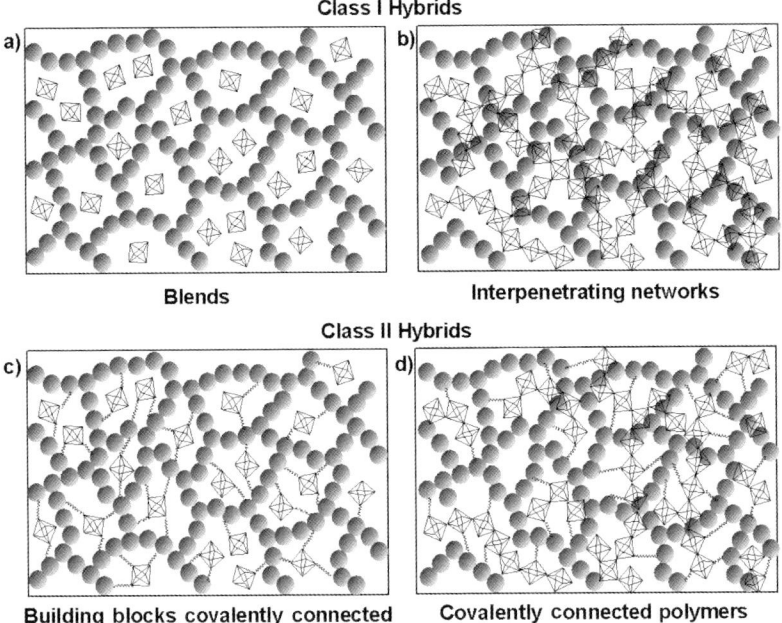

Scheme 1.2 The different types of hybrid materials.

because large molecular building blocks for hybrid materials, such as large inorganic clusters, can already be of the nanometer length scale. Commonly the term nanocomposites is used if discrete structural units in the respective size regime are used and the term hybrid materials is more often used if the inorganic units are formed *in situ* by molecular precursors, for example applying sol–gel reactions. Examples of discrete inorganic units for nanocomposites are nanoparticles, nanorods, carbon nanotubes and galleries of clay minerals (Fig. 1.2). Usually a nanocomposite is formed from these building blocks by their incorporation in organic polymers. Nanocomposites of nanoparticles are discussed in more detail in Chapter 2 and those incorporating clay minerals in Chapter 4.

1.1.4
Advantages of Combining Inorganic and Organic Species in One Material

The most obvious advantage of inorganic–organic hybrids is that they can favorably combine the often dissimilar properties of organic and inorganic components in one material (Table 1.3). Because of the many possible combinations of components this field is very creative, since it provides the opportunity to invent an almost unlimited set of new materials with a large spectrum of known and as yet unknown properties. Another driving force in the area of hybrid materials is the possibility to create multifunctional materials. Examples are the incorporation of

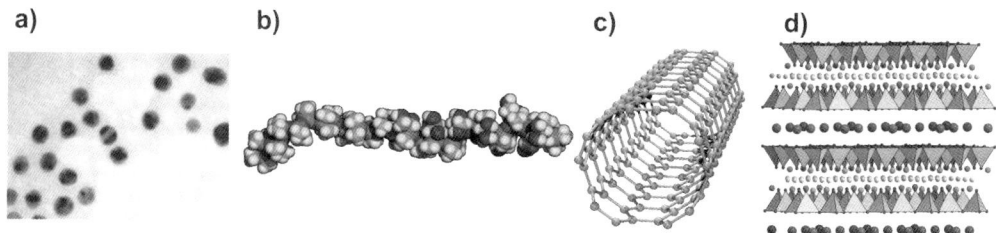

Fig. 1.2 Inorganic building blocks used for embedment in an organic matrix in the preparation of inorganic-organic nanocomposites: a) nanoparticles, b) macromolecules, c) nanotubes, d) layered materials.

Table 1.3 Comparison of general properties of typical inorganic and organic materials.

Properties	Organics (polymers)	Inorganics (SiO$_2$, transition metal oxides (TMO))
Nature of bonds	covalent [C—C], van der Waals, H-bonding	ionic or iono-covalent [M—O]
T_g	low (−120 °C to 200 °C)	high (≫200 °C)
Thermal stability	low (<350 °C–450 °C)	high (≫100 °C)
Density	0.9–1.2	2.0–4.0
Refractive index	1.2–1.6	1.15–2.7
Mechanical properties	elasticity plasticity rubbery (depending on T_g)	hardness strength fragility
Hydrophobicity	hydrophilic	hydrophilic
Permeability	hydrophobic ±permeable to gases	low permeability to gases
Electronic properties	insulating to conductive redox properties	insulating to semiconductors (SiO$_2$, TMO) redox properties (TMO) magnetic properties
Processability	high (molding, casting, film formation, control of viscosity)	low for powders high for sol–gel coatings

inorganic clusters or nanoparticles with specific optical, electronic or magnetic properties in organic polymer matrices. These possibilities clearly reveal the power of hybrid materials to generate complex systems from simpler building blocks in a kind of LEGO © approach.

Probably the most intriguing property of hybrid materials that makes this material class interesting for many applications is their processing. Contrary to pure solid state inorganic materials that often require a high temperature treatment for their processing, hybrid materials show a more polymer-like handling, either because of their large organic content or because of the formation of crosslinked inorganic networks from small molecular precursors just like in polymerization reactions. Hence, these materials can be shaped in any form in bulk and in films. Although from an economical point of view bulk hybrid materials can currently only compete in very special areas with classical inorganic or organic materials, e.g. in the biomaterials sector, the possibility of their processing as thin films can lead to property improvements of cheaper materials by a simple surface treatment, e.g. scratch resistant coatings.

Based on the molecular or nanoscale dimensions of the building blocks, light scattering in homogeneous hybrid material can be avoided and therefore the optical transparency of the resulting hybrid materials and nanocomposites is, dependent on the composition used, relatively high. This makes these materials ideal candidates for many optical applications (Chapter 9). Furthermore, the materials' building blocks can also deliver an internal structure to the material which can be regularly ordered. While in most cases phase separation is avoided, phase separation of organic and inorganic components is used for the formation of porous materials, as described in Chapter 5.

Material properties of hybrid materials are usually changed by modifications of the composition on the molecular scale. If, for example, more hydrophobicity of a material is desired, the amount of hydrophobic molecular components is increased. In sol–gel materials this is usually achieved if alkyl- or aryl-substituted trialkoxysilanes are introduced in the formulation. Hydrophobic and lipophobic materials are composed if partially or fully fluorinated molecules are included. Mechanical properties, such as toughness or scratch resistance, are tailored if hard inorganic nanoparticles are included into the polymer matrix. Because the compositional variations are carried out on the molecular scale a gradual fine tuning of the material properties is possible.

One important subject in materials chemistry is the formation of smart materials, such as materials that react to environmental changes or switchable systems, because they open routes to novel technologies, for example electroactive materials, electrochromic materials, sensors and membranes, biohybrid materials, etc. The desired function can be delivered from the organic or inorganic or from both components. One of the advantages of hybrid materials in this context is that functional organic molecules as well as biomolecules often show better stability and performance if introduced in an inorganic matrix.

1.1.5
Interface-determined Materials

The transition from the macroscopic world to microscopic, nanoscopic and molecular objects leads, beside the change of physical properties of the material itself, i.e. the so called quantum size effects, to the change of the surface area of the objects. While in macroscopic materials the majority of the atoms is hidden in the bulk of the material it becomes vice versa in very small objects. This is demonstrated by a simple mind game (Fig. 1.3). If one thinks of a cube of atoms in tight packing of $16 \times 16 \times 16$ atoms. This cube contains an overall number of 4096 atoms from which 1352 are located on the surface (~33% surface atoms); if this cube is divided into eight equal $8 \times 8 \times 8$ cubes the overall number is the same but 2368 atoms are now located on the surface (~58% surface atoms); repeating this procedure we get 3584 surface atoms (~88% surface atoms). This example shows how important the surface becomes when objects become very small. In small nanoparticles (<10 nm) nearly every atom is a surface atom that can interact with the environment. One predominant feature of hybrid materials or nanocomposites is their inner interface, which has a direct impact on the properties of the different building blocks and therefore on the materials' properties. As already explained in Section 1.1.3, the nature of the interface has been used to divide the materials in two classes dependent on the strength of interaction between the moieties. If the two phases have opposite properties, such as different polarity, the system would thermodynamically phase separate. The same can happen on the molecular or nanometer level, leading to microphase separation. Usually, such a system would thermodynamically equilibrate over time. However in many cases in hybrid materials the system is kinetically stabilized by network-forming reactions such as the sol–gel process leading to a spatial fixation of the structure. The materials formed can be macroscopically homogeneous and optically clear, because the phase segregation is of small length scale and therefore limited interaction with visible light occurs. However, the composition on the molecular or nanometer length scale can be heterogeneous. If the phase segregation

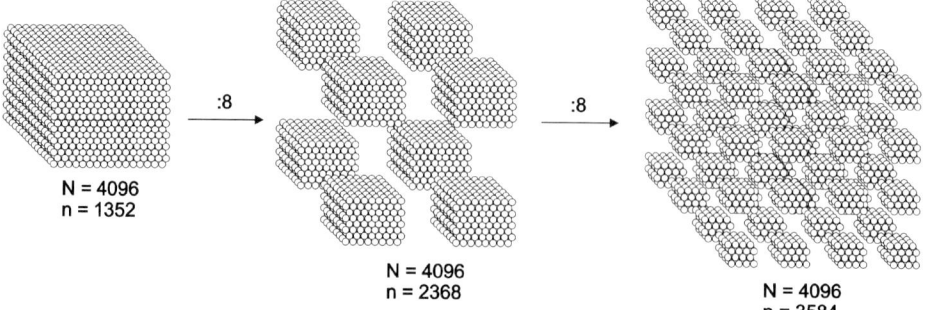

Fig. 1.3 Surface statistical consequences of dividing a cube with $16 \times 16 \times 16$ atoms. N = total atoms, n = surface atoms.

reaches the several hundred nanometer length scale or the refractive index of the formed domains is very different, materials often turn opaque. Effects like this are avoided if the reaction parameters are controlled in such a way that the speed of network formation is kept faster than the phase separation reactions.

The high surface area of nanobuilding blocks can lead to additional effects; for example if surface atoms strongly interact with molecules of the matrix by chemical bonding, reactions like surface reorganization, electron transfer, etc. can occur which can have a large influence on the physical properties of the nano-building blocks and thus the overall performance of the material formed. It is, for example, known that conjugated π-electron systems coordinated to the surface of titania nanoparticles can lead to charge transfer reactions that influence the color, and therefore the surface electronic properties of the particles.

Nanosized objects, such as inorganic nanoparticles, in addition show a very high surface energy. Usually if the surface energy is not reduced by surface active agents (e.g. surfactants), such particles tend to agglomerate in an organic medium. Thus, the physical properties of the nanoparticles (e.g. quantum size effects) diminish, and/or the resulting materials are no longer homogeneous. Both facts have effects on the final material properties. For example the desired optical properties of nanocomposites fade away, or mechanical properties are weakened. However, sometimes a controlled aggregation can also be required, e.g. percolation of conducting particles in a polymer matrix increases the overall conductivity of the material (see Chapter 10).

1.1.6
The Role of the Interaction Mechanisms

In Section 1.1.3 the interaction mechanism between the organic and inorganic species was used to categorize the different types of hybrid materials, furthermore of course the interaction also has an impact on the material properties. Weak chemical interactions between the inorganic and organic entities leave some potential for dynamic phenomena in the final materials, meaning that over longer periods of time changes in the material, such as aggregation, phase separation or leaching of one of the components, can occur. These phenomena can be avoided if strong interactions are employed such as covalent bonds, as in nanoparticle-crosslinked polymers. Depending on the desired materials' properties the interactions can be gradually tuned. Weak interactions are, for example, preferred where a mobility of one component in the other is required for the target properties. This is for example the case for ion conducting polymers, where the inorganic ion (often Li^+) has to migrate through the polymer matrix.

In many examples the interactions between the inorganic and organic species are maximized by applying covalent attachment of one to the other species. But there are also cases where small changes in the composition, which on the first sight should not result in large effects, can make considerable differences. It was, for example, shown that interpenetrating networks between polystyrene and sol–gel materials modified with phenyl groups show less microphase segregation than sol–gel materials with pure alkyl groups, which was interpreted to be an effect of π-π-interactions between the two materials.

In addition the interaction of the two components can have an influence on other properties, such as electronic properties if coordination complexes are formed or electron transfer processes are enabled by the interaction.

1.2
Synthetic Strategies towards Hybrid Materials

In principle two different approaches can be used for the formation of hybrid materials: Either well-defined preformed building blocks are applied that react with each other to form the final hybrid material in which the precursors still at least partially keep their original integrity or one or both structural units are formed from the precursors that are transformed into a novel (network) structure.

Both methodologies have their advantages and disadvantages and will be described here in more detail.

Building block approach As mentioned above building blocks at least partially keep their molecular integrity throughout the material formation, which means that structural units that are present in these sources for materials formation can also be found in the final material. At the same time typical properties of these building blocks usually survive the matrix formation, which is not the case if material precursors are transferred into novel materials. Representative examples of such well-defined building blocks are modified inorganic clusters or nanoparticles with attached reactive organic groups (Fig. 1.4).

Cluster compounds often consist of at least one functional group that allows an interaction with an organic matrix, for example by copolymerization. Depending on the number of groups that can interact, these building blocks are able to modify an organic matrix (one functional group) or form partially or fully crosslinked materials (more than one group). For instance, two reactive groups can lead to the formation of chain structures. If the building blocks contain at least three reactive groups they can be used without additional molecules for the formation of a crosslinked material.

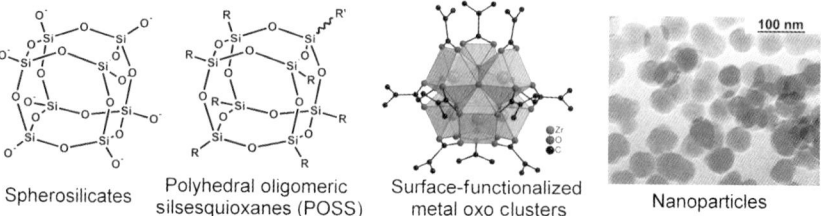

Fig. 1.4 Typical well-defined molecular building blocks used in the formation of hybrid materials.

Beside the molecular building blocks mentioned, nanosized building blocks, such as particles or nanorods, can also be used to form nanocomposites. The building block approach has one large advantage compared with the *in situ* formation of the inorganic or organic entities: because at least one structural unit (the building block) is well-defined and usually does not undergo significant structural changes during the matrix formation, better structure–property predictions are possible. Furthermore, the building blocks can be designed in such a way to give the best performance in the materials' formation, for example good solubility of inorganic compounds in organic monomers by surface groups showing a similar polarity as the monomers.

In situ formation of the components Contrary to the building block approach the *in situ* formation of the hybrid materials is based on the chemical transformation of the precursors used throughout materials' preparation. Typically this is the case if organic polymers are formed but also if the sol–gel process is applied to produce the inorganic component. In these cases well-defined discrete molecules are transformed to multidimensional structures, which often show totally different properties from the original precursors. Generally simple, commercially available molecules are applied and the internal structure of the final material is determined by the composition of these precursors but also by the reaction conditions. Therefore control over the latter is a crucial step in this process. Changing one parameter can often lead to two very different materials. If, for example, the inorganic species is a silica derivative formed by the sol–gel process, the change from base to acid catalysis makes a large difference because base catalysis leads to a more particle-like microstructure while acid catalysis leads to a polymer-like microstructure. Hence, the final performance of the derived materials is strongly dependent on their processing and its optimization.

1.2.1
In situ Formation of Inorganic Materials

Many of the classical inorganic solid state materials are formed using solid precursors and high temperature processes, which are often not compatible with the presence of organic groups because they are decomposed at elevated temperatures. Hence, these high temperature processes are not suitable for the *in situ* formation of hybrid materials. Reactions that are employed should have more the character of classical covalent bond formation in solutions. One of the most prominent processes which fulfill these demands is the sol–gel process. However, such rather low temperature processes often do not lead to the thermodynamically most stable structure but to kinetic products, which has some implications for the structures obtained. For example low temperature derived inorganic materials are often amorphous or crystallinity is only observed on a very small length scale, i.e. the nanometer range. An example of the latter is the formation of metal nanoparticles in organic or inorganic matrices by reduction of metal salts or organometallic precursors.

1.2.1.1 Sol–Gel Process

This process is chemically related to an organic polycondensation reaction in which small molecules form polymeric structures by the loss of substituents. Usually the reaction results in a three-dimensional (3-D) crosslinked network. The fact that small molecules are used as precursors for the formation of the crosslinked materials implies several advantages, for example a high control of the purity and composition of the final materials and the use of a solvent based chemistry which offers many advantages for the processing of the materials formed.

The silicon-based sol–gel process is probably the one that has been most investigated; therefore the fundamental reaction principles are discussed using this process as a model system. One important fact also makes the silicon-based sol–gel processes a predominant process in the formation of hybrid materials, which is the simple incorporation of organic groups using organically modified silanes. Si—C bonds have enhanced stability against hydrolysis in the aqueous media usually used, which is not the case for many metal–carbon bonds, so it is possible to easily incorporate a large variety of organic groups in the network formed. Principally $R_{4-n}SiX_n$ compounds (n = 1–4, X = OR', halogen) are used as molecular precursors, in which the Si—X bond is labile towards hydrolysis reactions forming unstable silanols (Si—OH) that condensate leading to Si—O—Si bonds. In the first steps of this reaction oligo- and polymers as well as cyclics are formed subsequently resulting in colloids that define the sol. Solid particles in the sol afterwards undergo crosslinking reactions and form the gel (Scheme 1.3).

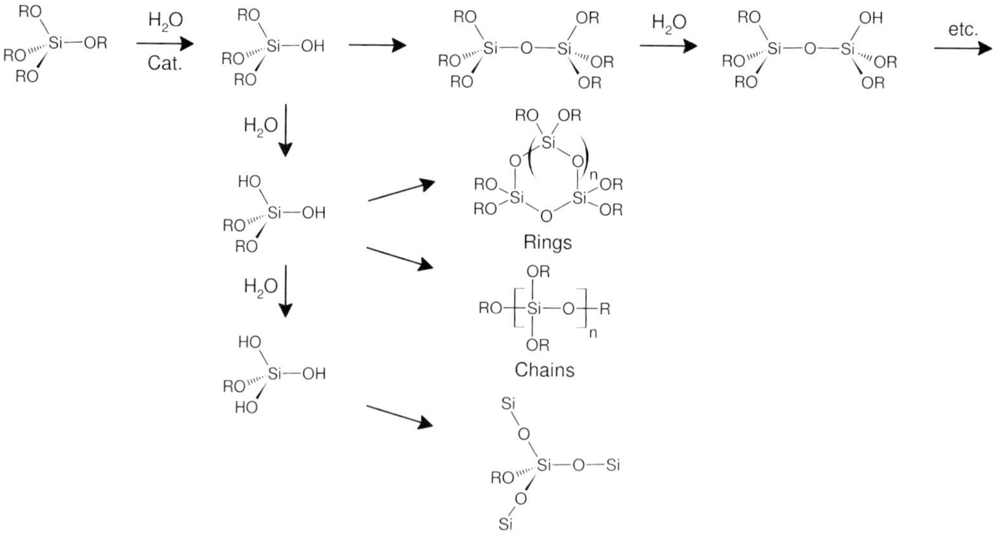

Scheme 1.3 Fundamental reaction steps in the sol–gel process based on tetrialkoxysilanes.

The process is catalyzed by acids or bases resulting in different reaction mechanisms by the velocity of the condensation reaction (Scheme 1.4). The pH used therefore has an effect on the kinetics which is usually expressed by the gel point

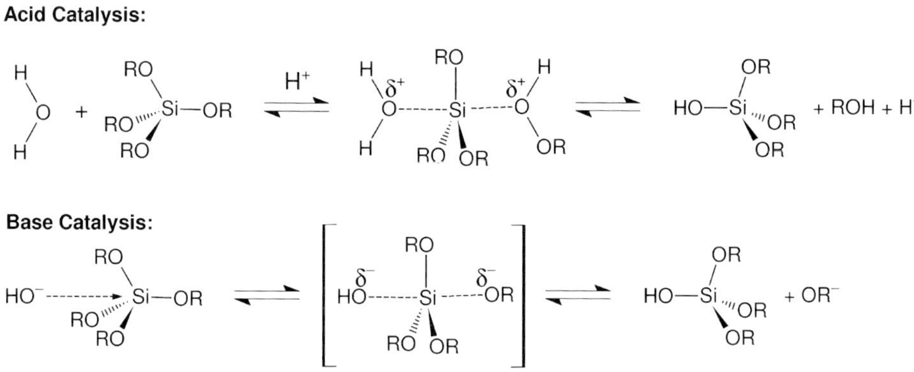

Acid Catalysis:

$$\text{H-O-H} + \text{RO-Si(OR)}_3 \rightleftharpoons \left[\text{H-O}^{\delta+}\cdots\text{Si(OR)}_3\cdots\text{O}^{\delta+}\text{-H} \right] \rightleftharpoons \text{HO-Si(OR)}_3 + \text{ROH} + \text{H}^+$$

with H^+ catalyst

Base Catalysis:

$$\text{HO}^- + \text{RO-Si(OR)}_3 \rightleftharpoons \left[\text{HO}^{\delta-}\cdots\text{Si(OR)}_3\cdots\text{OR}^{\delta-} \right] \rightleftharpoons \text{HO-Si(OR)}_3 + \text{OR}^-$$

Scheme 1.4 Differences in mechanism depending on the type of catalyst used in the silicon-based sol–gel process.

of the sol–gel reaction. The reaction is slowest at the isoelectric point of silica (between 2.5 and 4.5 depending on different parameters) and the speed increases rapidly on changing the pH. Not only do the reaction conditions have a strong influence on the kinetics of the reaction but also the structure of the precursors. Generally, larger substituents decrease the reaction time due to steric hindrance. In addition, the substituents play a mature role in the solubility of the precursor in the solvent. Water is required for the reaction and if the organic substituents are quite large usually the precursor becomes immiscible in the solvent. By changing the solvent one has to take into account that it can interfere in the hydrolysis reaction, for example alcohols can undergo *trans*-esterification reactions leading to quite complicated equilibria in the mixture. Hence, for a well-defined material the reaction conditions have to be fine-tuned.

The pH not only plays a major role in the mechanism but also for the microstructure of the final material. Applying acid-catalyzed reactions an open network structure is formed in the first steps of the reaction leading to condensation of small clusters afterwards. Contrarily, the base-catalyzed reaction leads to highly crosslinked sol particles already in the first steps. This can lead to variations in the homogeneity of the final hybrid materials as will be shown later. Commonly used catalysts are HCl, NaOH or NH_4OH, but fluorides can be also used as catalysts leading to fast reaction times.

The transition from a sol to a gel is defined as the gelation point, which is the point when links between the sol particles are formed to such an extent that a solid material is obtained containing internal pores that incorporate the released alcohol. However at this point the reaction has not finished, but condensation reactions can go on for a long time until a final stage is reached. This process is called ageing. During this reaction the material shrinks and stiffens. This process is carried on in the drying process, where the material acquires a more compact structure and the associated crosslinking leads to an increased stiffness. During the drying process the large capillary forces of the evaporating liquids in the porous structure take place which can lead to cracking of the materials. Reaction parameters such as drying rate, gelation time, pH, etc. can have a major influence on the cracking of the gels and have therefore to be optimized. Under some circumstances the destruction of the gel network can lead to the formation of powders instead of monoliths during materials formation.

Stress during the drying process can be avoided if the liquid in the pores is exchanged under supercritical conditions where the distinction between liquid and vapor no longer exists. This process leads to so-called highly porous aerogels compared with the conventionally dried xerogels.

As already mentioned above many parameters influence the speed of the sol–gel reaction; if a homogeneous material is required all parameters must be optimized. This is particularly true if hybrid materials are the target, because undesired phase separations of organic and inorganic species in the materials or between the network and unreacted precursors weaken the materials' properties. This can often even be observed by the naked eye if the material turns opaque.

The water to precursor ratio is also a major parameter in the sol–gel process. If tetraalkoxysilanes are used as precursors, two water molecules per starting compound are necessary to form completely condensed SiO_2. Applying a lower H_2O/Si ratio, would lead to an alkoxide containing final material.

1.2.1.2 Nonhydrolytic Sol–Gel Process

In this process the reaction between metal halides and alkoxides is used for the formation of the products (Scheme 1.5). The alkoxides can be formed during the process by various reactions. Usually this process is carried out in sealed tubes at elevated temperature but it can also be employed in unsealed systems under an inert gas atmosphere.

Non-hydrolytic sol–gel process:

M-Cl + M-OR $\longrightarrow$ M-O-M + R-Cl

In situ formation of alkoxides:

M-Cl + R-O-R $\longrightarrow$ M-OR + R-Cl

M-Cl + R-OH $\longrightarrow$ M-OR + HCl

Scheme 1.5 Mechanisms involved in the nonhydrolytic sol–gel process.

1.2.1.3 Sol–Gel Reactions of Non-Silicates

Metal and transition-metal alkoxides are generally more reactive towards hydrolysis and condensation reactions compared with silicon. The metals in the alkoxides are usually in their highest oxidation state surrounded by electronegative –OR ligands which render them susceptible to nucleophilic attack. Transition metal alkoxides show a lower electronegativity compared with silicon which causes them to be more electrophilic and therefore less stable towards hydrolysis in the sol–gel reactions. Furthermore, transition metals often show several stable coordination environments. While the negatively charged alkoxides balance the charge of the metal cation they generally cannot completely saturate the coordination sphere of

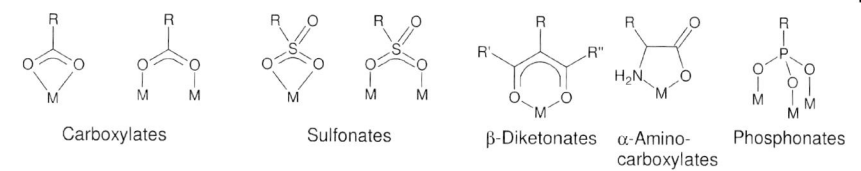

Fig. 1.5 Typical coordination patterns between bi- and multidentate ligands and metals that can be applied for the incorporation of organic functionalities in metal oxides.

the metals, which leads to the formation of oligomers via alkoxide or alcohol bridges and/or the saturation of the coordination environment by additional coordination of alcohol molecules, which also has an impact on the reactivity of the metal alkoxides. More sterically demanding alkoxides, such as isopropoxides, lead to a lower degree of aggregation and smaller alkoxides, such as ethoxides or n-propoxides, to a larger degree of aggregation. In addition, the length of the alkyl group in the metal alkoxides also influences their solubility in organic solvents, for example ethoxides often show a much lower solubility as their longer alkyl chain containing homologs.

As already mentioned M—C bonds in metal alkoxides are in most cases not stable enough to survive the sol–gel conditions. Therefore, contrary to the silica route, other mechanisms have to be employed if it is desired that hybrid inorganic–organic metal oxide materials be formed in a one-step approach. One solution to the latter problem is the use of organically functionalized bi- and multidentate ligands that show a higher bonding stability during the sol–gel reaction and, in addition, reduce the speed of the reaction by blocking coordination sites (Fig. 1.5).

1.2.1.4 Hybrid Materials by the Sol–Gel Process

Organic molecules other than the solvent can be added to the sol and become physically entrapped in the cavities of the formed network upon gelation. For this purpose the molecules have to endure the reaction conditions of the sol–gel process, namely the aqueous conditions and the pH of the environment. Hence, functional organic groups that can be hydrolyzed are not tolerated, but a partial tolerance for the pH can be obtained if the sol–gel reaction is carried out in a buffer solution. This is particularly necessary if biological molecules, such as enzymes, are to be entrapped in the gel. Physical entrapment has the disadvantage that sometimes the materials obtained are not stable towards phase separation or leaching because of differences in polarity. Chemical modification of organic compounds with trialkoxysilane groups can partially avoid such problems due to co-condensation during the formation of the sol–gel network and thus development of covalent linkages to the network. Trialkoxysilane groups are typically introduced by a platinum catalyzed reaction between an unsaturated bond and a trialkoxysilane (Scheme 1.6).

Scheme 1.6 Platinum catalyzed hydrosilation for the introduction of trialkoxysilane groups.

While the formation of homogeneous materials with a chemical link between the inorganic and organic component is in many cases the preferred route, there are cases where a controlled phase separation between the entrapped organic molecules and the sol–gel material is compulsory for the formation of the material, for example in the preparation of mesoporous materials (Chapter 5).

Besides the entrapment of organic systems, precursors with hydrolytically stable Si—C bonds can also be used for co-condensation reactions with tetraalkoxysilanes. In addition, organically functionalized trialkoxysilanes can also be used for the formation of 3-D networks alone forming so called silsesquioxanes (general formula R-SiO$_{1.5}$) materials. Generally a 3-D network can only be obtained if three or more hydrolyzable bonds are present in a molecule. Two such bonds generally result in linear products and one bond leads only to dimers or allows a modification of a preformed network by the attachment to reactive groups on the surface of the inorganic network (Fig. 1.6). Depending on the reaction conditions in the sol–gel process smaller species are also formed in the organotrialkoxysilane-based sol–gel process, for example cage structures or ladder-like polymers (Fig. 1.6). Because of the stable Si—C bond the organic unit can be included within the silica matrix without transformation. There are only a few Si—C bonds that are not stable against hydrolysis, for example the Si—C≡C bond where the Si—C bond can be cleaved by H$_2$O if fluoride ions are present. Some typical examples for trialkoxysilane compounds used in the formation of hybrid materials are shown in Scheme 1.7. Usually the organic functionalizations have a large influence on the properties of the final hybrid material. First of all the degree of condensation of a hybrid material prepared by trialkoxysilanes is generally smaller than in the case of tetraalkoxysilanes and thus the network density is also reduced.

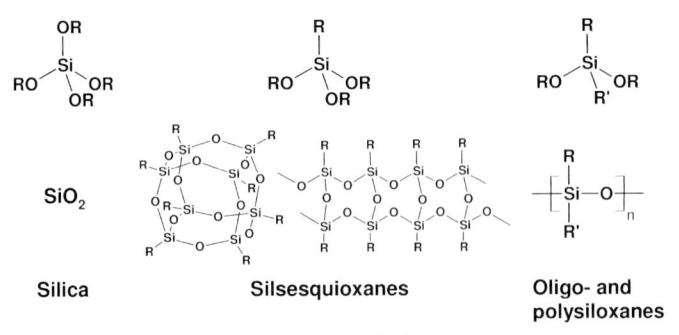

Fig. 1.6 Formation of different structures during hydrolysis in dependence of the number of organic substituents compared to labile substituents at the silicon atom.

In addition, the functional group incorporated changes the properties of the final material, for example fluoro-substituted compounds can create hydrophobic and lipophobic materials, additional reactive functional groups can be introduced to allow further reactions such as amino, epoxy or vinyl groups (Scheme 1.7). Beside molecules with a single trialkoxysilane group also multifunctional organic molecules can be used, which are discussed in more detail in Chapter 6.

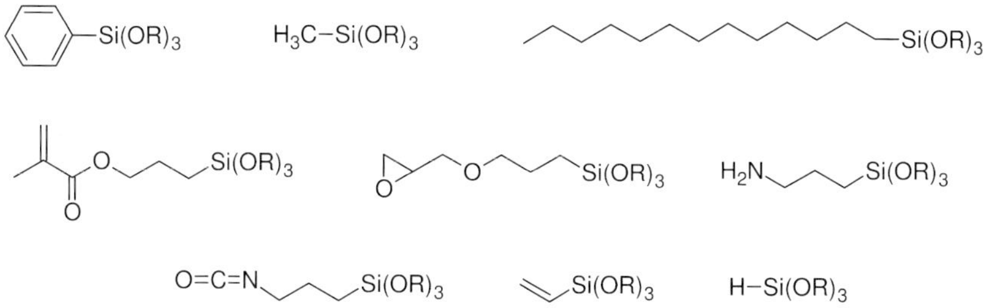

Scheme 1.7 Trialkoxysilane precursors often used in the sol–gel process.

More detailed discussions of the sol–gel process can be found in the cited literature.

1.2.1.5 Hybrid Materials Derived by Combining the Sol–Gel Approach and Organic Polymers

Compared with other inorganic network forming reactions, the sol–gel processes show mild reaction conditions and a broad solvent compatibility. These two characteristics offer the possibility to carry out the inorganic network forming process in presence of a preformed organic polymer or to carry out the organic polymerization before, during or after the sol–gel process. The properties of the final hybrid materials are not only determined by the properties of the inorganic and organic component, but also by the phase morphology and the interfacial region between the two components. The often dissimilar reaction mechanisms of the sol–gel process and typical organic polymerizations, such as free radical polymerizations, allow the temporal separation of the two polymerization reactions which offers many advantages in the material formation.

One major parameter in the synthesis of these materials is the identification of a solvent in which the organic macromolecules are soluble and which is compatible with either the monomers or preformed inorganic oligomers derived by the sol–gel approach. Many commonly applied organic polymers, such as polystyrene or polymethacrylates, are immiscible with alcohols that are released during the sol–gel process and which are also used as solvents, therefore phase separation is enforced in these cases. This can be avoided if the solvent is switched from the typically used alcohols to, for example, THF in which many organic polymers are soluble and which is compatible with many sol–gel reactions. Phase separation can also be avoided if the polymers contain functional groups that are more compatible with the reaction conditions of the sol–gel process or even undergo an interaction with the inorganic material formed. This can be achieved, for example by the incorporation of OH-groups that interact with, for example, hydroxyl groups formed during the sol–gel process or by ionic modifications of the organic polymer. Covalent linkages can be formed if functional groups that undergo hydrolysis and condensation reactions are covalently attached to the organic monomers. Some typically used monomers that are applied in homo- or copolymerizations are shown in Scheme 1.8.

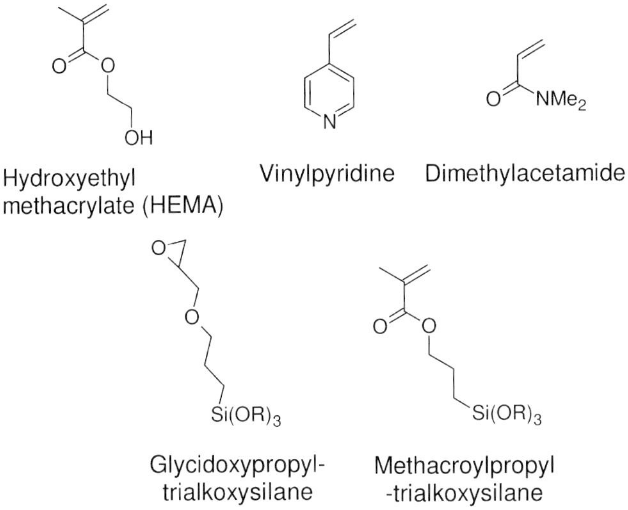

Scheme 1.8 Organic monomers typically applied in the formation of sol–gel/organic polymer hybrid materials.

1.2.2
Formation of Organic Polymers in Presence of Preformed Inorganic Materials

If the organic polymerization occurs in the presence of an inorganic material to form the hybrid material one has to distinguish between several possibilities to overcome the incompatibilty of the two species. The inorganic material can either have no surface functionalization but the bare material surface; it can be modified with nonreactive organic groups (e.g. alkyl chains); or it can contain reactive surface groups such as polymerizable functionalities. Depending on these prerequisites the material can be pretreated, for example a pure inorganic surface can be treated with surfactants or silane coupling agents to make it compatible with the organic monomers, or functional monomers can be added that react with the surface of the inorganic material. If the inorganic component has nonreactive organic groups attached to its surface and it can be dissolved in a monomer which is subsequently polymerized, the resulting material after the organic polymerization, is a blend. In this case the inorganic component interact only weakly or not at all with the organic polymer; hence, a class I material is formed. Homogeneous materials are only obtained in this case if agglomeration of the inorganic components in the organic environment is prevented. This can be achieved if the interactions between the inorganic components and the monomers are better or at least the same as between the inorganic components. However, if no strong chemical interactions are formed, the long-term stability of a once homogeneous material is questionable because of diffusion effects in the resulting hybrid material. Examples of such materials are alkyl chain functionalized silica nanoparticles that can be introduced into many hydrophobic polymers, the use of block copolymers containing a poly(vinyl pyridine) segment that can attach to many metal nanoparticles, or the use of hydroxyethyl methacrylates in the polymerization mixture together with metal oxide nanoparticles. In the latter example hydrogen bridges are formed between the polymer matrix and the particle surface. The stronger the respective interaction between the components, the more stable is the final material. The strongest interaction is achieved if class II materials are formed, for example with covalent interactions. Examples for such strong interactions are the use of surface-attached polymerizable groups that are copolymerized with organic monomers. Some examples of such systems are shown in the Chapters 2 and 3.

If a porous 3-D inorganic network is used as the inorganic component for the formation of the hybrid material a different approach has to be employed depending on the pore size, the surface functionalization of the pores and the stiffness of the inorganic framework. In many cases intercalation of organic components into the cavities is difficult because of diffusion limits. Several porous or layered inorganic materials have already been used to prepare hybrid materials and nanocomposites. Probably the most studied materials, class in this respect is that of two-dimensional (2-D) layered inorganic materials that can intercalate organic molecules and if polymerization between the layers occurs even exfoliate, producing nanocomposites. Contrary to intercalated systems the exfoliated hybrids only contain a small weight percentage of host layers with no structural order. The preparation of such materials is described in more detail in Chapter 4 but principally three methods for the formation of polymer–clay nanocomposites can be used:

1. Intercalation of monomers followed by *in situ* polymerization
2. Direct intercalation of polymer chains from solution
3. Polymer melt intercalation

The method applied depends on the inorganic component and on the polymerization technique used and will not be discussed in this introductory chapter.

Contrary to the layered materials, which are able to completely delaminate if the forces produced by the intercalated polymers overcome the attracting energy of the single layers, this is not possible in the case of the stable 3-D framework structures, such as zeolites, molecular sieves and M41S-materials. The composites obtained can be viewed as host–guest hybrid materials. There are two possible routes towards this kind of hybrid material; (a) direct threading of preformed polymer through the host channels (soluble and melting polymers) which is usually limited by the size, conformation, and diffusion behavior of the polymers and,

(b) the *in situ* polymerization in the pores and channels of the hosts. The latter is the most widely used method for the synthesis of such systems. Of course, diffusion of the monomers in the pores is a function of the pore size, therefore the pores in zeolites with pore sizes of several hundred picometers are much more difficult to use in such reactions than mesoporous materials with pore diameters of several nanometers. Two methods proved to be very valuable for the filling of the porous structures with monomers: one is the soaking of the materials in liquid monomers and the other one is the filling of the pores in the gas phase. A better uptake of the monomers by the inorganic porous materials is achieved if the pores are pre-functionalized with organic groups increasing the absorption of monomers on the concave surface. In principle this technique is similar to the increase of monomer absorption on the surface of silica nanoparticles by the surface functionalization with silane coupling agents.

Beside of well-defined 3-D porous structures, sol–gel networks are also inherently porous materials. Uniform homogeneous materials can be obtained if the solvent of the sol–gel process is a monomer for a polymerization. This can be polymerized in a second step after the sol–gel process has occurred. It is much more difficult to functionalize a dry porous xerogel or aerogel because here a stiff inorganic network has already formed and has to be filled again with organic monomers. Principally the same methods as in the case of the ordered 3-D networks can be used for this purpose. Infiltration of preformed polymers into sol–gel networks is as difficult as in the case of the well-ordered porous systems because of the difficulties connected with the slow diffusion of organic polymer chains into the porous inorganic network.

1.2.3
Hybrid Materials by Simultaneous Formation of Both Components

Simultaneous formation of the inorganic and organic polymers can result in the most homogeneous type of interpenetrating networks. Usually the precursors for the sol–gel process are mixed with monomers for the organic polymerization and both processes are carried out at the same time with or without solvent. Applying this method, three processes are competing with each other: (a) the kinetics of the hydrolysis and condensation forming the inorganic phase, (b) the kinetics of the polymerization of the organic phase, and (c) the thermodynamics of the phase separation between the two phases. Tailoring the kinetics of the two polymerizations in such a way that they occur simultaneously and rapidly enough, phase separation is avoided or minimized. Additional parameters such as attractive interactions between the two moieties, as described above can also be used to avoid phase separation.

One problem that also arises from the simultaneous formation of both networks is the sensitivity of many organic polymerization processes for sol–gel conditions or the composition of the materials formed. Ionic polymerizations, for example, often interact with the precursors or intermediates formed in the sol–gel process. Therefore, they are not usually applied in these reactions; instead free radical poly-

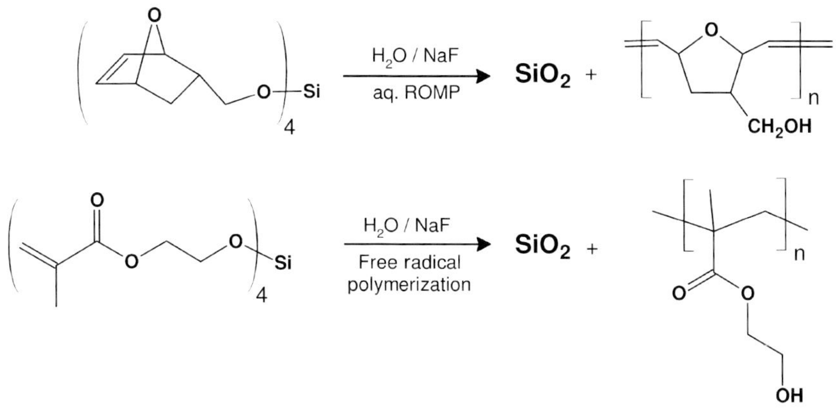

Fig. 1.7 Silicon sol-gel precursors with polymerizable alkoxides for ring opening metathesis polymerization (ROMP) or free radical polymerization.

merizations are the method of choice. This polymerization mechanism is very robust and can lead to very homogeneous materials. However, only selected, in particular vinyl, monomers can be used for this process. In addition, it is often also necessary to optimize the catalytic conditions of the sol–gel process. It is known, for example, that if the silicon sol–gel process is used basic catalysis leads to opaque final materials while the transparency can be improved if acidic conditions are used. This is most probably due to the different structures of the silica species obtained by the different approaches. While base catalysis leads to more particle-like networks that scatter light quite easily, acid catalysis leads to more polymer-like structures. Of course not only these parameters play a role for the transparency of the materials but also others such as the refractive index difference between organic polymer and inorganic species.

A very clever route towards hybrid materials by the sol–gel process is the use of precursors that contain alkoxides which also can act as monomers in the organic polymerization. The released alkoxides are incorporated in the polymers as the corresponding alcohol while the sol–gel process is carried out (Fig. 1.7). This leads to nanocomposites with reduced shrinkage and high homogeneity.

1.2.4
Building Block Approach

In recent years many building blocks have been synthesized and used for the preparation of hybrid materials. Chemists can design these compounds on a molecular scale with highly sophisticated methods and the resulting systems are used for the formation of functional hybrid materials. Many future applications, in particular in nanotechnology, focus on a bottom-up approach in which complex structures are hierarchically formed by these small building blocks. This idea is also one of the driving forces of the building block approach in hybrid materials.

Another point which was also already mentioned is the predictability of the final material properties if well-defined building blocks are used.

A typical building block should consist of a well-defined molecular or nanosized structure and of a well-defined size and shape, with a tailored surface structure and composition. In regard of the preparation of functional hybrid materials the building block should also deliver interesting chemical or physical properties, in areas like conductivity, magnetic behavior, thermal properties, switching possibilities, etc. All these characteristics should be kept during the material formation, for example the embedment into a different phase. Building blocks can be inorganic or organic in nature, but because they are incorporated into another phase they should be somehow compatible with the second phase. Most of the times the compatibility is achieved by surface groups that allow some kind of interaction with a second component.

1.2.4.1 Inorganic Building Blocks

Prime examples of inorganic building blocks that can keep their molecular integrity are cluster compounds of various compositions. Usually clusters are defined as agglomerates of elements that either exclusively contain pure metals or metals in mixture with other elements. Although the classical chemical understanding of a cluster includes the existence of metal–metal bonds, the term cluster should be used in the context of this book in its meaning of an agglomerate of atoms in a given shape. Regularly pure metal clusters are not stable without surface functionalization with groups that decrease surface energy and thus avoid coalescence to larger particles. Both coalescence and surface reactivity of clusters are closely related to that of nanoparticles of the same composition. Because of this similarity and the fact that the transition from large clusters to small nanoparticles is fluent, we will not clearly distinguish between them. While in commonly applied metal clusters the main role of the coordinating ligands is the stabilization, they also can serve for a better compatibilization or interaction with an organic matrix. Similar mechanisms are valid for binary systems like metal chalcogenide or multicomponent clusters. Hence, the goal in the chemical design of these systems is the preparation of clusters carrying organic surface functionalizations that tailor the interface to an organic matrix by making the inorganic core compatible and by the addition of functional groups available for certain interactions with the matrix. One major advantage of the use of clusters is that they are small enough that usual chemical analysis methods such as liquid NMR spectroscopy and, if one is lucky, even single crystal X-ray diffraction can be used for their analysis. The high ratio between surface groups and volume makes it possible to get important information of the bonding situation in such systems and makes these compounds to essential models for larger, comparable systems, such as nanoparticles or surfaces.

Two methods are used for the synthesis of such surface-functionalized molecular building blocks: either the surface groups are grafted to a pre-formed cluster ("post-synthesis modification" method) or they are introduced during the cluster synthesis ("*in-situ*" method).

Surface-functionalized metal clusters are one prominent model system for well-defined inorganic building blocks that can be used in the synthesis of hybrid materials. However, as with many other nanoscaled materials it is not possible to synthesize such pure clusters and to handle them without a specific surface coverage that limits the reactivity of the surface atoms towards agglomeration. From the aspect of the synthesis of hybrid materials this is no problem as long as the surface coverage of the cluster or nanoparticle contains the desired functionalities for an interaction with an organic matrix. A typical example of such a cluster is the phosphine-stabilized gold cluster of the type $Au_{55}(PPh_3)_{12}Cl_6$ which is prepared by reduction of $AuCl[PPh_3]$ with diborane. Blocking of surface sites by PPh_3 ligands guarantees the size limitation of these clusters and avoids their further growth. However, these phosphine-stabilized clusters are still unstable for example at elevated temperatures, which limits their applications. While triphenylphosphine only stabilizes the surface further functionalities can be included by an exchange of these capping agents. Ligand-exchange of the phosphine surface functionalization with alkyl- and arylthiols results in the corresponding thiol-stabilized cluster. This exchange leads to a complete substitution of the PPh_3 because of the better bonding capabilities of thiols to gold atoms compared with phosphines. A similar process is also used for the ligand exchange of stabilizing citrate groups on the surface of gold nanoparticles obtained by the reduction of $HAuCl_4$ by sodium citrate. The mild conditions during the exchange process (simple stirring in an organic solvent at room temperature), allows the functionalization of the clusters with different functionalized thiol ligands. Similar surface functionalizations can be carried out with other metal clusters and nanoparticles and with a variety of metal chalcogenide systems.

Beside pure metal clusters and nanoparticles an interesting class of materials are metal oxides, because they have interesting magnetic and electronic properties often paired with low toxicity. Simple easy-to-synthesize oxidic compounds are silicon-based systems such as silica particles or spherosilicate clusters, therefore these systems are often used as model compounds for the class of metal oxides, although they do not really represent the class of transition metal oxides that are probably more often used in technological relevant areas. Silica particles or spherosilicate clusters both have in common that the surface contains reactive oxygen groups that can be used for further functionalization (Fig. 1.8). Mono-functional polyhedral silsesquioxane (POSS) derivatives of the type $R'R_7Si_8O_{12}$ (R' = functional group, R = nonfunctional group) are prepared by reacting the incompletely condensed molecule $R_7Si_7O_9(OH)_3$ with $R'SiCl_3$. The eighth corner of the cubic closo structure is inserted by this reaction, and a variety of functional organic groups R' can be introduced, such as vinyl, allyl, styryl, norbornadienyl, 3-propyl methacrylate, etc (Fig. 1.8a). The incompletely condensed compounds $R_7Si_7O_9(OH)_3$ are obtained when certain bulky groups R (e.g. cyclopentyl, cyclohexyl, *tert.*-butyl) prevent the formation of the closo structures from $RSiX_3$ precursors and lead to the precipitation of open-framework POSS. These bulky substituents not only lead to open framework structures but also increase the solubility of the inorganic units in organic solvents. The closed cubic systems still show the high solubility which

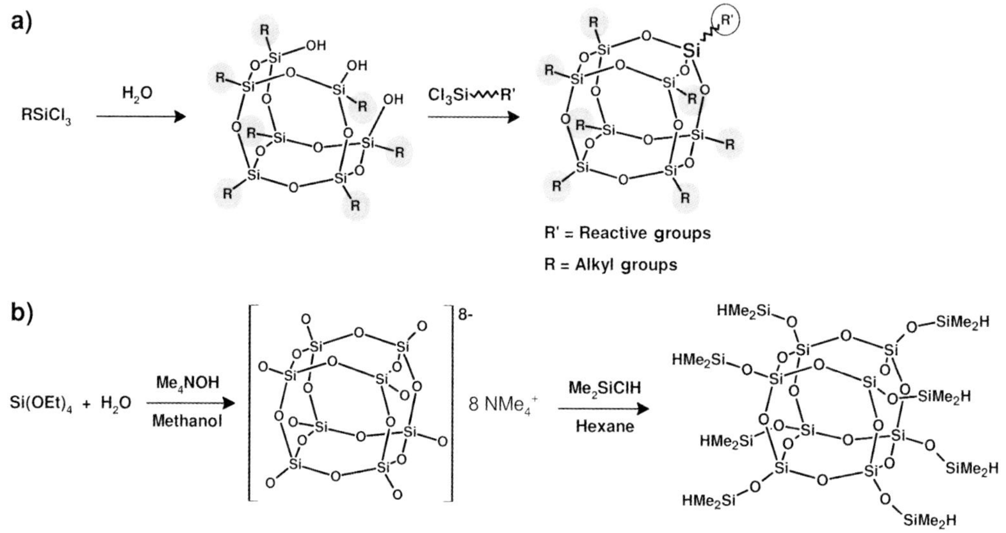

Fig. 1.8 Preparation of various well-defined silicon-based building blocks.

is necessary if the inorganic building blocks are to be incorporated in an organic environment for the functionalization of organic materials. The simple handling of these systems caused their boom in the preparation of hybrid materials. Other popular silsesquioxanes that have been prepared are the octahydrido- or the octavinyl-substituted molecules, which offer eight reactive sites. While the preparation of these systems is still very costly not least because of the low yields of the targeted products, another building block is much easier to obtain, namely spherosilicates. The polyhedral silicate species $[O-SiO_{3/2}]_n^{n-}$ are commonly prepared by hydrolysis of $Si(OR)_4$ in the presence of quaternary ammonium hydroxide (Fig. 1.8b). The best investigated compound is the cubic octamer (n = 8, "double four-ring", D4R). Spherosilicates are obtained from inorganic silica sources and can be considered the smallest piece of silica. The length of a Si—O—Si edge in the D4R structure is approximately 0.3 nm and the diameter of the cluster (Si—Si distance) 0.9 nm. Therefore the molecules can be considered sub-nanometer particles or – including the organic groups – particles in the low nanometer range. The anionic oxidic surface of the species $[SiO_{5/2}]_n^{n-}$ mimics the surfaces of larger silica particles, and therefore the polyhedral silicate clusters are also model systems for (nano)particles. After functionalization usually eight reaction sites are attached to these silica cores. Some typical reactions lead to the attachment of initiating or polymerizable groups at the corners and therefore the resulting clusters can be used as multifunctional initiators for polymerizations or as crosslinking monomers.

Recently the modification and embedment of transition-metal oxide clusters and particles has become more and more important, because of their catalytic, magnetic or electric properties. Chemical approaches different to those of silicate systems are often required for the attachment of organic groups to the surface of these compounds. The reason for these differences is the changed reactivity of these species, for example metal oxides often do not show highly nucleophilic oxygen atoms at their surface and charges are frequently delocalized over the whole cluster core. A variety of methods can be used for surface functionalization of pure clusters or nanoparticles depending on the reactivity of the surface, which often changes with parameters such as the pH value. In many cases only weak interactions are used to compatibilize the inorganic cluster or particle with an organic matrix, e.g. electrostatic interactions. Compatibilizing agents such as surfactants (e.g. amphiphilic molecules or block copolymers) are regularly used to increase the compatibility of the clusters or nanoparticles with an organic matrix. These molecules have two segments of which one undergoes interactions with the surface of the inorganic particle, for example by electrostatic or hydrogen bonding, and the other one, commonly a nonpolar block, interacts with the surrounding organic phase. However, functionalization of the cluster's surface by stable attachment of organic groups is preferred to a compatibilization with surfactants. Two methods can be employed for such a modification: either a cluster or particle is prepared in a first step and the surface is subsequently modified, or the surface functionalization is obtained *in situ* during the preparation.

Post-synthetic modification Post synthetic modification means that the cluster or nanoparticle is formed in a first step applying well-established procedures and the functionalization with organic groups is applied in a second step. Reactive surface functionalizations are required that allow a chemical reaction with the surface decorating molecules, for example nucleophilic substitution reactions. In the case of silica-based building blocks typically surface OH-groups are reacted with so-called silane coupling agents of the general composition $R_{4-n}SiX_n$ (n = 1–3; R = functional or nonfunctional organic group; X = halide or OR') to form stable covalent bonds as shown in Fig. 1.9. These molecules contain reactive Si—Cl or Si—OR groups that react with surface Si—OH groups to form stable Si—O—Si bonds. A plethora of such coupling agents is commercially available, containing various organic functional groups. If the desired group is not commercially accessible the molecule can easily be synthesized by a hydrosilation reaction. These compounds are used for the modification of silica particles or silica networks obtained in the sol–gel process. Other surface reactions can also be applied, for example the transformation of silanol groups into Si—Cl bonds followed by reactions with nucleophiles.

Prominent examples of nanoparticles that are functionalized by such methods are so-called Stöber particles which are monodispersed silica particles with

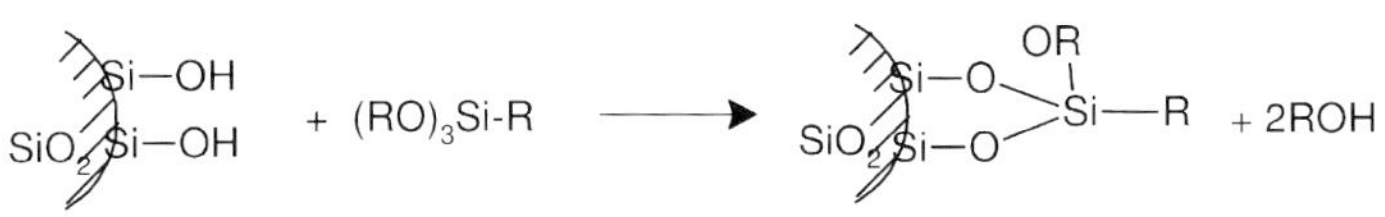

Fig. 1.9 A typical reaction of surface silanol-groups at the surface of an silica nanoparticle with a silane coupling agent.

diameters between 5 nm and 200 nm and low particle size distributions. Similarly to spherosilicates they are formed from $Si(OR)_4$ under alkaline conditions.

Only a few oxo clusters of other elements have similarly been substituted by functional silane coupling agents. The reason is that the surface functionalities in most clusters regularly lack the reactivity, e.g. basicity, of silicate surfaces, which is often based on the fact that negative charges are delocalized over the whole cluster. One of the few examples is the lacunary heterotungstate cluster $[SiW_{11}O_{39}]^{8-}$, in which the negative charge is rather located on two oxygen atoms. Derivatization of this cluster was obtained by reaction with various organotrichloro- or organotriethoxysilanes $RSiX_3$ (X = Cl or OEt) containing polymerizable groups R (R = allyl, vinyl, styryl or 3-methacryloxypropyl, $[(CH_2)_3C(O)OCMe=CH_2]$). The obtained anionic clusters have the composition $[SiW_{11}O_{35}(O_5Si_2R_2)]^{4-}$ containing two functional organic substituents per cluster unit.

Another way to attach functional groups to the surface of a preformed building block is the exchange of surface groups similar to the above mentioned ligand exchange on gold clusters. In such a reaction the charge and coordination number balance of the surface atoms must be retained, which is more complicated in the case of metal oxides with their heterogeneous surface than with the homogeneous surface of metal clusters and particles. Electrostatic or weak coordinative bonds between surface metals and organic groups are ideally suited for this reaction type. Many stabilized metal oxide species such as titanium oxo clusters can, for example, exchange surface alkoxide groups with other alkoxides.

In the case of metal oxo clusters or metal oxide particles containing metals in high oxidation states it is often difficult to attach functionalized surface molecules, like unsaturated bonds, during their synthesis because of the strong oxidizing conditions either of the cluster itself or the oxidative reaction conditions during their synthesis. Therefore such ligand exchange reactions offer a good way to form such functionalized molecules. Organic groups can also be attached to the cluster surface by bi- or multidentate (chelating or bridging) ligands via coordinative interactions, examples for anchor groups on metal oxide clusters or particles are carboxylates, sulfonates, phosphonates, β-diketonates, etc. These groups may carry organic functionalities, such as polymerizable double bonds. However, sometimes strong ligands can lead to the reorganization of a cluster surface or even to the degradation of transition-metal clusters or a partial degradation of the particles, which has been shown in several cases.

Beside the interaction types already mentioned, ionic interactions have also been employed for surface attachment of functional groups. Knowledge of the surface charges is necessary for this method. Charges of inorganic clusters or particles are sometimes influenced by variations of the reaction environment. Metal oxides, for example, can change their surface charge over a wide range depending on the metals and the pH value (isoelectric point). Typical anchor groups for the interaction with such charged surfaces are carboxylates, sulfonates, phosphonates as examples for anionic groups and ammonium groups as an example for cationic groups.

A special technique for the controlled formation of hybrid materials that relates on surface charges and their interaction with counterions is the so called layer-by-

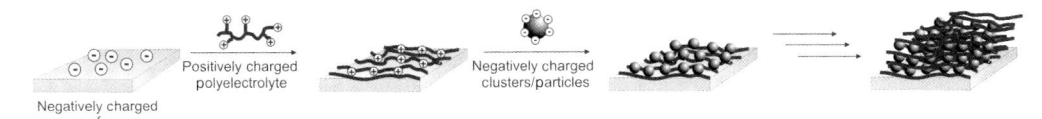

Fig. 1.10 Principle of layer-by-layer deposition.

layer (LbL) deposition. It allows the formation of inorganic–organic hybrid materials using the different charges on the surfaces of inorganic and organic polyelectrolytes. For example, anionic charges on a flat surface can be used to deposit cationically charged polyelectrolytes on it (Fig. 1.10). After deposition of the polymer the original surface charge is overcompensated and a layer of negatively charged inorganic building blocks, such as clusters or particles, can be deposited again. Afterwards again a layer of the polymer is deposited and so on. This technique enables the sequential deposition of oppositely charged building blocks. Because of its step-by-step character complex multilayer hybrid structures are readily accessible by this technique with control over layer thickness, composition and function. This method was also used for the surface functionalization of nanoparticles.

Functionalization of clusters and particles during their synthesis ("*in situ* functionalization") In the post-synthesis modification, functionalized building blocks are formed in two steps which are distinctly separate from each other: the inorganic core is formed first, and the functional organic groups are introduced later in a different reaction.

An alternative method is the formation of the inorganic building blocks in the presence of functional organic molecules (i.e. the functionalization of the clusters/particles occurs *in situ*). This is realized in the case of metal clusters or particles if the metal core is prepared in presence of ligands that control the size of the clusters and which contain the desired functionality. One limitation is that the organic groups have to withstand the reaction conditions of the cluster core formation. For example mild reductive reactions at room temperature are usually no restriction for the majority of potential surface functionalizations, but if elevated temperatures are employed for the synthesis of the inorganic building blocks, polymerizable groups in the ligands that undergo thermal polymerization, for example (meth)acrylic or styrene systems, should be avoided. The same is valid if unsaturated, easy to oxidize bonds are present in the synthesis of metal oxo clusters or particles containing metals in high oxidation states.

Silsesquioxanes with only one substitution pattern at each silicon atom are typical examples for the *in situ* formation of functionalized building blocks. As mentioned above they are prepared by the hydrolysis and condensation of trialkoxy- or trichlorosilanes, thus they contain inherently one functional group that is also present in the final material. Depending on the reaction procedure either ladder-like polymers or polyhedral silsesquioxanes (POSS) are obtained. The polyhedral compounds $[RSiO_{3/2}]_n$ can be considered silicon oxide clusters capped by the organic groups R.

Polyhedral silsesquioxanes $[RSiO_{3/2}]_n$ are obtained by controlled hydrolytic condensation of $RSiX_3$ (X = Cl, OR') in an organic solvent. There is a high driving force for the formation of polyhedral rather than polymeric compounds, particularly if the precursor concentration in the employed solvent is low (when the concentration is increased, increasing portions of network polymers may be formed). Which oligomers are produced and at which rate, depends on the reaction conditions, such as solvent, concentration of the monomer, temperature, pH and the nature of the organic group R. Chlorosilanes have a higher reactivity than the corresponding alkoxysilanes. In most cases, intractable mixtures of products are obtained except for those species that precipitate from the solution.

The best investigated silsesquioxane cages are the cubic octamers, $R_8Si_8O_{12}$. The compound $[HSiO_{3/2}]_8$ is, for example, prepared by hydrolysis of $HSiCl_3$ in a benzene / cc. H_2SO_4 mixture or by using partially hydrated $FeCl_3$ as the water source. It is a valuable starting compound for organically substituted derivatives as the Si—H functions of the silsesquioxane can be converted into organofunctional groups either by hydrosilation reactions (in the equation: X = Cl, OR', CN, etc.).

$$[HSiO_{3/2}]_8 + 8\ XCH_2CH=CH_2 \rightarrow [XCH_2CH_2CH_2SiO_{3/2}]_8$$

Only a few POSS with functional organic groups are insoluble enough to be obtained directly from the corresponding $RSiCl_3$ precursor by precipitation reactions. Examples are $(CH_2=CH)_8Si_8O_{12}$, $(p\text{-}ClCH_2C_6H_4)_8Si_8O_{12}$, $(R_2NCH_2CH_2)_8Si_8O_{12}$ or $(ClCH_2CH_2)_8Si_8O_{12}$. These compounds can also be transformed to other functional octa(silsesquioxanes). Examples include epoxidation of vinyl groups or nucleophilic substitution of the chloro group. Nonfunctional organic groups may be converted to functional organic by standard organic reactions. For example, the phenyl groups of the cubic silsesquioxane $Ph_8Si_8O_{12}$ were first nitrated and then reduced to give $(H_2NC_6H_4)_8Si_8O_{12}$. The systems prepared in this way can be used as building blocks for materials depending on their functional groups.

The polyhedral compounds $[RSiO_{3/2}]_n$ (POSS) or $[RO-SiO_{3/2}]_n$ discussed so far, formed by hydrolysis and condensation of a single precursor, are models for larger silica particles covered by organic groups and prepared from $RSi(OR')_3$ / $Si(OR')_4$ mixtures. The main parameter that controls the mutual arrangement of the $[SiO_4]$ and $[RSiO_{3/2}]$ building blocks is the pH. It was shown that upon sol–gel processing of $RSi(OR')_3$ / $Si(OR')_4$ mixtures (with nonbasic groups R) under basic conditions $Si(OR')_4$ reacts first and forms a gel network of agglomerated spherical nanoparticles. The $RSi(OR')_3$ precursor reacts in a later stage and condenses to the surface of the pre-formed silica nanoparticulate network. This kinetically controlled arrangement of the two building blocks from $RSi(OR')_3$ / $Si(OR')_4$ mixtures is another method to obtain surface-modified spherical Stöber particles (see above).

In situ functionalization is also a versatile route for many transition-metal clusters. In particular the synthesis of surface-functionalized early transition metal oxo clusters was studied by *in situ* processes. The main differences in these reactions are that surface functionalization is included in these systems, similar to the metal clusters, by functionalized ligands. For example a small zirconium oxo cluster is formed if zirconium alkoxides are mixed with functionalized carboxylic acids such as methacrylic acids. Under specific reaction conditions clusters of the type $Zr_6(OH)_4O_4(OMc)_{12}$ (OMc = methacrylate) are formed revealing an inorganic zirconium oxo core surrounded by a functional organic shell. The cluster core in this case, i.e. the arrangement of the eight-coordinate metal atoms and the oxygen atoms, corresponds to the basic structural unit of tetragonal zirconia. The cluster can therefore be considered the smallest possible piece of tetragonal zirconia stabilized by organic ligands. Clusters of different size and shape, and a different degree of substitution by organic ligands can be obtained by modifying the reaction conditions. The main parameters appear to be the metal alkoxide/carboxylic acid ratio and the kind of OR groups of the metal alkoxide. The surface-bound carboxylates in these clusters are loosely bound and they as well as alkoxides can be substituted by ligand exchange reactions quite easily.

Other bidentate ligands with polymerizable organic groups can be attached to the cluster surface by the *in situ* modification route as well (Fig. 1.11).

Interactions between metals or metal oxides cores to molecules that act as surface functionalizations are similar on a molecular scale and therefore do not usually change with the size of the core. Thus, the chemistry developed for isolated and structurally characterized metal and metal oxide clusters can also be applied for the functionalization of larger nanoparticles. This interface analogy offers the chance to study the chemistry on the molecular scale, which can be analyzed by conventional spectroscopic techniques much more easily, and transfer the obtained conclusions to the larger scale. There are many examples which show these similarities. Generally, as already mentioned above, the organic groups present in the reaction mixture and attached to the surface after the particle formation were mainly used to limit the growth of the derived particles by blocking reactive surface sites and guarantee a stable suspension in a specific solvent. Only recently have these groups been used to introduce a different surface characteristic to the surface of these building blocks or to add chemical functionalities to the surface.

Typical examples of capping agents for metals and II–VI semiconductor nanoparticles are alkylthiols for gold, CdS and related particles and tri-octyl phosphine and tri-octyl phosphine oxide for CdE (E = S, Se, Te). Oxide nanoparticles, such as iron oxides, are often stabilized by carboxylic acids. If these systems should be incorporated into a hydrophobic matrix long alkyl chains connected to the carboxylate groups are preferred, which favors fatty acid capping agents for such

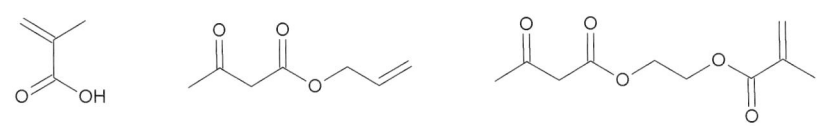

Fig. 1.11 Molecules that can act as bidentate ligands in the formation of surface-functionalized metal oxo clusters and metal oxide nanoparticles and which contain polymerizable groups for an attachment to an organic matrix.

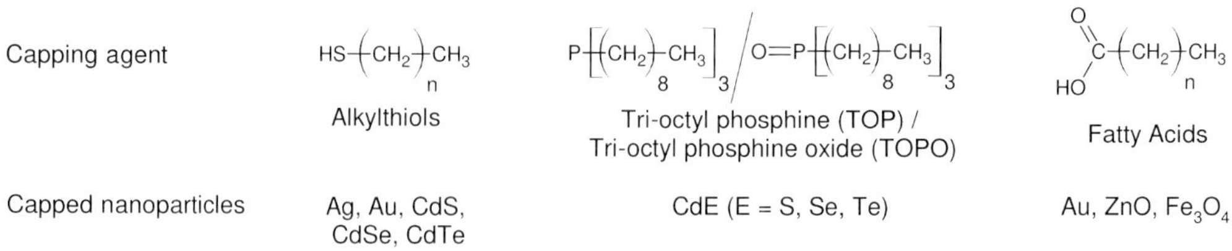

Capping agent	Alkylthiols: HS–(CH$_2$)$_n$–CH$_3$	Tri-octyl phosphine (TOP) / Tri-octyl phosphine oxide (TOPO): P[(CH$_2$)$_8$–CH$_3$]$_3$ / O=P[(CH$_2$)$_8$–CH$_3$]$_3$	Fatty Acids: HO–C(=O)–(CH$_2$)$_n$–CH$_3$
Capped nanoparticles	Ag, Au, CdS, CdSe, CdTe	CdE (E = S, Se, Te)	Au, ZnO, Fe$_3$O$_4$

Fig. 1.12 Typical capping agents used in nanoparticle preparation changing the surface properties of the nanoparticles to more hydrophobic.

applications. Generally attachment to the surface can occur by different interaction mechanisms and a variety of organic groups were applied for functionalization (Fig. 1.12). The selection of the capping agents is usually based on the composition of the particles and on their preparation route. The head groups of these capping agents can also be other functional molecules or polymers to graft them on the surface of the inorganic nanobuilding blocks. Mercapto-terminated linear polymers were for example used for a surface functionalization which provides a much better dispersion of the inorganic nanoparticles in various organic resins.

1.2.4.2 Organic Building Blocks

Beside the surface-functionalized inorganic building blocks described, of course organic building blocks can also be used for the formation of hybrid materials. Typical examples are oligo- and polymers as well as biological active molecules like enzymes. In principal similar methods have to be applied as in the formerly discussed example to increase the compatibility and the bonding between the two phases, therefore, a full description of the mechanisms is not necessary and would only lead to an enlargement of this introductory chapter. Hence, only a small selection of examples will be given here.

Small molecules The modification of inorganic networks with small organic molecules can be defined as the origin of hybrid materials. This is particularly true for sol–gel derived silicon-based materials. The mild conditions of the sol–gel process permit the introduction into the inorganic network of any organic molecule that consists of groups which do not interfere with these conditions, e.g. an aqueous and an acid or alkaline environment; it can then either be physically trapped in the cavities or covalently connected to the inorganic backbone (see Section 1.2.1.4). The latter is achieved by the modification of the organic molecules with hydrolysable alkoxysilane or chlorosilane groups. Phase separation is usually avoided by matching the polarity of the often hydrophobic molecules to that of the hydrophilic environment. If such a match can be obtained nearly every organic molecule can be applied to create hybrid materials. In recent years functional hybrids have been the particular focus of interest. Organic dyes, nonlinear optical groups, or switchable groups are only a small selection of molecules which have already been used to prepare hybrid materials and nanocomposites.

Macromolecules Oligo- and polymers as well as other organic macromolecules often show different solubilities in specific solvents compared with their monomers; most often the solubility of the polymers is much lower than that of the monomers. However, many formation mechanisms for hybrid materials and nanocomposites are based on solvent chemistry, for example the sol–gel process or the wet chemistry formation of nanoparticles. Therefore, if homogeneous materials are targeted, an appropriate solvent for both the inorganic and the organic macromolecules is of great benefit. For example many macromolecules are soluble in THF which is also a reasonable good reaction environment for the sol–gel process. Additional compatibilization is obtained if the polymers contain groups that can interact with the inorganic components. Similar mechanisms of interactions can be employed as already mentioned above, i.e. groups that interact via the formation of covalent bonds or others that compatibilize between the organic and inorganic components. Some typical monomers that form either homopolymers with a good interaction or that can be used to improve the interactions by copolymerization are shown in Fig. 1.13.

A particularly interesting group of macromolecules are block copolymers, consisting of a hydrophilic and a hydrophobic segment. They can be tailored in such a way that they can react with two phases that reveal totally different chemical characters and therefore, they are known as good compatibilizers between two components. These surfactants were for example often used for the modification of nanoparticles, where one segment interacts with the surface of the particle, and the other segment sticks away from this surface. In technology such systems are often used to overcome interfaces, for example when inorganic fillers are used for the modification of organic polymers. Novel controlled polymerization methods, such as atom transfer radical polymerization (ATRP) allow the preparation of block copolymers with a plethora of functional groups and therefore novel applications will soon be available.

Particles and particle-like structures Organic colloids formed from physically or chemically crosslinked polymers can also be used as building blocks for inorganic–organic hybrid materials and nanocomposites. The good control over their properties, such as their size, the broad size range in which they can be produced, from several nanometers to micrometers, accompanied by their narrow size distribution makes them ideal building blocks for many applications. Similarly to dendrimers, special interest in these systems is achieved after their surface modification, because of their self-assembling or simply by their heterophase

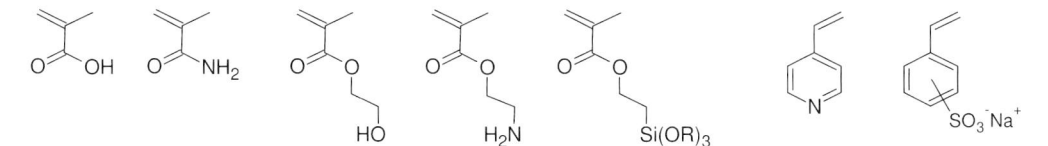

Fig. 1.13 Monomers that interact well with polar inorganic surfaces.

dispersion. Many examples for the use of such colloids are mentioned in other chapters of this book (templates for porous materials, precursors for core-shell nanoparticles), so this chapter will only provide some basic insight into this topic. As already mentioned above these latex colloids are formed in aqueous dispersions which, in addition to being environmentally more acceptable or even a mandatory choice for any future development of large output applications, can provide the thermodynamic drive for self-assembling of amphiphilics, adsorption onto colloidal particles or partitioning of the hybrid's precursors between dispersed nanosized reaction loci, as in emulsion or mini-emulsion free-radical polymerizations. For the use as precursors in inorganic–organic hybrid materials styrene or acrylate homo- and copolymer core latex particles are usually modified with a reactive comonomer, such as trimethoxysilylpropyl methacrylate, to achieve efficient interfacial coupling with silica environment during the sol–gel process.

Organic colloidal building blocks were in particular used for the preparation of 3-D colloidal crystals that were subsequently applied as templates in whose voids inorganic precursors were infiltrated and reacted to inorganic materials followed by removal of the colloids. Furthermore, discrete core-shell particles can also be produced consisting of an organic core and an inorganic shell. After removal of the organic core, for example by calcination, hollow inorganic spheres are obtained.

Another type of organic macromolecular building block is the hyperbranched molecules, so-called dendrimers. Dendrimers are highly branched regular 3-D monodisperse macromolecules with a tree-like structure. These macromolecules offer a wide range of unusual physical and chemical properties mainly because they have a well-defined number of peripheral functional groups that are introduced during their synthesis as well as internal cavities (guest–host systems). In particular the deliberate control of their size and functionality makes these compounds also interesting candidates as nanoscopic building blocks for hybrid materials. Spherically shaped dendrimers are, for example, ideal templates for porous structures with porosities that are determined by the radius of the dendrimeric building block. Generally approaches for surface functionalizations of these molecules are the modification with charged end-groups or the use of reactive organic groups. Triethoxysilyl-terminated dendrimers were, for example used as precursors for micro- and mesoporous hybrid dendrimer–silica xerogels and aerogels either as single precursors or in polycondensation reactions with tetraalkoxysilanes (Fig. 1.14).

The end groups of dendrimers can also be used for an interaction with metal clusters or particles and thus nanocomposites are formed often by simply mixing the two components. For example thiol-terminated phosphorus-containing dendrimers create supramolecular assemblies with Au_{55} clusters. A similar route was used to substitute alkoxide groups at titanium-oxo clusters against carboxylic acid or hydroxyl end-functionalized dendrimers forming 3-D networks. Multifunctionalized inorganic molecules can also act as the core of dendrimers. POSS and spherosilicate cages were, for example, used as the core units from which dendrons were either grown divergently or to which they were appended convergently.

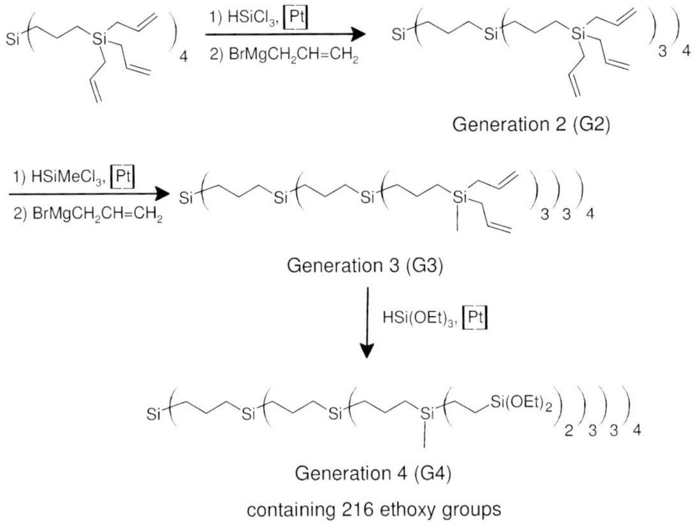

Fig. 1.14 Preparation of trialkoxysilyl-terminated carbosilane dendrimers.

Applying both silsesquioxanes and end-group functionalized dendrimers, both as multifunctional molecules crosslinked hybrid materials were obtained where well-defined inorganic molecules acted as crosslinking components.

Beside their role as crosslinking building blocks dendrimers can form hybrid materials by themselves. For example in their outer or inner shell precursors for nanoparticle synthesis can be attached to functional groups introduced during the synthesis and afterwards nanoparticles are grown within the branches of the dendrimers. One example is the complexation of metal ions in solution inside dendrimers that consist of groups, which can act as ligands, e.g. poly(amidoamine) (PAMAM) dendrimers. Subsequent chemical reactions, such as reductions, convert the entrapped metal salts to metal or semiconductor nanoparticles, which results in stable organic dendrimer-encapsulated inorganic nanoparticles.

1.3
Structural Engineering

An important area with respect to potential applications of hybrid materials and nanocomposites is the ability to design these materials on several length scales, from the molecular to the macroscopic scale. The importance of the design on the molecular scale was already expressed in the previous paragraphs. Because the processing of hybrid materials is more similar to that of organic polymers than to classical inorganic materials, such as ceramic or metal powders, based on the solvent-based chemistry behind the materials there is a variety of methods that can be adapted for their processing on the macroscopic scale. One has to distinguish between different applications to identify the best processing strategies. Thin films and coatings are generally formed by dip or spin coating, fibers are formed by spinning techniques and bulk materials are usually obtained if the often liquid precursors are simply poured into forms and curing is carried out. These techniques are used for the macroscopic processing of the materials, the molecular and microstructuring is obtained by a variety of other techniques that are mentioned below and throughout the chapters of this book.

One problem that has to be addressed in many of the macroscopic processing techniques in particular if the materials, such as the sol–gel network or the organic polymer in a hybrid are formed during the processing, is the problem of shrinkage. While in bulk materials this problem can be handled, it is more serious when films are applied on surfaces. At the interface between the coating and the support that does not follow the shrinkage considerable forces can appear that lead to inhomogeneities and cracks in the materials. As already mentioned this problem is particularly observed when the polymers or networks are formed during the processing, therefore often hybrid materials or nanocomposites are formed using preformed building blocks such as oligo- or polymers and clusters or nanoparticles to reduce this effect. However in many cases shrinkage can not totally avoided because the materials processing is usually based on solvents that are evaporated during processing.

Besides the macroscopic processing and the tailoring of the materials on a molecular scale, control of the nanometer structures is one of the major issues in the preparation of hybrid materials and nanocomposites. Many future technologies rely on hierarchically well-ordered materials from the micro- to the nanoscale. Applying the building block approach the structure of course is already engineered by the shape and distribution of the building blocks in the matrix. For example a homogeneous distribution of nanoparticles in an organic polymer has different properties than a material where the particles are agglomerated or where nanorods instead of nanoparticles are used. In addition to this building block based structural engineering the microstructure can often be influenced in hybrid materials similar to organic polymers, for example using lithographic techniques. However it is more difficult to create a structure on the nanometer level. Two different techniques can be identified for a structural control on the nanometer length scale: the top-down and the bottom-up approach. The top-down approach forms the nanostructures from larger objects by physical or chemical methods and is more engineering-related, while the bottom-up approach relies on self-organization (also called self-assembly) of molecules or nanometer sized compounds. The latter process is for many scientists the more elegant way to form large complex hierarchical structures. Think about the possibilities, for example if you have a liquid that contains your precursors and if you paint a support, it forms a hierarchical structure form the molecular over the nanometer to the macroscopic level over several hundred of square meters. This requires a detailed insight into many fundamental chemical and physico–chemical properties of the building blocks that should self assemble. In a bottom-up approach many reaction details have to be understood and controlled. Therefore, this process, although very promising, is yet only understood for quite small molecules and there are only some promising techniques that can already be applied for large scale technological applications. One of them is definitely the self-organization of block copolymers and their use in creating specific reaction environments. In this introductory chapter only a small selection of self-assembly mechanisms is presented, but the reader will find additional processes, explained in more detail in the following chapters.

Self-assembly is the major principle of the controlled formation of structural building blocks. It means nothing other than the spontaneous organization of unorganized systems to a complex structure, which is first of all an art to break and generate specific interactions and to work against entropy. There are several basic principles that have to be taken into account for a self-assembly process: the structure and shape of the building blocks, their interactions like the attractive and repulsive forces, their interactions with solvents, the environment where the reaction is carried out and diffusion processes. The balance between these principles is something which has to be set to create the desired structures. Two principles are employed in the synthesis of hybrid materials, the self-assembly of the building blocks of the material itself and the use of templates that can self-assemble and form a shape which is applied in the preparation of a material. Such templates are used in technology since a long time, for example as porogenes but it was only recently that they found a widespread use in the formation of hybrid materials.

Template-directed synthesis of hybrid materials Templates can fulfill various purposes, for example they can fill space and/or direct the formation of specific structures. Templates that fill space have been technologically used in materials processing for a long time; examples are porogenes for the formation of foam-like materials or single molecules that are employed in zeolite synthesis. In both examples the templates are removed after their use to form the pores in the material.

Templates for the synthesis of hybrid materials can be preformed structures such as dendrimers or nanoparticles that form 2-D or 3-D ordered structures. Furthermore the supramolecular self-organization of single molecules into larger 2-D and 3-D structures can also be employed as a template. One example is the application of amphiphilic surfactants that organize into micelles and more complicated 3-D arrangements such as hexagonal arranged rod-like structure, cubic interpenetrating networks or lamellar structures (Fig. 1.15). The latter are not

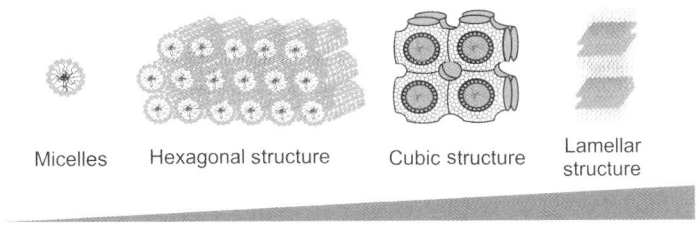

Fig. 1.15 Examples of ordered 3-D structures obtained by supramolecular self-assembly of surfactant molecules.

usually used for structural engineering because they collapse after removal of the template. The structures formed, that are dependent on parameters such as the concentration of the surfactants in the solvent, the temperature and sometimes the pH, consist of hydrophobic and hydrophilic regions and the interfaces between them which are applied in the formation of solid materials, primarily by sol–gel process (Fig. 1.16). As long as the organic surfactants are still incorporated in the formed inorganic matrix the materials can be considered as inorganic–organic hybrid materials. After removal of the template which commonly occurs by calcinations at temperatures above 450 °C, the materials are purely inorganic in nature. Probably the most prominent example of the use of single organic (macro)molecules and their 3-D assembly as templates for inorganic and hybrid materials is the use of surfactants in the formation of nano- and mesostructured materials porous materials (Chapter 5). Recently the silicate or metal oxide walls of the mesoporous materials were substituted by hybrid materials formed by bridged silsesquioxanes (Chapter 6) and so-called periodically mesoporous organosilicas (PMOs) were formed.

Colloids with narrow size distributions can order in 3-D objects so-called colloidal crystals. Most of the time latex or silica colloids are used for the preparation of such 3-D objects, because they can be prepared quite easily over different length scales and with the desired size distribution. They are crystallized in structurally well-ordered three-dimensional colloidal crystals that resemble the packing of atoms on a smaller scale. Because similar spheres can only fill the room by a close packing up to 74% the voids can be infiltrated by inorganic or organic reactive

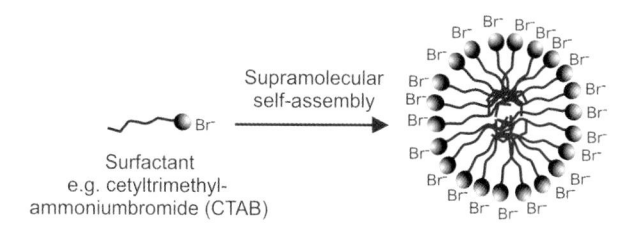

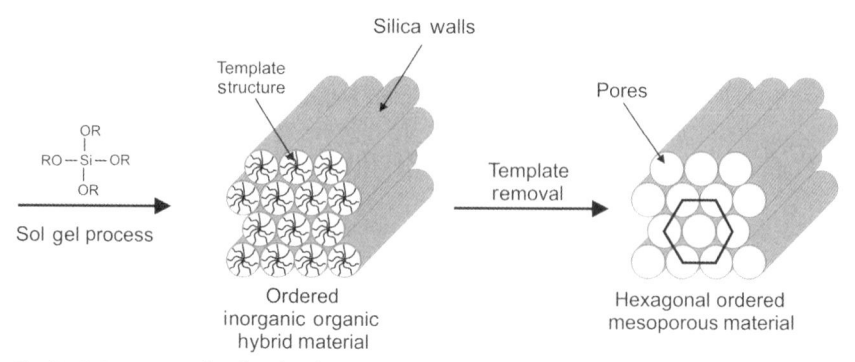

Fig. 1.16 Formation of well-ordered mesoporous materials by a templating approach.

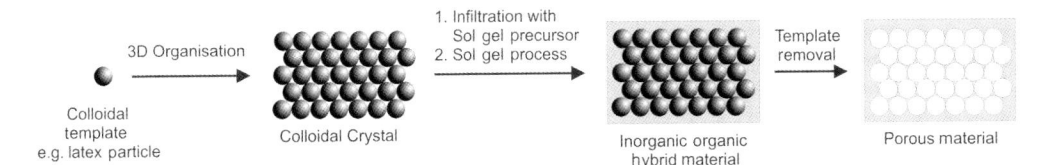

Fig. 1.17 Colloidal crystal templating.

species, such as materials precursors, clusters or smaller nanoparticles and a hybrid material is formed. Generally the templates are afterwards removed and porous materials are formed (Fig. 1.17).

Many other methods can be used for the controlled preparation of nanometer structures such as soft lithography, the use of nanoscale pores, for example in anodic alumina membranes, etc. are not mentioned here and the interested reader is referred to literature in the Bibliography.

1.4
Properties and Applications

There is almost no limit to the combinations of inorganic and organic components in the formation of hybrid materials. Therefore materials with novel composition–property relationships can be generated that have not yet been possible. Because of the plethora of possible combinations this introductory chapter can only present some selected examples. Many of the properties and applications are dependent on the properties of the precursors and the reader is therefore referred to the following chapters.

Based on the increased importance of optical data transmission and storage, optical properties of materials play a major role in many high-tech applications. The materials used can reveal passive optical properties, which do not change by environmental excitation, or active optical properties such as photochromic (change of color during light exposure) or electrochromic (change of color if electrical current is applied) materials. Both properties can be incorporated by building blocks with the specific properties, in many cases organic compounds, which are incorporated in a matrix. Hybrid materials based on silicates prepared by the sol–gel process and such building blocks reveal many advantages compared with other types of materials because silica is transparent and if the building blocks are small enough, does not scatter light, and on the other hand organic materials are often more stable in an inorganic matrix. One of the most prominent passive features of hybrid materials already used in industry are decorative coatings obtained by the embedment of organic dyes in hybrid coatings. Another advantage of hybrid materials is the increased mechanical strength based on the inorganic structures. Scratch-resistant coatings for plastic glasses are based on this principle. One of the major advantages of hybrid materials is that it is possible to include more than one function into a material by simply incorporating a second component with another property into the material formulation. In the case of scratch-resistant coatings, for example, additional hydrophobic or antifogging properties can be introduced. However, in many cases the precursors for hybrid materials and nanocomposites are quite expensive and therefore the preparation of bulk materials is economically not feasible. One of the advantages of hybrid materials, namely, their quite simple processing into coatings and thin films, can be one solution to this disadvantage. Applying such coatings to cheaper supports can be advantageous. Silica is preferred as the inorganic component in such applications because of its low optical loss. Other inorganic components, for example zirconia, can incorporate high refractive index properties, or titania in its rutile phase can be applied for UV absorbers. Functional organic molecules can add third order nonlinear optical (NLO) properties and conjugated polymers, conductive polymers can add interesting electrical properties. Nanocomposite based devices for electronic and optoelectronic applications include light-emitting diodes, photodiodes, solar cells, gas sensors and field effect transistors. While most of these devices can also be produced as fully organic polymer-based systems, the composites with inorganic moieties have important advantages such as the improvement of long-term stability, the improvement of electronic properties by doping with functionalized particles and the tailoring of the band gap by changing the size of the particles.

The enhancement of mechanical and thermal properties of polymers by the inclusion of inorganic moieties, especially in the form of nanocomposites, offers the possibility for these materials to substitute classical compounds based on metals or on traditional composites in the transportation industry or as fire retardant materials for construction industry.

Medical materials are also one typical application area of hybrid materials, as their mechanical properties can be tailored in combination with their biocompatibility, for example nanocomposites for dental filling materials. A high content of inorganic particles in these materials provides the necessary toughness and low shrinkage, while the organic components provide the curing properties combined with the paste-like behavior. Additional organic groups can improve the adhesion properties between the nanocomposites and the dentine.

Composite electrolyte materials for applications such as solid-state lithium batteries or supercapacitors are produced using organic–inorganic polymeric systems formed by the mixture of organic polymers and inorganic moieties prepared by sol–gel techniques. In these systems at least one of the network-forming species should contain components that allow an interaction with the conducting ions. This is often realized using organic polymers which allow an interaction with the ions, for example via coordinative or by electrostatic interactions. One typical example is proton conducting membranes which are important for the production of fuel cells The application of hybrid composites is interesting for these systems because this membrane is stable at high temperatures compared with pure organic systems.

These are only some applications for hybrid materials and there is a plethora of systems under development for future applications in various fields.

1.5
Characterization of Materials

The range of characterization methods used in the analysis of the composition, the molecular and nanometer structure as well as the physical properties of hybrid materials is quite large. Many of these methods are specific for particular materials' compositions, therefore a complete list of these techniques is out of the range of this chapter. Here only a small selection of techniques often used for the investigation of hybrid materials is explained. Compared with classical inorganic materials, hybrids are often amorphous so well-established characterization techniques such as X-ray diffraction are limited. Many methods used are more related to the characterization of amorphous organic polymers. The heterogeneous nature of hybrid materials means that generally a variety of analytical techniques has to be used to get a satisfactory answer to structure–property relationships.

NMR Liquid state NMR techniques are a well-known powerful tool in the characterization of solutions. The advantages of NMR spectroscopy is that it is a very sensitive technique for the chemical environment of specific nuclei and a plethora of nuclei can be investigated, which makes it also an interesting technique for solid materials. However, contrary to solution NMR where spectra usually consist of a series of very sharp lines due to an averaging of all anisotropic interactions by the molecular motion in the solution, this is different for a solid sample, where very broad peaks are observed due to anisotropic or orientation-dependent interactions. In principle high-resolution solid state NMR spectra can provide the same type of information that is available from corresponding solution NMR spectra, but special techniques/equipment are required, including magic-angle spinning (MAS), cross polarization (CP), special pulse sequences, enhanced probe electronics, etc. An additional advantage of the technique is that it is nondestructive. In this chapter we will not go into the details of this technique and the reader is referred to the corresponding literature. Here we will discuss the usefulness of the method on a selected example.

As mentioned, there are a variety of nuclei that can be used as probes in solid state NMR because of their NMR activity, many of them are also interesting for hybrid materials such as C, Si, Al, Sn, V, P, F and many others. By far the most investigated nucleus in the field of hybrid materials is ^{29}Si.

^{29}Si NMR is a powerful tool in the determination of the relative proportions of different silicon species in sol–gel derived materials. Therefore it offers insights into the kinetics of the process and the understanding of its fundamental parameters, such as precursor structure and reaction conditions. It is sensitive to the first and second nearest neighbors and therefore one can distinguish between different silicon atoms in the final material and their surroundings. Typical for hybrid materials is the nomenclature with letters and numbers. Four different species can be observed in hybrid materials derived by the sol–gel process depending on their substitution pattern at the silicon. In principle one can have one, two, three or four oxygen atoms surrounding the silicon atom and respectively the number of carbon atoms is reduced from three to zero. The abbreviations for these substitution patterns are M (C_3SiO), D (C_2SiO_2), T ($CSiO_3$) and Q (SiO_4). Depending how many Si atoms are connected to the oxygen atoms 4 to 1 a superscript is added to the abbreviation. In addition these silicon species have different chemical shifts which helps to distinguish between them. Some additional examples are shown in Chapter 6.

X-ray photoelectron spectroscopy (XPS) X-ray photoelectron spectroscopy (XPS) is a surface sensitive analytical tool used to examine the chemical compositions and electronic state of the surface of a sample. The sample is placed under high vacuum and is bombarded with X-rays which penetrate into the top layer of the sample (~nm) and excite electrons (referred to as photoelectrons). Some of these electrons from the upper approximately 5 nm are emitted from the sample and can be detected. The kinetic energy of these electrons is measured by an analyzer.

If monochromatic X-rays are used, the energy of an emitted photon imparts on an electron is a known quantity. The binding energy of the ejected electron can then be determined from:

$$E_{binding} = E_{photon} - E_{kinetic} - \Phi$$

where Φ is the work function of the material.

The energy of the core electrons is very specific for the element that the atom belongs to therefore the spectrum gives information on the elemental composition of the thin surface region. Shifts in the binding energies provide additional chemical information (e.g. the oxidation state of the element). The technique also offers the opportunity to use an ion bombardment to sputter the surface layer by layer and to obtain therefore a deep profile.

Electron microscopy Electron microscopy became one of the most important techniques to characterize the materials morphology on the nanometer and nowadays even on the atomic scale. This method provides a direct image of the sample and by coupling the microscopy with analytical techniques even elemental distribution and other properties can be resolved in these dimensions. Looking at electron micrographs one has always to be aware of some limiting points: (a) all specimens are in high vacuum and probably have another shape in a liquid or gel-like surrounding, (b) the images reveal only a small fragment of the whole sample raising the possibility that the region analyzed may not be characteristic of the whole sample, and (c) the sample is treated with a high energy electron beam and it was probably changed by this beam. Points (a) and (c) can partially be avoided using special technique such as cryogenic TEM where the sample is measured at low temperatures in a frozen state. In general electron microscopy results should always be confirmed by other techniques.

Scanning electron microscopy (SEM) Scanning electron microscopy (SEM) produces high resolution images of a sample surface. SEM images have a characteristic 3-D appearance and are therefore useful for judging the surface structure of the sample. The primary electrons coming from the source strike the surface and they are inelastically scattered by atoms in the sample. The electrons emitted are detected to produce an image. Beside the emitted electrons, X-rays are also produced by the interaction of electrons with the sample. These can be detected in a SEM equipped for energy dispersive X-ray (EDX) spectroscopy. Different detection modes can be applied such as the detection of backscattered electrons or the electron backscatter diffraction (EBSD) which gives crystallographic information about the sample.

The spatial resolution of the SEM techniques depends on various parameters, most of them are instrument related. Generally the resolution goes down to 20 nm to 1 nm, which is much lower than that of transmission electron microscopy (TEM) but SEM has some advantages compared with TEM. For example a quite large area of the specimen can be imaged, bulk materials can be used as samples and, as mentioned, a variety of analytical modes is available for measuring the composition and nature of the specimen.

Transmission electron microscopy (TEM) In transmission electron microscopy (TEM) images are produced by focusing a beam of electrons onto a very thin specimen which is partially transmitted by those electrons and carries information about the inner structure of the specimen. The image is recorded by hitting a fluorescent screen, photographic plate, or light sensitive sensor such as a CCD camera. The latter has the advantage that the image may be displayed in real time on a monitor or computer.

It is often difficult in TEM to receive details of a sample because of low contrast which is based on the weak interaction with the electrons. Particularly samples that have a high content of organic components often reveal this problem, which can partially be overcome by the use of stains such as heavy metal compounds. The dense electron clouds of the heavy atoms interact strongly with the electron beam.

However, sometimes the organic components of the sample are not detected because they decompose in the electron beam; this can be avoided using cryogenic microscopy, which keeps the specimen at liquid nitrogen or liquid helium temperatures (cryo-TEM).

Similar to SEM further information about the sample can be obtained by analytical TEM, for example the elemental composition of the specimen can be determined by analyzing its X-ray spectrum or the energy-loss spectrum of the transmitted electrons. Additionally if the material observed is crystalline, diffraction patterns are obtained that give information about the crystal orientation and very powerful instruments can even investigate the crystal structure.

Modern high-resolution TEM (HRTEM) goes down to a resolution <100 pm.

One of the major limitations of TEM is the extensive sample preparation, which makes TEM analysis a relatively time consuming process with a low throughput of samples.

Atomic force microscopy (AFM) Atomic force microscopy (AFM) became a very important tool in the analysis of surfaces and nanoobjects in recent years. The method is intriguingly simple; a cantilever with a sharp tip at its end, typically composed of silicon or silicon nitride with tip sizes on the order of nanometers is brought into close proximity of a sample surface. The van der Waals force between the tip and the sample leads to a deflection of the cantilever. Typically the deflection is measured applying a laser beam, which is reflected from the top of the cantilever into an array of photodiodes. If the tip is scanned at constant height, there would be a risk that the tip would collide with the surface, causing damage to the sample. Therefore, in most cases a feedback mechanism is employed to adjust the tip-to-sample distance to keep the force between the tip and the sample constant. Generally, the sample is mounted on a piezoelectric holder in all three space directions (x and y for scanning the sample, z for maintaining a constant force). The resulting map of $s(x,y)$ represents the topography of the sample.

Primary modes of operation for an AFM are contact mode, non-contact mode, and dynamic contact mode. In the contact mode operation, the force between the tip and the surface is kept constant during scanning by maintaining a constant deflection. In the non-contact mode, the cantilever is externally oscillated at or close to its resonance frequency. The oscillation gets modified by the tip–sample interaction forces; these changes in oscillation with respect to the external reference oscillation provide information about the sample's properties. In dynamic contact mode, the cantilever is oscillated such that it comes in contact with the sample with each cycle, and then enough force is applied to detach the tip from the sample.

AFM technique has several advantages over other nanoanalysis tools, such as the electron microscope. Contrary to other methods AFM provides a 3-D surface profile of a sample. Furthermore, samples viewed by AFM do not require any special treatment such as high vacuum and it is a non destructive technique, which is not the case for electron microscope images because here high energy electron beams often destroy the organic parts of a hybrid sample.

However, there are also limits of this method; for example, AFM can only show a maximum height on the order of micrometers and a maximum area of around 100 by 100 micrometers and the scanning speed is quite low compared with SEM.

X-ray diffraction X-ray diffraction is regularly used to identify the different phases in a polycrystalline sample. Two of its most important advantages for analysis of hybrid materials are that it is fast and nondestructive. When the positions and intensities of the diffraction pattern are taken into account the pattern is unique for a single substance. The X-ray pattern is like a fingerprint and mixtures of different crystallographic phases can be easily distinguished by comparison with reference data. Usually electronic databases such as the Inorganic Crystal Structure

Database (ICSD) are employed for this comparison. The major information one gets from this method is the crystalline composition and the phase purity. In the case of semicrystalline or amorphous materials broad humps are observed in the diffractogram. Therefore the degree of crystallinity can be qualitatively estimated.

If the crystallites of the powder are very small the peaks of the pattern will broaden. From this broadening it is possible to determine an average crystallite size by the Debye–Scherrer equation:

$$d = \frac{k\lambda}{B\cos\theta}$$

where k is a factor which is usually set to 0.9, λ is the wavelength of the X-ray radiation, B is the broadening of the diffraction line measured at half of its maximum intensity (radians) and θ is the Bragg angle. An error for the crystallite size by this fomular can be up to 50% which is appropriate.

Small-angle X-ray scattering (SAXS) SAXS is a technique where the source for the scattering of the X-rays is the inhomogeneities in the sample within the nanometer range. SAXS patterns are recorded at very low angles (typically <3–5°). In this angular range, information about the shape and size of the inhomogeneities, which can be, for example, clusters or nanoparticles in an organic matrix, and their distances is obtained. Furthermore pores can also be defined as inhomogeneities therefore SAXS is also used to characterized regularly ordered porous materials. The resolution of SAXS experiments is strongly dependent on the equipment. The highest resolution is obtained at beamlines at synchrotrons.

Thermal analysis techniques Thermogravimetric analysis (TGA) studies the weight changes of samples in relation to changes in temperature. TGA is commonly employed with respect to hybrid materials and nanocomposites to investigate the thermal stability (degradation temperatures), the amount of inorganic component, which usually stays until the end of the measurement due to its high thermal resistance, and the level of absorbed moisture or organic volatiles in these materials. Typically TGA plots show the weight lost in relation to the temperature and typical ranges that can be distinguished are the lost of moisture and absorbed solvents up to 150 °C, the decomposition of organic components between 300 and 500 °C. Usually the measurements are carried out under air or an inert gas.

Differential scanning calorimetry (DSC) is a thermoanalytical technique that compares the difference in the amount of heat required to increase the temperature of a sample and a reference with a well-defined heat capacity measured as a function of temperature. Both the sample and the reference are maintained at the same temperature throughout the experiment. Generally, the temperature program for a DSC analysis is designed such that the sample holder temperature increases linearly as a function of time. The basic principle underlying this technique is that, when the sample undergoes a physical transformation such as a phase transition or thermal decomposition, more or less heat is required compared with the reference, to maintain both at the same temperature. Whether more or less heat must flow to the sample depends on whether the process is exothermic or endothermic. Melting points of solids, for example, require more heat flow to the sample therefore these processes are endothermic, while thermal decompositions, for example oxidation processes are mostly exothermic events. An exothermic or endothermic event in the sample results in a deviation in the difference between the two heat flows to the reference and results in a peak in the DSC curve (plot of heat flow against temperature). The difference in heat flow between the sample and reference also delivers the quantitative amount of energy absorbed or released during such transitions. This information can be obtained by integrating the peak and comparing it with a given transition of a known sample.

Similar to TGA the experiments can be carried out under oxygen and other atmospheres such as inert gas.

1.6
Summary

Hybrid materials represent one of the most fascinating developments in materials chemistry in recent years. The tremendous possibilities of combination of different properties in one material initiated an explosion of ideas about potential materials and applications. However, the basic science is sometimes still not understood, therefore investigations in this field in particular to understand the structure–property relationships are crucial. This introductory chapter is intended to give an overview of critical issues in the synthesis and to guide the reader to the other chapters in this book, which focus on more specialized topics. This introduction has shown the importance of the interface between the inorganic and organic materials which has to be tailored to overcome serious problems in the preparation of hybrid materials. Different building blocks and approaches can be used for their preparation and these have to be adapted to bridge the differences of inorganic and organic materials. Beside the preparation of hybrid materials, their nano- and microstructure formation, processing and analysis is important. A variety of techniques can be used for these issues and several of them have been introduced to the reader.

Bibliography

Sol–gel process
J. D. Wright, N. Sommerdijk, *Sol-Gel Materials: Their Chemistry and Biological Properties*, Taylor & Francis Group, London, **2000**.
U. Schubert, N. Hüsing, A. Lorenz, *Hybrid Inorganic-Organic Materials by Sol-Gel Processing of Organofunctional Metal Alkoxides. Chem. Mater.* **1995**, *7*, 2010–2027.
C. J. Brinker, G. W. Scherer, *Sol-Gel Science: The Physics and Chemistry of Sol-Gel Processing*, Academic Press, London, **1990**.

Hybrid materials

Special issue of *J. Mater. Chem.* **2005**, *15*, 3543–3986.

P. Gómez-Romero, C. Sanchez (Eds.) *Functional Hybrid Materials*, Wiley-VCH, Weinheim, **2004**.

G. Schottner, *Hybrid Sol-Gel-Derived Polymers: Applications of Multifunctional Materials*. *Chem. Mater.* **2001**, *13*, 3422–3435.

G. Kickelbick, U. Schubert, *Inorganic Clusters in Organic Polymers and the Use of Polyfunctional Inorganic Compounds as Polymerization Initiators*, *Monatsh. Chem.* **2001**, *132*, 13.

K.-H. Haas, *Hybrid inorganic-organic polymers based on organically modified Si-alkoxides*. *Adv. Eng. Mater.* **2000**, *2*, 571–582.

C. Sanchez, F. Ribot, B. Lebeau, *Molecular design of hybrid organic-inorganic nanocomposites synthesized via sol-gel chemistry*. *J. Mater. Chem.* **1999**, *9*, 35–44.

K. G. Sharp, *Inorganic/organic hybrid materials*. *Adv. Mater.* **1998**, *10*, 1243–1248.

Nanocomposites

Special issue of *J. Nanosci. and Nanotechn.* **2006**, *6*, 265–572.

R. M. Laine, *Nanobuilding blocks based on the [OSiO1.5]x (x = 6, 8, 10) octasilsesquioxanes*. *J. Mater. Chem.* **2005**, *15*, 3725–3744.

A. Usuki, N. Hasegawa, M. Kato, *Polyme-clay nanocomposites*. *Adv. Polym. Sci.* **2005**, *179*, 135–195

B. C. Sih, M. O. Wolf, *Metal nanoparticle-conjugated polymer nanocomposites*. *Chem. Commun.* **2005**, 3375–3384.

C. Sanchez, B. Julian, P. Belleville, M. Popall, *Applications of hybrid organic-inorganic nanocomposites*. *J. Mater. Chem.* **2005**, *15*, 3559–3592.

P. Gómez-Romero, K. Cuentas-Gallegos, M. Lira-Cantu, N. Casan-Pastor, *Hybrid nanocomposite materials for energy storage and conversion applications*. *J. Mater. Sci.* **2005**, *40*, 1423–1428.

G. Kickelbick, *Concepts for Incorporation of Inorganic Building Blocks into Organic Polymers on a Nanoscale*. *Progr. Polym. Sci.* **2003**, *28*, 83–114.

Special issue of *Chem. Mater.* **2001**, *13*, 3059–3910.

Porous materials

A. Sayari, S. Hamoudi, *Periodic Mesoporous Silica-Based Organic-Inorganic Nanocomposite Materials*. *Chem. Mater.* **2001**, *13*, 3151–3168.

F. Hoffmann, M. Cornelius, J. Morell, M. Fröba, *J. Nanosci. Nanotechnol.* **2006**, *6*, 265–288.

B. Hatton, K. Landskron, W. Whitnall, D. Perovic, G. A. Ozin, *Past, Present, and Future of Periodic Mesoporous Organosilicas – The PMOs*. *Acc. Chem. Res.* **2005**, *38*, 305–312

J-L. Shi, Z.-L. Hua, L.-X. Zhang, *Nanocomposites from ordered mesoporous materials*. *J. Mater. Chem.* **2004**, *14*, 795–806.

R. Gangopadhyay, A. De, *Conducting Polymer Nanocomposites: A Brief Overview*. *Chem. Mater.* **2000**, *12*, 608–622.

C. J. Brinker, Y. Lu, A. Sellinger, H. Fan, *Evaporation-induced self-assembly. Nanostructures made easy*. *Adv. Mater.* **1999**, *11*, 579–585.

N. Hüsing, U. Schubert, *Aerogels – airy materials: chemistry, structure, and properties*. *Angew. Chem. Int. Ed.* **1998**, *37*, 22–45.

Building blocks

Macromolecules

D. Y. Godovsky, *Device applications of polymer-nanocomposites*. *Adv. Polym. Sci.* **2000**, *153*, 163–205.

J. Wen, G. L. Wilkes, Garth L. *Organic/Inorganic Hybrid Network Materials by the Sol-Gel Approach*. *Chem. Mater.* **1996**, *8*, 1667–1681.

Dendrimers

G. J. de A. A. Soler-Illia, L. Rozes, M. K. Boggiano, C. Sanchez, C.-O. Turrin, A.-M. Caminade, J.-P. Majoral, *New mesotextured hybrid materials made from assemblies of dendrimers and titanium(IV)-oxo-organo clusters*. *Angew. Chem. Int. Ed.* **2000**, *39*, 4249–4254.

J. L. Hedrick, T. Magbitang, E. F. Connor, T. Glauser, W. Volksen, C. J. Hawker, V. Y. Lee, R. D. Miller, *Application of complex macromolecular architectures for advanced microelectronic materials*. *Chem. – Europ. J.* **2002**, *8*, 3308–3319.

J. W. Kriesel, T. D. Tilley, *Carbosilane dendrimers as nanoscopic building blocks for hybrid organic-inorganic materials and catalyst supports*. *Adv. Mater.* **2001**, *13*, 1645–1648.

P. R. Dvornic, C. Hartmann-Thompson, S. E. Keinath, E. J. Hill, *Organic-Inorganic Polyamidoamine (PAMAM) Dendrimer-Polyhedral Oligosilsesquioxane (POSS) Nanohybrids*. Macromolecules **2004**, *37*, 7818–7831.

R. W. J. Scott, O. M. Wilson, R. M. Crooks, *Synthesis, Characterization, and Applications of Dendrimer-Encapsulated Nanoparticles*. J. Phys. Chem. B **2005**, *109*, 692–704.

Particles

V. Castelvetro, C. De Vita, Cinzia. *Nanostructured hybrid materials from aqueous polymer dispersions*. Adv. Coll. Interf. Sci. **2004**, *108–109*, 167–185.

E. Bourgeat-Lami, *Organic/inorganic nanocomposite colloids* in Encyclopedia of Nanoscience and Nanotechnology (Ed. H. S. Nalwa), American Scientific Publishers, Stevenson Ranch **2004**.

G. Kickelbick, L. M. Liz-Marzán, *Core-Shell Nanoparticles* in Encyclopedia of Nanoscience and Nanotechnology (Ed. H. S. Nalwa), American Scientific Publishers, Stevenson Ranch, **2004**.

Engineering of nanomaterials

G. A. Ozin, A. C. Arsenault, *Nanochemistry – A Chemical Approach to Nanomaterials*, RSC Publishing, London, **2005**.

2
Nanocomposites of Polymers and Inorganic Particles
Walter Caseri

2.1
Introduction

Polymers are ubiquitous and indispensable in daily life. Depending on the field of application, significant properties of polymers are, for example, low density, corrosion resistance, or electrical or thermal insulating behavior. In addition, the processing of polymers proceeds rapidly and consumes relatively low energy. However, polymers are applied not only on their own but often together with inorganic components which are enclosed in the polymer. Such multiphase systems belong to the class of composites, i.e. to materials which are composed of two or more different basic materials. The chemical nature of the inorganic component is manifold. Calcium carbonate is most commonly employed, but other salt-like compounds, metals, or compounds based on an extended structure of covalently bound atoms (e.g. carbon black or silicon dioxide) are also used. Reasons for using polymers not only as neat substances but also in combination with inorganic moieties are reduction of cost (e.g. calcium carbonate is drastically cheaper than any polymer) or an improvement of materials properties with respect to the neat polymer. In this context, the inorganic component is added to polymers, e.g. in order to increase stiffness, tensile strength, hardness, or abrasion resistance; to improve dimensional stability, heat resistance, or fire resistance; or to modify optical properties such as color or gloss. For some applications, the electrical and thermal insulating behavior of polymers is disadvantageous and this drawback can be improved by the addition of conducting inorganic materials.

In polymer composites, the polymer forms a continuous phase (also called matrix) in which the inorganic component is embedded. The latter is commonly present as spherical, cubic or plate-like particles or as fibers. In fact, not only the chemical nature but also the shape of the particles can influence properties of composites. Fibers such as glass fibers or carbon fibers (typical diameters in the order of 10 µm) especially improve mechanical properties. A pronounced influence of the shape of the inorganic component is observed in the electrical conductivity of composites where the polymer acts as an insulator and the inorganic material as a conductor, as will be referred to in Chapter 10.

Hybrid Materials. Synthesis, Characterization, and Applications. Edited by Guido Kickelbick
Copyright © 2007 Wiley-VCH Verlag GmbH & Co. KGaA, Weinheim
ISBN: 978-3-527-31299-3

Polymeric composites containing particle-like entities are also called filled polymers and consequently the incorporated objects are denoted as filler. A common range of inorganic filler sizes is of the order of 0.5–50 µm, although fillers consisting of primary particles with smaller diameters have also been used for a long time. Nonetheless, composites which contain inorganic particles up to a size of several tens of nanometers, attracted particular attention just in the last decade. Due to the small size of the inorganic particles ("nanoscale"), they are frequently termed nanosized particles or nanoparticles (formerly the designations colloids or ultrafine particles were more common), and the corresponding composites are therefore called nanocomposites. The term nanocomposite is also applied to polymer matrices that contain inorganic objects with nanoscale sizes in only one or two dimensions (nanoplatelets or nanofibers, respectively). A selection of polymers used as matrices for nanocomposites is presented in Table 2.1. Most investigations on polymer nanocomposites have been devoted to matrices containing rather spherical (or cubic) or plate-like particles. Since nanocomposites of the latter type are dealt with in a subsequent chapter, emphasis is placed in the following to nanocomposites comprising particles with more or less regular shape, although composites with rod-like nanoparticles have also been described, for instance with carbon nanotubes. Due to their electrical conductivity in combination with their rod-like shape, such species may introduce electrical conductivity in insulating polymer matrices at very low content of inorganic material. Carbon nanotubes are often expected to be particularly suited for the reinforcement of polymers, however, to date the performance of the resulting materials has been at most moderate, especially compared to their costs.

When decreasing the size of particles with rather uniform shape in three dimensions, optical and electrical characteristics and also other attributes can change or improve, as will be evident from the implementations below. Table 2.2 shows properties that have been introduced in polymer matrices by inorganic particles. Some of these properties (e.g. dichroism, absorption of UV light accompanied by full transmittance of visible light, or superparamagnetism) cannot be obtained with uniformly shaped particles in the micrometer range; other features (e.g. catalysis, photoconductivity, or vapor-dependent volume expansion caused by adsorption of organic vapors at particle surfaces) are expected to be improved by the use of nanoparticles compared to their counterparts in the micrometer range; and some properties can become unique for composites if coexistence of high translucence is required since high translucence is characteristic for individually dispersed very small particles, as discussed below. It will also become obvious below that encapsulation of nanoparticles in a surface-bound shell of organic molecules is frequently beneficial. This shell can be decisive for the dispersibility of the particles in the polymer matrix but also for the creation of a firm interface between particles and polymer matrix.

One should be aware that the size distribution of nanoparticles embedded in polymers is usually not monodisperse (i.e. the particles diameters are not identical), except in the case of small clusters which were built up by stepwise chemical reactions. Hence, the term "monodisperse particles", which is sometimes found in the literature, is often misleading since the particles appear frequently

Table 2.1 Examples of polymers that have been applied in nanocomposites.

Polymer	Structure	Remarks
Poly(acrylic acid)	(–CH$_2$—CHR–)$_n$, R=COOH	Water-soluble[a], coordination sites to inorganic species
Poly(acryl amide)	(–CH$_2$—CHR–)$_n$, R=CONH$_2$	Water-soluble[a]
Poly(aniline)	(–C$_6$H$_4$—NH–)$_n$[b]	Electrically conductive matrix
Poly(ethylene)	(–CH$_2$—CH$_2$–)$_n$	Formation of orientated structures by drawing, particularly attractive technical polymer
Poly(ethyleneoxide)	(–CH$_2$—CH$_2$—O–)$_n$	Water-soluble[a]
Poly(methyl methacrylate)	[c]	Soluble in many organic solvents[d]
Poly(propylene)	(–CH$_2$—CHR–)$_n$, R=CH$_3$	Particularly attractive technical polymer
Poly(styrene)	(–CH$_2$—CHR–)$_n$, R=C$_6$H$_5$	Soluble in many organic solvents[d]
Poly(vinyl alcohol)	(–CH$_2$—CHR–)$_n$, R=OH	Water-soluble[a], coordination sites to inorganic species, formation of orientated structures by drawing
Poly(vinyl carbazole)	[c]	Photoconductive matrix
Poly(2-vinyl pyridine), Poly(4-vinyl pyridine)	[c]	Soluble in aqueous acids[a], coordination sites to inorganic species
Poly(N-vinyl pyrrolidone)	[c]	Water-soluble[a]

a Suited for the preparation of nanocomposites via *in-situ*-formation of the inorganic particles from ionic presursors.
b Basic structure, poly(aniline) exists in different states of oxidation or protonation, respectively
c See scheme below.
d Suited for the preparation of nanocomposites with surface-modified particles.

polydisperse (i.e. the particles show different diameters) upon careful analysis. In such cases, the expression "particles with narrow size distribution" would be more correct. The reported particle diameters typically refer to number average particle diameters d_{na}, defined as

Poly(methyl methacrylate)

Poly(vinyl carbazole)

Poly(2-vinyl pyridine)

Poly(4-vinyl pyridine)

Poly(N-vinyl pyrrolidone)

Table 2.2 Examples of inorganic particles that introduce selected properties in polymeric matrices.

Discipline	Property	Example
Optics	Color	Au, Ag
	Dichroism	Au, Ag (parallel arrays)
	Absorption of both visible and UV light	Carbon black
	Absorption of UV light at full transmittance of visible light	TiO_2, ZnO
	Iridescence	SiO_2 (ordered lattice)
	Ultrahigh refractive index	Si, PbS
	Ultralow refractive index	Au
Electrical conductivity	Enhancement of electrical conductivity of polymers	Carbon black
	Photoconductivity (in combination with a conducting polymer matrix)	PbS, TiO_2, CdS, HgS
	Vapor-dependent electrical conductivity (using a conductive polymer matrix)	Cu
Miscellaneous	Superparamagnetism	Co, γ-Fe_2O_3
	Catalytic activity	CdS, Pd, Pt, Rh, Ir
	Reinforcement of elastomers	Carbon black, SiO_2
	Retardation of polymer decomposition by heat	$Al(OH)_3$
	Vapor-dependent volume expansion caused by vapor adsorption at particle surfaces	Au

$$d_{na} = \frac{\sum_i n_i d_i}{\sum_i n_i}$$

where n_i is the number of particles with diameter d_i. However, physical properties such as the refractive index of nanocomposites can depend on the volume fraction of the particles. Therefore, the volume-weighted average diameter d_{va} might be more expedient in such cases, where

$$d_{va} = \sqrt[3]{\frac{\sum_i n_i d_i^3}{\sum_i n_i}}$$

Fortunately, the number-weighted and the volume-weighted average diameters of the particles differ commonly by only 20–30%, which usually does not influence the results considerably. If very high volume fractions of particles are attempted, possible limitations in the volume fractions should be considered. For monodisperse spherical particles, the volume fraction cannot exceed a value of 0.74 for geometric reasons: even if the spheres are packed as tight as possible, there is still

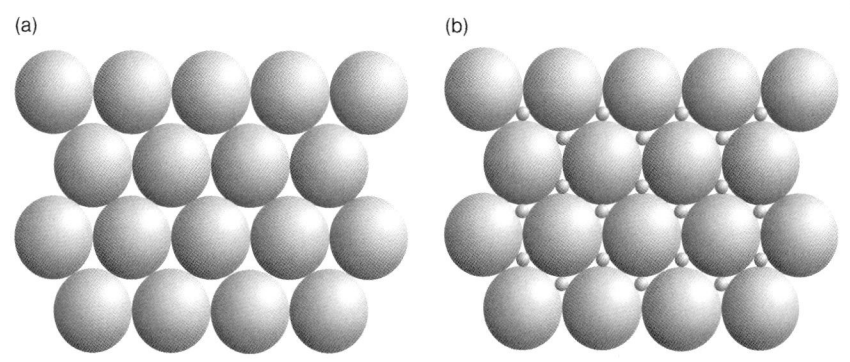

Fig. 2.1 Schematic two-dimensional representation of the densest packing of spheres: (a) monodisperse spheres, (b) polydisperse spheres with two different kinds of diameters whereby the smaller spheres fit into the spaces of the densest packing of the larger spheres.

some hollow space between them (Fig. 2.1). For polydisperse spherical particles, the maximum volume fraction might rise above that of monodisperse particles because smaller particles could fit in the free space of the densest packing of the larger particles (see Fig. 2.1). By contrast, monodisperse cubic particles can be packed without free space between the particles, i.e. the volume fraction of such particles can adopt any value in nanocomposites. However, in reality cubic particles are expected to show a certain polydispersity which leads to a less tight packing of the cubes, i.e. the upper limit of the volume fraction of polydisperse cubic particles is below that of monodisperse particles.

2.2
Consequences of Very Small Particle Sizes

Properties of nanocomposites can differ from those of analogous composites with larger particles because physical constants can depend on the particle size. This arises in particular when physical characteristics are based on the correlated interactions of numerous atoms, which has to be taken into account for metals and semiconductors. Namely, the energy levels of individual metal atoms (e.g. Au) or basic formula units of semiconductors (e.g. CdS) split upon formation of larger entities (e.g. Au_n or $(CdS)_n$) into more and more components with increasing number of atoms until they reach the quasi-continuous band structure of the bulk solid. For instance, individual sodium atoms exhibit sharply located energy levels as reflected in the optical spectrum by sharp absorption lines around 589 nm. In Na_3, the larger number of energy levels causes numerous but still resolved lines in the optical spectrum, while the myriad energy levels in Na_8 are hardly resolved and lead only to a broad absorption band with a maximum around 490 nm. This band shifts to ca. 570 nm for sodium particles with a diameter around 10 nm, and finally the spectrum of a sodium film (thickness 10 nm) does not show an absorption maximum anymore but a continuous increase in the visible wavelength range. Similarly, the optical absorption spectrum of CdS particles of ca. 4 nm diameter showed a well-pronounced maximum at 368 nm in contrast to the spectrum of bulk CdS which was characterized by a clearly different course (Fig. 2.2). Thus, UV/vis absorption spectra of metal or semiconductor nanoparticles show commonly shoulders or absorption maxima, which are characteristic for a certain diameter of the respective material. The particle-size-dependent UV/vis spectra imply that the color, as the most obvious optical property, may depend on the size of metallic or semiconductor particles (cf. Fig. 2.3, and examples that arise in the following sections), as well as the band gap which amounts, e.g. for CdSe particles of 3.9 nm diameter to 2.22 eV (absorption edge 558 nm) compared to 1.74 eV (absorption edge above 700 nm) for bulk CdSe. Generally, optical properties of particles begin to change significantly below a certain diameter, which is specific for each substance. For selected semiconductors, this diameter amounts to roughly 2 nm for CdSe, 2.8 nm for GaAs, 20 nm for PbS, and 46 nm for PbSe, as illustrated by the refractive index of PbS, which was estimated by extrapolation of refractive indices of gelatin–PbS nanocomposites with different PbS contents (Fig. 2.4). The refractive index around 4 for PbS particles with sizes above ca. 20 nm, which was close to that of the bulk material (4.3 at 619.9 nm and 1300), began to decrease when the PbS diameters fell below 15–20 nm. For diameters of 4 nm, the estimated refractive indices for PbS were only in the range of 2.

Properties such as the color or maximum absorption wavelength of nanoparticles can vary not only as the size of the primary particles but also as the distance to neighboring particles changes (cf. Fig. 2.3). Accordingly, individually dispersed

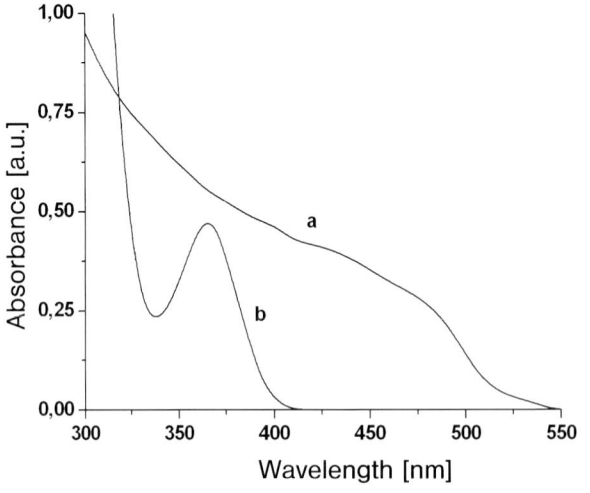

Fig. 2.2 UV/vis spectra of CdS. (a) bulk, (b) particles of a diameter of ca. 4 nm; sketch according to data of *J. Mater. Chem.* **1996**, *6*, 1643 and N. Herron, Y. Wang, H. Eckert, *J. Am. Chem. Soc.* **1990**, *112*, 1322.

a

Different particle diameters

b

Different distances between particles

c

Different orientation of linear particle assemblies with respect to the direction of linearly polarized light

Fig. 2.3 Origin of color differences of nanocomposites for a given metal or semiconductor. The double arrow in Example C, which is devoted to dichroism, indicates the polarization direction of the polarized light. The absorption maxima in optical spectra of the structures on the right side arise at higher wavelengths than those at the left side.

red silver or gold nanoparticles became blue at narrow distances between adjacent particles, and accordingly the absorption maximum in UV/vis spectra shifted to higher wavelengths upon decreasing the distance between the particles. This is of particular interest in polymer nanocomposites which contain uniaxially oriented

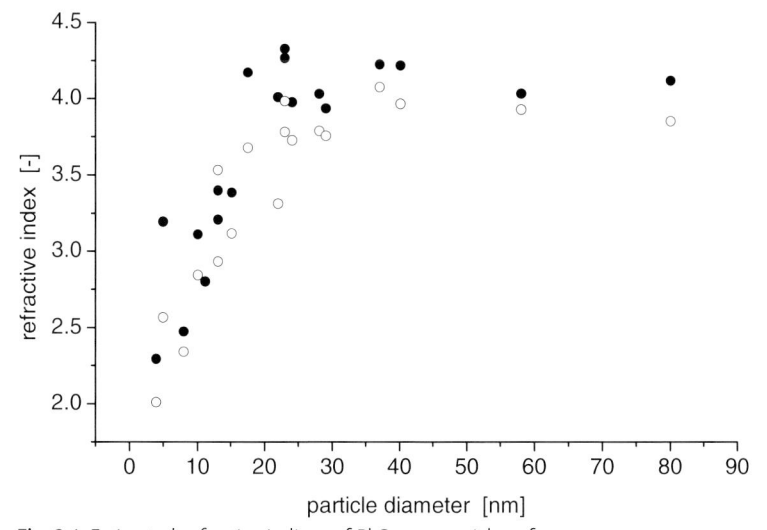

Fig. 2.4 Estimated refractive indices of PbS nanoparticles of different diameters at 1500 nm (open circles) and at 632.8 nm (filled circles), according to *J. Phys. Chem.* **1994**, *98*, 8992.

linear particle assemblies. At appropriate dimensions of such assemblies, polarized light interacts in a different way at the parallel and perpendicular orientations of the polarization plane with respect to the long axis of the particle assemblies, since the cooperative optical effects of adjacent metal particles under the influence of electromagnetic radiation are more pronounced for parallel orientation (provided that the short axis of the particle assemblies is small enough). As a consequence, the maximum absorption wavelength is higher for parallel than for perpendicular orientation between polarization and particle assembly direction and therefore different colors arise at parallel and perpendicular orientation, which is called dichroism (Fig. 2.3).

Since light scattering by particles with dimensions far below the wavelengths of visible light, i.e. far below 400 nm, is markedly reduced or even not relevant anymore, nanocomposites frequently appear translucent while larger particles provide opacity in composites of considerable thickness and particle content (unless the refractive index of the particles is very close to that of the polymer). Opacity also arises if nanoparticles are considerably agglomerated. The intensity loss of light caused by scattering depends in nanocomposites with randomly dispersed spherical particles on the particle radius r and refractive index n_p of the particles, the refractive index n_m of the polymer matrix, the volume fraction ϕ_p of the particles, the wavelength λ of the incident light, and the thickness x of the nanocomposite. The intensity loss can be estimated with the equation

$$\frac{I}{I_0} = e - \left[\frac{32\Phi_p x \pi^4 r^3 n_m^4}{\lambda^4} \left| \frac{\left(\frac{n_p}{n_m}\right)^2 - 1}{\left(\frac{n_p}{n_m}\right)^2 + 2} \right|^2 \right]$$

whereat I is the intensity of the transmitted and I_0 of the incident light. The transmittance I/I_0 becomes virtually 0 when the incident light is most efficiently scattered and 1 if no scattering occurs. The latter arrives if the refractive indices of polymer and particles are identical, independent on the other parameters. The transmittance of composites is shown in Fig. 2.5 as a function of the particle diameter d (note that $r = d/2$ in the above formula) for $x = 0.5$ mm, $\phi_p = 0.1$, and $\lambda = 589.3$ nm (Na$_D$ line). A refractive index of the matrix of 1.5 was selected, which is located in the usual range of organic polymers (Table 2.3). As refractive indices of the particles, values of common fillers were selected from Table 2.3, namely of BaSO$_4$ ($n_p = 1.64$), ZnO ($n_p = 2.00$), TiO$_2$ (rutile, $n_p = 2.62$), and in addition of the high refractive index compound PbS ($n_p = 3.91$) which has been applied particularly in nanocompsites (see below). With those parameters, the intensity loss by scattering becomes pronounced at particle diameters above ca. 5–25 nm, depending on the refractive index difference between particles and matrix, and the transmittance is close to 0 at particle diameters of 25–125 nm. It is obvious that the particle diameter plays a decisive role for the transparence; particularly at

Table 2.3 Refractive indices (at a wavelength of 589.3 nm) of polymers and inorganic compounds (J. C. Seferis in *Polymer Handbook* (Eds: J. Brandrup, E. H. Immergut), Wiley & Sons, New York, **1989**, p. VI/451; P. E. Liley, G. H. Thomson, D. G. Friend, T. E. Daubert, E. Buck in *Perry's Chemical Engineers' Handbook* (Eds: R. H. Perry, D. W. Green), McGraw-Hill, New York, **1997**, Section 2 p. 2–7).

Substance	Refractive index
Poly(tetrafluoroethylene)	1.35–1.38
Poly(vinylidene fluoride)	1.42
Poly(dimethylsiloxane)	1.43
Poly(oxypropylene)	1.45
Poly(vinyl acetate)	1.47
Poly(methyl acrylate)	1.47–1.48
Poly(oxymethylene)	1.48
Poly(methyl methacrylate)	1.49
Poly(vinyl alcohol)	1.49–1.53
Poly(ethylene)	1.51–1.55
Poly(1,3-butadiene)	1.52
Poly(acrylonitrile)	1.52
Poly(acrylic acid)	1.53
Nylon 6,6	1.53
Poly(vinyl chloride)	1.54–1.55
Poly(chloroprene)	1.55–1.56
Poly(styrene)	1.59
SiO_2 (quartz)	1.54
$CaCO_3$ (calcite)	1.55
$Mg(OH)_2$ (brucite)	1.56
$K_2Al_6Si_6O_{22} \cdot 2H_2O$ (muscovite)	1.59
$BaSO_4$ (barite)	1.64
$CaCO_3$ (aragonite)	1.68
$MgCO_3$ (magnesite)	1.70
Al_2O_3 (corundum)	1.77
ZnO (zincite)	2.00
Fe_3O_4 (magnetite)	2.42
TiO_2 (anatase)	2.53–2.56
TiO_2 (rutile)	2.62
Fe_2O_3 (hematite)	3.04
PbS (galena)	3.91
MoS_2 (molybdenite)	4.7

pronounced refractive index differences between matrix and particles, and that composites with very small particles can be highly transparent even at high refractive index differences. Note that, as indicated above, the transmittance can also decrease by absorption. For very small colored particles, absorption is predominantly responsible for decrease in transmittance. Thus, the intensity loss of light transmitted through dispersions of gold particles with diameters below 50 nm

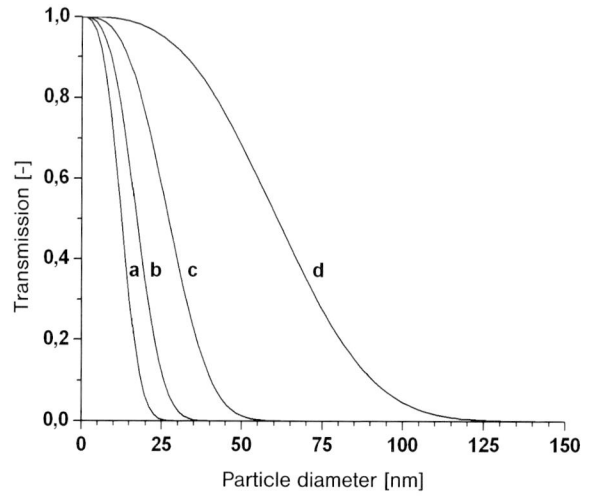

Fig. 2.5 Calculated transmission of composites as a function of the particle diameter at a wavelength of 589.3 nm (Na_D line) for a sample of 5 mm thickness loaded with 10% v/v particles. The refractive index of the matrix is 1.5 and that of the particles (a) 3.92, (b) 2.62, (c) 2.00, and (d) 1.64.

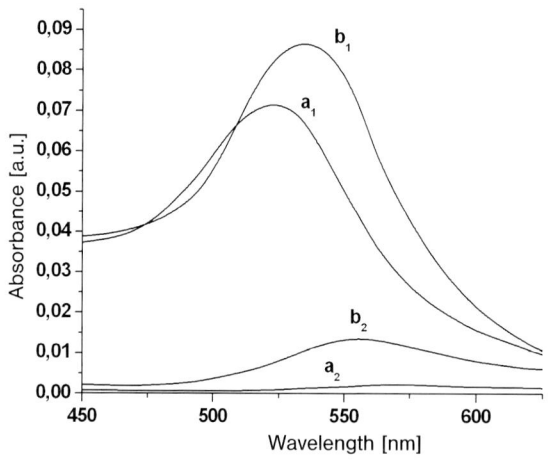

Fig. 2.6 Absorption (a_1) and scattering (a_2) by dispersed gold particles of ca. 20 nm diameter, and absorption (b_1) and scattering (b_2) by particles of 51 nm diameter; approximate representation according to Z. Chem. Ind. Kolloide **1907**, 2, 129.

originates mainly from absorption of light (Fig. 2.6) while scattering of light becomes more prominent for larger particles.

The small size of nanoparticles is important not only for optical and electronic properties, including photocatalytic processes where the diffusion of electrons and holes was considered to be more efficient in small particles thus rendering nanoparticles more suited for related applications, but also for processes which become more efficient at high internal surface area (interface area), such as catalytic activity. The interface area of the particles in nanocomposites can be extremely large, which is illustrated by a simple model calculation. Assuming that the nanocomposites contain individually dispersed particles with sharply separated polymer–particle interfaces (the surface area exposed to the environment is not taken into account since it is usually much smaller than the surface area within the nanocomposites), the interface area of the particles per volume unit, S_V (in m^{-1}, i.e. in m^2 interface area per m^3 volume of composite) is given by

$$S_V = \frac{A_i}{V_0} = \sum_k \frac{n_k A_k}{V_0}$$

where V_0 is the volume of the nanocomposite, n_k the number of particles of size k, A_i the total interface area of the nanoparticles, and A_k the interface area of a single particle of size k. Since

$$n_k = \frac{\varphi_k V_0}{V_k}$$

with φ_k the volume fraction of the particles of size k and V_k the volume of a single particle of size k, S_V becomes for cubic or spherical particles

$$S_V = 6 \sum_k \frac{\varphi_k}{d_k}$$

where d_k is the edge length of the cubes or the diameter of the spheres, respectively. Thus at a given volume fraction of particles, S_V increases with decreasing particle size. Assuming, for simplicity, that the particles are monodisperse with a d_k of 10 nm and a φ_k of 0.001–0.5, the value of S_V becomes as high as $6 \cdot 10^5$–$3 \cdot 10^8$ m^{-1}. Interface areas of this magnitude are of advantage in applications such as catalysis since inorganic particles acting as catalysts are basically the more efficient the higher the contact area between the particles and the molecules to be converted in the catalytic reaction (consider, however, that the catalytic activity itself can also depend on the particle size). As another example, the large interface area in nanocomposites can favor an efficient separation of photoinduced charges, which enhances the output of photoconductive composites based on semiconducting polymers.

When dealing with materials with high interface areas, reflections on proper surfaces, i.e. surfaces exposed to the vacuum, are adequate. In contrast to a bulk phase, the force field in such a surface region is not homogeneous displaying an energy gradient perpendicular to the surface. While no net energy is expended in transporting matter reversibly within a bulk phase, energy is required to transport matter from the bulk phase to the surface region, i.e. energy is employed in the generation of surface area. Thus proper surfaces are basically in an energetically unfavorable state. This is represented by the surface free energy, which corresponds to the reversible work for the formation of a unit surface area. Compared to surfaces exposed to the vacuum, the free energy of internal surfaces in nanocomposites can be lowered by interactions between polymer and particles. Therefore, the interface free energy becomes maximal in absence of polymer–particle interactions. The sum of the maximum interface free energies of polymer and particles per volume of nanocomposite, $E_{V,max}$, is given by the expression (whereat the interface area of the polymer is equal to that of the inorganic particles and the surface area of the nanocomposite body itself is neglected)

$$E_{V,\max} = S_V \left(\gamma_f + \gamma_p \right)$$

with γ_f the surface free energy of the inorganic particles and γ_p the surface free energy of the polymer in the vacuum. The maximum interface free energy is dominated by the inorganic particles because surface free energies of inorganic materials are generally much higher (typically 0.5–2.5 J/m^2) than those of polymers (typically 0.01–0.05 J/m^2) (the surface free energy of very small particles is similar to that of smooth macroscopic surfaces if the intermolecular forces are dominated by van der Waals interactions, otherwise the surface free energy of very small particles and macroscopic surfaces can differ). Assuming values of $\gamma_f + \gamma_p$ between 0.5 and 2.5 J/m^2 and a nanocomposite containing 10% v/v of monodisperse particles of 10 nm diameter (i.e. $\phi_k = 0.1$ and $d_k = 10$ nm), $E_{V,max}$ adopts values as high as $3 \cdot 10^7$–$1.5 \cdot 10^8$ J/m^3. As a consequence, a more favorable thermodynamic state might be reached by lowering the interface free energy. This can happen in two ways: the interface area and concomitantly the high interface free energy of the inorganic component can be lowered by contacts between the inorganic particles themselves, i.e. by agglomeration of the particles. In fact, agglomeration of inorganic particles can give rise to problems in the preparation of nanocomposites. First of all, commercially available pristine inorganic nanoparticles (in particular SiO_2, TiO_2 and carbon black) are present as agglomerates which can hardly be divided into the individual primary particles upon processing to nanocomposites. As an alternative, the free energy of a nanocomposite can be decreased by interactions between particles and matrix polymer or between particles and an ultrathin layer of surrounding organic molecules, i.e. by encapsulation of the particles in a shell of strongly adhering organic compounds (core-shell structures, Fig. 2.7). So the outer surface of the particles consists of organic matter. Since the surface free energy of organic matter is typically one or two orders of magnitudes below that of inorganic matter, the agglomeration tendency of the particles in a polymer matrix is expected to be drastically reduced provided the chemical nature of the shell and the polymer allow a miscibility at all and if chemical interactions between the shells of adjacent particles are little pronounced. Thus, gold particles with a shell of 1-alkanethiolates (Fig. 2.8) disperse well in poly(ethylene) as long as the surface layer is intact. If, however, the surface layer is partially removed by thermal treatment, the gold cores get in contact and the particles start to agglomerate. In case of strong interactions between core and shell, the free energy which is gained by core-shell interactions can overcome the interface free energy between inorganic core and organic polymer. This might be possible for gold, copper or silver

Organic layer

Inorganic particle

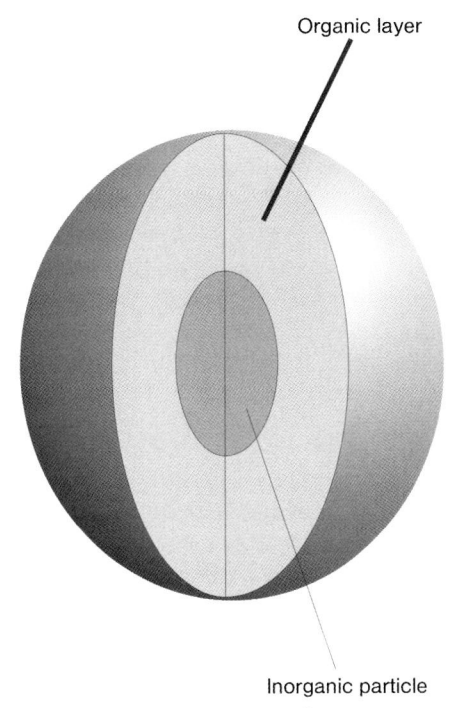

Fig. 2.7 Schematic illustration of an inorganic particle which is surrounded by a surface-bound layer of organic molecules (core-shell structure).

particles which are densely covered with 1-alkanethiolates, where the sulfur atoms interact strongly with the metal surfaces. The surface free energy of SiO_2 particles is frequently lowered by treatment with silanes (e.g. of the formula $(OR')_3SiR$ or Cl_3SiR) in presence of water, which leads to a strongly bound layer of crosslinked polysiloxane, as a result of hydrolysis of Si—OR' or Si—Cl bonds.

If the interactions between the components of a nanocomposite are not sufficient to compensate the interfacial free energies, the formation of nanocomposites is kinetically controlled, which is probably true for many cases. Once the

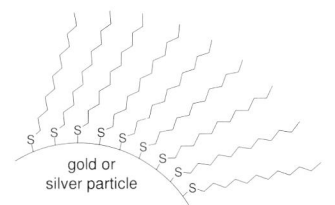

Fig. 2.8 Schematic illustration of a section of gold or silver particles covered by a surface-bound layer established by adsorption of 1-dodecanethiol. It is assumed that the thiol group is converted to a thiolate group after contact with the gold surface.

nanocomposites are formed, however, subsequent agglomeration of particles is usually negligible even over extended periods due to the restricted mobility of the particles in the solid polymer matrix. For instance, in nanocomposites composed of an ethylene–methacrylic acid copolymer and PbS (ca. 3% v/v), the particle sizes remained constant after prolonged storage of the nanocomposites in air or extended exposure to water, dichloromethane or hexane, and the period required for the agglomeration of the PbS particles was estimated to take 10^5 years at 25 °C. Also, nanocomposites of gold colloids embedded in amine-terminated poly(ethylene oxide) did not show perceptible changes for more than a year, and the particle-induced dichroism in nanocomposites with silver and poly(ethylene) did not fade significantly for 6 years.

Although the physical concept of the surface free energy helps in understanding certain basic phenomena, a more detailed view of the behavior of particles embedded in polymers is provided when chemical interactions are included. Finally, the dispersion behavior of the particles is a result of the balance between the particle–particle, polymer–polymer and particle–polymer interactions. An organic shell can crucially influence this balance. Thus, an organic shell can induce uncommon effects, as in the case of platinum particles of 1–2 nm diameter which arranged into spherical superstructures in presence of ammonium O,O'-dialkyldithiophosphates which were supposed to coat the platinum particles (Fig. 2.9). The influence of the ammonium O,O'-dialkyldithiophosphates on the interactions between the components of the composite is evident from the fact that superstructures were not observed in absence of a surface-modifying agent. As another example, in aqueous solutions of appropriate ratios of poly(ethylene oxide) and the surface active agent sodium dodecylsulfate, *in-situ*-prepared PbS particles of 2–4 nm diameter assembled in nanotube superstructures (Fig. 2.10). The walls of those objects consisted of alternating layers of PbS particles and organic matter.

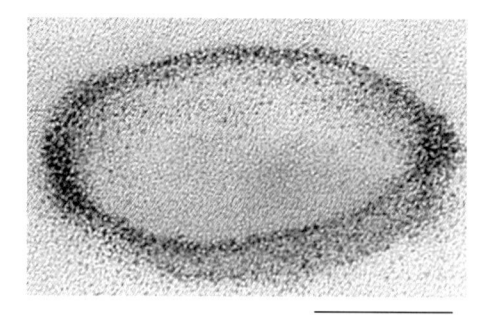

Fig. 2.9 Transmission electron micrograph of a superstructure of platinum nanoparticles embedded in poly(styrene) in presence of ammonium O,O'-dioctadecyldithiophosphate, (J. Phys. Chem. B **2001**, *105*, 7399).

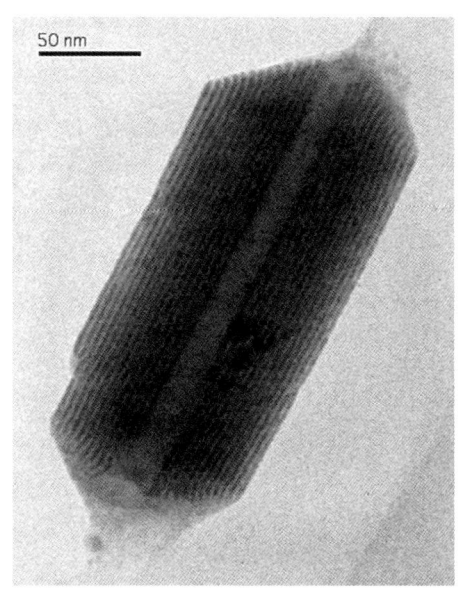

Fig. 2.10 Nanotube evolved from a mixture containing PbS nanoparticles, sodium dodecylsulfate and poly(ethylene oxide). The walls of the nanotube consist of layers of PbS particles separated by layers of organic matter (*Nano Lett.* **2003**, *3*, 569).

2.3
Historical Reports on Inorganic Nanoparticles and Polymer Nanocomposites

Since nanoparticles of metals or semiconductors are frequently colored, they can act as colorizing agents for polymers. This implementation was established originally in inorganic matrices. For instance, even in the 9th century nanoparticles prepared from silver, copper and iron salts were used to generate an iridescent glaze in pottery (luster). Metal nanoparticles have also been used to dye glass windows at least since the middle of the 16th century (so-called gold ruby glasses). Inorganic nanoparticles were embedded in polymers in 1833: gold was dissolved in aqua regia followed by evaporation of a part of the volatile contents at elevated temperature until a solid skin was observed on the surface of the aqueous solution. When the solution cooled down to room temperature, the reaction mixture became entirely solid. The resulting hygroscopic matter, which contained gold salts, was rapidly dissolved in water and the solution thus obtained was mixed with a solution of dissolved $SnCl_2$ and gum arabic. The $SnCl_2$ acted as a reducing agent for the gold ions, which yielded red gold nanoparticles. Subsequent addition of ethanol caused precipitation of gum arabic together with gold nanoparticles (coprecipitation). The same principle was applied in 1899 for the preparation of nanocomposites of gum and silver. Silver ions were reduced in an aqueous solution which also contained gum, and nanocomposites were obtained by coprecipitation upon addition of ethanol.

At the end of the 19th century, natural polymers (plant and animal fibrils) were treated with solutions of gold or silver salts (such as silver nitrate, silver acetate, or gold chloride) whereupon metal ions penetrated into the fibrils. By exposure to light, these ions were subsequently reduced to metal nanoparticles within the polymer matrix. The resulting samples exhibited a pronounced dichroism. For example, the color of gold-containing spruce wood fibers turned from red to blue green at parallel and respectively perpendicular orientation of the polarization direction of the light with respect to the fiber axis. By 1927, dichroic fibers had been prepared with 17 pure chemical elements, in particular Cu, Ag, Au, Hg, Os, Rh (designated with the symbol Rd in the original literature), Pd, Pt, P, As, Sb, Bi, S, Se Te, Br, and I, i.e. not only with metals. The colors observed in the dichroic samples were found to depend not only on the incorporated chemical element, but also on the particle size, e.g. the color of perpendicular and parallel orientation between polarization plane and fiber direction changed in ramie fibers from straw yellow to indigo blue for gold particles of 8.5 nm diameter and from claret red to green at 12.3 nm diameter. It was recognized early that this dichroism was caused by particles which assembled in uniaxially oriented spaces of the fibers to linear aggregates (Fig. 2.11). The *in-situ*-formed particles were investigated in more detail two or three decades later by X-ray diffraction. The evaluation of the line widths (Scherrer's equation) revealed particle diameters of silver and gold crystallites in dichroic samples between 5 and 14 nm, and the ring-like

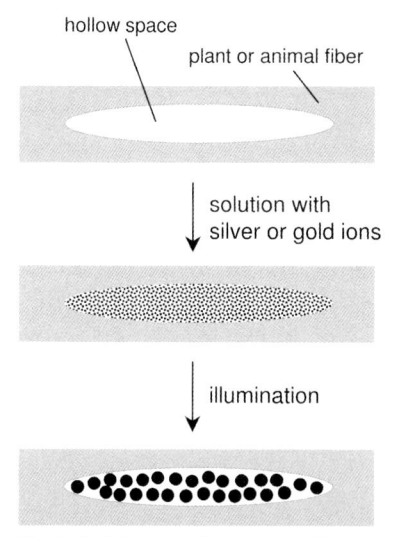

Fig. 2.11 Schematic illustration of the preparation of dichroic plant and animal fibers around 1900. The fibers, which contained oriented anisotropic hollow spaces, were impregnated with a solution of metal salts, and the corresponding metal ions were subsequently reduced and converted to nanoparticles by illumination.

diffraction patterns in oriented fibers indicated that the individual primary crystallites themselves were not oriented, which supported the view that the dichroism was not a consequence of an orientation of the individual particles themselves, but of an orientation of particle assemblies.

It may appear amazing that gold was used as the inorganic component in early studies on nanocomposites. However, this metal was well represented in initial reports on nanoparticles. According to Ostwald, a pioneer in nanoparticle research, Richter was the first who found experimental evidence for the presence of extremely small gold particles, namely in 1802, followed by, for instance, Faraday, Zsygmondy and Gutbier (silver nanoparticles) in the next 100 years. The final demonstration of the existence of nanoparticles was adduced by particle size measurements with the ultramicroscope in 1903. These instruments allowed the visualization of particles with dimensions below the resolution limit of optical microscopes with the help of light scattering. Emphasis was put on the investigation of glass–gold nanocomposites, and gold particle diameters were disclosed down to the resolution limit (ca. 6 nm for gold). Diameters of gold particles obtained with ultramicroscopy were confirmed by Scherrer in 1918 by evaluation of line widths in X-ray diffraction patterns. He also showed with X-ray diffraction that the gold lattices of nanoparticles and the bulk substance were identical even for particles as small as 4–5 unit cell lengths. Nonetheless, nanoparticles of substances other than gold (or silver) had also been synthesized already 100–200 years ago, for example platinum by reduction of $PtCl_4^{2-}$ with hydrazine in aqueous solution, SiO_2 by hydrolysis of tetramethoxysilane, aluminum oxides by hydrolysis of aluminum chloride, iron oxides by hydrolysis of iron chloride, and TiO_2 nanoparticles in the course of the discovery of titanium, by hydrolysis of titanium tetrachloride or by treatment of a gel-like titanate with a small amount of hydrochloric acid followed by dialysis. With regard to the use of nanoparticles in composites, particular attention has to be paid to SiO_2 produced by flame hydrolysis of $SiCl_4$ around 1940 and carbon black, which was shown to consist of nanoparticles by electron microscopy at the same time. Both SiO_2 nanoparticles and carbon black have been applied as reinforcing fillers for elastomers. Moreover, since carbon black absorbs UV and visible light efficiently, light-induced polymer degradation processes are markedly retarded by carbon black nanoparticles, and it was reported already in 1950 that carbon black extended the weathering life of poly(ethylene) from 6 months to 20 years. The distinct electrical conductivity of carbon black is another property, which has found application for decades in polymer composites. Since polymers are usually insulators, filling with carbon black can introduce electrical conductivity up to 1–10 S/cm.

2.4
Preparation of Polymer Nanocomposites

In order to prepare nanocomposites, either the particles or the polymers or both components can be synthesized *in situ* (i.e. not applied as isolated species) or used in the final state (Fig. 2.12). Mixing of the final or the precursor components can proceed, for instance, in solution or in a polymer melt. In such processes, inorganic or organometallic precursor molecules can also be bound to polymer molecules. The conversion of the precursors to the final inorganic particles can take place in solutions, in polymer melts or in polymeric solids. The chemical reactions for the generation of the inorganic nanoparticles are often simple. For example, salt-type nanoparticles are frequently synthesized by reaction of soluble precursor salts, e.g. Cu_2S particles from $[Cu(CH_3CN)_4]CF_3SO_3$ and Li_2S in N-methylpyrrolidone or PbS particles from $Pb(CH_3COO)_2$ and H_2S in water. Hydrolysis of precursors is also a common method for nanoparticle synthesis, e.g. SiO_2 particles are obtained by hydrolysis of tetraethoxysilane. In contrast to powders of bare inorganic particles, particles surrounded by an organic shell, which can be gained as solids or viscous substances, may disperse as individual primary particles in water, organic solvents or even directly in polymer melts, as indicated above. A number of such surface-modified inorganic particles are available, for example, gold, silver, platinum or CdS nanoparticles with a shell of organic thiols, where the thiol groups are attached to the particle surfaces. The surfaces of inorganic particles can also be modified with reactive organic groups that allow the attachment of matrix polymer molecules to the surfaces of the inorganic particles, thus leading to a firm polymer–particle interface. As an alternative to chemical methods, nanoparticles were also produced by evaporation of bulk materials (e.g. silver or gold).

Nanocomposites may be obtained basically not only as powders or coherent bodies without particular shape but also as films or fibers, using the common methods of polymer processing. Thus, films can be manufactured from a liquid mixture of the composite components by casting followed by evaporation of the

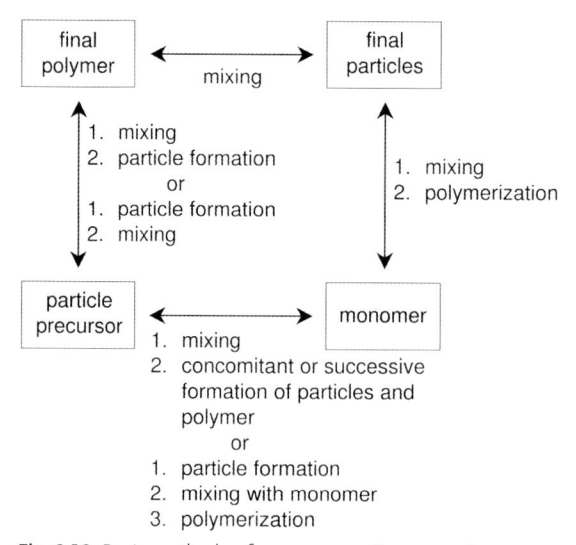

Fig. 2.12 Basic methods of nanocomposite preparation.

volatile parts or by insertion of a solid nanocomposite in a hot press, and fibers by conventional spinning processes including electrospinning.

2.4.1
Mixing of Dispersed Particles with Polymers in Liquids

A common method for nanocomposite preparation is based on the mixing of polymer and dispersed nanoparticles in a liquid (Fig. 2.13). If the particles are synthesized *in situ*, the polymer can be present already during the synthesis. In this case, coordination of precursors, e.g. metal ions or organometallic compounds, with polymer molecules may influence not only the final particle size and shape but also the particle distribution in the polymer matrix. The interactions between precursor and polymer can also be of ionic nature, as with cationic polyelectrolytes and $[PtCl_6]^{2-}$ or $[PdCl_4]^{2-}$ which are subsequently reduced to metal nanoparticles (Fig. 2.14).

Depending on the system, a polymer can destabilize or stabilize particles dispersed in a liquid medium. In the former case, the polymer and the particles precipitate together (coprecipitation) and the nanocomposite can be collected by filtration or decanting of the outstanding solution. Such procedures remove soluble reaction side products stemming from *in situ* synthesis of the particles. Spontaneous coprecipitation occurred, e.g. when PbS nanoparticles were prepared *in situ* in presence of poly(ethylene oxide). The particles in nanocomposites obtained by spontaneous coprecipitation can even be ordered in appropriate systems. When

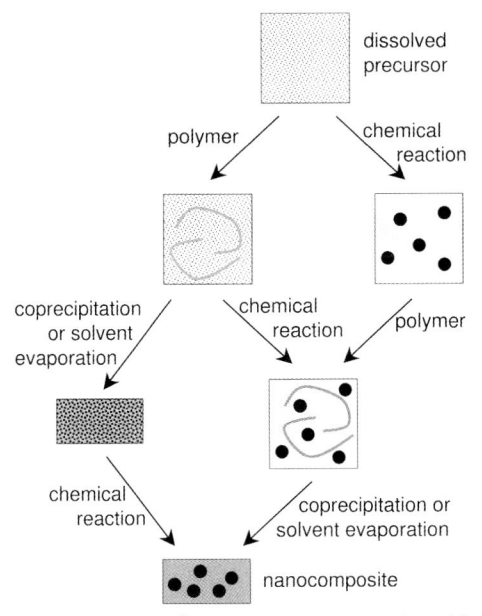

Fig. 2.13 Design of nanocomposite preparation with *in-situ*-prepared nanoparticles.

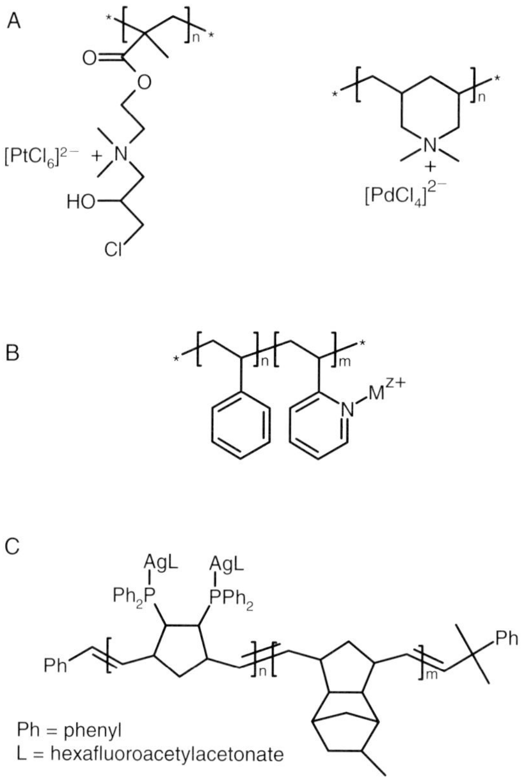

Fig. 2.14 A) Ionic interactions of cationic polyelectrolytes with anionic metal complexes prior to conversion to metal nanoparticles (cf. *Mater. Sci. Eng. C* **1998**, *6*, 155), B) coordination of metal ions (M^{z+} denotes Cu^{2+}, Co^{2+}, Ni^{2+}, Cd^{2+}, Zn^{2+}, or Ag^+) to a distinct block of a copolymer prior to conversion to metal sulfide nanoparticles (*New J. Chem.* **1998**, (no volume), 685), C) coordination of an organometallic precursor to a distinct block of a copolymer prior to conversion of silver nanoparticles (*New J. Chem.* **1998**, 685).

gold particles modified at the surface with a mixed layer of 1-octanethiol and 11-mercaptoundecanoic acid were merged with poly(amidoamine) (PAMAM) dendrimers, acid–base interactions between amine groups of the polymer and carboxylic acid group of the particle shell (Fig. 2.15) resulted in a rather ordered arrangement of the particles in the matrix, the distances between the particles depending on the size of the dendrimer molecules. If, on the other hand, a polymer stabilizes a particle dispersion, coprecipitation is often induced by addition of a liquid in which the polymer is insoluble. For instance, a nanocomposite of Cu_2S and poly(aniline) was prepared by addition of ethanol to the stabilized Cu_2S–polymer dispersion in *N*-methylpyrrolidone.

As an alternative to coprecipitation, nanocomposite films can be prepared from stabilized particle–polymer dispersions by spin coating or by casting the dispersions into a flat dish followed by solvent evaporation. If the particles were prepared

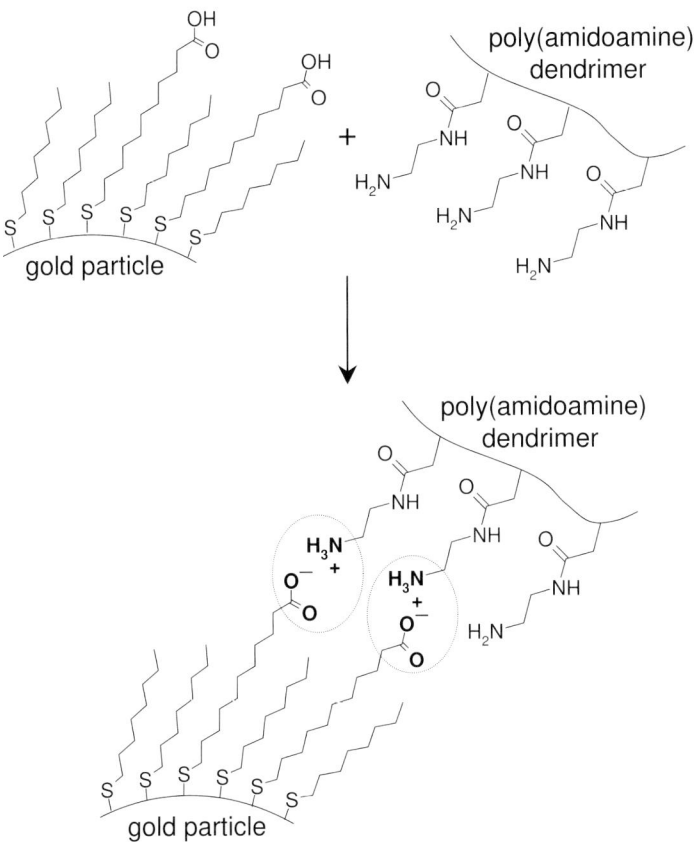

Fig. 2.15 Formation of charged moieties (dotted oval) by interactions of surface-modified gold particles with poly(amidoamine) (PAMAM) dendrimers (rather globular polymers whose exterior is densely covered with amine groups) (*J. Polym. Sci. A* **1992**, *30*, 2241).

in situ under formation of nonvolatile byproducts, the nanocomposites prepared by casting or spin coating typically also contain these byproducts as impurities. If this is undesirable, attention has to be paid to the exclusive formation of volatile byproducts like, e.g. acetic acid which results from the conversion of lead(II) acetate with H_2S to PbS nanoparticles. Problems with the incorporation of reaction side products from nanoparticle synthesis do not arise when dispersions of previously isolated core-shell nanoparticles are used, such as alkanethiolate-modified silver or gold nanoparticles which were dispersed in polymer solutions and subsequently processed to films by spin coating or casting and subsequent solvent evaporation (Fig. 2.16). Ultrathin nanocomposite films with controlled layer thickness can be generated by alternating immersion of a substrate in dispersions or solutions of oppositely charged particles and polymers, respectively. In this

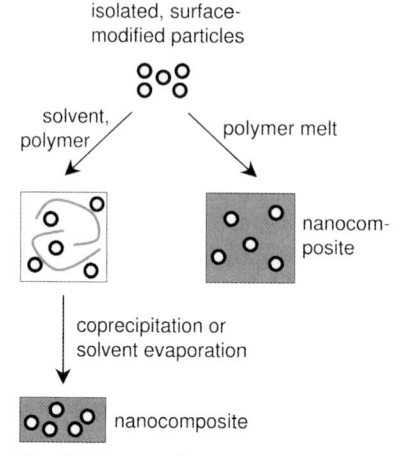

Fig. 2.16 Design of nanocomposite preparation with surface-modified nanoparticles.

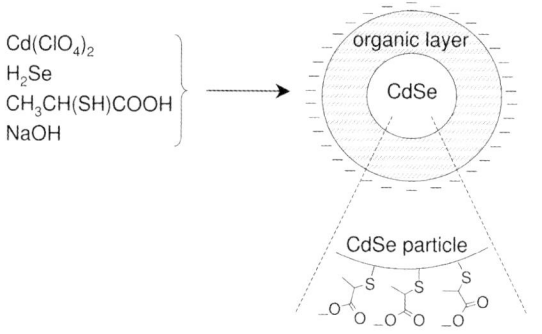

Fig. 2.17 In situ formation of CdSe nanoparticles modified with surface-bound carboxylate groups (J. Phys. Chem. B **1998**, 102, 4096).

manner, the layer thickness increases with the number of dipping cycles. For example, CdSe particles modified with strongly bound carboxylate groups were prepared *in situ* (Fig. 2.17). Upon immersion of a piece of flat glass or silicon, coated with a layer of poly(ethyleneimine), into the CdSe dispersion, the negatively charged CdSe particles adsorbed, probably with sodium ions as counterions (Fig. 2.18). After removal of the substrates and exposure to a solution of poly(allylamine hydrochloride), the positively charged polymer was attracted by the negative charges at the CdSe shell, leading to a deposition of a layer of the polymer. Subsequently, a new CdSe layer deposited after treatment with CdSe dispersion, and the adsorption cycles could be repeated many times, leading to a material with the nanoparticles and the polymer strongly interacting by electrostatic forces.

Occasionally, nanocomposites were also obtained by diffusion of nanoparticles into a swollen polymer matrix. This method, however, is suited only for rather

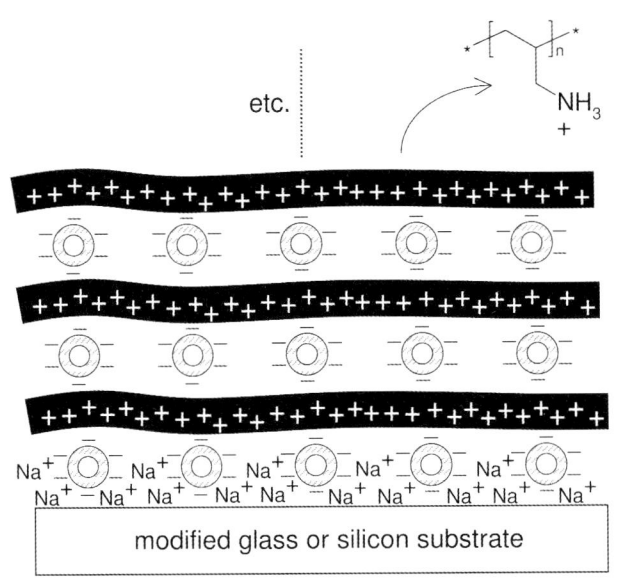

Fig. 2.18 Nanocomposite formed by alternating immersion (layer by layer deposition) of a modified glass or silicon substrate in a dispersion of *in-situ*-prepared CdSe nanoparticles modified with surface-bound carboxylate groups (see Fig. 2.17) and an aqueous solution of poly(allylamine hydrochloride). Excess negative charges of CdSe particles at the substrate might be compensated by sodium ions present in the CdSe dispersion (*J. Phys. Chem. B* **1998**, *102*, 4096).

insoluble polymers with good swelling behavior. Since the particle incorporation throughout the specimen becomes slower with increasing thickness of the matrix, thin films may be particularly suited for the application of this method. Thus, nanocomposites of CdTe and poly(aniline) (PANI) were prepared by exposure of a PANI film of ca. 100 nm thickness to an aqueous dispersion of thioglycerol-modified CdTe particles for 1 h.

2.4.2
Mixing of Particles with Monomers Followed by Polymerization

Particles, which are either prepared *in situ* or applied in the final state as core-shell particles, can be dispersed in suited monomer solutions or even in neat monomers. As an example, CdS nanoparticles with surface-bound 4-hydroxyphenol or 2-mercaptoethanol were prepared from $CdCl_2$ and Na_2S in reverse micellar systems. The particles were collected by centrifugation. Most likely, hydroxyl groups were exposed to the environment, which is expected to favor particle dispersion in polar solvents. Accordingly, the particles were dissolved in dimethyl sulfoxide. Upon addition of ethylene glycol and toluenediyl-2,4-diisocyanate, a poly(urethane) established *in situ*, whereupon the corresponding poly(urethane) –CdS nanocomposite was obtained by coprecipitation with water.

If appropriate modifications are made at the surfaces of the inorganic particles, the *in-situ*-generated polymer can be connected to the particle surface. This was attempted, for instance, with *in-situ*-prepared SiO_2 particles (Fig. 2.19). In order to establish a SiO_2 surface which enables the binding of polymer molecules, the hydroxyl groups at the pristine SiO_2 surfaces were converted with 3-(trimethoxysilyl)propyl methacrylate under formation of surface-bound siloxane groups and release of methanol. The organic surface layer not only implemented an appropriate chemical reactivity but also reduced the agglomeration tendency of the particles, which were dispersed in the monomers methyl methacrylate or methyl acrylate. A photoinitiator dissolved in the monomer enabled polymerization by irradiation with light. It was assumed that, apart from the polymerization in the bulk, at least a fraction of surface-bound methacrylate groups reacted to yield polymer molecules, which were attached to SiO_2 surfaces. Moreover, an initiator can also be bound to surfaces. Consequently, in presence of a suited monomer, polymerization is initiated from the surface under concomitant binding of the growing polymer to the surface. For instance, a peroxide initiator was attached first to SiO_2 nanoparticles, and after mixing the modified particles with methyl methacrylate at 70 °C the peroxide group decayed under formation of radicals which initiated the formation of surface-bound poly(methyl methacrylate) (Fig. 2.20).

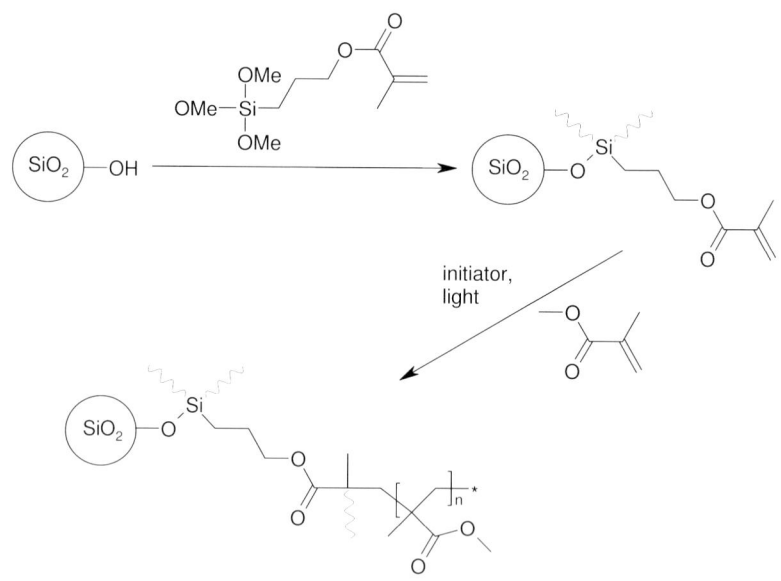

Fig. 2.19 Schematic illustration of a reaction sequence which was suggested to yield poly(methyl methacrylate)–SiO_2 nanocomposites containing polymer molecules which are bound to particle surfaces (see *Chem. Mater.* **1994**, *6*, 362). For simplicity, only one reactive site per SiO_2 particle is considered in the illustration.

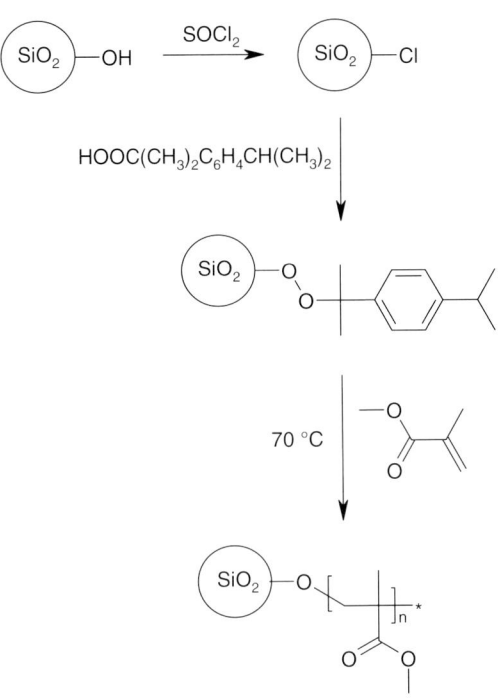

Fig. 2.20 Modification of a SiO$_2$ surface with a peroxide initiator which subsequently induced polymerization of methyl methacrylate upon heating to 70 °C, finally resulting in surface-bound poly(methyl methacrylate) (J. Polym. Sci. A **1992**, 30, 2241).

2.4.3
Nanocomposite Formation by means of Molten or Solid Polymers

In the simplest case, nanocomposites are prepared by mixing of inorganic particles in the final state into polymer melts. As explained above, this method provides random particle dispersion in the polymer matrix primarily if the particles are encapsulated by a shell of organic molecules whose chemical nature is compatible with that of the polymer. When the particles are prepared *in situ*, the final step of nanocomposite formation can proceed under the action of molten or solid polymers. Polymer melts were the starting point for the generation of metal nanoparticles which emerged by thermal decomposition of a precursor. When solutions of, e.g. Cu(HCOO)$_2$, Ti(C$_6$H$_5$CH$_2$)$_4$, or Fe(CO)$_5$ came in contact with polymer melts at 200–260 °C, elemental copper, titanium or iron, respectively, formed under concomitant evaporation of the solvent, release of volatile reaction byproducts and mixing of particles and polymer. The exposure of polymer melts to metal vapors can also result in the formation of polymer nanocomposites with well-dispersed particles. However, nanoparticle formation by thermal decomposition of an inorganic or organometallic precursor has been applied more frequently in a solid polymer matrix. Thus, in a composite film consisting of copper(II) formate and poly(2-vinylpyridine), prepared from a methanol solution followed by solvent evaporation, the copper ions were reduced in the solid polymer matrix upon heating to metallic copper nanoparticles. The reaction side products (H_2 and CO_2) were gaseous and therefore favorably escaped from the polymer matrix. Precursors in a solid polymer matrix can be transformed to nanoparticles not only by thermal decomposition but also by chemical reaction with agents in the gaseous or dissolved state. For instance, the precursor diethyl zinc was incorporated in sheets of a statistical copolymer of ethylene and vinyl acetate (EVA) by diffusion from a diethyl zinc solution in hexane. After removal of the sheets from the solution, the diethyl zinc hydrolyzed rapidly by water present in the ambient atmosphere, yielding zinc oxide particles (diameter ca. 10 nm). Again, the reaction side product of the formation of the zinc oxide particles (ethane) was volatile and therefore left the nanocomposite.

Coordination of metal ions to ligand sites of polymers can lead to controlled formation of nanoparticles. In particular in microphase-separated block copolymers, precursor molecules may interact predominantly with one specific block, and therefore this block will harbor the final nanoparticles sprout from the precursor by reaction in the solid material. As a consequence of the microphase separation, a structured distribution of the particles can even result. For instance, when solids of poly(styrene-*block*-2-vinylpyridine) containing Cu^{2+}, Co^{2+}, Ni^{2+}, Cd^{2+}, Zn^{2+}, or Ag^+ salts were prepared from solution, coordination of metal ions to pyridine moieties was suggested (see Fig. 2.14) and as consequence the corresponding metal sulfide nanoparticles which emerged after treatment with gaseous H_2S finally arose only in the ordered domains of the 2-vinylpyridine blocks. As another example, in films of poly(styrene-*block*-ethylene oxide) micelles previously treated with Li[AuCl$_4$], Li$^+$ ions coordinated only to the ethylene oxide units under concomitant binding of the oppositely charged [AuCl$_4$]$^-$ ions. Annealing of the films lead to the formation of a single gold particle in each poly(ethylene oxide) core. As a consequence, the gold particles were ordered in a hexagonal structure. Finally, organometallic polymers can also be used as precursors for nanocomposites. For example, nanocomposites with silver particles arranged in lamellar structures were prepared from microphase-separated diblockcopolymers where one block was functionalized with phosphine units (see Fig. 2.14). The phosphorous atoms served as ligands for silver(I), which was additionally coordinated by hexafluoroacetylacetonate. Silver particles emerged in the end upon thermal treatment of the solid polymers.

2.4.4
Concomitant Formation of Particles and Polymers

For nanocomposite generation, both the inorganic particles and the polymer can be prepared *in situ*. In presence of monomers, which can also be used as a solvent for the particle precursors, nanocomposites can emerge upon polymerization and concomitant *in situ* formation of the nanoparticles. For instance, heat or radiation can induce both polymerization and nanoparticle formation. As an example, when a solution of tris(styrene)platinum(0) dissolved in styrene was heated to 60 °C in the presence of a radical initiator, the metal complex decomposed into platinum nanoparticles and concomitantly the styrene polymerized to poly(styrene), and a nanocomposite of poly(acryl amide) containing nickel or cobalt nanoparticles was prepared by irradiation with γ-rays starting from nickel(II) sulfate or cobalt(II) sulfate in an aqueous solution containing the monomer acryl amide. In the latter case, the γ-rays initiated both the reduction of the metal(II) ions to metal(0) atoms and the polymerization of acrylamide. The polymerization was supposed to be induced by radicals which formed under the action of the γ-irradiation, e.g. H˙ or OH˙ radicals which may arise by radiolysis of water molecules. The *in situ* manufacture of both polymer and particles was also performed by simultaneous evaporation (co-evaporation) of an inorganic species and a monomer. While the monomer converts into a polymer film after contact with a substrate, the inorganic vapors decompose grow to nanoparticles in the concomitantly growing polymer matrix. As an example, paracyclophane was evaporated and the vapors were heated to 650–700 °C which caused a conversion of the *p*-cyclophane to *p*-xylylene. The *p*-xylylene vapors were coevaporated with palladium, tin, or copper vapors in presence of a quartz slide which was cooled to −196 °C. The hot vapors condensed on the cold quartz surface, and when the quartz with the deposited solids was warmed to room temperature, the *p*-xylylene polymerized to poly(*p*-xylylene), which is otherwise difficult to process, and the inorganic vapors yielded the respective metal nanoparticles. The evaporation of inorganic substances for the preparation of nanocomposites is not restricted to metals; for instance PbS was also used in presence of *p*-cyclophane in order to create nanocomposites as described above.

2.5
Properties and Applications of Polymer Nanocomposites

2.5.1
Properties

As indicated above, optical and electrical properties can differ in nanocomposites compared to composites containing larger particles of the same volume fraction. For instance, below a certain size, which is specific for each compound, the electrical conductivity of semiconductors and conductors decreases because the bandgap increases. As also mentioned above, the high specific interface area in nanocomposites can be of advantage in photoconducting processes where inorganic and polymeric semiconductors are combined. Photoconductivity was found, e.g. in nanocomposites of poly(aniline) (PANI) and PbS, CdS or TiO_2, in films of poly(N-vinylcarbazole) (PVK) and CdS or HgS, or in films of poly(p-xylylene) (PPX) and PbS. The detailed origin of the photoconductivity may depend on the nanocomposite. In the PANI-TiO_2 nanocomposites, electrons seem to be transferred from PANI to TiO_2 upon irradiation, and the generated charges move subsequently to the respective electrodes. The photoconductivity in the PVK-CdS films was concluded to be initiated with the absorption of a photon by a CdS particle while PVK was responsible for the charge transport. Remarkably, the activation energy in PPX–PbS nanocomposites was rather low (less than 0.1 eV) and therefore it was suggested that the photoelectric behavior of those materials was determined most likely by photoinduced tunneling transfer of electrons between PbS particles.

With a view to magnetic properties of nanoparticles, the emphasis is on monodomain particles, whereat a domain is a group of aligned spins that act in a concerted way. In bulk magnetic materials, domains are separated by domain walls of certain thickness and formation energy, and the motion of those domain walls is involved in the reversibility of the magnetization effect. In the bulk, the formation of domain walls is energetically favorable while domain wall motion is not relevant in monodomain particles. If the particles are very small, the spins become sensitive to the thermal energy and thus arrange in a disordered state when an external magnetic field, which previously caused alignment of the spins, is switched off. Such superparamagnetic systems are characterized by zero remanent magnetization (absence of hysteresis) and zero coercive force. In order to keep individual monodomain particles separated from each other, they can be embedded in block-copolymers whereat the nanoparticle surfaces are compatible only with one block which is immiscible with the other block and forms islands which contain not more than one particle. Corresponding superparamagnetic nanocomposites were prepared, for example, with cobalt or γ-Fe_2O_3 nanoparticles. Superparamagnetic nanocomposites were also obtained with conductive polymers such as poly(pyrrole) or poly(aniline). Such materials combine a superparamagnetic with an electrically conducting or semiconducting behavior.

As elaborated above, nanocomposites with randomly dispersed particles are characterized by significantly reduced light scattering compared to related composites comprising larger particles. If the particles in the polymer matrix are, however, ordered in a regular lattice, remarkable optical effects can emerge for particles with diameters at the upper limit of the nanosize region, as a result of diffraction of light at the lattice planes according to Bragg's equation. As a consequence, iridescence is observed. Corresponding materials were prepared with concentrated dispersions (35–45% w/w) of spherical SiO_2 particles of narrow size distribution and diameters on the order of 100–200 nm in the monomers methyl methacrylate or methyl acrylate. The SiO_2 particles were enclosed in a reactive shell which allowed the binding of the polymerizing monomers to the particles, according to Fig. 2.19. After polymerization, scanning electron micrograph (SEM) images revealed indeed a hexagonal pattern of the silicon dioxide particles in the nanocomposites, and iridescent materials developed with a distinct absorption maximum in UV/vis spectra as expected from Bragg's equation. As another consequence of Bragg's equation, the maximum absorption wavelength depended on the angle between the incident light and the sample surface; a sample with an absorption maximum of 490 nm at perpendicular incidence of light showed a value of 472 nm at an incidence angle of 68°.

Generally, the incorporation of inorganic particles in elastomeric polymers increases the strength but decreases the extensibility of the matrix. This is also the case in composites comprising surface-modified SiO_2 nanoparticles which were most likely bound to polymer molecules of a poly(methyl acrylate) matrix. Two types of such composites were prepared, one with randomly dispersed and the other with agglomerated SiO_2 particles. As evident from Table 2.4, the presence of SiO_2 resulted in an increase in Young's modulus by a factor of 10–130, whereat the composites with the aggregated particles exhibited higher moduli than those with the randomly dispersed particles. The extensibilities, which are represented by the elongation at break, were markedly smaller in the composites than in the neat polymer. The toughness, which is described by the energy required to break the sample, i.e. the area under a stress–elongation curve, decreased in most cases (Table 2.4). Again, the samples with aggregated particles provided more favorable values than those with randomly dispersed particles. Interestingly, when SiO_2 particles were arranged in the polymer in a regular lattice, no noteworthy differences in Young's modulus, elongation at break or toughness were found when compared to materials with randomly dispersed particles, indicating that the order of individually dispersed particles in the matrix is not of high importance for mechanical properties at least in those systems. One should be aware, however, that the reinforcement of thermoplastic polymers by nanoparticles can be little pronounced. For instance, poly(ethylene) is not reinforced considerably by

Table 2.4 Mechanical properties of a poly(methyl methacrylate)—SiO_2 nanocomposite containing different contents of randomly dispersed or agglomerated SiO_2 particles (Chem. Mater. **1997**, 9, 2442).

SiO_2 content (w/w)	State in dispersion	Young's modulus (MPa)	Elongation at break (%)	Toughness (J/m^3 × 10^7)
0	n.a.	2.07	1419	1.54
0.35	random	20.3	120	1.01
	agglomerated	29.9	254	2.24
0.40	random	30.1	98	1.13
	agglomerated	68.6	184	1.80
0.45	random	126	46	0.36
	agglomerated	262	127	1.02

carbon black as well as EVA (ethylene–vinyl acetate copolymer) by Al(OH)$_3$ nanoparticles whose surface was modified with alkyl groups in order to enhance the compatibility of the particles at least with the poly(ethylene) segments in EVA.

Thermal studies of nanocomposites were performed in particular with clay minerals as the inorganic component, which are referred to extensively in Chapter 4. Thermal properties such as the glass transition temperature of polymers can increase, decrease or remain unchanged in presence of inorganic particles. The reasons for the different behavior are not clear at the moment. Yet similar systems can show a different behavior, as in the case of the previously mentioned nanocomposites of methacrylate-modified SiO$_2$ and poly(methyl methacrylate) or poly(methyl acrylate), respectively. While the glass transition temperature of poly(methyl methacrylate) clearly increased in presence of SiO$_2$, that of poly(methyl acrylate) remained essentially unchanged or even decreased slightly.

2.5.2
Applications

2.5.2.1 Catalysts

As indicated above, the large specific surface (or interface) area of catalytically active nanoparticles enhances the efficiency of heterogeneous catalytic reactions (hereby, the catalyst is present as a solid and the reacting agents are in the liquid, dissolved or gaseous state). Heterogeneous catalyses are widely applied in technical processes since they offer, for instance, a simple separation of the catalyst from the reaction mixture, which is often difficult to achieve in systems based on homogeneous catalysis. Catalysis is of technical importance, e.g. in hydrogenation of olefins or acetylenes (Fig. 2.21). Such reactions also proceed with nanocomposites containing catalysts based on palladium, platinum, rhodium or iridium embedded in poly(vinyl alcohol), poly(vinyl acetate) or block-copolymers. Selective

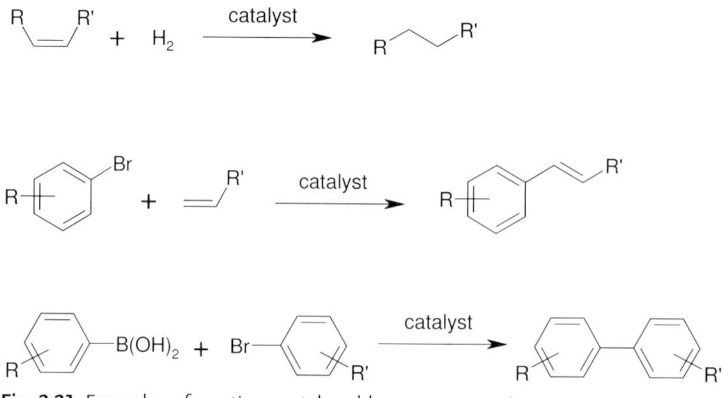

Fig. 2.21 Examples of reactions catalyzed by nanocomposites (top: hydrogenation of olefins, middle: Heck reaction, bottom: Suzuki reaction).

hydrogenation in molecules with different reaction sites is also possible. For instance, the selective hydrogenation of dehydrolinalool (oct-6-ene-1-yne-3-ol), where a C≡C and a C=C bond can basically react with hydrogen, is of industrial importance in fragrance chemistry. The C≡C bond was hydrogenated rather than the C=C bond with a selectivity of 99.5% with palladium nanoparticles embedded in poly(styrene)-*block*-poly(4-vinylpyridine). Even higher selectivities (up to 99.8%) were achieved with bimetallic Pd/Au, Pd/Pt or Pd/Zn nanoparticles. It was suggested that a coordination of pyridine units to the metal surface contributed to the high selectivity.

Numerous other reactions can also be catalyzed with nanocomposites, such as the Heck reaction or the Suzuki reaction (Fig. 2.21). Also, the oxidation of L-sorbose into 2-keto-L-gulonic acid, an intermediate stage in the synthesis of ascorbic acid (vitamin C), was catalyzed by platinum nanoparticles in crosslinked poly(styrene), or a poly(thiourethane) containing benzylthiol- or phenylthiol-modified CdS nanoparticles caused the generation of hydrogen from 2-propanol under illumination while bulk CdS did not show a significant photocatalytic activity. When the starting compounds and the reaction products are in the gaseous state, catalytic membranes can be used to keep the starting materials separate from the reaction products, as in the case of a nanocomposite based on an aromatic poly(amide imide) containing 15% w/w bimetallic Pd/Ag (77/23 w/w) nanoparticles, which was effective in the hydrogenation of nitrous oxide.

In spite of the successful examples of catalytic active nanocomposites one should be aware that the enclosure of nanoparticles in polymers is a general drawback because the polymer retards the diffusion of the reagents to and from the catalytic active centers. In order to minimize this effect, nanocomposite materials which are riddled with open pores should be preferred. Also, it should be considered that an organic shell frequently decreases the efficiency of nanoparticles, most likely because the organic layer prevents a contact between the active surface sites and the molecules to be converted.

2.5.2.2 Gas Sensors

Gas sensors find increasing interest in the monitoring of vapors of halogenic or, more general, organic compounds, e.g. for toxic or environmental reasons. Indeed, nanocomposites might be helpful in the detection of such vapors. For instance, poly(aniline)–copper nanocomposites turned out to be useful for the detection of chloroform vapors. Since poly(aniline) shows, in contrast to most polymers, significant electrical conduction, the chloroform caused a change in conductivity which increased with increasing chloroform content up to a saturation value. When the samples were removed from the chloroform vapors and exposed to dry air, the chloroform left the nanocomposite and the electrical conductivity adopted the initial value again. Poly(aniline) itself did not show a comparable conductivity effect, which implies that the copper particles adsorbed reversibly chloroform vapors. The concentration of gases in a nanocomposite was also determined by thickness measurements. If the gaseous species adsorb reversibly at the internal surfaces of the inorganic particles, the thickness of the composite can increase considerably because the high specific interface area (i.e. interface area per volume unit) of the incorporated nanoparticles allows for the adsorption of a large quantity of gaseous molecules. Consequently, the thickness of nanocomposites consisting of a fluorinated polymer and gold increased upon exposure to vapors of carbon tetrachloride, isopropanol or acetone. The extent of this increase depended on the gaseous substance and the gold content. In the example of acetone, the thickness of the nanocomposite increased linearly with the volume fraction of gold, which indicates that the thickness expansion is indeed not simply related to dissolution of the organic molecules in the polymer matrix but rather to an adsorption of the vapors at the gold particles. These particles probably adhere little to the polymer since otherwise the adsorption of the gaseous molecules might be prevented.

2.5.2.3 Materials with Improved Flame Retardance

A disadvantage of polymers compared to inorganic materials is their more or less pronounced burning tendency at high temperatures, which may arise in particular in the course of accidents. As a consequence, the flame retardance of technical polymers is improved by addition of inorganic additives, in particular $Al(OH)_3$ and, to a less extent, $Mg(OH)_2$. Upon heating, these hydroxides decompose to metal oxides under release of water. The flame retardance effect of these agents bases on their energy consumption due to endothermal decomposition in an appropriate temperature range, release of an inert gas (water) which dilutes the combustion gases, and formation of an oxide layer which thermally insulates the polymer. It appears to be likely that a higher specific interface area of the particles enhances the flame retardance efficiency because the release of water is expected to proceed fastest at the interface region. Indeed, it was found, for instance, that $Al(OH)_3$ nanoparticles were more powerful in dropping the heat release rate in composites of EVA (ethylene–vinyl acetate copolymer) than a common grade of $Al(OH)_3$.

2.5.2.4 Optical Filters

As already indicated above, nanoparticles have been used to colorize matrix materials for centuries. Since light scattering in the resulting nanocomposites is commonly irrelevant, colored but translucent materials emerge which therefore act as translucent color filters. As an example, color filters consisting of poly(N-vinylpyrrolidone) and silver particles appeared translucent yellow or red, depending on the silver particle diameters which ranged between 6 and 12 nm. More pronounced color variations at different particle diameters were observed in poly(vinyl alcohol)–gold nanocomposites. When viewed in transmitted light, the nanocomposites appeared pink at 16 nm, purple at 43 nm, and blue at 79 nm particle diameter. Under oblique observation, the composites comprising particles with 79 nm diameter changed their color to brown (Tyndall effect).

Since the long-term stability of technical polymers is frequently affected by UV radiation, UV stabilizers are commonly added. In colorless materials, these stabilizers consist usually of organic compounds. However, UV-absorbing, visually transparent nanoparticles might also serve as UV-absorbing entities, and nanocomposites which are fully transparent for visible light but efficiently absorb UV radiation can act as UV filters. Gratifyingly, ZnO and TiO_2 absorb UV radiation over a broad wavelength range but are transparent in the spectrum of visible light. Accordingly, nanocomposites with TiO_2 and poly(vinyl alcohol) or ZnO and a statistical copolymer of ethylene and vinyl acetate (EVA) were transparent to visible light but absorbed UV light to a high extent close to 400 nm. Further, visually transparent nanoparticles of ZnO showed a UV-protective effect of poly(propylene) or poly(ethylene) matrices. Upon irradiation with UV light, discoloration of poly(propylene) or carbonyl formation in poly(ethylene) was substantially reduced by 1–2% w/w ZnO. Remarkably, the thermal stability of low-density poly(ethylene) was also improved by the presence of ZnO nanoparticles.

2.5.2.5 Dichroic Materials

Dichroic materials are of interest in common twisted-nematic liquid crystal displays (LCDs). If one of the polarizers is substituted by a dichroic sheet, the display becomes bicolored. Impressive dichroic effects in nanocomposites were caused by parallel oriented, linear assemblies of nanoparticles. Besides the dichroic fibrils mentioned above, dichroic materials were obtained more recently by processing of nanocomposites on the basis of synthetic polymers. When poly(ethylene) films containing gold or silver nanoparticles were heated to 120 °C, the polymer became soft which allowed drawing with moderate expenditure of energy. The resulting films showed pronounced dichroism. The observed colors depended on the type of metal (gold or silver) and the particle diameters of a given metal. For example, nanocomposites with gold changed their color from blue to red amber when turning the angle ϕ between the polarization direction of the incident light and the drawing direction of the films from 0° to 90°, and materials with silver, depending on the particle size, from red to yellow (particle diameter 4.5 nm) or purple to amber (diameter 10 nm), respectively. In fact, it was demonstrated that bicolored LCDs can indeed be established with such nanocomposites. The dichroism was also reflected in UV/vis spectra recorded with polarized light, which strongly depended on ϕ (Fig. 2.22). The absorption maximum at $\phi = 0°$ arose at higher wavelengths than that at $\phi = 90°$. Isosbestic points, i.e. points at which the absorbance was independent on ϕ at a particular wavelength (as evident from Fig. 2.22) appeared frequently in the spectra of the dichroic materials. Transmission electron microscope (TEM) images suggested that most pronounced dichroism was obtained when the initial nanocomposite films were prepared under conditions which provoked the formation of spherical particle assemblies in the polymers which deformed into linear assemblies upon drawing under the action of the shear forces. As a consequence of the directive drawing process, these assemblies oriented themselves uniaxially in the drawing direction.

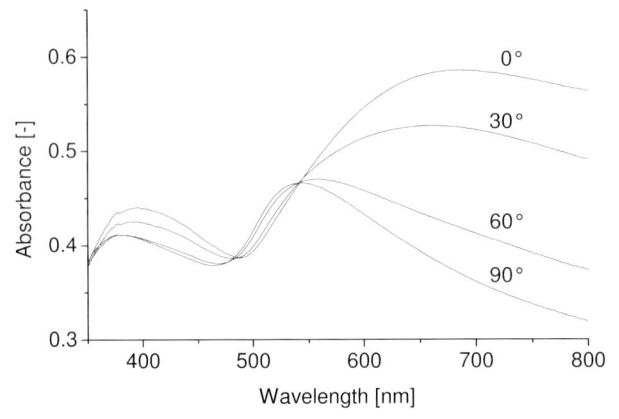

Fig. 2.22 UV/vis spectra of a drawn nanocomposite consisting of poly(ethylene) and 1-dodecanethiol–modified gold nanoparticles recorded with linear polarized light at different angles between the polarization and the drawing direction (0° corresponds to parallel orientation of polarization and drawing direction) (*Appl. Optics* **1999**, *38*, 658; *J. Mater. Sci.* **1999**, *34*, 3859; *Adv. Mater.* 1999, *11*, 223).

2.5.2.6 High and Low Refractive Index Materials

The refractive index is a key feature in many optical applications, such as lenses or waveguides. The refractive indices of organic polymers are typically in the range of 1.3 and 1.7 while inorganic materials can possess refractive indices far below 1 (e.g. gold) or above 3 (e.g. PbS or silicon) over a broad wavelength range. Hence, it may be expected that the refractive index of nanocomposites containing particles with extreme refractive index can fall outside of the typical range of that of organic polymers. Further, it should be possible to adjust the refractive index of composite materials by the particle content. This was indeed demonstrated with poly(ethylene oxide), a poly(thiourethane) or gelatin which were charged with PbS nanoparticles, with gelatin–silicon nanocomposites and gelatin–gold nanocomposites containing a high content (up to 50% v/v) of inorganic particles. Refractive indices of composites with PbS or silicon around 3 and of gelatin–gold composites down to 1.0 were found. The refractive index of the nanocomposites composed of gelatin and PbS, silicon or gold seemed to depend rather linearly on the volume content of the particles (Fig. 2.23). Linear relations of refractive indices and volume fractions of inorganic particles were also reported for materials composed of poly(methyl methacrylate) and silicon dioxide or of a poly(thiourethane) and titanium dioxide. Of course, some deviation from the linear relationships is possible due to the limited experimental precision of the determination of the refractive indices and the volume fractions of the particles. According to theory, a linear relation between refractive index and volume content of filler could be expected for many composites but nonlinear relations cannot be excluded.

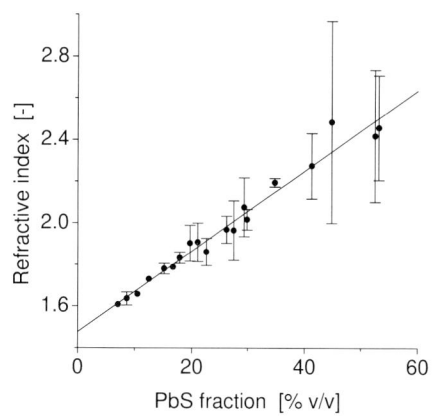

Fig. 2.23 Refractive index of a gelatin–PbS nanocomposite as a function of the volume fraction of PbS, according to *Polym. Adv. Technol.* **1993**, 4, 1.

2.6
Summary

Polymer nanocomposites as dealt with in this section consist of a continuous polymer matrix which contains inorganic particles of a size below ca. 100 nm at least in one dimension. The largest variety of chemically different inorganic components has been described with more or less spherical or cubic particles. Although such materials have been reported already long ago (e.g. with SiO_2, TiO_2, carbon black or gold nanoparticles), polymer nanocomposites found particular attention only in the last 15 years. The increasing interest in nanocomposites is based on the fact that they can exhibit properties which are more pronounced or even differ from those of the analogous composites with larger particles. Such effects occur especially due to size-dependent physical properties of inorganic particles or of the high internal surface area. As a consequence of these characteristics, materials properties such as optical transparency, color (including dichroism), photoconductivity, iridescence, catalytic activity or superparamagnetism, can be of particular use in nanocomposites when compared to analogous composites with larger particles. Thus, the properties of nanocomposites are considered to provide applications in the areas of, e.g. optics, catalysis, electronics, or magnetism.

It is of considerable importance for many materials properties (in particular optical or magnetic properties) that the particles are dispersed as individual primary particles and not as agglomerated particles (secondary particles). In order to avoid agglomeration, the nanoparticles are frequently synthesized *in situ* or applied as surface-modified particles (or core-shell particles where an inorganic core is surrounded by a shell composed of organic molecules). The preparation of nanocomposites with randomly dispersed particles is markedly facilitated by the use of nanoparticles with appropriately modified surface since such particles enable a good dispersion of the primary particles in polymer solutions or melts.

In addition, the use of particles with appropriately modified surfaces comprising reactive organic molecules allows the binding of the particles to suited polymers which leads to strongly connected polymer–particle systems.

Bibliography

Introduction
M. Terrones, *Int. Mater. Rev.* **2004**, *49*, 325.
B. Pukánsky, *Eur. Polym. J.* **2005**, *41*, 645.

Consequences of very small particle sizes
U. Kreibig, M. Vollmer, *Optical Properties of Metal Clusters*, Springer, Berlin, **1995**.
A. Chevreau, B. Phillips, B. G. Higgins, S. H. Risbud, *J. Mater. Chem.* **1996**, *6*, 1643.
N. Herron, Y. Wang, H. Eckert, *J. Am. Chem. Soc.* **1990**, *112*, 1322.
X.-D. Ma, X.-F. Qian, J. Yin, H.-A. Xi and Z.-K. Zhu, *J. Colloid Interface Sci.* **2002**, *252*, 77.
L. L. Beecroft, C. K. Ober, *Chem. Mater.* **1997**, *9*, 1302.
T. Kyprianidou-Leodidou, W. Caseri, U. W. Suter, *J. Phys. Chem.* **1994**, *98*, 8992.
F. Kirchner, Ph.D. thesis, University of Leipzig, imprinted by Bernhard Vopelius, Jena, **1903**.
J. C. Maxwell Garnett, *Phil. Trans. Royal Soc. London A* **1906**, *205*, 237.
J. C. Maxwell Garnett, *Phil. Trans. Royal Soc. London A* **1904**, *203*, 385.
R. J. Nussbaumer, *Ph.D. Thesis ETH Zürich, No. 15516*, Zürich, **2004**.
G. Mie, *Z. Chem. Ind. Kolloide* **1907**, *2*, 129.
Y. Dirix, C. Darribère, W. Heffels, C. Bastiaansen, W. Caseri, P. Smith, *Appl. Optics* **1999**, *38*, 658.
W. Mahler, *Inorg. Chem.* **1988**, *27*, 435.
S. Deki, K. Sayo, T. Fujita, A. Yamada, S. Hayashi, *J. Mater. Chem.* **1999**, *9*, 943.
M. Gianini, W. R. Caseri, U. W. Suter, *J. Phys. Chem. B* **2001**, *105*, 7399.
E. Leontidis, M. Orphanou, T. Kyprianidou-Leodidou, F. Krumeich, W. Caseri, *Nano Lett.* **2003**, *3*, 569.

Historical reports on inorganic nanoparticles and polymer nanocomposites
I. Borgia, I. Brunetti, I. Mariani, A. Sgamellotti, F. Cariati, P. Fermo, M. Mellini, C. Viti, G. Padeletti, *Appl. Surf. Sci.* **2002**, *185*, 206.
A. Cornejo, *Z. Chem. Ind. Kolloide* **1913**, *12*, 1.
A. Neri, *The Art of Glass, WHEREIN Are shown the wayes to make and colour Glass, Pastes, Enamels, Lakes, and other Curiosities, English translation from the Italian original text*, printed by A. W., London, **1662**, facsimile reproduction by UMI Books on Demand, Ann Arbor (Michigan), **2002**.
Lüdersdorff, *Verhandlungen Verein. Beförderung Gewerbefleiss. Preussen* **1833**, *12*, 224.
K. Stoeckl, L. Vanino, *Z. Phys. Chem. Stöchiom. Verwandtschaftslehre* **1899**, *30*, 98.
H. Ambronn, *Königl. Sächs. Ges. Wiss.* **1896**, *8*, 613.
A. Frey, *Jahrbuch wiss. Botanik* **1927**, *67*, 597.
A. Frey-Wyssling, *Protoplasma* **1937**, *27*, 372.
W. Ostwald, *Z. Chem. Ind. Kolloide* **1909**, *4*, 5.
M. Faraday, *Experimental Researches in Chemistry and Physics*, Taylor and Francis, Red Lion Court, London **1859**.
R. Zsigmondy, *Ann. Chem. (Liebig's Ann.)*, **1898**, *301*, 29.
A. Gutbier, *Z. anorg. Chem.* **1902**, *32*, 347.
H. Siedentopf, R. Zsigmondy, *Ann. Phys., vierte Folge (Drude's Ann.)* **1903**, *10*, 1.
P. Scherrer, *Nachr. Königl. Ges. Wiss. Göttingen, Math.-phys. Klasse* **1918**, 98.
E. Grimaux, *Compt. Rend. Acad. Sci.* **1884**, *98*, 1434.
T. Graham, *Ann. Chem. Pharm. (Liebig's Ann.)* **1862**, *121*, 1.
H. Rose, *Pogg. Ann. (Ann. Phys. Chem.)* **1844**, *61*, 507.
T. Graham, *Ann. Chem. Pharm. (Liebig's Ann.)* **1865**, *135*, 65.
M. Ettlinger, *Schriftenreihe Pigmente (4th Edition)* **1989**, *56*, 2.
F. Lyon, K. A. Burgess in *Encyclopedia of Polymer Science and Engineering, Vol. 2*

(Eds.: H. F. Mark, N. M. Bikalees, C. G. Overberger, G. Menges, J. I. Kroschwitz), Wiley & Sons, New York, **1985**, p. 623.

Preparation of polymer nanocomposites
Mixing of dispersed particles with polymers in liquids
A. B. R. Mayer, *Mater. Sci. Eng. C* **1998**, *6*, 155.

M. Weibel, W. Caseri, U. W. Suter, H. Kiess, E. Wehrli, *Polym. Adv. Technol.* **1991**, *2*, 75.

B. L. Frankamp, A. K. Boal, V. M. Rotello, *J. Am. Chem. Soc.* **2002**, *124*, 15146.

D. Yu. Godovsky, A. E. Varfolomeev, D. F. Zaretsky, R. L. N. Chandrakanthi, A. Kündig, C. Weder, W. Caseri, *J. Mater. Chem.* **2001**, *11*, 2465.

L. Zimmermann, M. Weibel, W. Caseri, U. W. Suter, *J. Mater. Res.* **1993**, *8*, 1742.

K. E. Gonsalves, G. Carlson, J. Kumar, F. Aranda, M. Jose-Yacaman, *ACS Symp. Ser.* **1996**, *622*, 151.

Y. Dirix, C. Bastiaansen, W. Caseri, P. Smith, *J. Mater. Sci.* **1999**, *34*, 3859.

M. Gao, B. Richter, S. Kirstein, H. Möhwald, *J. Phys. Chem. B* **1998**, *102*, 4096.

N. P. Gaponik, D. V. Talapin, A. L. Rogach, *Phys. Chem. Chem. Phys.* **1999**, *1*, 1787.

Mixing of particles with monomers followed by polymerization
T. Hirai, M. Miyamoto, I. Komosawa, *J. Mater. Chem.* **1999**, *9*, 1217.

H. B. Sunkara, J. M. Jethmalani, W. T. Ford, *Chem. Mater.* **1994**, *6*, 362.

J. M. Jethmalani, W. T. Ford, *Chem. Mater.* **1996**, *8*, 2138.

N. Tsubokawa, H. Ishida, *J. Polym. Sci. A* **1992**, *30*, 2241.

Nanocomposite formation by means of molten or solid polymers
T. Kyprianidou-Leodidou, P. Margraf, W. Caseri, U. W. Suter, P. Walther, *Polym. Advan. Technol.* **1997**, *8*, 503.

S. Deki, K. Sayo, T. Fujita, A. Yamada, S. Hayashi, *J. Mater. Chem.* **1999**, *9*, 943.

S. P. Gubin, *Colloids Surf. A* **2002**, *202*, 155.

A. M. Lyons, S. Nakahara, M. A. Marcus, E. M. Pearce, J. V. Waszczak, *J. Phys. Chem.* **1991**, *95*, ,1098.

J. F. Ciebien, R. T. Clay, B. H. Sohn, R. E. Cohen, *New J. Chem.* **1998**, (no volume), 685.

Concomitant formation of particles and polymers
M. Gianini, W. R. Caseri, U. W. Suter, *J. Phys. Chem. B* **2001**, *105*, 7399.

T. Hirai, M. Miyamoto, I. Komosawa, *J. Mater. Chem.* **1999**, *9*, 1217.

A. Sarkar, S. Kapoor, G. Yashwant, H. G. Salunke, T. Mukherjee, *J. Phys. Chem. B* **2005**, *109*, 7203.

A. Biswas, O. C. Aktas, J. Kanzow, U. Saeed, T. Strunskus, V. Zaporojtchenko, F. Faupel, *Mater. Lett.* **2004**, *58*, 1530.

N. Cioffi, I. Farella, L. Torsi, A. Valentini, A. Tafuri, *Sens. Actuators B* **2002**, *84*, 49.

S. A. Zavyalow, A. M. Pivkina, J. Schooman, *Solid State Ionics* **2002**, *147*, 415.

E. V. Nikolaeva, S. A. Ozerin, A. E. Grigoriev, E. I. Grigoriev, S. N. Chvalun, G. N. Gerasimov, L. I. Trakhtenberg, *Mater. Sci. Eng.* **1999**, *C8–9*, 217.

Properties and applications of polymer nanocomposites
Properties
F. Lyon, K. A. Burgess in *Encyclopedia of Polymer Science and Engineering, Vol. 2* (Eds.: H. F. Mark, N. M. Bikalees, C. G. Overberger, G. Menges, J. I. Kroschwitz), Wiley & Sons, New York, **1985**, p. 623.

D. Yu. Godovsky, A. E. Varfolomeev, D. F. Zaretsky, R. L. N. Chandrakanthi, A. Kündig, C. Weder, W. Caseri, *J. Mater. Chem.* **2001**, *11*, 2465.

H. B. Sunkara, J. M. Jethmalani, W. T. Ford, *Chem. Mater.* **1994**, *6*, 362.

J. M. Jethmalani, W. T. Ford, *Chem. Mater.* **1996**, *8*, 2138.

E. V. Nikolaeva, S. A. Ozerin, A. E. Grigoriev, E. I. Grigoriev, S. N. Chvalun, G. N. Gerasimov, L. I. Trakhtenberg, *Mater. Sci. Eng.* **1999**, *C8–9*, 217.

N. P. Gaponik, D. V. Sviridov, *Ber. Bunsenges. Phys. Chem.* **1991**, *101*, 1657.

W. Feng, E. Sun, A. Fujii, H. Wu, K. Niihara, K. Yoshino, *Bull. Chem. Soc. Jpn.* **2000**, *73*, 2627.

J. G. Winiarz, L. Zhang, M. Lal, C. S. Friend, P. N. Prasad, *J. Am. Chem. Soc.* **1999**, *121*, 5287.

J. G. Winiarz, L. Zhang, J. Park, P. N. Prasad, *J. Phys. Chem. B* **2002**,*106*, 967.

L. M. Bronstein, S. N. Sidorov, P. M. Valetsky, *Russ. Chem. Rev.* **2004**, *73*, 501.

T. Seto, H. Akinaga, F. Takano, K. Koga, T. Orii, M. Hirasawa, *J. Phys. Chem. B* **2005**, *109*, 13403.

Z. Tang, *Chemtech* **1999**, *29(11)*, 7.

R. Gangopadhyay, A. De, *Chem. Mater.* **2000**, *12*, 608.

Z. Pu, J. E. Mark, J. M. Jethmalani, W. T. Ford, *Chem. Mater.* **1997**, *9*, 2442.

M. Okoshi, H. Nishizawa, *Fire Mat.* **2004**, *28*, 423.

J. K. Pandey, K. R. Reddy, A. P. Kumar, R. P. Singh, *Polym. Degrad. Stab.* **2005**, *88*, 234.

Applications

1 Catalysts

T. Hirai, M. Miyamoto, T. Watanabe, S. Shiojiri, I. Komasawa, *J. Chem. Eng. Japan* **1998**, *31*, 1003.

J. F. Ciebien, R. T. Clay, B. H. Sohn, R. E. Cohen, *New J. Chem.* **1998**, (no volume), 685.

L. M. Bronstein, S. N. Sidorov, P. M. Valetsky, *Russ. Chem. Rev.* **2004**, *73*, 501.

R. Shenhar, T. B. Norsten, V. M. Rotello, *Adv. Mater.* **2005**, *17*, 657.

D. Fritsch, K.-V. Peinemann, *Catal. Today* **1995**, *25*, 277.

2 Gas sensors

N. Cioffi, I. Farella, L. Torsi, A. Valentini, A. Tafuri, *Sens. Actuators B* **2002**, *84*, 49.

S. Sharma, C. Nirkhe, S. Pethkar, A. A. Athawale, *Sens. Actuators B* **2002**, *85*, 131.

3 Materials with improved flame retardance

M. Okoshi, H. Nishizawa, *Fire Mat.* **2004**, *28*, 423.

4 Optical filters

G. Carotenuto, *Appl. Organomet. Chem.* **2001**, *15*, 344.

R. J. Nussbaumer, W. Caseri, P. Smith, T. Tervoort, *Macromol. Mater. Eng.* **2003**, *288*, 44.

R. J. Nussbaumer, W. Caseri, T. Tervoort, P. Smith, *J. Nanoparticle Res.* **2002**, *4*, 319.

T. Kyprianidou-Leodidou, P. Margraf, W. Caseri, U. W. Suter, P. Walther, *Polym. Adv. Technol.* **1997**, *8*, 505.

A. Ammala, A. J. Hill, P. Meakin, S. J. Pas, T. W. Turney, *J. Nanoparticle Res.* **2002**, *4*, 167.

J. I. Hong, K. S. Cho, C. I. Chung, L. S. Schadler, R. W. Siegel, *J. Mater. Res.* **2002**, *17*, 940.

5 Dichroic materials

Y. Dirix, C. Darribère, W. Heffels, C. Bastiaansen, W. Caseri, P. Smith, *Appl. Optics* **1999**, *38*, 658.

Y. Dirix, C. Bastiaansen, W. Caseri, P. Smith, *J. Mater. Sci.* **1999**, *34*, 3859.

Y. Dirix, C. Bastiaansen, W. Caseri, P. Smith, *Adv. Mater.* **1999**, *11*, 223.

6 High- and low-refractive index materials

M. Weibel, W. Caseri, U. W. Suter, H. Kiess, E. Wehrli, *Polym. Adv. Technol.* **1991**, *2*, 75.

L. Zimmermann, M. Weibel, W. Caseri, U. W. Suter, *J. Mater. Res.* **1993**, *8*, 1742.

J. G. Winiarz, L. Zhang, M. Lal, C. S. Friend, P. N. Prasad, *J. Am. Chem. Soc.* **1999**, *121*, 5287.

C. Lü, C. Guan, Y. Liu, Y. Cheng, B. Yang, *Chem. Mater.* **2005**, *17*, 2448.

F. Papadimitrakopoulos, P. Wisniecki, D. E. Bhagwagar, *Chem. Mater.* **1997**, *9*, 2928.

L. Zimmermann, M. Weibel, W. Caseri, U. W. Suter, P. Walther, *Polym. Adv. Technol.* **1993**, *4*, 1.

E. J. A. Pope, M. Asami, J. D. Mackenzie, *J. Mater. Res.* **1989**, *4*, 1016.

C. Lü, Z. Cui, C. Guan, J. Guan, B. Yang, J. Shen, *Macromol. Mater. Eng.* **2003**, *288*, 717.

J. C. Seferis, R. J. Samuels, *Polym. Eng. Sci.* **1979**, *19*, 975.

J. C. Seferis in *Polymer Handbook* (Eds.: J. Brandrup, E. H. Immergut), Wiley & Sons, New York, **1989**, p. VI/451.

3
Hybrid Organic/Inorganic Particles

Elodie Bourgeat-Lami

Abstract

Organic/inorganic hybrid particles with diameters ranging from ten nanometers up to several hundred nanometers are an important class of hybrid materials with potential applications in a variety of domains ranging from the encapsulation and controlled release of active substances to their utilization as fillers for the paint and coating industries. This review chapter discusses the different strategies and general concepts of synthesizing hybrid particles with defined shapes (core-shell, multinuclear, raspberry and hairy-like particles) and nanoscale dimensions. Synthetic techniques are mainly based on physicochemical routes or polymerization methods. The physicochemical route involves interaction of preformed macromolecules and/or nanoparticles with particles templates, whereas in the chemical route, the mineral and organic phases are generated *in situ* in the presence of organic or inorganic particles, respectively. The simultaneous reaction of organic and inorganic precursors to produce single-phase hybrid nanoparticles will also be considered. This chapter gives a general overview of the different techniques and briefly mentions potential applications of such systems.

3.1
Introduction

Although colloidal particles received little attention until the end of the 19th century, man has observed and made used of their properties since the earliest days of civilization. The alchemists of medieval times, in their search for the elixir of life, reported the transmutation of base precious metals into gold or silver and highlighted the amazing changes in color that accompanied the overall process.[1] However, it was not until 1856 that Faraday made the first systematic study of colloidal gold and tentatively reported factors that could be responsible for the

---

1) The color is due to the wavelength dependence of scattering and absorption.

Hybrid Materials. Synthesis, Characterization, and Applications. Edited by Guido Kickelbick
Copyright © 2007 Wiley-VCH Verlag GmbH & Co. KGaA, Weinheim
ISBN: 978-3-527-31299-3

stability of these dispersions. The term "colloid" (in Greek *kolla* = glue) was first introduced by Thomas Graham (1805–1869) at the beginning of the 19th century to distinguish between substances that diffused through a semipermeable collodion membranes and those (such as glue, gelatine or starch) that did not. Today, there are many examples of systems for which this simple classification is inadequate. Indeed, colloids cover a broad range of seemingly different materials and are currently defined as two-phase systems in which particles of colloidal size (typically within the range 10 nm–1 µm) of any nature (e.g. solid, liquid or gas, organic or inorganic) are dispersed in a continuous phase of a different composition or state. Hence, colloids are not limited to solid particles and also include suspensions of gas in liquids (foams), liquids or solids in gas (aerosols) or liquids in liquids (emulsions). However, for simplicity, only suspensions of solids in liquids will be considered in this article. Each particle in a colloidal suspension (or a sol) consists of a large number of molecules or "molecular aggregates" stabilized in solution by chemical or electrochemical means. Due to their small size and stabilization, they remain dispersed for a long time and also have the property to scatter light (so-called *Tyndall* effect). However, particles whose size is larger than a critical size (e.g. around 1 µm) settle in solution at rates that depends on their specific gravity. Note that it is not necessary for all three dimensions to lie below 1 µm: platelets and fibers for which only one and two dimensions, respectively, are in the colloidal range may also be classified as colloidal systems (Fig. 3.1). Another particularity of colloids is their high surface area. Indeed, owing to their small size, colloids are almost all surface! Not only are surface phenomena accentuated in colloidal materials but physical confinement due to boundaries effects also creates strong size-dependent properties. Because of their large surface area, colloids are also prone to adsorb large amounts of chemicals, which is largely responsible

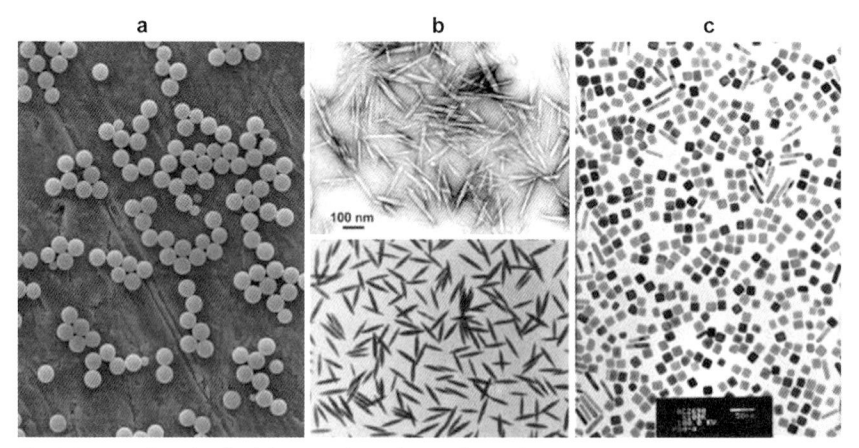

Fig. 3.1 Examples of colloidal materials: a) latex spheres, b) rods (top: cellulose whiskers and bottom: iron oxide particles) and c) gold cubes together with rods. (Reprinted with permission from: *J. Am. Chem. Soc.* **2004**, 126, 8648–8649.)

for their stability in solution and also allows introducing a variety of molecules on their surface. Thus, properties of colloidal materials can be tuned at will by controlling particle size and surface characteristics.

Despite their long history, the scientific study of colloids is a relatively recent development that was mainly stimulated by the demand of industry for colloidal materials of high quality and well defined characteristics. Indeed, colloidal particles are the major components of a lot of industrial products such as foods, inks, paints, coatings, pharmaceutical and cosmetic preparations, photographic films and rheological fluids and play a key role in all these technologies. This progression in the quality of commercial products is also the consequence of significant progress in physical chemistry and the concomitant development of sophisticated modern instrumental techniques. Since the pioneering work of Michael Faraday and Thomas Graham, increasing interest has been devoted to colloids and colloidal processes. The remarkable scientific progress which has been done in inorganic chemistry now allows the production of large quantities of inorganic colloidal dispersions using very simple procedures under mild conditions. These colloidal systems are also perfectly defined in shape, size, composition and surface properties. Polymeric materials can also be elaborated in the form of colloidal spheres (known as latexes) using free radical polymerization procedures. Synthetic latexes have raised increasing interest in the last century and large quantities of commodity polymers (polyvinyl chloride, styrene–butadiene or polychloroprene rubbers. . . .) are manufactured as aqueous dispersions.

Although most colloidal systems are either organic or inorganic, colloids can also be both organic *and* inorganic. Interest in organic/inorganic (O/I) particles mainly arose from the necessity of combining and controlling several properties in a single structure. Indeed, minerals and polymers display radically different properties and can find consequently significantly different applications. By combining the two materials, it appears possible to achieve a unique combination of properties and obtain totally new synergetic behaviors. On the one hand, O/I colloids benefits from the processing and handling advantages of organic polymers. In addition, the latter can be made optically transparent and can be selected to impart distinguishable chemical or physical properties such as conducting, bioactive or electroluminescent characteristics. On the other hand, inorganic materials have fascinating electronic, optical, magnetic or catalytic properties which are complementary to those of pure organic materials. The incorporation of inorganic fillers into polymers (such as clay or silica for instance) additionally contributes to a significant improvement of the thermal and mechanical properties of the resulting composite material.

Organic/inorganic (O/I) hybrid particles can be defined as colloidal particles that contain both organic and inorganic domains. The organic and inorganic components can form either two clearly distinguishable macroscopic phases such as in *composite particles*, or exhibit some degree of phase mixing at the molecular level such as in *hybrids*. However, in practice, the distinction between composite and hybrid is more often subtle and the appellation *O/I particles* will be used in the following to designate both composite and hybrid particles. O/I particles can be

3 Hybrid Organic/Inorganic Particles

classified into three groups according to the method which has been used for their synthesis (Fig. 3.2):

1. O/I colloids can be constructed by assembling preformed organic and inorganic components (e.g. in the form of particles or macromolecules) which elementary units (or bricks) constitute the building blocks of the resulting hybrid colloid,
2. O/I colloids can be produced *in situ* by polymerizing organic (vs. inorganic) precursors in the presence of preformed inorganic (vs. organic) particles and,
3. O/I particles can be obtained by simultaneously reacting organic and inorganic molecular precursors.

The properties of O/I particles depend not only on the chemical nature of the organic and inorganic components but also on their morphology and on the spatial arrangement of the different phases which, in turn, are controlled by the synthet-

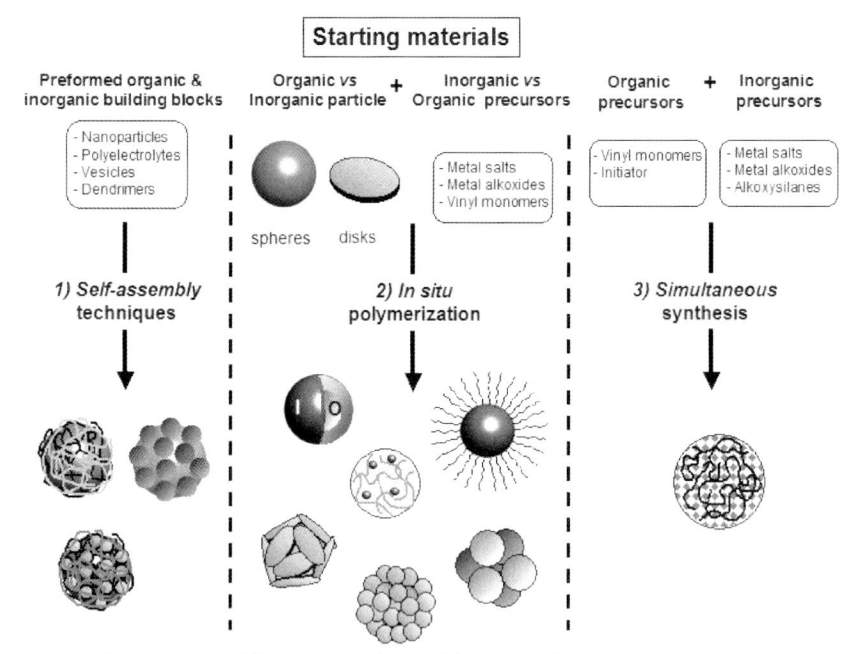

Fig. 3.2 The three main different approaches to fabrication of O/I particles. Route 1: self-assembly of preformed organic and inorganic components, route 2: *in situ* formation of organic polymers or inorganic domains in the presence of inorganic *versus* organic colloids, and route 3: simultaneous reaction of organic and inorganic molecular precursors.

ic procedure and the experimental conditions. For example, if the organic polymer is located at the outer particle surface (in the so-called core-shell morphology), it may protect the core from environmental aggressions or provide functional groups to improve interactions with the surrounding medium (e.g. a better colloidal stability or sensing properties, for instance). On the contrary, when the polymer is surrounded by the mineral and thus plays the role of a template, hollow particles can be produced by subsequent removal of the core, which structures are of particular interest in encapsulation technologies, drug delivery or as pigments for the paint industry. In addition, polymer particles, specifically designed such as to complex inorganic precursors, are potential nanoreactors to control the shape, size and size distribution of *in situ* generated metal or semiconductor particles. Thus, by choosing the appropriate route and by varying the reaction conditions, it is possible in principle to generate an infinite variety of O/I particles with distinct morphologies and functionalities. Last, it is worth mentioning that O/I particles can not only be used as colloidal suspensions, but also be processed as thin films or assembled into three-dimensional (3-D) colloidal crystals, which structures have attracted intense interest in the recent literature. However, despite the growing interest in such macroscopic devices, in the following, only colloidally stable O/I particulate systems will be described, keeping in mind that such nanostructured colloids could be used as precursors for the construction of higher hierarchical materials.

This chapter highlights recent research in the area of hybrid polymer/inorganic particles with well-defined shapes and nanostructures. These include core-shell, hairy-like, multipod-like and other exotic morphologies. The chapter starts with a brief description of some of the most important methods of synthesizing particles. A distinction will be made between organic particles (e.g. polymer colloids but also, vesicular, dendritic and block copolymer structures), and inorganic particles (e.g. metal oxide, metallic and non oxide semiconductor particles). The discussion will be limited to colloidal systems. The second part of the chapter reports physicochemical methods to assemble organic and inorganic building blocks into hybrid particles. The third part describes synthetic routes to elaborate organic/inorganic colloids *in situ* starting from either preformed mineral or polymer templates. These templates can be either particles or supramolecular structures. The emphasis is put on the techniques that have been developed to control the surface chemistry of the templating materials and on the morphology of the resulting composite particles. Finally, the last part of the chapter provides examples of hybrid colloid formation by the simultaneous reaction of organic and inorganic precursors. This approach aims at producing particles with a more intimate mixing of the organic and inorganic components, which strategy should enlarge the range of properties and applications of hybrid colloids.

Only basic aspects are reported in this chapter. Readers who want to know more about this particular class of hybrid particulate materials are referred to reviews and text books that cover in more depth this exiting and broad area of research.

3.2
Methods for Creating Particles

3.2.1
Polymer Particles

Polymeric particles are mostly produced through heterogeneous polymerization processes. Heterophase polymerization systems can be defined as two-phase systems in which the resulting polymer and/or starting monomer are in the form of a fine dispersion in an immiscible liquid medium defined as the "polymerization medium", "continuous phase" or "outer phase". Even if oil-in-water (o/w) systems are greatly preferred on an industrial scale, water-in-oil (w/o) systems may also be envisaged for specific purposes. Heterogeneous polymerization processes can be classified in suspension, dispersion, precipitation, emulsion, miniemulsion and microemulsion techniques according to interdependent criteria which are the initial state of the polymerization mixture, the kinetics of polymerization, the mechanism of particle formation and the size and shape of the final polymer particles (Table 3.1, Fig. 3.3).

3.2.1.1 Oil-in-water Suspension Polymerization
It may be roughly described as a bulk polymerization in which the reaction mixture is suspended as droplets in the aqueous continuous phase. Therefore, the initiator, monomer and polymer must be insoluble in water. The suspension mixture is prepared by addition of a solution of initiator in monomer to the pre-heated aqueous suspension medium. Droplets of the organic phase are formed and maintained in suspension by the use of (a) vigorous agitation throughout the reaction and (b) hydrophilic macromolecular stabilizers dissolved in water (e.g. low molar mass polymers such as poly(vinyl alcohol), polyvinylpyrrolidone, hydroxymethyl-

Table 3.1 The different types of heterogeneous polymerization systems. Adapted from: R.G. Gilbert, *Emulsion Polymerization. A mechanistic approach*, Academic Press, New York, **1995** with permission.

Type	Continuous phase	Droplet size	Initiator	Particle diameter
Emulsion	water	~1–10 µm	water-soluble	50–600 nm
Precipitation	water	no droplets	water-soluble	100 nm–10 µm
Suspension	water	~1–10 µm	oil-soluble	>1 µm
Dispersion	organic (poor solvent for polymer formed)	No droplets	oil-soluble	100 nm to larger than 1 µm
Miniemulsion	water	~50–300 nm	water or oil-soluble	50–300 nm
Microemulsion	water	10 nm	water-soluble	10–50 nm

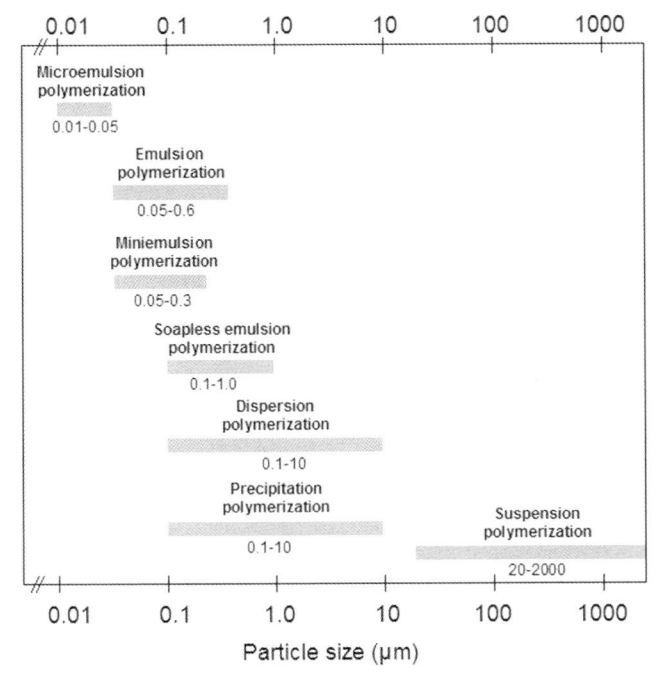

Fig. 3.3 Classification of heterophase polymerization processes according to the particle size range of the resulting colloidal particles. (Redrawn from: R. Arshady in *Preparation and Chemical Applications*, MML Series 1 (Ed. R. Arshady) Citus Book, London, **1999**, pp. 85–124.)

Particle size (μm) ranges shown:
- Microemulsion polymerization: 0.01–0.05
- Emulsion polymerization: 0.05–0.6
- Miniemulsion polymerization: 0.05–0.3
- Soapless emulsion polymerization: 0.1–1.0
- Dispersion polymerization: 0.1–10
- Precipitation polymerization: 0.1–10
- Suspension polymerization: 20–2000

cellulose, etc.). Each droplet acts as a small bulk polymerization reactor for which the normal kinetics apply. Polymer is produced in the form of beads whose average diameter is close to that of the initial monomer droplets (0.01 to 2 mm) even if inadvertent droplet breaking and coalescence widen the bead size distribution. Polymer beads are easily isolated by filtration provided they are rigid and not tacky. Therefore, the suspension polymerization process is unsuitable for preparing polymers that have low glass transition temperatures. It is widely used for styrene, methyl methacrylate and vinyl chloride monomers for instance.

3.2.1.2 Precipitation and Dispersion Polymerizations

In precipitation polymerization, the reaction mixture is initially homogeneous like in solution polymerization, but it is a precipitant for the polymer. Thus the initially formed macromolecules collapse and coagulate to create particle nuclei, which gradually flocculate into irregularly shaped and polydisperse particles. Such a process concerns for instance the synthesis of polytetrafluoroethylene in water or polyacrylonitrile in bulk. In the case of dispersion polymerization, the polymerization medium is not a precipitant but a poor solvent for the resulting polymer. Thus, the macromolecules swell rather than precipitate and the polymerization proceeds largely within these individual particles leading to more monodisperse products. To ensure their stability, macromolecular stabilizers have to be used as in suspension polymerization. Lastly, another characteristic of dispersion polymerization reactions is the diameter of the polymer particles (in the range 0.5–10 µm) which is generally much larger than in emulsion polymerization although small polymer particles (100–500 nm) can also be obtained in the presence of reactive stabilizers or block copolymers.

3.2.1.3 Oil-in-water Emulsion Polymerization

Emulsion polymerization is another heterogeneous process of great industrial importance and allows the elaboration of aqueous colloidally-stable dispersions of polymer particles, known as latexes. In "conventional" emulsion polymerization, the polymer particles are formed by starting from an insoluble (or scarcely soluble) monomer emulsified by the aid of a surfactant above its critical micelle concentration (CMC). The monomer is originally distributed between coarse emulsion droplets, surfactant micelles and the water phase where a small proportion of monomer (depending on its solubility) is molecularly dissolved. Unlike in suspension polymerization, the initiator is soluble in water and it leads to a strongly different particle formation mechanism. Polymerization thus starts in the aqueous phase by the formation of free radicals through the initiator thermolysis and the addition of the first monomer units. These oligomeric radical species (oligoradicals) are rapidly captured by the monomer-swollen micelles, where propagation is supported by absorption of monomer diffusing from the monomer droplets through the aqueous phase to maintain equilibrium. Therefore, stabilized nuclei are produced leading to primary particles, growing gradually until the monomer is completely consumed. The size of these particles is determined by the number of primary latex particles formed and the time during which they grow. The polymer particles generally have final diameters in the range 50–600 nm, i.e. considerably smaller than for suspension polymerization. In emulsifier-free polymerizations, the polymerization is carried out in the same way as described above, except that no surfactant is used. Nucleation occurs by oligoradical precipitation into unstable nuclei which collide to form larger particles. Polymerization takes place mainly within these monomer-swollen particles and particles grow similarly to conventional emulsion polymerization. One of the important features of emulsion polymerization is also the ability to control particle morphology (e.g. formation of core-shell particles and other equilibrium morphologies), by successive additions of different monomers. Polymers prepared by emulsion polymerization are used either directly in the latex form or after isolation by coagulation or spray-drying of the latex. They are used as binder in paints, adhesives, paper coating, carpet backing, water-based inks non woven textiles and related domains. They are also used as support for medical diagnostics. Critical parameters in all these applications are the particle size, the presence of reactive end groups for covalent bonding of targeted molecules (in biomedical applications) or for adhesion to a given substrate (in coatings) and the stability of the colloidal suspension.

3.2.1.4 Oil-in-water Miniemulsion Polymerization

The miniemulsion polymerization may be roughly described as a suspension polymerization leading to polymer particles in the range of submicronic sizes. Indeed, particles are obtained by direct conversion of small monomer droplets without serious exchange kinetics being involved. Nevertheless, due to the small droplets size, the initiator can be either oil- or water-soluble. In a first step, miniemulsion droplets of 50–300 nm are formed by shearing (high pressure homogenizer or ultrasound) a system containing the dispersed phase, the continuous phase, a surfactant and an hydrophobe playing the role of an osmotic pressure agent for preventing the interdroplet mass transfer phenomenon, known as Ostwald ripening. Therefore, polymer particles are obtained by direct conversion of monomer droplets and their final size can be thoroughly controlled by the shearing conditions. The advantages of miniemulsion polymerization are mainly associated to its versatility and applicability to nonradical polymerizations and the encapsulation of resins, liquid and preformed particles.

3.2.1.5 Oil-in-water Microemulsion Polymerization

Microemulsion polymerization is usually performed in mixtures of water, monomer and surfactant in large concentrations in the presence of a water-soluble initiator. A co-surfactant (generally alcohols, amines, or other amphiphilic molecules) is introduced in the microemulsion formulation. Under such conditions, extremely small and stable microemulsion droplets are formed into which polymerization can take place as in miniemulsion polymerization. However, unlike emulsions or miniemulsions, microemulsions are thermodynamically stable and form spontaneously upon contact of the ingredients. Because of the small size of the droplets, microemulsions are usually transparent. The resulting particles are also very small, typically in the range 10–50 nm in diameter and contain only a few polymer chains.

3.2.2
Vesicles, Assemblies and Dendrimers

As mentioned in the introduction, colloidal materials are not limited to solid particles and include macromolecular aggregates (such as polymeric micelles), dendrimers and lamellar vesicles.

3.2.2.1 Vesicles

Surfactant vesicles (also called liposomes) are important class of lamellar bilayer structures extensively used as model membranes for artificial cells. They are non-equilibrium aggregates mostly kinetically stabilized resulting from the interaction of phospholipid molecules. Liposomes are either unilamellar or multilamellar depending on the number of bilayers that compose the aggregate (Fig. 3.4). A large number of methods and a wide range of amphipathic lipids are available for preparation of liposomes of different sizes and void volumes and it is beyond the scope of this chapter to describe all possibilities. Because of their lamellar structure,

Unilamellar vesicle   **Multilamellar vesicle**

Phopholopid molecule

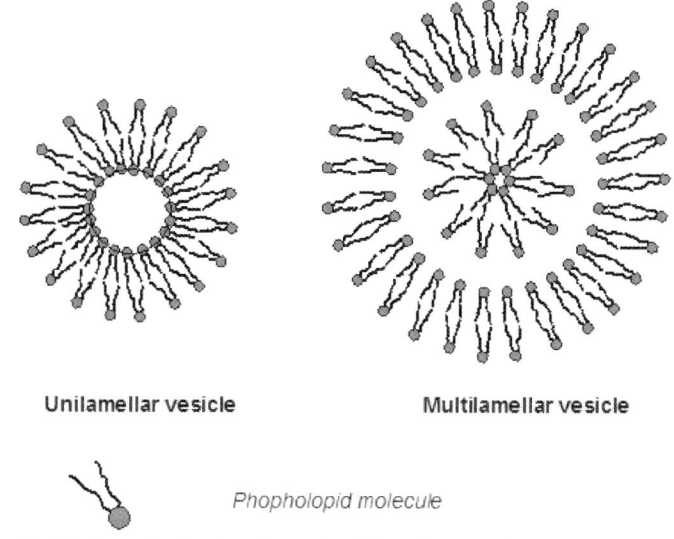

Fig. 3.4 Example of unilamellar and multilamellar vesicular structures.

vesicles constitute an important class of capsule materials with applications in drug delivery or as red cell substitutes. However, because of their low phase transition temperature and imperfections of the bilayer structure (such as branching or unsaturation in the alkyl chain), liposomes are not very stable, resulting in poor retention of encapsulated drugs. Various methods have thus been developed in order to improve the long term stability of vesicular structures including for instance the use of lipids that are functionalized by polymerizable groups. Like natural phospholipids, block copolymers can also form vesicular structures in aqueous solution (so called polymersomes or polymeric liposomes). These materials are more stable than conventional liposomes due to the larger size and the lower dynamic of the polymer molecules.

3.2.2.2 Block Copolymer Assemblies

Block copolymers are characterized by two or more chemically distinct polymer chains or blocks. They are obtained by living polymerization techniques such as anionic, cationic, group-transfer, ring-opening metathesis and "controlled" free radical polymerizations. They can also be elaborated by coupling of preformed end-functionalized homopolymers. Significant progress has been done in recent years in this field and a wide range of polymers of various architectures (stars, diblocks, triblocks, grafts, etc.) and different solubility and functionality can be now produced by these techniques (Fig. 3.5). One particular property of block copolymers is their capacity to phase separate in selective solvents. The ability of amphiphilic diblock copolymers to self-assemble into colloidal size aggregates has been studied for several decades and will not be reviewed here. Suffice it to say that

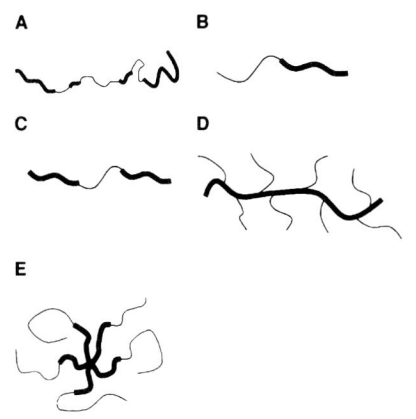

Fig. 3.5 Representation of different block copolymer architectures: (A) random copolymer, (B) diblock copolymer, (C) triblock copolymer, (D) graft copolymer and (E) star copolymer. (Reproduced from: W. Loh, Block Copolymer Micelles in *Encyclopedia of Surface and Colloid Science*, Marcel Dekker, New York, **2002**, pp. 803–813 with permission.)

block copolymers can adopt a variety of morphologies depending on their concentration, the polarity of the solvent, the chain length, the chemical structure and the composition of the different blocks.

3.2.2.3 Dendrimers

Dendrimers are 3-D globular highly branched macromolecules with a centrosymmetric architecture which are obtained by an iterative sequence of reaction steps. They are usually characterized by: a) a core or initiating central unit, b) an interior volume made of the different generations and c) an outward structure in interaction with the surrounding solution and located at the periphery of the molecule. Dendrimers can be grown either by a divergent or a convergent method. In the divergent technique, the structure is built up by reacting the core molecule with a multifunctional monomer to form the first generation. Growth then proceeds by subsequent addition of monomer molecules to the end groups of preceding generations (Fig. 3.6a). In the convergent method, each new generation is added to a constant number of active sites starting from the periphery and terminating at the focal point or end group (Fig. 3.6b). Dendrimers contain a predetermined number of functional groups located either in the internal cavities, in the interior branches or at the surface. Owing to their high density of functional groups, dendrimers constitute a special class of highly reactive molecules of particular interest as sensor materials, as nanosized containers for drug encapsulation and controlled release, as carriers for heterogeneous catalysis or as confined reaction vessels for nanoparticles synthesis.

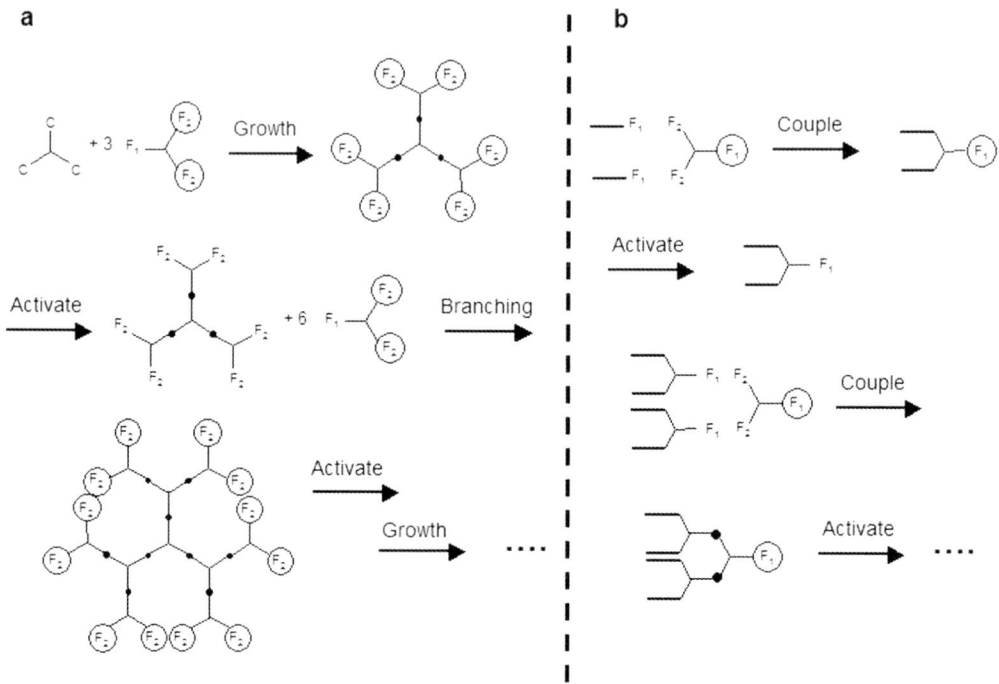

Fig. 3.6 Schematic illustration of convergent (a) and divergent (b) growth synthesis of dendrimers with a AB$_2$ type monomer. (Adapted from: I. Gitsov, K. R. Lambrych, Dendrimers. Synthesis and Applications in *Dendrimers, Assemblies, Nanocomposites*, MML Series 5 (Eds R Arshady, A. Guyot), Citus Book, London, **2002**, Chap. 2, pp. 31–68.)

3.2.3
Inorganic Particles

Colloidal mineral particles can be divided into metal oxides, metals and non oxide semiconductors. The following section briefly describes some of the methods that are commonly used for their synthesis.

3.2.3.1 Metal Oxide Particles

Metal oxide colloids are usually prepared by controlled hydrolysis/precipitation of organometallic precursors or metal salts from homogeneous solutions according to the so-called sol–gel process (see Chapter 6 of this book). The most commonly used precursors are metal alkoxides of the type $M(OR)_z$ where M = Si, Ti, Sn, Zr, etc, and R stands for an alkyl group. The precursor is usually dissolved in an organic solvent (e.g. typically an alcohol) and acidic or basic catalysts are used to accelerate hydrolysis and condensation reactions. A typical illustration of particles preparation by the sol–gel process is the synthesis of silica spheres through the reaction of tetraethylorthosilicate (TEOS) in a basic solution of water and ethanol according to the so-called Stöber process. In a typical procedure, TEOS is introduced in a mixture of alcohol, ammonia and water. Hydrolysis and condensation reactions yield dense (compact) monodisperse silica spheres whose diameter is controlled by the reactant concentrations. However, the breadth of the distribution broadens and the particles become less spherical when the TEOS concentration is increased up to around 0.2 M. The largest size that can be achieved with a narrow size distribution is ~0.7 μm.

Forced hydrolysis and controlled release of ions Metal oxide colloids can also be obtained by addition of a strong base to metal salt solutions. This procedure is highly sensitive to experimental parameters such as the pH, the nature of the ions and the salt concentration, which sensitivity prevents the generation of monodisperse particles. To control the kinetics of metal ion hydrolysis, two methods have been developed: forced hydrolysis and controlled release of ions. In the forced hydrolysis method, the aqueous salt solution is aged at temperatures typically between 80 °C and 100 °C for different periods of time. Deprotonation of coordinated water molecules is greatly accelerated with increasing temperature which ensures the surpersaturation to be reached rapidly resulting in the formation of a large number of small nuclei. A variety of metal oxide particles of controlled chemical composition, morphology and structure can be obtained by this technique as reported in many papers and reviewed. Instead of deprotonating hydrated metal ions, the hydrolysis of cations can be controlled by a slow release of hydroxide ions from organic molecules like urea of formamide. For example, heating a solution of yttrium and europium chlorides in the presence of urea allows the liberation of hydroxide ions which released ions cause the precipitation of Y_2O_3:Eu nanoparticles.

3.2.3.2 Metallic Particles

The most relevant method of metal particle formation is by chemical reduction of metal salts in solution. Since a large fraction of the constituent atoms of metallic clusters are present at their surface, raw metals are inherently unstable. Metallic particles must be thus stabilized either electrostatically or sterically using coordinating (capping) agents that allow preventing nanoparticles agglomeration. Typical preparation procedures can be illustrated by taking the example of gold nanoparticles synthesis. The earliest method of gold formation relies on sodium citrate reduction of chloroauric acid at 100 °C in aqueous solution. In a typical procedure, the metal salt is mixed with sodium citrate in highly diluted water solution and boiled for several hours while maintaining a constant suspension volume by the continuous addition of water. The procedure yields approximately 20 nm diameter gold nanoclusters stabilized by an electrical double layer responsible for electrostatic (coulombic) repulsion. In parallel to the aqueous route, non aqueous methods of colloidal synthesis that require steric stabilization have also developed. Steric stabilization is accomplished by adsorbing polymers, molecular surfactants or coordinating molecules (hereinafter referred to as monolayer-protected clusters

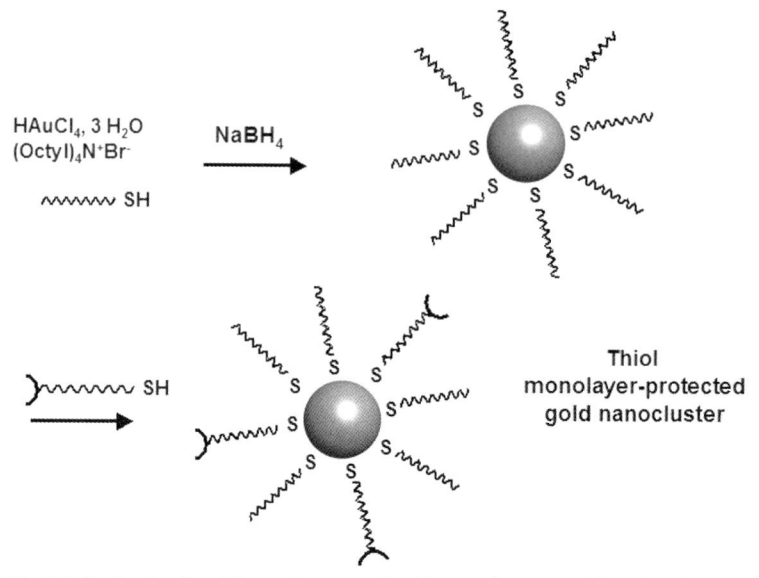

Fig. 3.7 Synthesis of multilayered-protected gold nanoclusters and ligand exchange.

(MPCs)) which role is to provide a barrier against aggregation and control nanoparticles size.[2] Numerous other papers report the synthesis of gold nanoclusters using cationic surfactants that allows transport of the inorganic salt precursor from water to organic solutions. Basically, the synthesis involves the phase transfer of chloroaurate ions from aqueous to organic solution in a two-phase liquid/liquid system followed by reduction with sodium borohydride in the presence of alkane thiol stabilizing ligands. Importantly, the ligand can be further replaced by another stronger capping agent which allows tuning the surface properties of the metal nanocluster by choosing appropriate ligands (Fig. 3.7). The particles can be thus isolated, stored as dried powders and subsequently redispersed in a suitable solvent which can be polar or apolar depending on the nature of the capping agent.

Table 3.2 provides a non-exhaustive list of precursors, reducing agents and stabilizers (capping agents) commonly used in the synthesis of metal colloids. While polyvinyl alcohol or sodium polyacrylate are well adapted for stabilization in polar media, alkanethiols, mercapto alcohols or mercaptocarboxylic acids have been shown to be highly suitable capping agents for a variety of metallic nanoclusters in organic solutions including silver, platinum, iridium and palladium.

---

2) In some sense, metal protected nanoclusters could be regarded as hybrid particles with a mineral core coated with an organic protecting layer. However, in our classification, these particles are considered to be surface stabilized clusters rather than true hybrids although this demarcation is somehow arbitrary.

Table 3.2 Examples of precursor molecules, reducing agents and polymeric stabilizers involved in the synthesis of metal colloids.

Precursors	Chemical formula
	$PdCl_2$
Hydrogen hexachloro platinate IV	H_2PtCl_6
Silver nitrate	$AgNO_3$
Silver tetraoxyl chlorate	$AgClO_4$
Chloroauric acid	$HAuCl_4$
Rhodium chloride	$RhCl_3$
Reducing agents	
Sodium citrate	$Na_3C_6H_5O_7$
Hydrogen	H_2
Sodium carbonate	Na_2CO_3
Palladium chloride	$NaOH$
Hydrogen peroxide	H_2O_2
Sodium tetrahydroborate	$NaBH_4$
Lithium tetrahydroaluminate	$LiAlH_4$
Lithium tetrahydroborate	$LiBH_4$
Ammonium ions	NH_4^-
Polymeric stabilizers	
Poly(vinyl pyrrolidone)	PVP
Polyvinyl alcohol	PVA
Polyethyleneimine	PEI
Sodium polyphosphate	NaPP
Sodium polyacrylate	NaPA
Poly (N-isopropyl acrylamide)	PNIPAM
Surfactants	
Cetyl trimethyammonium bromide	CTAB
Tetraoctylammonium bromide	TOAB
Didodecyldimethylammonium bromide	DDAB
Capping agents–passivators	
Alkanethiols	C_nSH
Octylamine	$C_{18}H_{37}NH_2$
Trioctylphosphine	TOP

3.2.3.3 Semiconductor Nanoparticles

Non-oxide semiconductor particles are commonly produced by pyrolysis of organometallic precursors in a hot coordinating solvent that mediates nanoparticle growth and stabilizes the inorganic surface. A rapid injection of the reagents in the hot reaction vessel ensures a short burst of homogeneous nucleation (caused by an abrupt surpersaturation and a sharp decrease in temperature) and prevents renucleation by the the fast depletion of the reagents from the suspension medium. Typical high boiling point coordinating solvents are tri-*n*-octyl phosphine (TOP) or tri-*n*-octyl phosphine oxide (TOPO). In the following, the case of cadmium selenide (CdSe) nanoparticles synthesis is used as an example to illustrate the overall process. Dimethyl cadmium (Me$_2$Cd) was used as the cadmium source and trioctyl phosphide selenide (TOPSe) as the selenium source. Typically, two stock solutions of Me$_2$Cd and TOPSe into TOP are added to a reaction flack containing TOPO and maintained at 300 °C under an argon atmosphere. The rapid introduction of the reagent mixture produces a yellow/orange solution characteristic of CdSe nanocrystals formation. The suspension medium is finally maintained at 230–260 °C by gradually increasing the temperature during the aging period. The resulting suspension is purified to remove by-products and excess reagents.

Nanoparticles of metal sulfides are usually synthesized by reacting a soluble metal salt (as for instance Cd(ClO$_4$)$_2$) and H$_2$S (or Na$_2$S) in the presence of an appropriate stabilizer such as sodium metaphosphate according to the sulfidation reaction:

$$Cd(ClO_4)_2 + Na_2S \rightarrow CdS \downarrow + 2NaClO_4$$

The growth of the CdS nanoparticles in the course of reaction is arrested by an abrupt increase in pH of the solution. A variety of semiconducting nanocrystals such as cadmium sulfide or cadmium telenide, have been elaborated following these procedures. The size of the nanocrystals depends on a large number of experimental conditions such as the nature and concentration of the precursor, the nature of the solvent, the temperature and the reaction time.

3.2.3.4 Synthesis in Microemulsion

An extension of the use of coordinating agents is the synthesis of mineral particles in water-in-oil microemulsions. Indeed, inverse micelles are nanoreactors in which inorganic precursors are dissolved and precipitated. A variety of metal oxide particles (including TiO$_2$, SiO$_2$ and magnetic colloids), mixed metal-oxides, metal carbonates, and a number of other ultrafine particulate materials (CdS, Pd, Pt, Au) have been synthesized in reverse microemulsion systems. Typically, the organometallic and/or metal salt precursors are dissolved inside the water pools of the reverse spherical micelles, and are allowed to react via droplet collision and rapid inter-micellar exchange of their water content. This process is mainly controlled by diffusion and is also dependent on the nature of the surfactant. Two distinct procedures are described. In the first, the catalyst is already contained in the water phase and the precursor, dissolved in the continuous medium progressively enters the water pools of the reverse microemulsion. In a second route, the precursor and the catalyst are prepared as two separate microemulsions, and mixed together to allow the reaction to proceed. Because rapid exchange is taking place between micelles, nanoparticles synthesized in reverse micelles most often do not have narrow size distributions. Table 3.3 summarizes the most commonly used inorganic particles and their preparation methods.

Table 3.3 Most commonly used inorganic particles and their preparation methods.

Inorganic particles	Chemical formula	Technique	Precursors	Stabilizers
Silica, zinc oxide, titanium dioxide, iron oxide	SiO_2, ZnO, TiO_2, Fe_2O_3, Fe_3O_4	Sol-gel, hydrothermal hydrolysis	Alkoxides, metal salts	Surface charges
Gold, Silver, Palladium, Platinium, rhodium	Au, Ag, Pd, Pt, Rh	Chemical reduction	Metal salts and metal complexes (metallorganic)	Organothiol capping agents, surfactants
Sulfides, Selenides	CdS, PbS, CdSe	Hydrothermal synthesis	Metal salts and metal complexes	Passivating

3.3
Hybrid Nanoparticles Obtained Through Self-assembly Techniques

It has been known for some time that complex ordered architectures can form spontaneously by self-assembly. Self-assembly can be briefly described as the spontaneous 2D or 3-D organization of molecules, clusters, aggregates or nanoparticles. Typical examples of self-assembled systems either in solution or on solid supports are ligand-stabilized metal nanoparticles, block copolymer assemblies, ordered 2D arrays of colloidal particles deposited on planar substrates or 3-D super lattice colloidal crystals. The following section will concentrate only on the assembly of unlike particles or particles and polymers in solution to prepare complex colloidal structures with well defined geometries. As the main driving forces of assembly techniques are electrostatic attraction and molecular recognition, the two techniques will be discussed separately.

3.3.1
Electrostatically Driven Self-assembly

3.3.1.1 Heterocoagulation

The interaction between particles of different characteristics (sizes, charges and chemical composition) is generally known as heterocoagulation. This kind of process is common in nature and frequently encountered in many industrial and biological processes such as oil recovery, mineral flotation or treatment of wastewaters. Attractive interactions between positively charged particles and negatively charged colloids also provide a possible mechanism for the formation of raspberry-like O/I particles made of an inorganic core surrounded by smaller heterocoagulated polymer particles or *vice versa*. The interaction is mainly driven by electrostatic attraction and generally involves particles of opposite charges (Fig. 3.8).

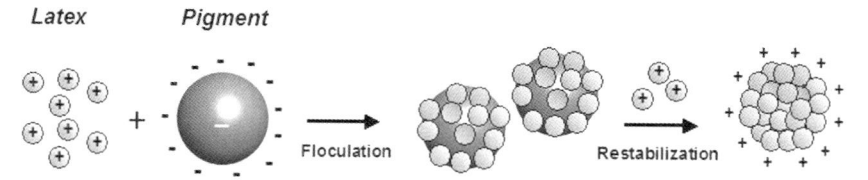

Fig. 3.8 Schematic picture illustrating the heterocoagulation of small positively charged latex particles onto larger negatively charged inorganic particles.

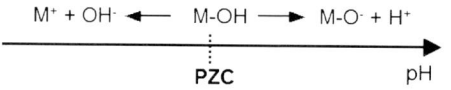

Fig. 3.9 Ionization reactions of mineral oxide surfaces as a function of pH.

Most inorganic particles and more particularly oxide particles develop a surface charge in aqueous solutions. The surface charge of mineral oxide particles is pH dependent. Protons and hydroxyl ions adsorb on the surface of the oxide and protonate or deprotonate the MOH bonds according to:

$$M-OH + H^+ \rightarrow M-OH_2^+ \qquad (1)$$

$$M-OH + H^- \rightarrow M-O^- + H_2O \qquad (2)$$

The ease with which protons are added or removed from the surface (e.g. the acidity of the MOH group) depends on the nature of the metal. The pH at which the surface charges changes from positive to negative is called the point of zero charge (PZC). Thus, the surface charge is positive at pH > PZC and negative at pH < PZC (Fig. 3.9).

The PZC values of a series of oxide particles are reported in Table 3.4. While a PZC value lower than 7 indicates an acidic surface, a PZC higher than 7 indicates a basic surface.

Table 3.4 Point of zero charge of selected oxides (reproduced with permission from: C. J. Brinker, G. W. Scherer, *Sol-Gel Science – The Physics and Chemistry of Sol-Gel Processing*, Academic Press, New York, 1990).

Metal oxide	PZC
MgO	12
FeOOH	6,7
Fe_2O_3	8,6
Al_2O_3	9,0
Cr_2O_3	8,4
SiO_2	2,5
SnO_2	4,5
TiO_2	6,0

PZCs are experimentally determined by acid–base titration. However, in practise, characterization of surface charge is most often carried out by the measurement of their electrophoretic mobility. When a colloidal suspension is submitted to an electric field, each particle with closely associated ions is attracted towards the electrode of opposite charge. The velocity of the particle is commonly referred to as its electrophoretic mobility (U_E). U_E is dependent on the strength of electric field, the dielectric constant of the medium, the viscosity of the medium and the zeta potential which corresponds to the electric potential at the boundary between the ionic particles and the surrounding solution (slipping plane). With the knowledge of U_E, we can thus obtain the zeta potential of the particles by application of the Henry equation.

$$U_E = \frac{2\varepsilon\zeta f(K_a)}{3\eta}$$

where:

ζ: zeta potential,
U_E: electrophoretic mobility,
ε: dielectric constant,
η: viscosity and,
$f(K_a)$: Henry's function.

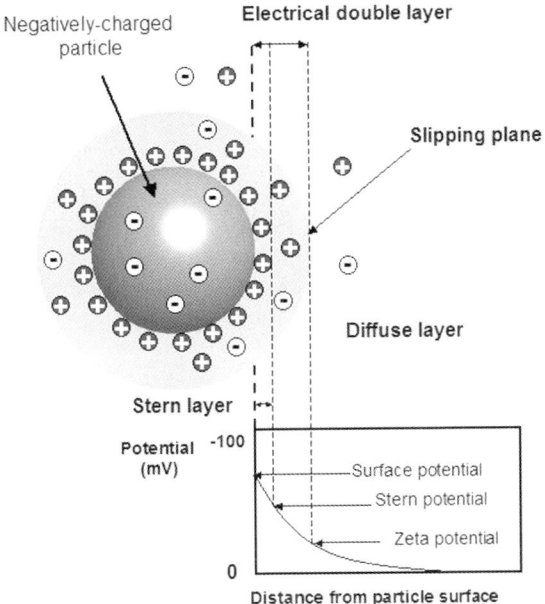

Electrophoretic determination of zeta potential is most commonly made in aqueous media in the presence of moderate electrolyte concentrations. $f(K_a)$ in this case is equal to 1.5, and the Henry equation is then referred to as the Smoluchowski approximation. Therefore, calculation of zeta potential from the mobility is straightforward for systems that fit the Smoluchowski model, that is to say for particles larger than 0.2 µm dispersed in electrolytes containing more than 10^{-3} molar salt.

Variation of zeta potential as a function of pH allows determining the isoelectric point (IEP) of the mineral which corresponds to the pH value at which the net electric charge of the particles is zero (i.e. there is no charge outside the slipping plane). Zeta potential therefore reflects the effective charge on the particles. It is worth noticing that as the PZC and the IEP are determined by different methods, they do not always necessarily coincide except when the measurements are performed at very low ionic strengths.

Table 3.5 Examples of ionic monomers used in the synthesis of functional polymer particles through emulsion polymerization. Reproduced from: C. Pichot, T. Delair, Functional Nanospheres by Emulsion Polymerization in *Microspheres, Microcapsules & Liposomes*, MML Series 1 (Eds.: R Arshady), Citus Book, London, **2002**, Chap.5, pp. 125–163, with permission.

Ionic group	Chemical name
$-SO_3^-, M^+$	Sodium styrene sulfonate
	Sodium 2-sulfoethyl or propyl methacrylate
	2, Acrylamide, 2-methylpropane sulfonate
$-Py^+, X^-$	4-Vinyl pyridinium
	1-Methyl-4-vinylpyridinium bromide (or iodide)
	1,2-dimethyl-5-vinylpyridinium methyl sulfate
$-N^+(CH_3)_3, Cl^-$	(N-trimethyl-N-ethyl methacrylate) ammonium chloride
	3-Methacrylamidinopropyl trimethyl ammonium chloride
$-N^+R_3, Cl^-$	2-Dimethylaminoethyl methacrylate hydrochloride
	Vinylbenzyltrimethylammonium chloride
	Vinylbenzylamine hydrochloride
$-SC^+(NH_2)NH_2Cl^-$	Vinylbenzylisothiouronium chloride

Polymer colloids are also characterized by a surface charge. The surface charge of polymer particles may have different origins including ionic initiator, ionic surfactant and ionic functional monomers (Table 3.5). Again, electrophoresis is used to evaluate the isoelectric point of polymer colloids by measuring the variation of the surface potential as a function of pH.

Owing to its major technological implications, the fundamental aspects associated to the stability behavior of mixed colloidal dispersions, including the mechanisms and kinetics of the heterocoagulation process, have been extensively studied both experimentally and theoretically for a long time. Systematic studies have been performed using different types of inorganic and polymer colloids such as metal oxides and polymer latexes. By changing the pH, both the sign and the surface potential of the colloids can be finely tuned, making it possible to evaluate the effects of these determinant parameters on the interaction of both sets of particles.

For example, preformed amphoteric latex particles were adsorbed on the surface of titanium dioxide pigments. The latex particles were synthesized in the presence of a zwitterionic emulsifier, N, N'-dimethyl n-lauryl betaïne (LNB) at pH 7.0, and showed an isoelectric point in the range of pH 7–8. Strong interactions were observed between pH 3 and pH 8, where the latexes were positively charged while titanium dioxide particles were negatively charged. As evidenced by turbidity measurements, the mixed heterocoagulated suspensions were destabilized upon addition of an increased number of latex particles due to the neutralization of the surface charge of the pigment, but restabilization occurred with further addition of the latexes. Similarly, cationic polystyrene latexes were adsorbed onto spherical rutile titanium dioxide particles. It was shown that the ionic strength of the suspension medium had a great influence on the adsorption behavior. More latex particles were heterocoagulated on the TiO_2 surface when the electrolyte concentration was increased due to the diminution of the electrostatic repulsion between neighboring adsorbed particles. It should be noticed that not only can polymer latexes be self-assembled at the surface of inorganic colloids, but dendrimers and vesicles can also be electrostatically adsorbed on mineral particles surfaces according to the heterocoagulation mechanism. The heterocoagulation technique also allows small metal oxide particles to be deposited onto larger polymer particles. For example, monodisperse hydrophilic magnetic polymer latexes have been synthesized using a two-step procedure. In a first step, magnetic iron oxide particles were adsorbed onto various cationic latexes (e.g. polystyrene, core shell poly (styrene–N-isopropylacrylamide: NIPAM) and poly (NIPAM)) via electrostatic interactions. In a second step, the iron oxide-coated polymer latexes were encapsulated by poly (NIPAM) after separation of the excess free iron oxide particles in a magnetic field. In a similar procedure, magnetic latex particles were prepared by mixing amphoteric polymer latexes with different sizes and nanosized magnetic $NiOZnOFe_2O_3$ particles. Latex particles have also been coated by CdTe nanocrystals using a related strategy.

3.3.1.2 Layer-by-layer Assembly

Since Decher and co workers demonstrated in 1997 that uniform polymer films could be deposited onto mineral substrates by the sequential adsorption of polyanions and polycations using the so-called layer-by-layer (L*b*L) assembly technique, interest in polyelectrolyte assembly has increased considerably. This technique has been recently extrapolated to colloidal particles to generate colloidally stable multilayered composite particles (Fig. 3.10).

Typically, the polyion multilayer is formed by adsorbing a polyelectrolyte solution onto particles of apposite surface charges, removing the excess polyions by rinsing and repeating the procedure. Charge overcompensation occurs after every adsorption step which allows adsorption of the subsequent oppositely charged layers. Changes in surface charge after each sequence of adsorption is evidenced by zeta potential measurements. Fig. 3.11 displays a list of some of the most common polyelectrolytes used in the preparation of LbL assemblies involving colloidal templates.

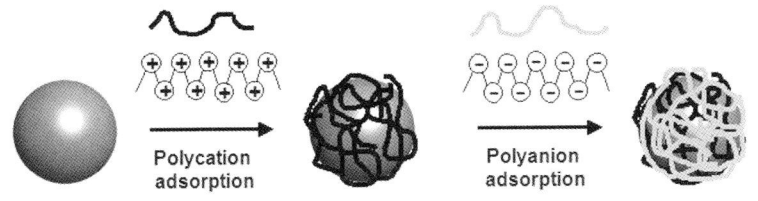

Fig. 3.10 Schematic illustration of alternate adsorption of negatively and positively charged polymers onto colloidal templates.

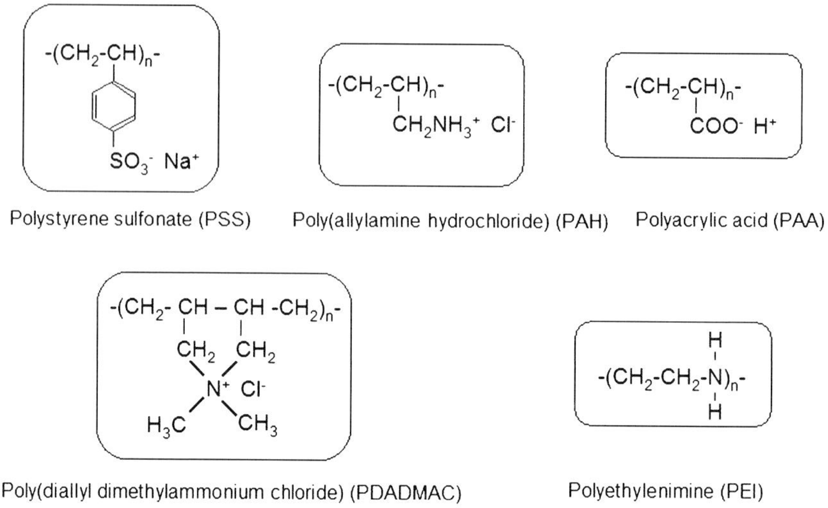

Fig. 3.11 Structure of common polyelectrolytes used for the preparation of LbL colloidal assemblies.

Polystyrene sulfonate (PSS); Poly(allylamine hydrochloride) (PAH); Polyacrylic acid (PAA); Poly(diallyl dimethylammonium chloride) (PDADMAC); Polyethylenimine (PEI)

Beyond the use of polymers, the technique has been recently extended to the layer-by-layer construction of complex colloidal systems involving both polymer electrolytes and mineral particles. The technique is similar to that reported previously and consists in the step-wise, layer-by-layer neutralization and subsequent resaturation of the surface charge of the colloid by alternate polyelectrolyte/nanoparticles coatings. By using this approach, a variety of inorganic particles such as metals, semiconductors, metal oxides, silicates or mixtures of these materials have been deposited onto latex spheres as illustrated in Fig. 3.12 for silica.

For instance, colloidal clay nanosheets have been adsorbed onto cationic polystyrene latexes as a thin crystalline layer. Tetramethoxysilane was used as inorganic precursor to consolidate the coating and increase shell stability. The polymer template was removed in a next step to generate hollow silicate capsules. Hollow titania shells have been produced in a similar way by alternate deposition of polyethylenimine and titania nanosheets on polymer latexes and removal of the organic template by heat or UV treatment of the core-shell nanocomposite particles. Micrometer-sized polystyrene beads have also been coated with zeolite A nanocrystals using a cationic polymeric agent as binder between the core and shell materials. Contrary to metal oxide colloids that carry a pH-dependent surface charge, metallic and semiconductor particles are uncharged in their native state. Therefore, in this particular case, the choice of the stabilizing or coordinating molecules used in the synthesis in solution is essential in the creation of surface charges.

Advantages of this LbL coating procedure is the ability to precisely control the thickness of the multilayered assembly by varying the number of coating cycles, the ability to incorporate a large variety of polyelectrolytes and inorganic particles

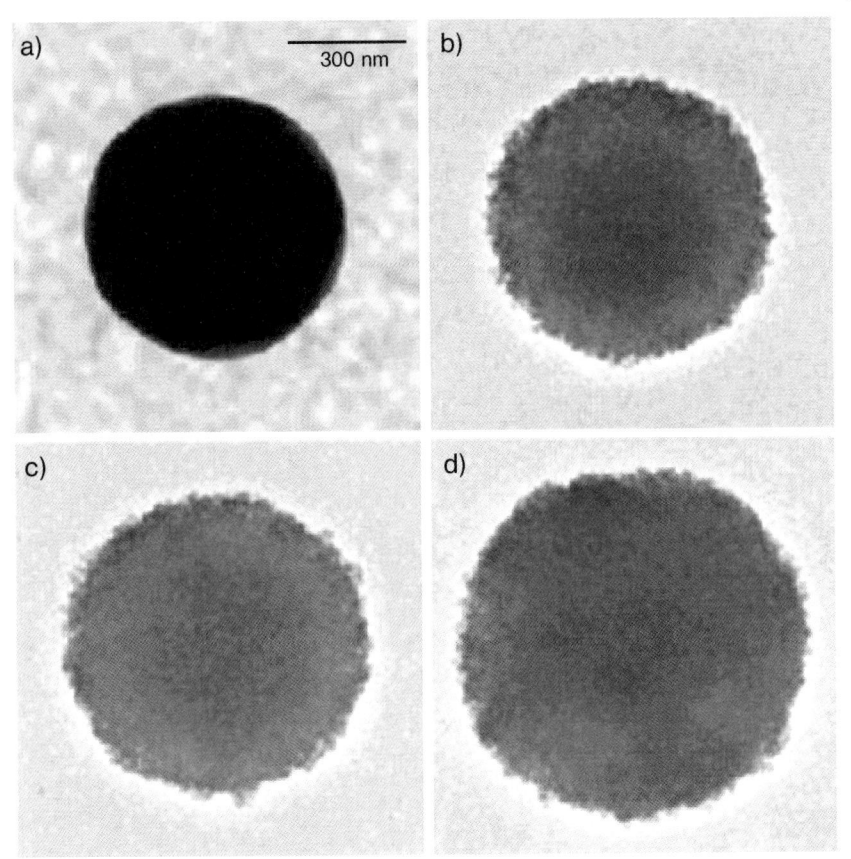

Fig. 3.12 TEM micrograph of silica/ PDADMAC multilayer shell deposited onto 640 nm polystyrene latex particles. (a) Bare polystyrene latex particles and (b–d) polystyrene particles coated with one, two and four SiO_2/polyelectrolyte bilayers, respectively. (Reproduced from: F. Caruso, H. Lichtenfeld, M. Giersig, H. Möhwald, Electrostatic self-assembly of silica nanoparticle-polyelectrolyte multilayers on polystyrene latex spheres. *J. Am. Chem. Soc.* **1998**, *120*, 8523–8524, with permission.)

of different chemical and physical functionalities in any desired order and the possibility to control particles shape by an appropriate choice of the precursor colloid used as template. At last, it is worth mentioning that not only does this method allow the formation of O/I composite particles but it also permits synthesizing hollow spheres by removal of the colloidal template.

3.3.2
Molecular Recognition Assembly

Controlling and tuning interaction between particles has always been a relevant challenge both experimentally and theoretically. Systems can be specifically designed to undergo self colloidal organization by using bifunctional mediating molecules bearing reactive groups on both ends capable of bonding particles together. Biologically programmed assembly of nanoparticles has been first described by Mirkin and Alivisatos. They showed that complementary DNA antigens could be used to self-assemble nanoparticles. The concept was recently applied to gold colloids by Mann and co-workers who used antigen/antibody recognition assembly to induce the reversible aggregation of the inorganic nanoparticles and produce a conjugated hybrid material with long-range interconnectivity. Recent examples also include the elaboration of colloid/colloid, dendrimer/colloid and polymer/colloid composite superstructures. However, one major difficulty of the technique is to control the size and the morphology of the aggregate structure. Indeed, the recognition assembly process most often yields to extended network formation instead of individual aggregates.

Based on this same general principal of molecular recognition assembly, it was shown that amine-functionalized polystyrene particles could be assembled for instance into highly ordered 2D monolayers on the surface of glutaraldehyde-activated silica microspheres (Fig. 3.13). Core-shell composite particles consisting of a silica core and a polystyrene shell were obtained in a subsequent step by heating the assembled colloids above the Tg of polystyrene.

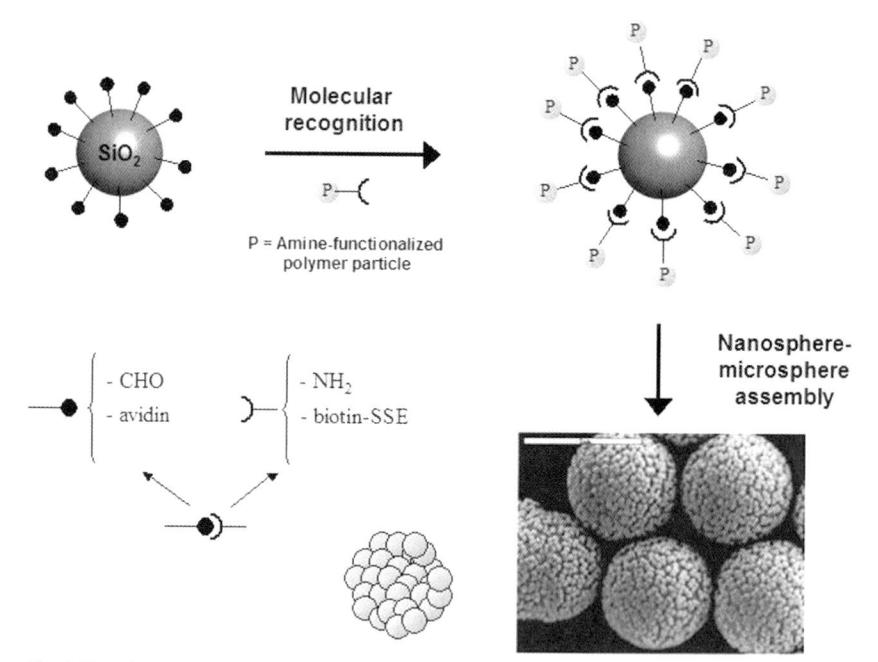

Fig. 3.13 Schematic representation of the colloidal assembly of dissimilar particles in a binary suspension via chemical and biospecific interactions. (Adapted from: M. S. Fleming, T. K. Mandal, D. R. Walt, Nanosphere-microsphere assembly: methods for core-shell materials preparation, *Chem. Mater.* **2001**, *13*, 2210–2216, with permission.)

3.4
O/I Nanoparticles Obtained by *in situ* Polymerization Techniques

3.4.1
Polymerizations Performed in the Presence of Preformed Mineral Particles

Fillers such as fumed silica, metals and mineral fibers have been incorporated into polymers for more than a century. However, in many applications, these inorganic fillers display a poor miscibility with the polymer matrix which often yields to phase segregation and degradation of the final (optical or mechanical) properties. If the polymer could be introduced on the surface of the inorganic particles so that the mineral became the core and the polymer the shell, optimal disposition of the inorganic particles within the polymeric film could be achieved resulting therefore in optimal light scattering. It becomes thus possible this way to make composite materials that maintain their optical clarity while exhibiting enhanced mechanical properties.

To allow efficient formation of polymer on the mineral surface, it is common to introduce groups that are reactive in the polymerization process (e.g. monomers, initiators or chain transfer agents). These groups may be either chemically bound or physically adsorbed on the surface. Moreover, they can be introduced in a separate step before polymerization or during polymerization. The polymerization reaction can also be conducted in different ways. In the following, a distinction will be done between polymerizations performed in a good solvent for the polymer (e.g. in solution) and those performed in a precipitating medium (e.g. through heterophase polymerization) (Fig. 3.14). An overview of the various methods that

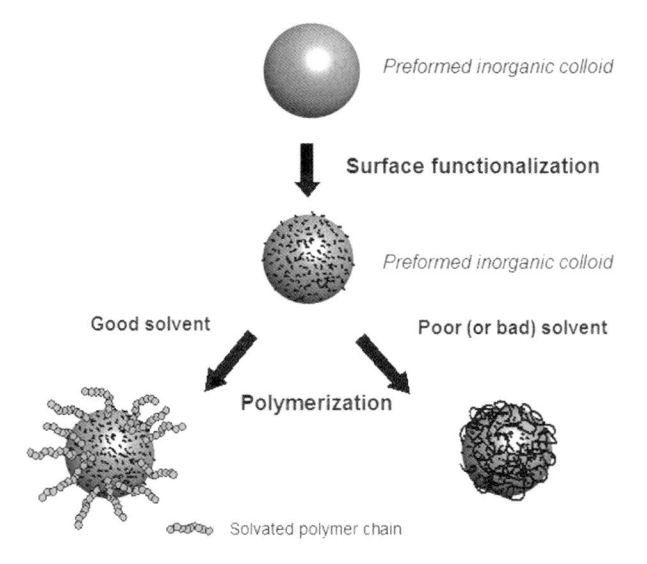

Fig. 3.14 General strategy used to grow polymers at the surface of inorganic particles.

allow modification of mineral particle surfaces and the subsequent growth of polymer chains either in solution or in multiphase systems is given in this section.

3.4.1.1 Surface Modification of Inorganic Particles

Grafting of organosilane and organotitanate coupling agents Silane and titanate coupling agents have been used for decades in order to provide enhanced adhesion between a variety of inorganic substrates and organic resins. They are organometallic derivatives of the type R_nMX_{4-n} where M is a metal (Si or Ti), X is a chloride or an alkoxy group and R is an organic group that can bear different functionalities (Table 3.6).

Organosilane compounds are known to react with hydroxylated surfaces to form mono or multilayer coverages depending on the experimental conditions (namely, the nature of the organosilane molecule and the amount of water). Organosilanes can be either deposited from organic solvents, aqueous alcoholic solutions or water. The halogen or alkoxy groups of the organosilane molecules hydrolyze in contact with water and the resulting silanols form hydrogen bonds with neighboring hydrolyzed silanes and with surface hydroxyls. Siloxane bonds are formed with release of water. The suspensions are made free of excess materials either by dialysis or by successive centrifugation/redispersions cycles in alcohols.

Electrostatic interaction and complexation chemistry are other possible ways to change the surface properties of minerals. Such reactions are highly sensitive to the nature of the solute (which may sometimes compete for complexation), the pH of the solution (which determines the surface charge of the particles and, thus controls its interaction with ionic compounds), and the surface area of the mineral particles. Adsorption is usually investigated through the construction of adsorption isotherms. Adsorption isotherms give information on the extent and

Table 3.6 Examples of organosilane and organotitanate coupling agents commonly used in the surface modification of metal oxide particles.

Coupling agent	Chemical structure	Symbol
vinyl trimethoxysilane	$CH_2{=}CHSi(OCH_3)_3$	VTMS
3-trimethoxysilyl propyl methacrylate	$CH_2{=}C(CH_3)COOCH_2CH_2CH_2Si(OCH_3)_3$	MPS
3-trimethoxysilyl propane thiol	$HSCH_2CH_2CH_2Si(OCH_3)_3$	MPTS
amino propyl trimethoxysilane	$H_2NCH_2CH_2CH_2Si(OCH_3)_3$	APS
glycidoxy propyl trimethoxysilane	$CH_2(O)CHCH_2O(CH_2)_3Si(OCH_3)_3$	GPMS
diisopropyl methacryl isostearoyl titanate	$((H_3C)_2HCO)_2Ti(OCOCH_2C_{16}H_{33})(OCOC(CH_3){=}CH_2)$	KR7
trimethacryl isopropyl titanate	$(CH_2{=}C(CH_3)COO)_3TiOCH(CH_3)_2$	/
diisopropyl diisostearoyl titanate	$((H_3C)_2HCO)_2Ti(OCOCH_2C_{16}H_{33})_2$	KR TTS

the nature of the interaction and may exhibit different shapes depending on the nature of the sorbent, the suspension medium and the physicochemical properties of the system. As described later, mineral particles strongly interact with ethylene oxide-based surfactants or macromonomers and oppositely charged monomer or initiator molecules.

3.4.1.2 Polymerizations in Multiphase Systems

Polymer encapsulation of metal oxide particles One frequently encountered strategy for synthesizing O/I particles by *in situ* polymerization in multiphase systems is the grafting of a methacrylate silane molecule (MPS, Table 3.6) that allows anchoring of the growing polymer chains on the mineral surface during the earlier stages of polymerization and enables O/I particles formation in a very efficient way. A variety of silica/polymer composite particles have been elaborated by this technique using emulsion, dispersion or miniemulsion polymerization processes resulting in different particle morphologies (Fig. 3.15).

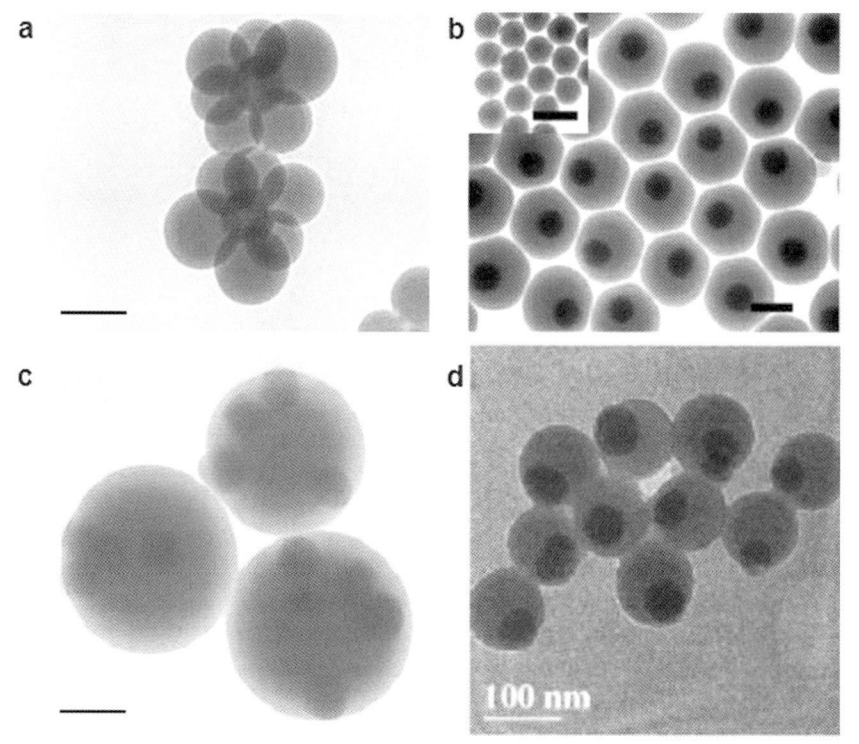

Fig. 3.15 TEM images of silica/polystyrene composite particles through emulsion (a–b), dispersion (c–d) and miniemulsion polymerization (e–f) using MPS as silane coupling agent.

3 Hybrid Organic/Inorganic Particles

Composite particle morphologies mainly depend on three parameters:
1. The MPS grafting density,
2. The silica particles size and concentration which in turn determine the overall surface area available for capture of the growing free radicals and,
3. The experimental conditions used for polymerization (e.g. the nature of the initiator, the surfactant or the monomer and their respective concentrations).

In emulsion polymerization, the mechanism of nanocomposite particle formation can be described as follows (Fig. 3.16). The initiator starts to decompose in the water phase giving rise to the formation of radicals. These radicals propagate with aqueous phase monomers until they undergo one of the following fates: a) aqueous phase termination or b) entry into a micelle or precipitation (depending on the surfactant concentration), creating somehow a new particle. Aqueous-phase oligomers of all degrees of polymerization can also undergo frequent collision with the surface of the silica seed particles, and have therefore a high probability to

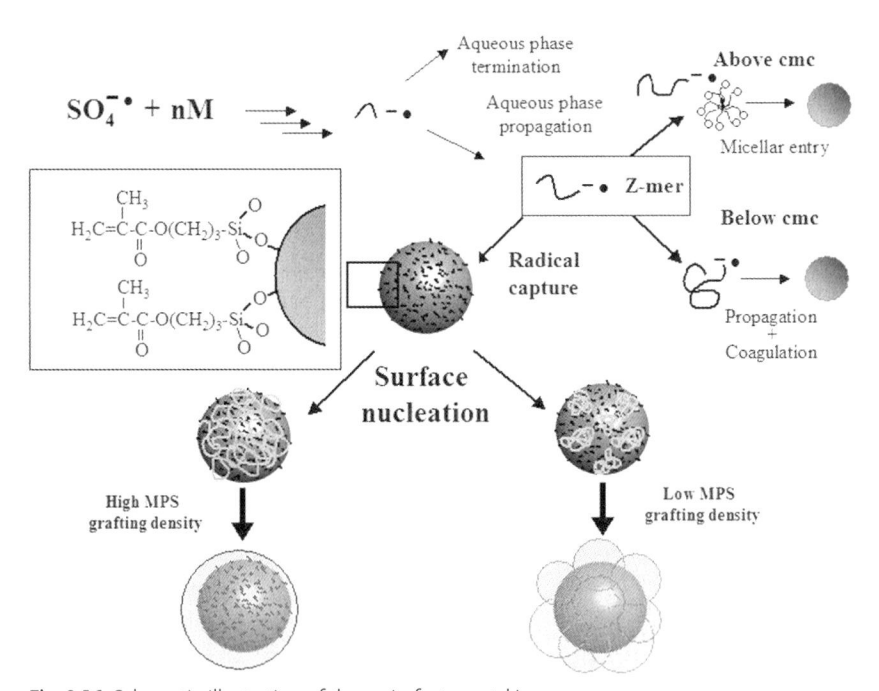

Fig. 3.16 Schematic illustration of the main features taking place during the formation of the silica/polymer nanocomposite particles through emulsion polymerization using MPS as silane coupling agent. (Reproduced from: E. Bourgeat-Lami, M. Insulaire, S. Reculusa, A. Perro, S. Ravaine, E. Duguet, J. Nanosci. Nanotechnol. **2006**, 6, 432–444 with permission.)

copolymerize with the double bonds of silica, thus generating chemisorbed polymer chains in the early stages of polymerization. These discrete loci of adsorption are preferred to adsorb further oligomers or radicals compared with the bare seed surface. As a result, these discrete loci of adsorption become discrete loci of polymerization. Provided that new polymer particle formation is not promoted (which in turn depends on the number of double bonds and the overall surface area), polymerization will exclusively take place around silica. If the MPS grafting density is sufficiently high, a large number of primary particles are captured by the seed surface and a shell is formed around silica (Fig. 3.15b). The shell may result from the collapsing of the growing polymer chains on the functionalized silica surface or from the coalescence of freshly nucleated neighboring primary particles, this last issue being promoted by the close proximity of these precursor particles and the correspondingly low surface energy. For a too low MPS concentration, in contrast, the polymer chains form segregated domains around the silica particles as the high interfacial energy (due to the presence of unreacted silanol groups), does not promote spreading of the polymer chains on the surface nor interparticle coalescence (Fig. 3.15a). Therefore, the affinity of the growing polymer spheres for the silica surface, and hence the final morphology of the composite particles can be tuned on demand by varying the amount of double bonds attached to the surface and the respective sizes of the inorganic core and growing polymer nodules.

A variety of O/I composite particles with diverse functionalities have been elaborated by this technique using different types of nanoparticles and organic polymers as illustrated in a recent work on the synthesis of multilayered gold–silica–polystyrene core shell particles through seeded emulsion polymerization. In this article, silica-coated gold colloids were encapsulated by polystyrene using MPS as silane coupling agent according to the procedure just described for silica. These particles were subsequently transformed into hollow spheres by chemical etching of the silica core in acidic medium.

Apart from the use of silanes, another strategy to O/I colloids formation through heterophase polymerization consists in promoting monomer or initiator adsorption on the mineral surface. As mentioned above, some conveniently selected molecules may spontaneously adsorb on inorganic surfaces through electrostatic or complexation chemistry. In some cases, monomer adsorption can also be promoted by the presence of adsorbed surfactant molecules on the surface (so-called admicellization/adsolubilization behavior). A schematic illustration of this concept is shown in Fig. 3.17 while Table 3.7 provides a non-exhaustive list of monomers, initiators, surfmers or macromonomers which are concerned by this general strategy.

Adsorption of surfactants onto solid particles may involve complex mechanisms and had been the subject of a huge number of works. Mineral oxide/surfactant interactions are promoted by charge/charge attractions, hydrogen-bonding or hydrophobic interactions. Depending on the nature of the surfactant and the kind of interaction, the amphiphilic molecule may either stabilize the inorganic particles or induce significant aggregation. It has been demonstrated for instance that positively charged particles flocculate upon addition of anionic surfactants, and

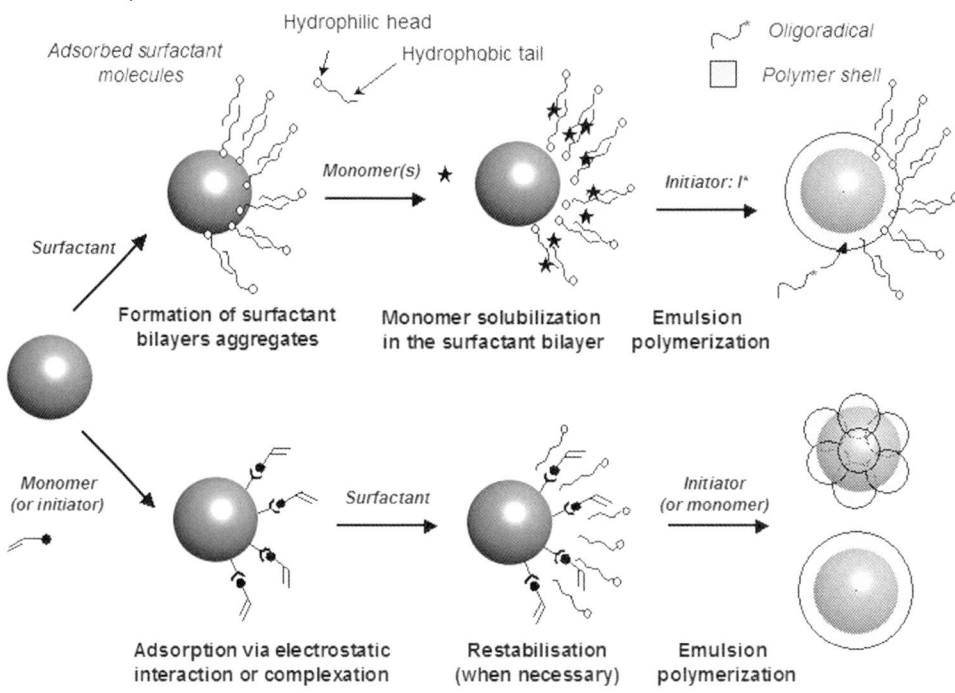

Fig. 3.17 Principle of encapsulation through surfactant, monomer and initiator adsorption. (a) polymerization in adsorbed surfactant bilayers; (b) surface polymerization induced by initiator or monomer adsorption.

that on further addition, the particles redisperse due to formation of surfactant bilayers. Adsorbed surfactants may adopt a variety of structures as for instance hemimicelles, admicelles or vesicles, depending on the respective surfactant/ surfactant and surfactant/mineral interactions. As a direct consequence of the assembly process of surfactants on mineral surfaces, a hydrophobic interlayer can form on the inorganic particles into which monomers can solubilize and polymerization subsequently proceed (Fig. 3.17). The concept of admicellization/polymerization in adsorbed surfactant assemblies (so-called ad-polymerization) has been well described in case of planar substrates but only few examples have been reported concerning colloidal systems. Organofunctional titanate molecules carrying hydrophobic groups have been used for instance to promote adsorption of sodium dodecyl sulfate on amorphous titanium dioxide. This allowed the formation of surfactant-surrounded pigments which proved to be convenient seed particles in the subsequent construction of an organic polymer layer on their surface. Positively charged iron oxide pigments stabilized with adsorbed sodium dodecyl sulfate bilayers were also reported to be encapsulated by this technique. More recently, polymer-coated silver nanoparticles have been elaborated through emulsion

Table 3.7 Functional monomers, initiators, surfmers and macromonomers used during the synthesis of O/I composite particles through emulsion polymerization.

	Nomenclature	Chemical structure
INITIATOR	2,2'-azo(bis) isobutyramidine dihydrochloride (AIBA)	
MONOMERS	Pyrrole, aniline	
	4-vinyl pyridine	
SURFMERS - MACROMONOMERS	N-[(ω-methacryloyl)-ethyl] trimethyl ammonium chloride	
	N-dimethyl-N-[(ω-methacryloyl)-ethyl] alkyl ammonium chloride	
	N-(decadecyl styrene) trimethyl ammonium chloride	
	N-[(ω-methacryloyl)-decadecyl] trimethyl ammonium chloride	
	Polyethylene oxide monomethylether mono methacrylate	$CH_3O-(CH_2-CH_2O)_n-CO(CH_3)=CH_2$
	ST—PVP: 3, poly(N-vinyl pyrrolidone) Styrene	

polymerization. The silver colloids were previously modified with oleic acid which readily adsorbed on the surface. A uniform and thin layer of poly(styrene/methacrylic acid) copolymers was formed on the hydrophobized inorganic seed particles providing a protective organic and functional shell to the metal colloid. It is relevant to mention here that oleic acid derivatives are also of great benefit in the encapsulation reaction of colloidal magnetic particles through emulsion polymerization. The surface of the magnetite colloid is first coated by a monolayer of sodium monooleate, and then stabilized by adsorption of a second layer of sodium dodecyl benzene sulfonate (SDBS) surfactant. Thermosensitive magnetic immunomicrospheres were prepared according to this technique by reacting styrene, N-isopropylacrylamide and methacrylic acid comonomers in the presence of the double layer surfactant-coated ferrofluid at 70 °C and using potassium persulfate as initiator.

As shown in Fig. 3.17, monomers can also be directly adsorbed on the particle surface. Adsorption is most often promoted by charge/charge interactions through acid–base mechanisms, and involves in this case the use of basic (as for instance 4-vinylpyridine, quaternary ammonium methacrylate salts) or acidic (e.g. acrylic acid, methacrylic acid) monomers depending on the PZC of the mineral particles. A variety of silica–polymer colloidal nanocomposites have been elaborated according to this strategy by copolymerizing 4-vinylpyridine (4VP) with methyl methacrylate, styrene, n-butyl acrylate or n-butyl methacrylate monomers. The comonomer feed composition was chosen such as to afford either hard or soft film-forming materials. Owing to the inherent strong acid–base interaction between the basic 4VP monomer and the acidic silica surface, nanocomposite particles with "current-bun morphologies" or silica-rich surfaces were obtained. The resulting films presented a high gloss and a good transparency as well as unusually low water uptake. These water-borne colloidal nanocomposites could find applications in the coating industry as fire-retardant or abrasion-resistant materials.

Silica/polypyrrole and silica/polyaniline nanocomposite colloids have been synthesized using a related approach. The silica nanoparticles participate to stabilization of the polymeric suspension and are mainly located at the composite particles surface that display a raspberry-like morphology characterized by silica beads glued together into the conducting composite latexes (Fig. 3.18).

Fig. 3.18 Schematic representation of the formation of polypyrrole-inorganic oxide nanocomposite colloids by dispersion polymerization of pyrrole in aqueous medium (Reprinted from: S. P. Armes, S. Gottesfeld, J. G. Beery, F. Garzon, S. F. Agnew, Conducting Polymer-Colloidal Silica Composites, Polymer 1991, 32, 2325–2330, with permission.)

Poly(pyrrole) and poly(N-methylpyrrole)-gold composite particles have also been produced by aqueous solution reduction of the corresponding monomer in the presence of gold colloids. The conductive polymer-gold composite particles were next converted to hollow polymeric nanocapsules by chemical etching of the colloidal gold template.

Alternatively, the adsorbed molecules can combine the property of a surfactant with that of a monomer (so-called polymerizable surfactant). For example, quaternary alkyl salts of dimethylaminoethyl methacrylate (C_nBr) surfactants were used to promote polymer encapsulation of porous silica particles in aqueous suspension. The polymerizable surfactant adsorbed on the silica surface in a bilayer fashion (Fig. 3.19). The CnBr amphiphilic molecule was either homopolymerized or copolymerized with styrene adsolubilized in the reactive surfactant bilayer. High encapsulation efficiencies were readily obtained by this technique.

In a similar approach, a series of head- and tail-type surface active cationic monomers were adsorbed on the surface of colloidal silica. The adsorbed monomers were shown to spontaneously polymerize in THF or chloroform at 60°C and even 40°C giving rise to the formation of small polymer plots on the surface.

Macromonomers (e.g. linear macromolecules with polymerizable group at one end) can also be adsorbed on mineral surfaces. Dispersion polymerization of styrene performed in presence of a styrene-terminated poly(4-vinylpyridine) (ST-PVP) macromonomer was shown to give polystyrene-coated silica particles. Similarly, it has been demonstrated that rapsberry-like silica/polystyrene colloidal nanocomposites can be elaborated in emulsion polymerization. They showed that addition of a small amount of a monomethylether mono methylmethacrylate poly(ethyleneoxide) macromonomer allowed the formation of nanometric polystyrene latex particles on the surface of submicronic silica particles through an *in situ* nucleation and growth process (Figs 3.20 and 3.21).

Cationic initiators, as for instance 2,2′-azobis(2-amidinopropane) dihydrochloride (AIBA) strongly adsorb on negatively charged surfaces. Interaction of the initiator with the inorganic surface can be finely tuned by changing the pH of the

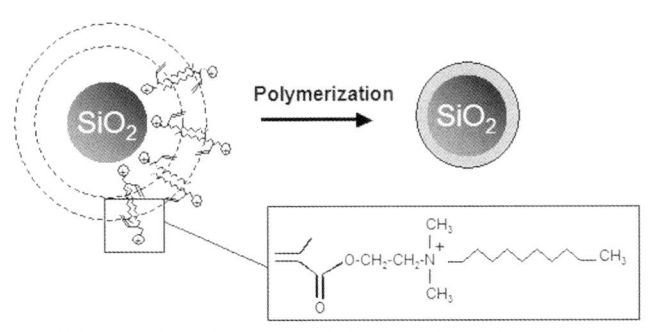

Fig. 3.19 Principle of silica encapsulation through emulsion polymerization by adsorption of a surface active monomer.

Macromonomer: **Ma-PEO**

$CH_3O-(CH_2-CH_2O)_x-\underset{\underset{O}{\|}}{C}-\underset{\underset{CH_3}{|}}{C}=CH_2$

Non ionic surfactant: **NP$_{30}$**

C_9H_{19}-⌬-$(O-CH2-CH2)_{30}$-OH

Ma-PEO adsorption

1) NP$_{30}$
2) Styrene
3) KPS (70°C)

Emulsion polymerization

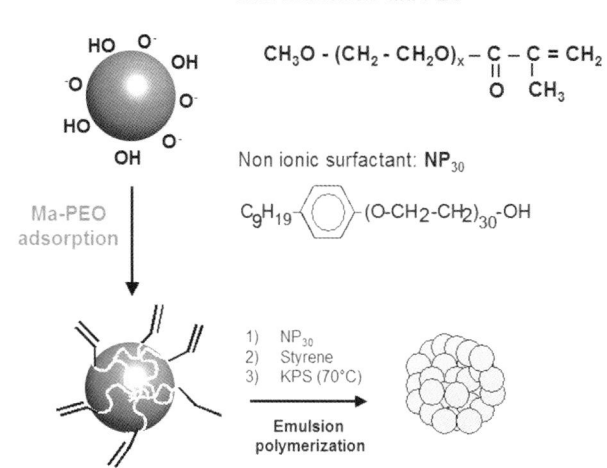

Fig. 3.20 Schematic representation of the macromonomer-mediated assembly process of polymer latexes onto colloidal silica nanoparticles.

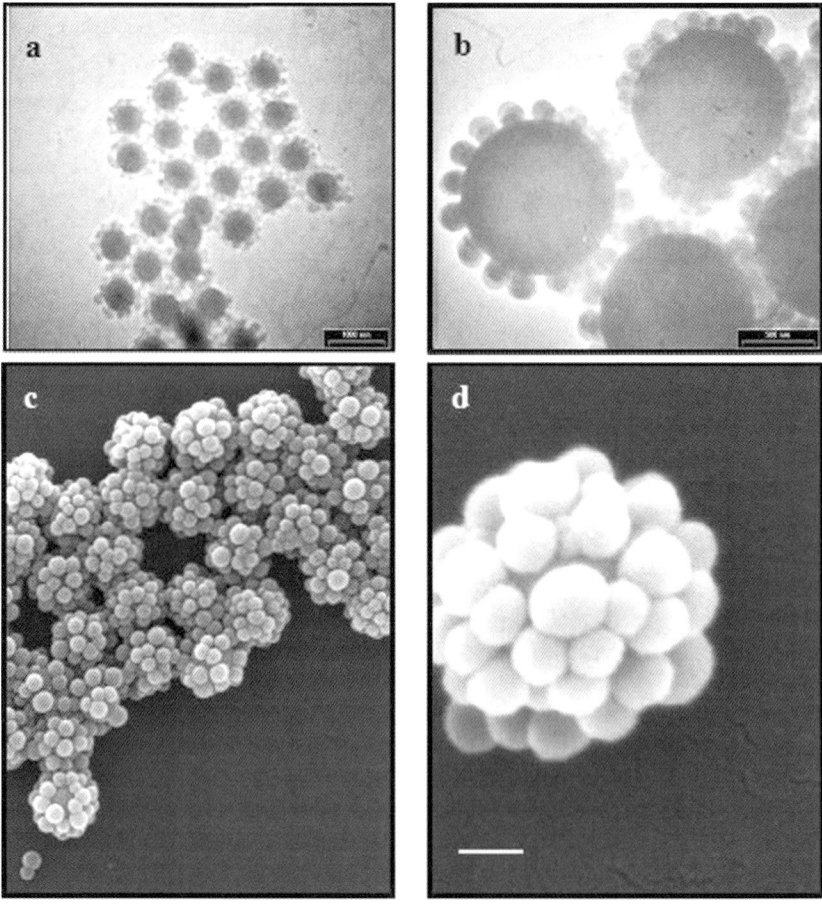

Fig. 3.21 TEM (a, b) and SEM (c, d) images illustrating the macromonomer-mediated self-assembly process of colloidal polystyrene particles onto submicronic silica spheres through emulsion polymerization. (a, c): Dp SiO$_2$ = 500 nm; and (b, d): Dp SiO$_2$ = 1 μm. Note the homogeneous distribution of the polymer particles on the silica surface. (Reprinted from: S. Reculusa, C. Poncet-Legrand, S. Ravaine, C. Mingotaud, E. Duguet and E. Bourgeat-Lami, *Chemistry of Materials*, with permission from American Chemical Society.)

surrounding medium. AIBA was used for instance to initiate the polymerization reaction of styrene in presence of titanium dioxide pigments. High encapsulation efficiencies were achieved when the terminal ionic group of the polymer chain and the surface charge of the inorganic pigment were oppositely charged. AIBA adsorption was shown to proceed by means of ion-exchange and the attached initiator allowed the subsequent growth of polymer chains on the surface. When AIBA is used in combination with suitable amounts of surfactant, positively charged latexes are generated *in situ* during polymerization, and concurrently heterocoagulated on the mineral surface. Depending on the diameter of the silica beads, either strawberry-like or core-shell morphologies can be produced by this technique. AIBA was also used to polymerize vinyl monomers from several layered silicate substrates. The elaboration of clay-based composite particles is the focus of the following section.

Polymer/clay nanocomposite particles   In addition to spherical particles, anisotropic fillers such as clays or carbon nanotubes have retained major attention in recent literature. Indeed, because of their high aspect ratio, platelet-shaped clay particles, a few nanometers thick and several hundred nanometers long, allow a substantial improvement in strength, modulus and toughness while retaining optical transparency. Additional benefits are enhanced tear, radiation and fire resistance as well as a lower thermal expansion and permeability to gases. As reviewed in Chapter 4 of this book, one major issue of the elaboration of clay-based nanocomposites is the exfoliation of the clay layers within the polymer matrix. Three methods are currently reported: a) melt intercalation, b) exfoliation/adsorption and c) *in situ* polymerization.

Although numerous studies have been devoted to *in situ* intercalative polymerization in solution or in bulk, only a limited number of contributions have dealt with the synthesis of clay/polymer nanocomposites through heterophase polymerization. Intercalated nanocomposites based on MMT and various polymers or copolymers have been elaborated through conventional emulsion polymerization using bare non-modified clay suspensions as seeds. Confinement of the polymer chains in the interlayer gallery space was evidenced by DSC and TGA measurements and was suspected to originate from ion-dipole interactions between the organic polymers and the MMT surface. Unfortunately, as the composite particles were precipitated, no information was given on their morphology. However, the clay being used as supplied, it is very unlikely that special interactions were taking place between the exfoliated clay layers and the growing latex particles in the diluted suspension medium. It can be anticipated rather that the polymer particles were physically entrapped between the clay layers consequently to flocculation and drying of the composite suspension as schematically represented in Fig. 3.22. For steric and energetic considerations, the polymer latex particles could no longer move from the interlayer space resulting in polymer chains confinement at the vicinity of the clay surface.

Work involving the use of organically-modified clay particles in heterophase polymerization is rather scarce. To the best of our knowledge, only two reports

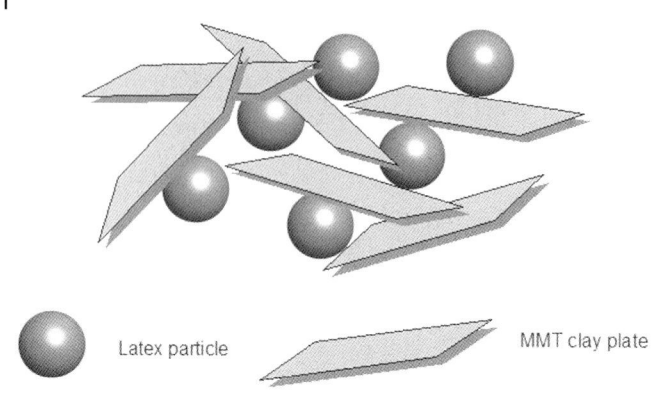

Fig. 3.22 Suspected morphology of polymer/MMT composite materials produced through conventional emulsion polymerization without any pre-treatment of the clay particles.

combine the emulsion or suspension polymerization approaches and ion exchange reaction. In one of these reports, 2,2′-azo(bis) isobutyramidine dihydrochloride (AIBA) is immobilized in the clay interlayer region to yield exfoliation of MMT in the PMMA matrix through suspension polymerization. In another relevant work, it is demonstrated that exfoliated structures can be obtained by post-addition of an aqueous dispersion of layered silicates (either MMT or laponite) into a polymethyl methacrylate latex suspension produced in the presence of suitable cationic compounds (cationic initiator, monomer or surfactant). Since the latex particles were cationic and the clay platelets anionic, strong electrostatic forces were developed at the polymer/clay interface but again no mention was made of particles' morphology.

Successful formation of nanocomposite particles was recently evidenced in a recent work on Laponite. Laponite is a synthetic clay similar in structure to Hectorite. Advantages of using Laponite instead of Montmorillonite is the dimension of the crystals (e.g. 1 nm thick and 40 nm large), which is of the same order of magnitude as the diameter of polymer latexes. Following strategies similar to those mentioned previously for spherical fillers, polystyrene-co-butylacrylate/Laponite composite particles have been synthesized through emulsion polymerization. The clay particles were first modified by incorporating reactive groups on their surface. This was performed either by exchanging the sodium ions by suitable organic cations (AIBA and MADQUAT, respectively, Table 3.7) or by reacting methacryloyloxy alkoxysilanes with hydroxyl groups located on the clay edges. The organoclay was next suspended in water, which process required the use of high shear devices or chemicals (peptizing agents) in order to assist in redispersion. Fig. 3.23 depicts the different steps involved in the synthesis of the polymer/laponite latex particles.

Finally, the emulsion polymerization reaction was accomplished in a conventional way using potassium persulfate or 2,2′-azobis cyanopentanoic acid as initiators

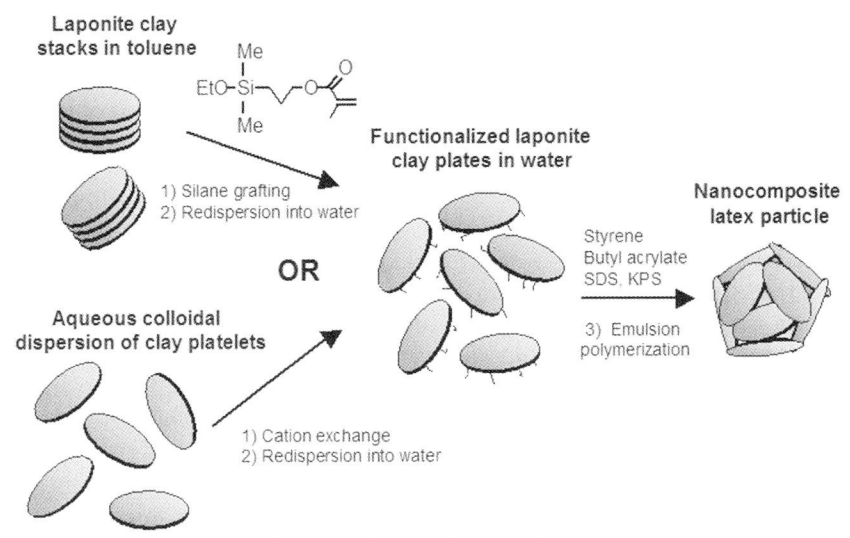

Fig. 3.23 Schematic picture illustrating the procedure used for synthesis of polymer /Laponite composite particles.

and sodium dodecyl sulfate as surfactant. Stable composite latexes with diameters in the range 50–150 nm were successfully produced provided that the original clay suspension was stable enough. The clay plates were found to be located at the external surface of the polymer latex particles as illustrated on the TEM pictures of Fig. 3.24. It is worthwhile to notice that similar morphologies were obtained

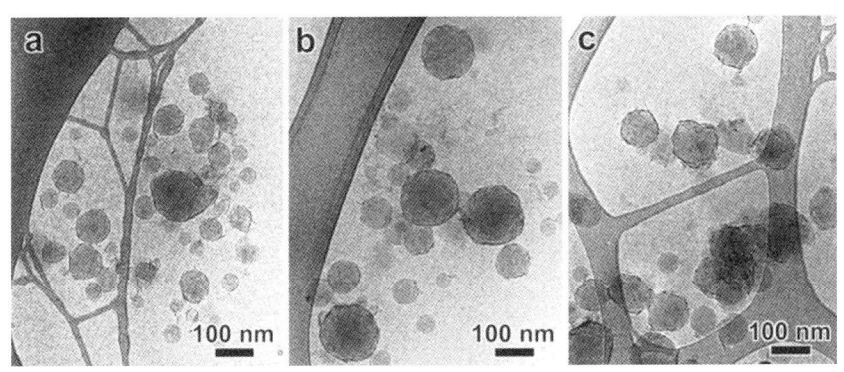

Fig. 3.24 Cryoelectron microscopy images of poly(styrene-co-butyl acrylate)/laponite nanocomposite particles prepared through emulsion polymerization using (a) γ-MPS, (b) AIBA and (c) MADQUAT as reactive compatibilizers. The nanoparticles are seen embedded in a film of vitreous ice. The thin dark layer covering the surface of the polymer particles corresponds to 1 nm-thick diffracting clay platelets that are oriented edge-on with respect to the direction of observation.

whatever the reactive compatibilizer introduced on the clay surface (e.g. either the organosilane, the cationic initiator and the cationic monomer).

3.4.1.3 Surface-initiated Polymerizations

The graft-from and the graft-to techniques Apart from the formation of dense coatings, core-shell O/I particles can also be elaborated by templating inorganic colloids with polymer brushes in solution. There are two general methods used for attachment of polymers to nanoparticle surfaces: the graft-to and the graft-from techniques. In the graft-to technique, a functional group of a preformed polymer is reacted with active sites on the inorganic surface. Covalent attachment of the polymer chains thus requires the macromolecules to be first modified by conveniently selected end groups. For instance, isocyanate-capped and triethoxysilyl-terminated polyethylene oxide (PEO) oligomers were synthesized and grafted further to the surface of silica particles by reaction of the end groups of PEO with the silanol groups of silica. In the graft-from technique, polymers are grown directly *from* the inorganic surface which has been previously functionalized with the appropriate initiator or catalyst. The grafting reaction can be done in various ways through anionic, cationic or free radical processes while a variety of colloidal materials such as silica, semiconductor, metallic nanoparticles and clays have been used as templating materials. Basically, the general synthetic strategy involves the covalent attachment of the initiator molecule, the controlled agent or the catalyst on the inorganic surface and the subsequent growth reaction of the polymer chains from the anchored molecules.

Controlled radical polymerization from inorganic particle surfaces Among the different techniques, controlled radical polymerization (CRP) has found an increased interest in recent literature due to its simplicity and versatility compared to ionic processes. Indeed, CRP can tolerate water, air and some impurities and is applicable to a broad spectrum of monomers. Moreover, one key advantage of CRP in comparison to conventional free radical processes is the possibility to synthesize well-defined polymers which can be grown with the desired thickness and composition. Owing to the narrow molecular weight polydispersity of the polymer chains, the grafted particles can self-organize into 2D arrays with controlled interparticle distances function of the degree of advancement of the reaction.

CRP is usually divided into three categories: atom transfer radical polymerization (ATRP), reversible radical addition, fragmentation and transfer (RAFT) and nitroxide mediated polymerization (NMP). All three techniques permit the polymer molecular weight, the polydispersity and the polymer architecture to be accurately controlled owing to a reversible activation/deactivation process of the growing macroradicals into dormant species as schematically represented below:

$$\text{MINERAL}{-}CA + P_n^{\bullet} \ (M) \rightleftarrows \text{MINERAL}{-}CA{-}P_n$$

CA = initiator (controlled agent)

Propagating radical = active species

Dormant species

Covalent grafting of the polymer chains requires the initiator (for NMP and ATRP) or the chain transfer agent (for RAFT) to be chemically attached to the mineral surface. Table 3.8 provides a list of conventional initiators, macroinitiators and transfer agents which have been developed in the recent literature for this purpose.

The atom transfer radical polymerization (ATRP) technique has been extensively studied and applied to the grafting of polymers from the surface of silica nanoparticles. Typical ATRP initiators for such systems are halide-functionalized alkoxysilanes of the type reported in Table 3.8. The grafting is performed as described previously for organosilane molecules by reacting the alkoxysilyl-terminated halide initiator with the hydroxyl groups of the mineral surface. Not only silica but also aluminum oxide particles, clay, magnetic colloids, gold and photoluminescent cadmium sulfide nanoparticles have been functionalized by ATRP. In case of gold, for instance, an initiator carrying a 2-bromoisobutyryl group and a thiol functionality has been specifically synthesized to allow both complexation of the nanoparticles and initiation of the polymer chains from their surface (Table 3.8). When cast from solution, the resulting nanocomposite film materials exhibit hexagonal ordering of the inorganic cores and properties arising from the inorganic component (Fig. 3.25). For example, silica-coated photoconductive CdS nanoparticles were used as inorganic particles and the resulting materials were shown to retain the photoluminescent properties of the core. However, no mention was made about the colloidal stability of these systems. The ATRP has also been recently reported to work with efficacy in the absence of solvent and in aqueous media using hydrophilic water-soluble acrylic monomers. Lastly, it is worth mentioning that hollow polymeric microspheres have been produced through ATRP by templating silica microspheres with poly(benzyl methacrylate), and subsequently removing the core by chemical etching.

The so-called nitroxide-mediated polymerization (NMP) can also be advantageously used to initiate the polymerization of vinyl monomers from inorganic surfaces. Although a library of nitroxide and nitroxide-based alkoxyamine compounds has been recently reported in the literature for the free radical polymerization of a variety of monomers, the extrapolation of the NMP technique to the grafting of inorganic surfaces requires the development of adequate surface-active initiators

Table 3.8 Chemical structures of the macro-initiators involved in the CRP polymerization of a variety of monomers from nanoparticulate inorganic surfaces.

Conventional free radical polymerization	Mineral/monomer
≡O–Si–[cyclohexene with COOR and COOH substituents]	Silica / methyl methacrylate, styrene, acrylonitrile
≡Si–OOR	Silica / methyl methacrylate
≡Si–$(CH_2)_3$–O–C(=O)–$(CH_2)_3$–C(CH_3)(CN)–N=N–C(CH_3)(CN)–CH_3	Silica gel / styrene
≡R–NH–C(=O)–$(CH_2)_2$–C(CH_3)(CN)–N=N–C(CH_3)(CN)–$(CH_2)_3$–C(=O)–OH	Silica gel / styrene
≡Si(OEt)$_2$–$(CH_2)_3$–O–CH_2–CH(OH)–CH_2–O–C(=O)–$(CH_2)_2$–C(CH_3)(CN)–N=N–C(CH_3)(CN)–$(CH_2)_2$–COOH	Silica gel / styrene

Nitroxide-mediated polymerization	Mineral / monomer
≡O–Si–(CH$_2$)$_n$–O–CH$_2$–[phenyl]–CH(O–N(tBu)–CH(iPr)–phenyl)	Colloidal silica / styrene, maleic anhydride
≡Si(OEt)$_2$–$(CH_2)_{10}$–C(=O)–O–CH_2–CH(phenyl)–O–N(tBu)–CH(P(=O)(OEt)$_2$)	Silica gel / styrene

Table 3.8 Continued

Atom-transfer radical polymerization	Mineral / monomer
⫽O–Si–(CH$_2$)$_3$–O–C(=O)–CH(CH$_3$)–Br , CuBr : dNbipy	Silica / styrene
⫽O–Si–(CH$_2$)$_3$–O–C(=O)–C(CH$_3$)$_2$–Br , CuBr : dNbipy	Silica / styrene, SStNa, DEA, NaVBA, DMA*
⫽O–Si–(CH$_2$)$_3$–O–C(=O)–C(CH$_3$)$_2$–Br , CuBr : dNbipy	CdS / SiO$_2$ / MMA**
⫽O–Si–(CH$_2$)$_2$–C$_6$H$_4$–Cl , CuCl : dNbipy	Silica / styrene
⫽O–Si–(CH$_2$)$_{11}$–O–C(=O)–CH(Ph)–Cl , CuCl : dNbipy	Silica / styrene
HO–C(=O)–CH$_2$–CH$_2$–Cl , CuCl : dNbipy	MnFe$_2$O$_4$ / styrene
⫽O–C(=O)–CH(Br)(CH$_3$) , CuBr/PMDETA	Alumine / MMA**
Br$^-$ (CH$_3$)$_3$N$^+$–(CH$_2$)$_{11}$–O–C(=O)–C(CH$_3$)$_2$–Br , CuBr : HMTETA	Montmorillonite / MMA**
HS–(CH$_2$)$_{11}$–O–C(=O)–C(CH$_3$)$_2$–Br , CuBr : Me$_6$tren	Gold / butyl acrylate Gold / MMA**

Reversible addition-fragmentation chain transfer (RAFT)	Mineral / monomer
⫽Si–(CH$_2$)$_{11}$O–C(=O)–(CH$_2$)$_2$–C(CH$_3$)(CN)–N=N–C(CH$_3$)(CN)–CH$_3$ + Ph–CH(CH$_3$)–S–C(=S)–Ph	Silica gel / styrene

* SStNa: sodium styrene sulfonate, DEA: 2-(diethyl amino ethyl) methacrylate. NaVBA: sodium 4, vinyl benzoate. DMA: 2-(dimethyl amino ethyl) methacrylate. ** MMA: methyl methacrylate

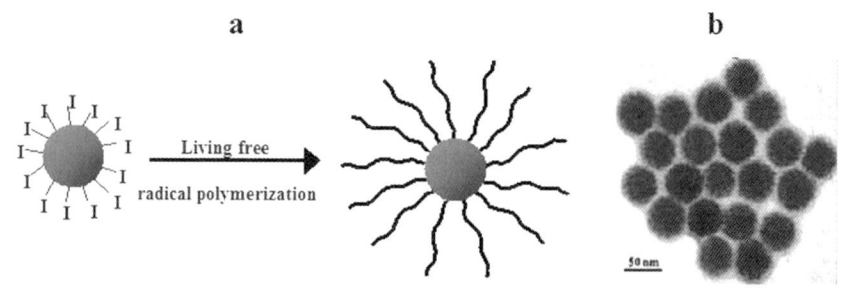

Fig. 3.25 (a) Reaction scheme for the synthesis of polymer-grafted inorganic particles via controlled radical polymerization using chemically anchored CRP macroinitiators. (b) TEM illustration of the CdS/SiO$_2$/PMMA hybrid nanoparticles produced by atom transfer radical polymerization initiated from the CdS/SiO$_2$ nanoparticles surface. (Adapted with permission from *Chem. Mater.* **2001**, *13*, 3920–3926. Copyright 2001 Am. Chem. Soc.)

and has been much less explored. Reactive unimolecular alkoxyamine initiators carrying trichlorosilyl or triethoxysilyl end-groups for further attachment onto mineral substrates have been synthesized, and employed with success for instance in the growth reaction of polymer chains with controlled molecular weights and well-defined architectures from the surface of silica particles. One of the prime advantages of these unimolecular systems is the possibility to accurately control the structure and concentration of the initiating species. However, a major drawback is the multi-step reaction required for synthesis of the functional alkoxyamine. Thus, bimolecular systems have been developed. In this strategy, the NMP process is initiated from an azo or a peroxide initiator attached to the mineral particles. While successful, this approach still involves a two-step chemical reaction to synthesize the functional azoic or peroxidic initiator. Therefore, a versatile one-step synthetic strategy has been recently reported (Fig. 3.26) that allows elaborating surface alkoxyamine compounds by reacting simultaneously a polymerizable silane, a source of radical and *N-tert*-butyl-*N*-[1-diethylphosphono-(2,2-dimethylpropyl)] nitroxide used as spin trap.

As for ATRP, the NMP technique can also be advantageously used for the designed construction of nanoparticles and nanomaterials with new shapes and structures. Following this line, shell-crosslinked polymeric capsules have been elaborated in a multistep procedure by templating colloidal silica with polymeric compounds and crosslinking the polymer shell (Fig. 3.27). Micrometric silica beads were first modified by grafting on their surface a chlorosilane alkoxyamine initiator (see Table 3.8). Copolymers were then grown from the surface-attached initiator using an appropriate amount of sacrificial "free" alkoxyamine. The copolymer chains were designed to carry maleic anhydride functional groups for further crosslinking reactions. A diamine crosslinker was added in a third step to effect interchain coupling via the formation of a bisimide. The inorganic silica template was finally removed in a last step by chemical etching. In an alternative strategy, styrene monomer was copolymerized with 4-vinylbenzocyclobutene, and

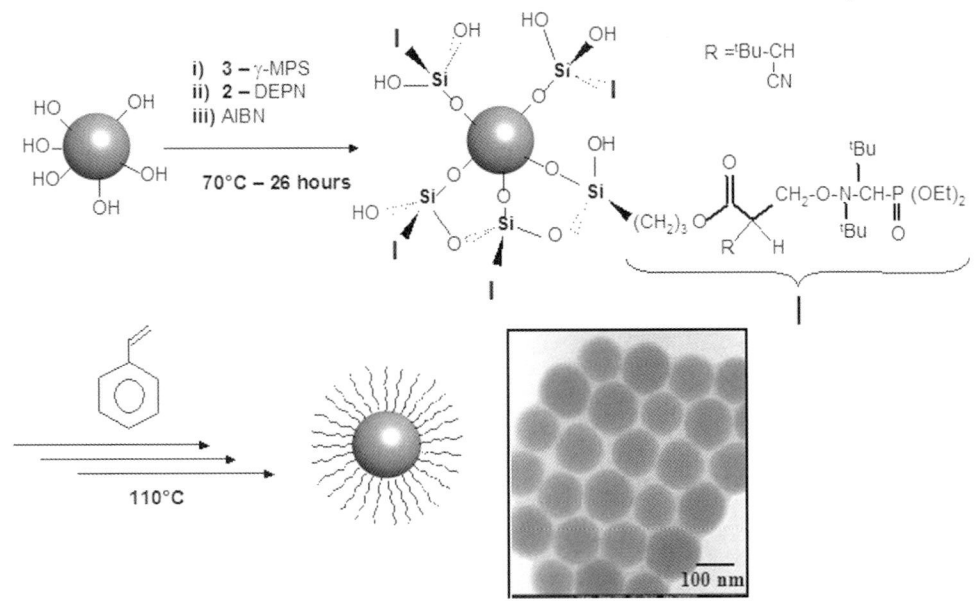

Fig. 3.26 Reaction scheme for one-step covalent bonding of a DEPN-based alkoxyamine initiator onto silica particles and subsequent grafting of polystyrene from the functionalized silica surface.

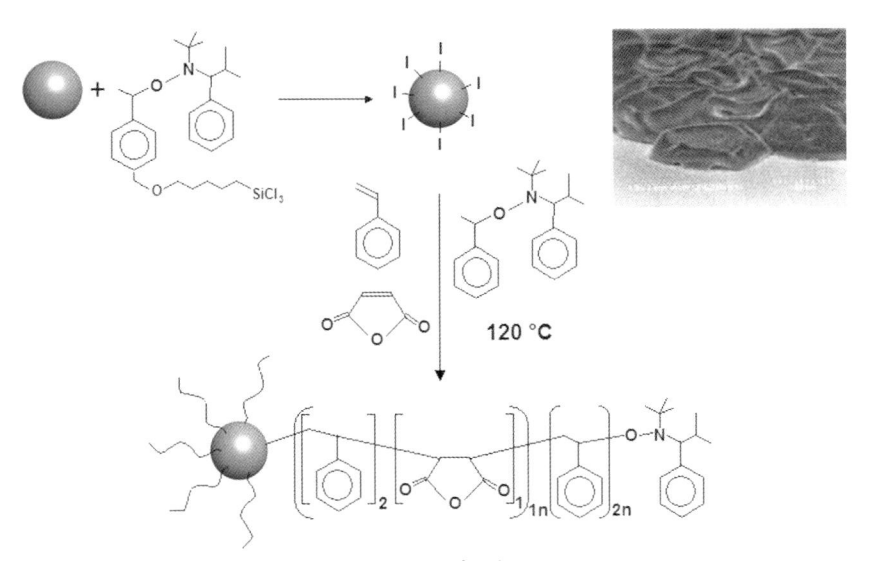

Fig. 3.27 Schematic scheme for the preparation of maleic anhydride-functionalized silica particles and SEM picture of the resulting polymeric capsules obtained shell reticulation and removal of the inorganic core by chemical etching.

the resulting nanocomposite core/shell particles were heated at 200 °C for thermal crosslinking.

3.4.2
In situ Formation of Minerals in the Presence of Polymer Colloids

3.4.2.1 Polymer Particles Templating

Sol–gel nanocoating Core-shell particles have attracted much research attention in recent years because of the great potential in protection, modification and functionalization of the core particles with suitable shell materials to achieve specific physical or chemical performances. For instance, the optical, electrical, thermal, mechanical, magnetic and catalytic properties of polymer particles can be finely tuned by coating them with a thin mineral shell. A general approach for preparation of polymer core/inorganic shell particles consists in performing a sol–gel polycondensation in the presence of polymer latex particles used as templates. Hollow particles can be obtained in a subsequent step by thermal or chemical degradation of the templating colloid as illustrated in Fig. 3.28.

In a similar way as for the coating of mineral particles with polymers, the surface of the organic colloidal sphere must be functionalized by grafting or adsorption of appropriate compounds that can enhance the coupling (and thus deposition) of the inorganic precursor on the particles surface. These molecules are either groups capable to undergo a chemical reaction with the inorganic precursor or ionic molecules capable to promote electrostatic attraction of ionic precursors. In one of these methods, cationic groups have been introduced onto the surface of polystyrene latex particles. The positive charges on the surface ensured quick deposition of the titania precursors on the seed particles in the early beginning of the sol–gel reaction. Very thin (in the range typically a few nanometers up to 50 nm), and smooth coatings were thus produced in a one-step method. Crystalline hollow spheres were further obtained by calcination of the TiO_2-coated particles at elevated temperatures. Increasing the temperature up to 600 °C yielded

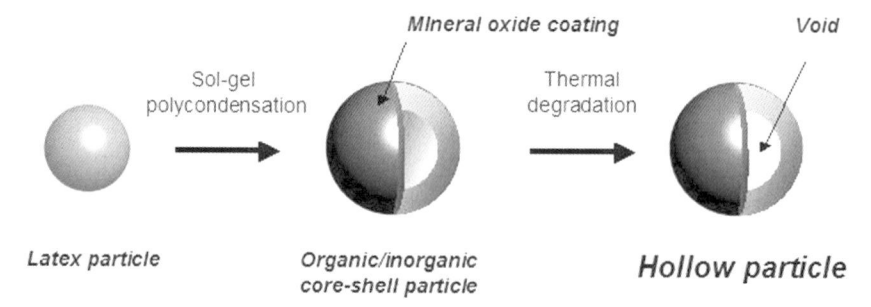

Fig. 3.28 Schematic picture illustrating *in situ* coating of organic particulate templates with a mineral oxide shell and subsequent formation of hollow spheres.

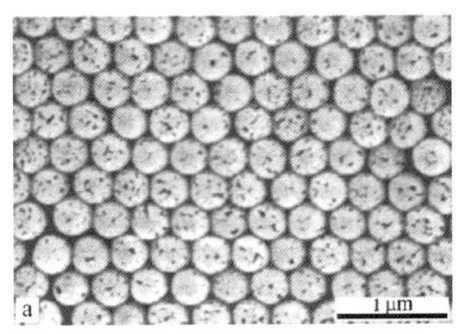

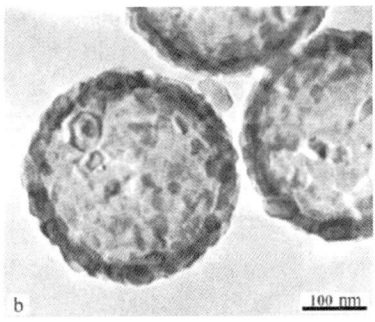

Fig. 3.29 Scanning (a) and transmission (b) electron microscopy images of hollow titania spheres obtained by calcination of polystyrene/TiO_2 core/shell particles at 600°C under air. (Reproduced from A. Imhof, Preparation and characterization of titania-coated polystyrene spheres and hollow titania shells, *Langmuir* **2001**, *17*, 3579–3585, with authorization.)

hollow crystalline anatase titania particles whereas the rutile form of TiO_2 was obtained by calcining at 900–100 °C (Fig. 3.29).

In another recent example, silanol groups were introduced on the surface of polystyrene (PS) latex particles using MPS as a functional (co)monomer. The presence of the silanol groups on the polymer surface enabled the subsequent growth of a silica shell on the functionalized PS seed particles by addition of tetraethoxysilane and ammonia to the colloidal suspension either in water or in a mixture of ethanol and water (Fig. 3.30) without renucleation. That no separate silica particles were formed in this work indicates strong affinity of the sol–gel precursor for the polymer colloid. Burning of the latex core resulted in the formation of hollow nanometer sized silica capsules. A clear advantage of this method is that the nature of the polymeric core, the particles size and the shell thickness can be finely tuned by conventional polymer colloid chemistry. The technique was thus successfully applied to the synthesis of core-shell latexes with a soft polybutylacrylate core and a rigid silica shell which soft/hard particles could find applications as nanofiller for impact resistance improvement.

In addition to polymer latexes, vesicles can also be used as templating materials for transcription into inorganic capsules. The transcriptive synthesis approach is identical to the colloidal templating strategy described previously. For instance, cationic dioctadecyldimethylammonium vesicles were shown to provide effective receptors for silica growth due to electrostatic interaction of the alkoxysilane precursors with the surfactant molecules. The so-produced "petrified" vesicles were stable to dehydratation and could be visualized by conventional TEM without additional staining agents.

Coating with metallic and semiconductor particles The immobilization of fine metal colloids onto nanoparticle surfaces has received a lot of attention in recent years

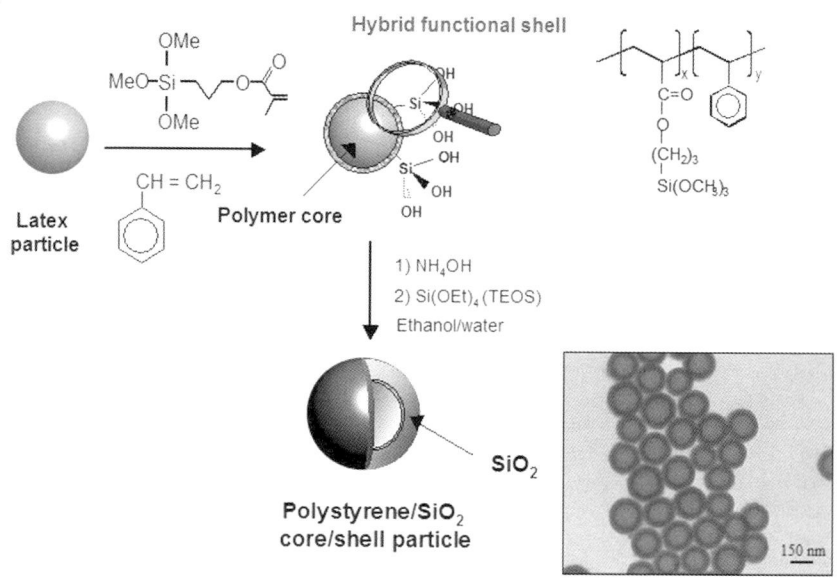

Fig. 3.30 Synthetic scheme for the formation of silica/coated polymer latexes and the resulting hollow silica nanoparticles using SiOH-functionalized latex particles as colloidal templates and TEM image of poly(butylacrylate)/SiO$_2$ core-shell colloids.

because of the potential use of metal-decorated particles in optic, electronic and heterogeneous catalysis. A variety of methods has been successfully reported for the coating of colloidal templates with metallic nanoparticles. Two approaches can be distinguished. In the first method, the nanoparticles are precipitated *in situ* onto the colloidal templates by the reaction of the metal salt precursors previously adsorbed on their surface through ion exchange or complexation chemistry whereas in the second method, preformed metal colloids are adsorbed onto colloidal templates of opposite charges through electrostatic interaction as extensively reported in the previous section. In both methods, the colloidal templates must contain surface groups with strong affinity for the metal precursors and/or the nanoparticles. Functional groups such as carboxylic acid (—COOH), hydroxy (—OH), thiol (—SH) and amine (—NH$_2$) derivatives can be easily introduced into polymer latexes by copolymerizing their corresponding monomers. The surface-complexed metal salts are then directly transformed into metal colloids by the addition of reducing agents (Fig. 3.31). Following this route, palladium rhodium, nickel, cobalt, silver and gold nanoparticles have been successively anchored onto the surface of a series of functional polymer microspheres. The resulting composite colloids were shown to display high catalytic activity in for instance the hydrogenation of alkenes.

In alternative procedures, the coating can also be produced by the controlled hydrolysis of the metal salts into metal oxide followed by reduction of the oxide into the corresponding metal. Submicrometer-sized composite spheres of yttrium and

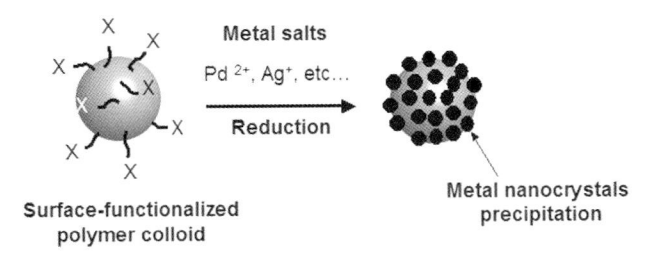

3.4 O/I Nanoparticles Obtained by in situ Polymerization Techniques

Fig. 3.31 Schematic representation of metal particles formation at the surface of polymer latexes through chemical reduction of metal salts. TEM image of Pd particles precipitated on the surface of a carboxylated polystyrene latex. (Redrawn from: P. H. Wang, C.-Y. Pan, Ultrafine palladium particles immobilized on polymer microspheres, *Colloid Polym. Sci.* **2001**, *279*, 171–177, with authorization.)

X = - COOH, CN, NH$_2$, SH, OH, etc…

zirconium compounds and hollow metallic spheres have been prepared this way by coating cationic polystyrene latex particles with basic yttrium carbonate and basic zirconium sulfate, respectively, followed by calcination of the coated latexes at elevated temperatures. Uniform coatings of copper and iron oxide compounds have been formed in a similar procedure by aging at high temperature aqueous solutions of the metal salt in presence of urea, poly(*N*-vinyl pyrrolidone) (PVP) and anionic polystyrene latexes. The coating was shown to proceed by *in situ* heterocoagulation of the precipitating metal colloids on the organic seed surface. Voids were produced in a subsequent step by complete thermal oxidative decomposition of the polymer core.

Following procedures similar to those described previously for metals, polymer microspheres were coated with semiconductor nanocrystals. Semiconductor particles can be advantageously used in coating applications to provide specific optical response to the material. Monodisperse nanocomposite particles with inorganic CdS nanocrystals sandwiched between a PMMA core and a P(MMA-*co*-BA) outer copolymer shell layer have been prepared to this purpose. The particles are obtained by emulsion polymerization in three steps (Fig. 3.32). In a first step, polymer latexes are used as host matrices for CdS nanocrystals formation. To do so, monodisperse poly(methyl methacrylate-*co*-methacrylic acid) (PMMA–PMAA) latex particles were ion exchanged with a Cd(ClO$_4$)$_2$ solution. The Cd^{2+} ions thus introduced into the electrical double layer were further reduced into CdS nanoclusters by addition of a Na$_2$S solution. The CdS-loaded nanocomposite particles were subsequently recovered by a film forming polymer shell by reacting methyl methacrylate and butyl acrylate monomers. The resulting colloidal nanocomposites were finally assembled in 3-D periodic arrays consisting of rigid PMMA–PMAA/CdS core particles regularly distributed within the soft polymer matrix.

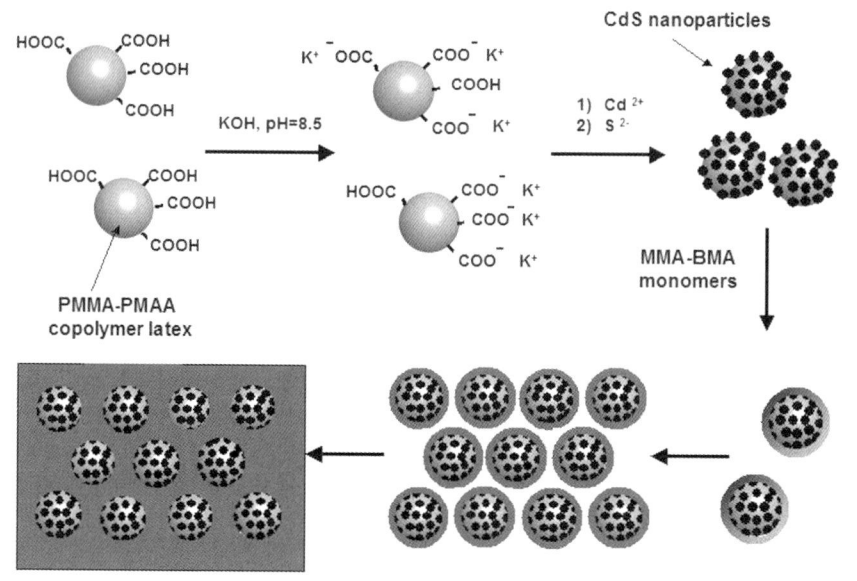

Fig. 3.32 Schematic representation of the synthesis of PMMA-co-PMAA/CdS/PMMA-co-BuA multilayered hybrid particles with a periodic structure. (Redrawn and adapted from: J. Zhang, N. Coombs, E. Kumacheva, A new approach to hybrid nanocomposite materials with a periodic structure, *J. Am. Chem. Soc.* **2002**, *124*, 14512–14513.)

3.4.2.2 Block Copolymers, Dendrimers and Microgels Templating

There have been many studies of *in situ* precipitation of metals, oxides, and sulfides into various elastomeric, glassy, and semi-crystalline polymers. In general, *in situ* growth of particles within a polymer matrix allows for greater control of particle size, orientation, distribution and crystal phase or morphology. Such reactions can also be performed into the confined space of organic particles or diblock copolymer micelles used as nanoreactors as thoroughly reviewed in recent literature. Briefly, the nanoreactor principle consists in an ion exchange followed by a reaction step of either reduction, oxidation or sulfidation (Fig. 3.33) precipitating zerovallent metallic, metal oxide or metal sulfide semiconducting particles.

Block copolymer templating As mentioned previously, self-assembled nanoscale morphologies can be advantageously used as templates to control the nucleation, growth and distribution of inorganic particles. Dilute solutions of block copolymers mostly form spherical micelles in water, the interior of which have been used as nanoreactors to nucleate metal and semiconductor colloids. In a general strategy, a salt precursor solution is loaded into the core of the micellar aggregates and reduced into metal nanoparticles. Semiconductor colloids are prepared in a similar way by addition of H_2S to the metal precursor-loaded micellar solution leading to the formation of quantum-size nanoparticles. For instance, pH-sensitive core-

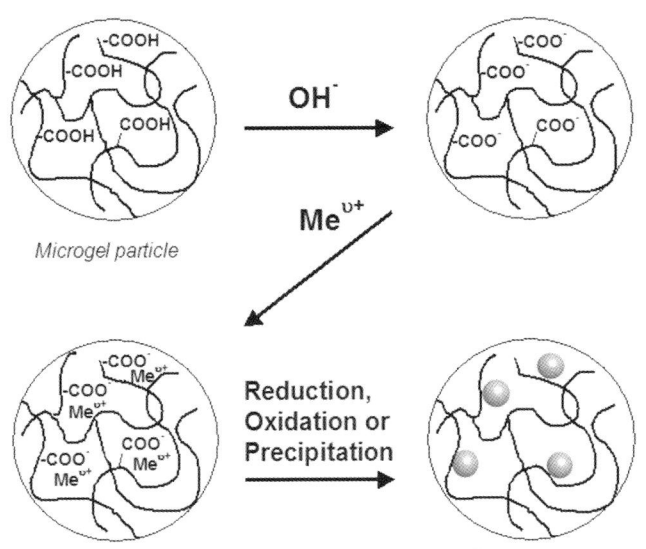

Fig. 3.33 Schematic representation of the principle of *in situ* metallation of polymer colloids and assemblies.

shell-corona (CSC) polystyrene-*b*-poly(2-vinyl pyridine)-*b*-poly(ethylene oxide) triblock copolymer micelles were loaded with $HAuCl_4$ gold salt. The metal salt was transformed in a next step into metal colloids through $NaBH_4$ reduction. Owing to preferred interaction between the protonated P2VP block of the terpolymer and the metal ions, precipitation of gold nanoparticles took place into the P2VP outerlayer of the triblock micelle. Palladium colloids have been synthesized in a similar way within poly-4-vinyl pyridine-*b*-polystyrene (P4VP-*b*-PS) diblock copolymer micelles using palladium acetate $(Pd(OAc)_2)$ as the palladium source.

Not only metals but also metal oxide particles can be prepared in block copolymer mesophases (Table 3.9). For instance, iron oxide magnetic nanoparticles have been precipitated inside the core of the PI-*b*-PCEMA-*b*-PtBA triblock micelles. For that purpose, the triblock nanospheres were first rendered water dispersible by hydroxylation of the polyisoprene block. The PtBA block was then converted into polyacrylic acid (PAA) by hydrolysis. The resulting polymeric triblock nanospheres were loaded with Fe^{2+} metal ions by exchange of the PAA protons and the iron oxide nanoparticles were finally precipitated by addition of NaOH.

Microgel colloids used as templates This section reports the use of polymer gel colloids (also called microgels) as reactors for the controlled precipitation of minerals. Microgel particles are polymer latex particles which swell in a good solvent environment for the polymer matrix. The swelling properties of these particles are of particular interest for encapsulation purposes. The polymeric gel can entrap inorganic precursors (metal ions, metal alkoxides ...), and be used as a host matrix for mineral formation. In a typical procedure, the hydrogel microsphere is impregnated with metal salt or metal oxide precursors which are reacted *in situ* to

Table 3.9 Examples of inorganic precursors and colloids that can be prepared in block copolymers assembly. Adapted from: S. Forster, M. Antonietti, Adv. Mater. **1998**, 10, 195–217, with permission from Wiley-VCH Verlag.

Precursor	Colloid
$FeCl_2/FeCl_3$	Fe_2O_3
$Cd(ClO_4)_2$, $CdMe_2$	CdS
$Pb(ClO_4)_2$, $PbEt_4$, $PbCl_2$	PbS
$Cu(OAc)_2$	CuS
$FeCl_2$	FeS
$PbCl_2$	Pb
$NiCl_2$	Ni
AgOAc, $AgClO_4$, $AgNO_3$	Ag
$Rh(OAc)_2$	Rh
$HAuCl_4$, $LiAuCl_4$, $AuCl_3$	Au

afford organic colloids with entrapped inorganic particles. The procedure allows a variety of precursors and host colloids to be used provided that there exist significant interactions between both components, and that the guest molecules can efficiently enter the gel structure of the colloidal template. For instance, poly(N-isopropylacrylamide) (NIPAM) hydrogels were shown to entrap iron salts, giving rise to the formation of iron oxide particles embedded into the polymer gel. In case of poly(NIPAM), the thermosensitive and swelling properties of the template are of particular interest to provide new materials with tunable properties. One could imagine for instance to elaborate stimuli-responsive controlled-release hybrid colloids that could liberate their active inorganic content upon changing the temperature, pH or salt concentration of the surrounding medium.

Along with metal oxide colloids, metallic palladium nanoparticles have also been synthesized into the internal volume of polymer microgel by exchanging the internal ions by Pd^{2+} cations followed by chemical reduction. A good control over the size and shape of the metal particles was achieved by changing the composition of the microgel particles. The metal-loaded microgel were heated at pH 4 to expel the water from the microsphere interior and coated in a last step with an hydrophobic polymeric shell to irreversibly entrap the metallic particles. Such hybrid spheres are suitable building block for photonic crystals applications. By carefully controlling the reaction conditions, and by varying the nature of the reducing agent, a variety of morphologies can be obtained. Periodic structures of polyacrylic/silver colloids have been elaborated in a similar way using Ag^+ ions as precursors. The method obviously opens a new avenue for producing optically responsive materials with a controlled periodicity.

In another related approach, polymer microspheres were used as host to control the nucleation and growth of cadmium sulfide semiconductor particles. The host polymer contained chelating groups capable to stabilize the Cd^{2+} cations, and

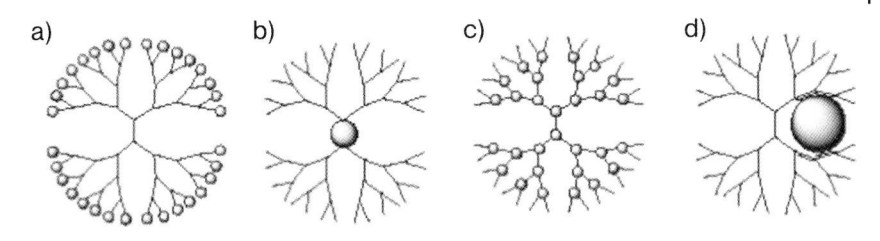

Fig. 3.34 Schematic illustration of different types of nanoparticles-loaded dendrimers. The inorganic particles are located at the periphery (a), in the core (b), in the interior branches (c) or in the inner cavities of the dendritic structure (d). (Reprinted from: K. Kaneda, M. Ooe, M. Murata, T. Mizugaki, K. Ebitani, Dendritic Nanocatalysts in *Encyclopedia of Nanoscience and Nanotechnology*, Marcel Dekker, New York, **2004**, pp. 903–911.)

thus control their subsequent reaction with HS⁻ ions. CdS nanocrystals were prepared *in situ* within the chelate polymer beads which proved to play a determinant role in the control of the CdS crystals formation and characteristics (particle sizes and dispersion state).

Dendrimer templating The unique architecture of dendrimers also provides special opportunities for the growth of particles in the confined volume of the dendritic structure that plays the role of a nanoreactor (Fig. 3.34). Indeed, the surface functional groups of dendrimers can be easily modified with various ligands capable of binding metal complexes. Active sites can be introduced specifically at the surface, the core or the branches of dendrimers in a controlled manner. For instance, monodisperse gold particles with diameters of about 1 nm have been obtained by reduction of metallic salt with UV irradiation in the presence of dendrimers. In a typical procedure, the dendrimer structure is loaded with the aqueous salt which is reduced further to yield metallic nanoparticles with a narrow size distribution. The dendritic structure plays both the role of a template and a stabilizer of the nanoparticles. A variety of transition metal particles including cupper, palladium, platinum and silver have been prepared within the internal cavities of dendrimers and the sequestered metals were found to be stabilized against agglomeration. For instance, poly(amidoamine) dendrimers have shown to display an effective protective action during synthesis of gold nanoparticles.

3.5
Hybrid Particles Obtained by Simultaneously Reacting Organic Monomers and Mineral Precursors

3.5.1
Poly(organosiloxane/vinylic) Copolymer Hybrids

Composite materials where the organic and inorganic components are intimately intertwined within one another at the molecular level are important class of materials which properties are controlled by the functionality and connectivity of the molecular precursors. Such materials are usually produced by the sol–gel technique and processed as thin films, powders, gels or monoliths. But, surprisingly, there are only few examples of nanoparticles synthesis by reacting simultaneously organic and inorganic precursor molecules to form O/I interpenetrated networks (IPN). An example of morphology that can be produced by this strategy is shown on Fig. 3.35 which represents a gel-like colloidal particle made of an organic–inorganic interpenetrated network. It is expected that the properties of those hybrid colloids will be significantly different than a simple combination of the properties of the two components. Typical examples of this general approach are provided in this section.

The combination of various polymers and copolymers with inorganic structures, like silica and silsesquioxanes, to yield inorganic particles doped with organic polymers or vitreophilic polymer colloids can be readily conducted in multiphase media. As a matter of fact, silica networks and structured silicate for instance are easily obtained by hydrolysis and condensation of tetrafunctional $(Si(OR)_4)$ or trifunctional $(R'_n Si(OR)_{4-n})$ alkoxysilanes in various dispersion systems. In addition, the polymerization reaction of a variety of acrylic monomers and comonomers can be carried out into these systems as well. So, provided that the rate of both reactions are not too much different and that a coupling agent is used to link the inorganic network and the organic polymer, hybrid colloids with interpenetrated organic–inorganic networks could be formed. Microemulsion for instance is a convenient system for both metal oxides and polymer latexes synthesis. On one hand,

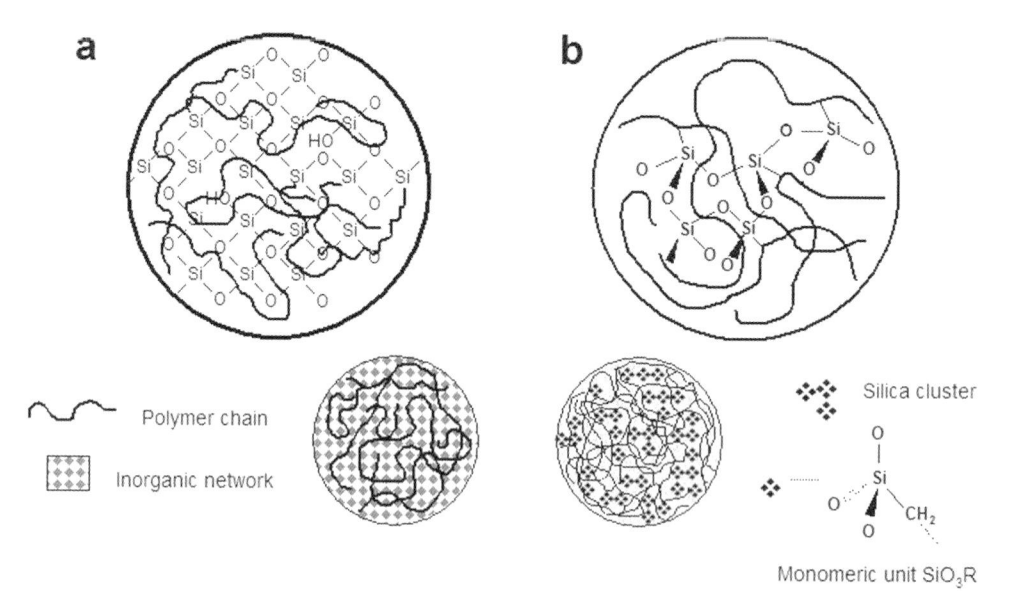

Fig. 3.35 Schematic representation of nanocomposite colloids with organic/inorganic interpenetrated network.

alcohols are usually used as short chains cosurfactants (SCC) in conjunction with sodium dodecyl sulfate to stabilize the microemulsion. On the other hand, the sol–gel reaction can take place into alcohol–water mixtures. Consequently, the sol–gel reaction of TEOS and the polymerization of acrylic monomers can be performed simultaneously in microemulsion systems in which the continuous phase is a mixture of alcohol (typically methanol) and water and the organic phase is constituted of TEOS and the acrylic monomer. In a typical recipe, the inorganic precursor, the organic monomer (methyl methacrylate or vinyl acetate) and the coupling agent are added simultaneously, and interpenetrated networks can be obtained by adjusting the kinetics of the organic and inorganic reactions. Typical coupling agents are organoalkoxysilane molecules with a terminal double bond reactive in free radical polymerization processes of the type described previously (as for instance v-MPS). A crosslinker may be additionally introduced in the formulation to promote the formation of the polymer network. The formation of simultaneous interpenetrated polymer–inorganic networks (SIPIN) resulted in an increase of the glass transition temperature and improved thermal resistance of the organic network due to the presence of the inorganic phase.

MPS molecule can also be reacted directly with acrylic and styrene monomers to produce functional self-cross-linkable hybrid copolymer latexes with interpenetrated organic/inorganic networks via emulsion or miniemulsion polymerizations. The polymer latex particles are synthesized in batch or in semi-batch. In the semi-batch reactions, the silane molecule is introduced as a shot after consumption of part of the acrylic and styrene monomers. Core-shell colloids with a polymer core and an hybrid shell were thus produced by this technique. The core-to-shell ratio could be easily adjusted by addition of the silane molecule at different seed conversions. The silane concentration was varied from 5 to 40 wt.% with respect to the monomer without significant influence on particles size and particles stability except when non ionic surfactants were used as stabilizers. Film forming copolymer latexes were also produced by this technique by reacting MPS, styrene and butyl acrylate monomers. The composite films were fully transparent up to 15% MPS content suggesting an homogeneous distribution of the silane units within the organic/inorganic network and the absence of macroscopic phase separation. The resulting materials were characterized by dynamic mechanical spectroscopy and were shown to display significantly improved mechanical properties in comparison to their polymeric counterparts.

Not only can tetrafunctional and trifunctional alkoxsilanes be reacted with vinylic monomers, but polysiloxanes with difunctional repeating units can also be incorporated into polymer latexes via the emulsion copolymerization reaction of silicon monomers namely octomethyl tetracyclosiloxane and methacryloxy propyl trimethoxysilane, and a series of acrylic compounds (e.g. MMA, butyl acrylate (BuA), acrylic acid (AA) and N-hydroxyl methyl acrylamide). In order to obtain stable latexes, the copolymerization reaction must be carried out under specific experimental conditions. It was shown for instance that stable monodisperse particles were formed only when the monomers were pre-emulsified in water and added dropwise at 85 °C into an aqueous solution of the initiator. The silane coupling agent was used to control particles morphology and provide a successful incorporation of the silicone polymers into the acrylic latexes. The films produced from the hybrid latexes were shown to display improved water resistance and a higher gloss.

3.5.2
Polyorganosiloxane Colloids

Organoalkoxysilanes of the type described previously ($R_nSi(OR')_{4-n}$) can also be processed separately as fine particles by emulsion polymerization in the presence of benzethonium chloride surfactant and a base catalyst. A series of spherical elastomeric micronetworks (so-called organosilicon microgels) of narrow size distribution were produced by this technique. Typical examples of alkoxysilane derivatives which have been used in these syntheses are listed in Table 3.10. From the mechanistic point of view, the inorganic polymerization reaction can be more regarded as a polycondensation in microemulsion than as a conventional emulsion polymerization process. Particle sizes were principally governed by the ratio of surfactant to monomer concentration, and in a limited range, the final microgel diameters could be well described by the theory of µ-emulsion although the suspensions were not fully transparent. The main interest of the technique is the possibility to synthesize highly functionalized nanoparticles with nearly uniform diameters in the range typically 10–40 nm. The cross-linking density of the

Table 3.10 Examples of organoalkoxysilanes involved in the preparation of organosilicon microgel colloids.

Organoalkoxysilanes	Chemical structure	Abbreviation
Tetraethoxysilane	$(CH_3CH_2O)_4Si$	TEOS
Methacryloxypropyl trimethoxysilane	$(CH_3O)_3Si(CH_2)_3OCOC(CH_3)=CH_2$	γ-MPS
Triethoxysilane	$(CH_3CH_2O)_3SiH$	TMS
Vinyl trimethoxysilane	$(CH_3O)_3SiCH=CH_2$	VMS
Allyl trimethoxysilane	$(CH_3O)_3SiCH_2CH=CH_2$	AMS
Mercaptopropyl trimethoxysilane	$(CH_3O)_3Si(CH_2)_3SH$	MPTMS
Mercaptopropyl triethoxysilane	$(CH_3CH_2O)_3Si(CH_2)_3SH$	MPTES
Methyl trimethoxysilane	$(CH_3O)_3SiCH_3$	MMS
Dimethyl dimethoxysilane	$(CH_3O)_2Si(CH_3)_2$	DMMS
Trimethy methoxysilane	$(CH_3O)Si(CH_3)_3$	TMMS
Dihydridotetramethyl disiloxane	$HSi(CH_3)_2OSi(CH_3)_2H$	DTDS
Hexamethyldisilazane	$(CH_3)_3SiNHSi(CH_3)_3$	HMDS
Cyanatopropyl triethoxysilane	$(CH_3CH_2O)_3Si(CH_2)_3CN$	CPTMS
Chlorobenzyl trimethoxysilane	$(CH_3O)_3SiPh(CH_2)Cl$	CMS
Phenyl trimethoxysilane	$(CH_3O)_3SiPh$	PMS
Glycidoxypropyl triethoxysilane	$(CH_3CH_2O)_3Si(CH_2)_3OCH_2CHOCH_2$	GLYMO
Aminopropyl trimethoxysilane	$(CH_3O)_3Si(CH_2)_3NH_2$	APTMS
n-Octadecyl trimethoxysilane	$(CH_3O)_3Si(CH_2)_{17}CH_3$	C18-TMS

microgel particles could be finely tuned by using a mixture of trifunctional (T) trialkoxysilane and difunctional (D) dialkoxysilane molecules. Particle sizes were shown to increase with increasing content of D-units suggesting intraparticle gelation. Other critical parameters influencing colloidal stability were the dispersion concentration, the amount of catalyst, the temperature and the monomer addition rate. The particles could be additionally rendered chemically inert towards further interparticle condensation reactions, and hydrophobic by end-capping of the —SiOH groups into —SiOSi(CH$_3$)$_3$ or —SiOSi(CH$_3$)$_2$H moieties. The curing process was performed by addition of trimethyl methoxysilane or dihydridotetramethyldisiloxane, respectively. The resulting colloids could be easily solubilized or redispersed into organic solvents making it possible to grow polystyrene chains from their surface by the hydrosilylation reaction of vinyl-terminated polystyrene macromonomers with the SiH functional groups. Those model systems have been shown to be particularly suitable for the elaboration of thermodynamically stable homogeneous mixtures of the hybrid colloids with linear polymeric chains.

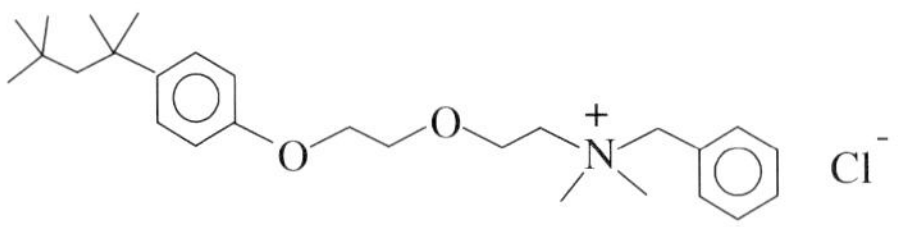

Scheme 3.2 Benzethonium chloride surfactant.

In addition to copolycondensates, well defined core/shell structures can also be prepared by the subsequent addition of different types of monomers on particle seeds. The cores were made for instance of linear chains of polydimethylsiloxane while the shell was crosslinked by reacting different trialkoxysilane precursors. One important feature of these core-shell colloids is that specific reactivities can be imparted to the cores by the reaction of convenient functional monomers. These topologically entrapped functional molecules are confined reactive sites that can undergo further chemical reactions. According to this principle, organic dye molecules have been selectively attached to microgel particles by the reaction of the dye labels with chlorobenzyl functional groups previously incorporated in the internal part of the colloid. In a similar procedure, functionalized μ-network gel particles were used as nanoreactors to entrap metal clusters. The microstructure of the host microgel particles was designed such as to contain reducing SiH moities in their core. Metal ions were subsequently loaded into the microgel particles by diffusion of the metal salt solution (H(AuCl$_4$)) through the shell. Metal salt reduction was taking place in situ by means of the confined SiH reactive sites. The aqueous dispersion of the metal cluster entrapped colloids was finally transferred to organic solvents by reaction of SiOH with monoalkoxy silanes as reported above.

An original approach that combines templating techniques and organoalkoxysilane chemistry has been recently reported. In this method, silica core/mesoporous shell (SCMS) colloids have been synthesized by reacting a mixture of triethoxysilane and n-octadecyl trimethoxysilane (C18-TMS, see Table 3.10) on the surface of nanometer-sized silica particles produced by the Stöber process. Calcination of the C18-TMS porogen molecules resulted in the formation of inorganic spheres which surface porosity can be finely tuned by the concentration of the organoalkoxysilane precursor incorporated into the shell. Carbon capsules have been elaborated following a related strategy by templating the mesopores of the SCMS colloid with a phenolic resin followed by carbonization. Hollow spheres were produced in a last step by dissolution of the silica core. More recently, the technique was extended to the synthesis of complex shell-in-shell nanocomposite particles and gold loaded nanocapsules using silica-coated gold colloid as the seed instead of pure silica spheres. Silica-coated gold particles have also been used as templates for the subsequent overgrowth of a polyorganosiloxane shell on their surface. The process involves the hydrolysis and condensation reaction of a mixture of functional alkoxysilanes on the surface of gold colloids previously rendered vitreophilic by complexing 3-mercapto propyl trimethoxysilane on their surface. Gold/polysiloxane core/shell nanoparticles with various functionalities (allyl, phenyl mercapto, amino, cyano...), and a controlled shell thickness have been successfully prepared by this technique.

3.6
Conclusion

This review highlighted the preparation of O/I particles. The relatively recent advances in the synthesis of these particles has paved the way to a huge range of new materials with outstanding properties. In this article, we discussed various preparation methods using either *ex situ* or *in situ* techniques. In *ex situ* techniques, preformed organic and inorganic building blocks are assembled into nanocomposite colloids through electrostatic attraction, complexation or acid–base chemistry. We have shown that the method is highly sensitive to parameters such as the suspension pH or the ionic strength which, in turn, control the attraction potential of both sets of particles or particles and polymers. *In situ* techniques are divided into two groups. In a first group, organic (vs inorganic) polymerizations are performed in the presence of inorganic (vs organic) colloids. When minerals are used as seeds, suitable interactions of the growing polymer with their surface is provided by the previous reaction and/or adsorption of coupling agents such as for instance organosilane molecules reactive in the polymerization process, macromonomers, ionic initiators or ionic monomers. Polymerization can be indifferently performed in multiphase systems (e.g. through emulsion, miniemulsion, dispersion or suspension polymerizations) or in solution. While core-shell, raspberry-like and other exotic morphologies are obtained in the former route, the later method provides access to hairy colloids characterized by mineral particles surrounded by a hairy protecting polymer shell. The overall strategy allows an accurate control over the composite particles morphology and also affords the opportunity to precisely design the surface properties of the polymer-coated mineral particles by selecting appropriate functional monomers.

Metallic particles, semiconductors and metal oxides can also be generated at the surface of polymer colloids used as templates. Again, the seed particles surface must carry suitable functionalities to promote interaction (and thus deposition) of the inorganic precursor. Not only can the surface of polymer particles be modified by inorganic particles but metal salts and metal complexes can also be sequestered into the internal space of microgels, dendrimers or block copolymers assemblies. Inorganic nanoparticles are obtained in a subsequent step by chemical reduction or any other conventional way. This nanoreactor strategy allows control of the nucleation, growth and distribution of the inorganic particles which are protected against agglomeration.

We discussed in a previous section *in situ* chemical preparation of hybrid colloids by simultaneously reacting organic monomers and inorganic precursors. A typical illustration of this strategy is the synthesis of polysiloxane-based or silicone-based latexes. These latexes are of particular interest in coating applications as they allow significant improvements in film mechanical properties, and wetting behavior. Moreover, the presence of reactive silanols within the hybrid latex particles provides self-cross-linking ability to the resulting copolymer film.

In summary, it is clear that O/I nanoparticles represent a huge domain of material and colloidal science that brings together all fundamental aspects of both organic and inorganic particles synthesis, properties and applications. This article attempted to summarize the most important issues, emphasizing the important role of interactions in the elaboration of nanocomposite colloids. As we tried to be selective in the topics and examples described, only a few of the many potential achievements in the field have been covered in this article, but we hope to have given the reader sufficient information to develop their own expertise and create new organic–inorganic hybrid particles and nanocomposites with outstanding characteristics and properties.

Bibliography

Colloidal science

"Definitions and classification of colloids", can be found under http://www.iupac.org/reports/2001/colloid, **2001**.

D. H. Everett, *Basic principles of colloid science*, Royal Society of Chemistry, London, **1988**.

W. B. Russel, D. A. Saville, W. R. Schowalter, *Colloidal Dispersions*, Cambridge University Press, Cambridge, **1989**.

R. J. Hunter, *Introduction to Modern Colloid Science*, Oxford University Press, Oxford, **1993**.

Reviews on hybrids and nanocomposites

C. H. M. Hofman-Caris, Polymers at the surface of oxide nanoparticles, *New J. Chem.* **1994**, *18*, 1087–1096.

A. D. Pomogalio, Hybrid polymer/inorganic nanocomposites, *Russ. Chem. Rev.* **2000**, *69*, 53–80.

E. Bourgeat-Lami, Organic/inorganic nanocomposites by multiphase polymerization in *Dendrimers, Assemblies and Nanocomposites*, MML Series 5 (Eds R. Arshady, A. Guyot), Citus Books, London, **2002**, Chap. 5, pp. 149–194.

E. Bourgeat-Lami, Organic-inorganic nanostructured colloids, *J. Nanosci. Nanotechnol.* **2002**, *2*, 1–24.

E. Bourgeat-Lami, Organic/inorganic nanocomposite colloids in *Encyclopedia of Nanoscience and Nanotechnology*, Vol. 8 (Ed. H. S. Nalwa), American Scientific Publishers, Los Angeles, **2004**, pp. 305–332.

V. Castelvetro, C. De Vita, Nanostructured hybrid materials from aqueous polymer dispersions, *Adv. Colloid Interface Sci.* **2004**, *108–109*, 167–185.

Polymer latexes – synthesis – properties and applications

I. Piirma, *Emulsion Polymerization*, Academic Press, New-York, **1982**.

G. W. Poehlein, R. H. Ottewill, J. W. Goodwin, *Science and Technology of Polymer Colloids*, Vol. II, Martinus Nijhoff, Boston MA, **1983**.

R. G. Gilbert, *Emulsion Polymerization. A mechanistic approach*, Academic Press, New-York, **1995**.

R. Arshady in *Preparation and Chemical Applications*, MML Series 1 (Ed. R. Arshady) Citus Book, London, **1999**, pp. 85–124.

K. Tauer in *Encyclopedia of Polymer Science and Technology*, 3rd edition (Ed. H. F. Mark), Wiley & Sons, Inc., **2004**.

M. Antonietti, K. Tauer, *Macromol. Chem. Phys.* **2003**, *204*, 207–219.

K. Tauer in *Colloids and Colloid Assemblies*, (Ed. F. Caruso), Wiley VCH, Weinheim, **2004**, pp. 1–51 and references therein.

K. E. J. Barret, *Dispersion Polymerisation in Organic Media*, Wiley, London, **1975**.

C. Larpent in *Colloidal Polymers. Synthesis and Characterization*, Surfactant Science Series, Vol. 115 (Ed. A. Elaissari), Marcel Dekker, New York, **2003**, pp. 145–187.

J. M. Asua, *J. Polym. Sci. Part A: Polym. Chem.* **2004**, *42*, 1025–1041.

K. O. Calvert, *Polymer Latices and their Applications*, Applied Science Publishers, London, **1982**.

D. Urban, K. Takamura, *Polymer Dispersions and their Industrial Applications*, Wiley-VCH, Weinheim, **2002**.

Vesicles, dendrimers and block copolymer assemblies

I. Gitsov, K. R. Lambrych, Dendrimers. Synthesis and Applications in *Dendrimers, Assemblies, Nanocomposites*, MML Series 5 (Eds R. Arshady, A. Guyot), Citus Book, London, **2002**, Chap. 2, pp. 31–68.

K. Kaneda, M. Ooe, M. Murata, T. Mizugaki, K. Ebitani, Dendritic Nanocatalysts in *Encyclopedia of Nanoscience and Nanotechnology*, Marcel Dekker, New York, **2004**, pp. 903–911.

I. U. Hamley, *The Physics of Block Copolymers*, Oxford University Press, Oxford, UK, **1998**.

G. Riess, P. Dumas, G. Hurtrez, Block Copolymer Micelles and Assemblies in *Dendrimers, Assemblies, Nanocomposites*, MML Series 5 (Eds R. Arshady, A. Guyot), Citus Book, London, **2002**, Chap. 3, pp. 69–110.

W. Loh, Block Copolymer Micelles in *Encyclopedia of Surface and Colloid Science*, Marcel Dekker, New York, **2002**, pp. 803–813.

N. Hadjichristidis, S. Pispas, G. Floudas, *Block Copolymers. Synthetic Strategies, Physical Properties, and Applications*, Wiley Interscience, New York, **2003**.

Mineral nanoparticles synthesis

E. Matijevic in *Science of Ceramic Chemical Processing* (Eds L. L. Hench, D. R. Ulrich), Wiley, New York, **1986**, pp. 463–481; E. Matijevic, *Acc. Chem. Res.* **1981**, *14*, 22–29; E. Matijevic in *Ultrastructure Processing of Ceramics, Glasses and Composites* (Eds L. L. Hench, D. R. Ulrich), Wiley, New York, **1984**, pp. 334–352; T. Sugimoto, *Adv. Colloid Interf. Sci.* **1987**, *28*, 65–108.

G. Cao, *Nanostructures & Nanomaterials – Synthesis, Properties & Applications*, Imperial College Press, London, **2004**.

J. P. Jolivet, *De la Solution à l'Oxyde*, Interéditions/CNRS éditions, Paris, **1994**.

V. H. Perez-Luna, K. Aslan, P. Betala, Colloidal Gold, in *Encyclopedia of Nanoscience and Nanotechnology*, Vol. 2 (Ed. H. S. Nalwa), American Scientific Publishers, Los Angeles, **2004**, pp. 27–49 and references therein.

J. P. Wilcoxon, Quantum dots made of metals: preparation and characterization in *Encyclopedia of Nanoscience and Nanotechnology*, Marcel Dekker, New York, **2004**, pp. 3177–3202 and references therein.

M. Brust, J. Fink, D. Bethell, D. J. Schiffrin, C. J. Kiely, Synthesis and reaction of functionalized gold nanoparticles, *J. Chem. Soc. Chem. Commun.* **1995**, *16*, 1655–1658.

M. Brust, C. J. Kiely, Some recent advances in nanostructure preparation from gold and silver particles: a short topical review, *Colloids and Surfaces, A, Physicochemical and Engineering Aspects*, **2002**, *202*, 175–186.

Heterocoagulation

E. Bleier, E. Matijevic, Heterocoagulation. I. Interaction of monodispersed chromium hydroxide with polyvinyl chloride latex, *J. Colloid Interf. Sci.* **1976**, *55*, 510–523.

H. Sasaki, E. Matijevic, E. Barouch, Heterocoagulation. VI. Interactions of monodispersed hydrous aluminum oxide sol with polystyrene latex, *J. Colloid Interf. Sci.* **1980**, *76*, 319–329.

K. Kato, M. Kobayashi, K. Esumi, K. Meguro, Interaction of pigments with polystyrene latex prepared using a zwitterionic emulsifier, *Colloids and Surfaces* **1987**, *23*, 159–170.

N. J. Marston, B. Vincent, N. G. Wright, The synthesis of spherical rutile titanium dioxide particles and their interaction with polystyrene latex particles of opposite charge, *Prog. Colloid Polym. Sci.* **1998**, *109*, 278–282.

Layer-by-layer assembly

G. Decher, Fuzzy nanoassemblies: toward layered polymeric multicomposites, *Science* **1997**, *277*, 1232–1237.

F. Caruso, H. Lichtenfeld, M. Giersig, H. Möhwald, Electrostatic self-assembly of silica nanoparticle-polyelectrolyte multilayers on polystyrene latex spheres, *J. Am. Chem. Soc.* **1998**, *120*, 8523–8524.

S. Susha, F. Caruso, A. L. Rogach, G. B. Sukhorukov, A. Kornowski, H. Möhwald, M. Giersig, A. Eychmüller, H. Weller H, Formation of luminescent spherical core-shell particles by the consecutive adsorption of polyelectrolyte and CdTe(S) nanocrystals on latex colloids, *Colloid Surface A* **2000**, *163*, 39–44.

F. Caruso F, Hollow capsule processing through colloidal templating and self-assembly, *Chem. Eur. J.* **2000**, *6*, 413–419.

Recognition assembly

I. Lee, H. Zheng, M. F. Rubner, P. T. Hammond, Controlled cluster size in patterned particles arrays via directed adsorption on confined surfaces, *Adv. Mater.* **2002**, *14*, 572–577.

P. Alivisatos, K. P. Johnsson, X. Peng, T. E. Wilson, C. J. Loweth, M. Bruchez, P. G. Schultz, Organization of Nanocrystal Molecules using DNA, *Nature* **1996**, *382*, 609–611; C. J. Loweth, W. B. Caldwell, X. G. Peng, A. P. Alivisatos, P. G. Schultz, DNA-based assemblies of gold nanocrystals, *Angew. Chem. Int. Ed.* **1999**, *38*, 1808–1812.

B. L. Frankamp, O. Uzun, F. Ilhan, A. K. Boal, V. M. Rotello, Recognition-Mediated Assembly of Nanoparticles into Micellar Structures with Diblock Copolymers, *J. Am. Chem. Soc.* **2002**, *124*, 892–893.

S. Mann, W. Shenton, M. Li, S. Connolly, D. Fitzmaurice, Biologically Programmed Nanoparticle Assembly, *Adv. Mater.* **2000**, *12*, 147–150.

M. S. Fleming, T. K. Mandal, D. R. Walt, Nanosphere-microsphere assembly: methods for core-shell materials preparation, *Chem. Mater.* **2001**, *13*, 2210–2216.

Encapsulation of pigments and fillers through emulsion polymerization

A. M. van Herk, Encapsulation of inorganic particles in *Polymeric Dispersions: Principles and Applications*, NATO ASI Series E: Applied Sciences, Vol. 335 (Ed. J. M. Asua), Kluwer Academic, Dordrecht, **1997**, pp. 435–450.

A. M. van Herk, A. L. German, Microencapsulated pigments and fillers in *Microspheres Microcapsules & Liposomes*, Vol. 1, *Preparation & Chemical Applications* (Ed. R. Arshady), Citus Books, London, **1999**, Chap. 17, pp. 457–486.

γ-MPS used as coupling agent

K. Shiratsuchi, H. Hokazono, Aqueous dispersion of core-shell type composite particles with colloidal silica as the cores and with organic polymers as the shells and production method thereof, US Patent 5 856 379 (1999).

K. Zhang, H. Chen, X. Chen, Z. Chen, Z. Cui, B. Yang, Monodisperse silica/polymer core/shell microspheres via surface grafting and emulsion polymerization, *Macromol. Mater. Eng.* **2003**, *288*, 380–385.

E. Bourgeat-Lami, J. Lang, Encapsulation of inorganic particles by dispersion polymerization in polar media. 2. Effect of silica size and concentration on the morphology of silica-polystyrene composite particles, *J. Colloid Interf. Sci.* **1999**, *210*, 281–289.

F. Corcos, E. Bourgeat-Lami, C. Novat, J. Lang, Poly (styrene-b-ethylene oxide) copolymers as stabilizer for the synthesis of silica-polystyrene core-shell particles, *Colloid Polym. Sci.* **1999**, *277*, 1142–1151.

S. Reculusa, C. Mingotaud, E. Bourgeat-Lami, E. Duguet, S. Ravaine, Synthesis of daisy-shaped and multipod-like silica/polystyrene nanocomposites, *Nano Lett.* **2004**, *4*, 1677–1682.

E. Bourgeat-Lami, M. Insulaire, S. Reculusa, A. Perro, S. Ravaine, E. Duguet, Nucleation of polystyrene latex particles in the presence of γ-methacryloxypropyl trimethoxysilane-functionalized silica particles, *J. Nanosci. Nanotechnol.* **2006**, *6*, 432–444.

S. Gu, J. Onishi, E. Mine, Y. Kobayashi, M. Konno, Preparation of multilayered gold-silica-polystyrene core-shell particles by seeded polymerization, *J. Colloid Interf. Sci.* **2004**, *279*, 284–287.

Admicellar polymerization

K. Esumi, Interactions between surfactants and particles: dispersion, surface modification and adsolubilization, *J. Colloid Interf. Sci.* **2001**, *241*, 1–17.

W. H. Waddell, J. H. O'Haver, L. R. Evans, J. H. Harwell, Organic polymer-surface modified precipitated silica, *J. Appl. Polym. Sci.* **1995**, *55*, 1627–1641.

L. Quaroni, G. Chumanov, Preparation of polymer-coated functionalized silver nanoparticles, *J. Am. Chem. Soc.* **1999**, *121*, 10642–10643.

Monomer and macromonomer adsorption

M. J. Percy, C. Barthet, J. C. Lobb, M. A. Khan, S. F. Lascelles, M. Vamvakaki, S. P. Armes, Synthesis and characterization of vinyl polymer-silica colloidal nanocomposites, *Langmuir* **2000**, *16*, 6913–6920; J. I. Amalvy, M. J. Percy, S. P. Armes, Synthesis and characterization of novel film-forming vinyl polymer/silica colloidal nanocomposites, *Langmuir* **2001**, *17*, 4770–4778.

S. P. Armes, S. Gottesfeld, J. G. Beery, F. Garzon, S. F. Agnew, Conducting Polymer-Colloidal Silica Composites, *Polymer* **1991**, *32*, 2325–2330.

K. Yoshinaga, F. Nakashima, T. Nishi, Polymer modification of colloidal particles by spontaneous polymerization of surface active monomers, *Colloid Polym. Sci.* **1999**, *277*, 136–144.

K. Nagai, Y. Ohishi, K. Ishiyama, N. Kuramoto, *J. Appl. Polym. Sci.* **1989**, *38*, 2183; K. Nagai, *Macromol. Symp.* **1994**, *84*, 29.

K. Yoshinaga, T. Yokoyama, T. Kito, *Polym. Adv. Technol.* **1992**, *4*, 38.

S. Reculusa, C. Poncet-Legrand, S. Ravaine, C. Mingotaud, E. Duguet, E. Bourgeat-Lami, Syntheses of Raspberrylike silica/polystyrene materials, *Chem. Mater.* **2002**, *14*, 2354–2359.

Initiator adsorption

Y. Haga, T. Watanabe, R. Yosomiya, *Die Angew. Makromol. Chem.* **1991**, *189*, 23.

J. L. Luna-Xavier, A. Guyot, E. Bourgeat-Lami, Synthesis and Characterization of Silica/Poly(methyl methacrylate) Nanocomposite Latex Particles using a Cationic Azo Initiator, *J. Colloid and Interf. Sci.* **2001**, *250*, 82–92.

J. L. Luna-Xavier, A. Guyot, E. Bourgeat-Lami, Preparation of nano-sized silica/poly(methyl methacrylate) composite latexes by heterocoagulation, *Polym. Int.* **2004**, *53*, 609–617.

Polymer/clay nanocomposite particles

X. Tong, H. Zhao, T. Tang, Z. Feng, B. Huang, Preparation and characterization of poly(ethyl acrylate)/bentonite nanocomposites by *in situ* emulsion polymerization, *J. Polym. Sci., Part A: Polym. Chem.* **2002**, *40*, 1706–1711.

Y. S. Choi, M. H. Choi, K. H. Wang, S. O. Kim, Y. K. Kim, I. J. Chung, Synthesis of exfoliated PMMA/Na-MMT nanocomposites via soap-free emulsion polymerization, *Macromolecules* **2001**, *34*, 8978–8985.

D. Wang, J. Zhu, Q. Yao, C. A. Wilkie, A comparison of various methods for the preparation of polystyrene and poly(methyl methacrylate) clay nanocomposites, *Chem. Mater.* **2002**, *14*, 3837–3843.

G. Chen, Y. Ma, Z. Qi, Preparation and morphological study of an exfoliated polystyrene/montmorillonite nanocomposite, *Scr. Mater.* **2001**, *44*, 125–128.

Y. K. Kim, Y. S. Choi, K. H. Wang and I. J. Chung, Synthesis of exfoliated PS/Na-MMT nanocomposites via emulsion polymerization, *Chem. Mater.* **2002**, *14*, 4990–4995.

D. C. Lee, L. W. Jang, Preparation and characterization of PMMA-clay hybrid composite by emulsion polymerization, *J. Appl. Polym. Sci.* **1996**, *61*, 1117–1122.

M. W. Noh, D. C. Lee, Synthesis and characterization of PS-clay nanocomposite by emulsion polymerization, *Polym. Bull.* **1999**, *42*, 619–626.

M. W. Noh, L. W. Jang, D. C. Lee, Intercalation of styrene-acrylonitrile copolymer in layered silicate by emulsion polymerization, *J. Appl. Polym. Sci.* **1999**, *74*, 179–188.

M. W. Noh, D. C. Lee, Comparison of characteristics of SAN-MMT nanocomposites prepared by emulsion and solution polymerization, *J. Appl. Polym. Sci.* **1999**, *74*, 2811–2819.

X. Huang, W. J. Brittain, Synthesis of a PMMA-layered silicate nanocomposite by suspension polymerization, *Polymer Preprints* **2000**, *41*, 521–522.

X. Huang, W. J. Brittain, Synthesis and characterization of PMMA nanocomposites by suspension and emulsion polymerization, *Macromolecules* **2001**, *34*, 3255–3260.

N. Negrete-Herrera, J.-M. Letoffe, J.-L. Putaux, L. David, E. Bourgeat-Lami, Aqueous dispersions of silane-functionalized laponite clay platelets. A first step toward the elaboration of water-based polymer/clay nanocomposites, *Langmuir* **2004**, *20*, 1564–1571.

N. Negrete-Herrera, S. Persoz, J-L. Putaux, L. David and E. Bourgeat-Lami, Preparation and characterization of polymer/laponite colloidal nanocomposites by emulsion polymerization, *J. Nanosci. Nanotechnol.* **2006**, *6*, 421–431.

Surface-initiated polymerizations

R. C. Advincula, Surface-initiated polymerization from nanoparticle surfaces, *J. Disp. Sci. Technol.* **2003**, *24*, 343–361.

N. Fery, R. Hoene, K. Hamann, Graft reaction of silicon surfaces, *Angew. Chem. Int. Ed. Engl.* **1972**, *11*, 337.

T. von Werne, T. E. Patten, Atom transfer radical polymerization from nanoparticles: a tool for the preparation of well-defined hybrid nanostructures and for understanding the chemistry of controlled/living radical polymerizations from surfaces, *J. Am. Chem. Soc.* **2001**, *123*, 7497–7505.

B. Gu, A. Sen, Synthesis of aluminium oxide/gradient copolymer composites by Atom Transfer Radical Polymerization, *Macromolecules* **2002**, *35*, 8913–8916.

C. R. Vestal, Z. J. Zhang, Atom transfer radical polymerization synthesis and magnetic characterization of $MnFe_2O_4$/polystyrene core/shell nanoparticles, *J. Am. Chem. Soc.* **2002**, *124*, 14312–14313.

Y. Wang, X. Teng, J.-S. Wang, H. Yang, Solvent-free atom transfer radical polymerization in the synthesis of Fe_2O_3@polystyrene core-shell nanoparticles, *Nano Lett.* **2003**, *3*, 789–793.

S. Nub, H. Böttcher, H. Wurm, M. L. Hallensleben, Gold nanoparticles with covalently attached polymer chains, *Angew. Chem. Int. Ed.* **2001**, *40*, 4016–4018.

K. Ohno, K. Koh, Y. Tsujii, T. Fukuda, Fabrication of ordered arrays of gold nanoparticles coated with high-density polymer brushes, *Angew. Chem., Int. Ed.* **2003**, *42*, 2751–2754.

S. C. Farmer, T. E. Patten, Synthesis of luminescent organic/inorganic polymer nanocomposites, *Polym. Mater. Sci. Eng.*, **2000**, *82*, 237–238; S. C. Farmer, T. E. Patten, Photoluminescent polymer/quantum dot composite nanoparticles, *Chem. Mater.* **2001**, *13*, 3920–3926.

C. Perruchot, M. A. Khan, A. Kamitsi, S. P. Armes, T. von Werne, T. E. Patten, Synthesis of well-defined polymer-grafted silica particles by aqueous ATRP, *Langmuir* **2001**, *17*, 4479–4481.

T. K. Mandal, M. S. Fleming, D. R. Walt, Production of hollow polymeric microspheres by surface-confined living radical polymerization on silica templates, *Chem. Mater.* **2000**, *12*, 3481–3487.

C. Bartholome, E. Beyou, E. Bourgeat-Lami, P. Chaumont, N. Zydowicz, Nitroxide-mediated polymerization of styrene initiated from the surface of silica nanoparticles. In-situ generation and grafting of alkoxyamine initiators, *Macromol.* **2005**, *38*, 1099–1106.

Inorganic particles formation at the surface of polymer colloids
Sol-gel nanocoating
F. Caruso, Nanoengineering of particle surfaces, *Adv. Mater.* **2001**, *13*, 11–22.

H. Shiho, N. Kawahashi, Titanium compounds as coatings on polystyrene latices and as hollow spheres, *Colloid Polym. Sci.* **2000**, *278*, 270–274.

S. Eiden, G. Maret, Preparation and characterization of hollow spheres of rutile, *J Colloid Interf. Sci.* **2002**, *250*, 281–284.

A. Imhof, Preparation and characterization of titania-coated polystyrene spheres and hollow titania shells, *Langmuir* **2001**, *17*, 3579–3585.

I. Tissot, C. Novat, F. Lefebvre, E. Bourgeat-Lami, Hybrid Latex Particles Coated with Silica, *Macromolecules* **2001**, *34*, 5737–5739; I. Tissot, J. P. Reymond, F. Lefebvre, E. Bourgeat-Lami, SiOH-functionalized polystyrene latexes: a step toward the synthesis of hollow silica nanoparticles, *Chem. Mater.* **2002**, *14*, 1325–1331.

D. H. W. Hubert, M. Jung, P. M. Frederik, P. H. H. Bomans, J. Meuldijk, A. L. German. Vesicle-directed growth of silica. *Adv. Mater.* **2000**, *12*, 1286–1290.

Coating with metallic and semiconductor particles
H. Tamai, S. Sakurai, Y. Hirota, F. Nishiyama, H. Yasuda, Preparation and characteristics of ultrafine metal particles immobilized on fine polymer particles, *J. Appl. Polym. Sci.* **1995**, *56*, 441–449.

C.-W. Chen, M.-Q. Chen, T. Serizawa, M. Akashi, In-situ formation of silver nanoparticles on poly(N-isopropylacrylamide)-coated polystyrene microspheres, *Adv. Mater.* **1998**, *10*, 1122–1126.

P. H. Wang, C.-Y. Pan, Ultrafine palladium particles immobilized on polymer microspheres, *Colloid Polym. Sci.* **2001**, *279*, 171–177.

S. Mecking, R. Thomann, Core-shell Microspheres of a catalytically Active Rhodium Complex Bound to a Polyelectrolyte-Coated Latex, *Adv. Mater.* **2000**, *12*, 953–956.

N. Kawahashi, E. Matijevic, Preparation and properties of uniform coated colloidal particles, V. Yttrium basic carbonate on polystyrene latex, *J. Colloid Interf. Sci.* **1990**, *138*, 534–542.

N. Kawahashi, H. Shiho, Copper and copper compounds as coatings on polystyrene particles and as hollow spheres, *J. Mater. Chem.* **2000**, *10*, 2294–2297.

J. Zhang, N. Coombs, E. Kumacheva, A new approach to hybrid nanocomposite materials with a periodic structure, *J. Am. Chem. Soc.* **2002**, *124*, 14512–14513.

***In situ* metallation of block copolymer micelles, polymer microgels and dendrimers**
L. M. Bronstein, Nanostructured Polymeric Nanoreactors for Metal Nanoparticle Formation, Chap. 4, **2005**, pp. 123–154; L. M. Bronstein, Nanostructured Polymers in *Encyclopedia of Nanoscience and Nanotechnology*, Vol. 7 (Ed. H. S. Nalwa), American Scientific Publishers, Los Angeles, **2004**, pp. 193–206.

L. M. Bronstein, Polymer Colloids and their Metallation in *Encyclopedia of Nanoscience and Nanotechnology*, Marcel Dekker, New York, **2004**, pp. 2903–2915.

Block copolymer templating
J. F. Gohy, N. Willet, S. Varshney, J. X. Zhang, R. Jérôme, *Angew. Chem. Int. Ed.* **2001**, *40*, 3214–3216.

S. Förster, M. Antonietti, Amphiphilic block copolymers in structure-controlled nanomaterial hybrids, *Adv. Mater.* **1998**, *10*, 195–217.

R. S. Underhill, G. Liu, Triblock nanospheres and their use as template for inorganic nanoparticles preparation, *Chem. Mater.* **2000**, *12*, 2082–2091.

Microgel templating
C. R. Mayer, V. Cabuil, T. Lalot, R. Thouvenot, Magnetic nanoparticles trapped in pH 7 hydrogels as a tool to characterize the properties of the polymeric network, *Adv. Mater.* **2000**, *12*, 417–420.

S. Xu, J. Zhang, C. Paquet, Y. Lin, E. Kumacheva, From hybrid microgel to photonic crystals, *Advanced Functional Materials* **2003**, *13*, 468–472.

S. Xu, J. Zhang, E. Kumacheva, Hybrid polymer-inorganic materials: multiscale hierarchy, *Composite Interfaces* **2003**, *10*, 405–421.

H. Yao, Y. Takada, N. Kitamura, Electrolyte effects of CdS nanocrystals formation in chelate polymer particles: optical and distribution properties, *Langmuir*, **1998**, *14*, 595–601.

Dendrimer templating

V. Chechik, R. M. Crooks, Monolayers of thiol-terminated dendrimers on the surface of planar and colloidal gold, *Langmuir* **1999**, *15*, 6364–6369.

Poly(organosiloxane/vinylic) copolymer hybrids

D. Donescu, M. Teodorescu, S. Serban, L. Fusulan, C. Petcu, Hybrid materials obtained in microemulsion from methyl methacrylate, methacryloxypropyl trimethoxysilane, tetraethoxysilane, *Eur. Polym. J.* **1999**, *35*, 1679–1686.

I. Marcu, E. S. Daniels, V. L. Dimonie, C. Hagiopol, M. S. El-Aasser, Incorporation of alkoxysilanes in model latex systems: vinyl copolymerization of vinyltriethoxysilane and n-butyl acrylate, *Macromolecules* **2003**, *36*, 326.

I. Marcu, E. S. Daniels, V. L. Dimonie, J. E. Roberts, M. S. El-Aasser, A miniemulsion approach to the incorporation of vinyltriethoxysilane into acrylate latexes, *Prog. Colloid Polym. Sci.* **2004**, *124*, 31–36.

Y. Wu, H. Duan, Y. Yu, C. Zhang, Preparation and Performance in Paper Coating of Silicone-Modified Styrene-Butyl Acrylate Copolymer Latex, *J. Appl. Polym. Sci.* **2001**, *79*, 333–336.

C. Y. Kan, D. S. Liu, X. Z. Kong, X. L. Zhu, Study on the Preparation and Properties of styrene-butyl acrylate – silicone copolymer latices, *J. Appl. Polym. Sci.* **2001**, *82*, 3194–3200.

Polyorganosiloxane colloids

F. Baumann, M. Schmidt, B. Deubzer, M. Geck, J. Dauth, On the preparation of Organosilicon µ-spheres: A Polycondensation in µ-Emulsion, *Macromolecules* **1994**, *27*, 6102–6105.

F. Baumann, B. Deubzer, M. Geck, J. Dauth, S. Sheiko, M. Schmidt, Soluble Organosilicon micronetworks with spatially confined reaction sites, *Adv. Mater.* **1997**, *9*, 955–958.

N. Jungmann, M. Schmidt, M. Maskos, J. Weis, J. Ebenhoch, Synthesis of Amphiphilic Poly(organosiloxane) Nanospheres with Different Core-Shell Architectures, *Macromolecules* **2002**, *35*, 6851–6857.

C. Roos, M. Schmidt, J. Ebenhoch, F. Baumann, B. Deubzer, J. Weis, Design and Synthesis of Molecular Reactors for the Preparation of Topologically Trapped Gold Clusters, *Adv. Mater.* **1999**, *11*, 761–766.

C. Graf, W. Schärtl, K. Fischer, N. Hugenberg, M. Schmidt, Dye-Labeled Poly(organosiloxane) Microgels with Core-shell Architecture, *Langmuir* **1999**, *15*, 6170–6180.

G. Büchel, K. K. Unger, A. Matsumoto, K. Tsutsumi, A novel pathway for synthesis of submicrometer-size solid core/mesoporous shell silica spheres, *Adv. Mater.* **1998**, *10*, 1036–1038.

S. R. Hall, S. A. Davis, S. Mann, Cocondensation of Organosilica Hybrid Shells Nanoparticle Templates: A Direct Synthetic Route to Functionalized Core-Shell Colloids, *Langmuir* **2000**, *16*, 1454–1456.

4
Intercalation Compounds and Clay Nanocomposites
Jin Zhu and Charles A. Wilkie

4.1
Introduction

Polymer nanocomposites are a newly emerging technology. Owing to the presence of the nano-dispersed filler in the polymer matrix, nanocomposites exhibit many enhanced properties, including stiffness, strength, barrier, chemical resistance, thermal stability and fire retardancy. For example, nanocomposites show improvement in both strength and toughness, which is impossible with conventional technologies.

There are likely several events that could be cited as the beginning of work in the area of polymer-clay nanocomposites; in this chapter it is suggested that the field actually began in the 1960s with the work of Blumstein. He carried out the polymerization of methyl methacrylate that was adsorbed onto montmorillonite clay and found that there was an oriented growth of the polymer. At this time, pre-nanotechnology, these were simply interesting materials. The next significant event occurred in the late 1980s at the Toyota Research Laboratories where it was for the first time recognized that nano-dimensional clay platelets were obtained in polyamide 6. One interesting result from the Toyota discovery was that the clay could be used at quite low levels, in the range of 2 to 8%, whereas the typical filler, glass or inorganic materials, were used at between 10 and 40%. The discovery is that the addition of 5% clay brings about a 40% increase in tensile strength, a 68% increase in tensile modulus, a 60% increase in flexural strength, a 126% increase in flexural modulus while the heat distortion temperature increase from 65 °C to 152 °C and the impact strength is lowered by only 10%. The tensile strength is the ability of a material to withstand the forces that tend to pull it apart while the tensile modulus is an indication of the relative stiffness of a material and can be determined from a stress-strain curve. A typical stress–strain curve is illustrated in Fig. 4.1. The flexural strength is the ability of the material to withstand a bending force applied perpendicular to its longitudinal axis while the flexural modulus is a measure of the stiffness during the first, or initial, part of the bending process. The heat distortion temperature (HDT) is defined as the temperature at which a

Hybrid Materials. Synthesis, Characterization, and Applications. Edited by Guido Kickelbick
Copyright © 2007 Wiley-VCH Verlag GmbH & Co. KGaA, Weinheim
ISBN: 978-3-527-31299-3

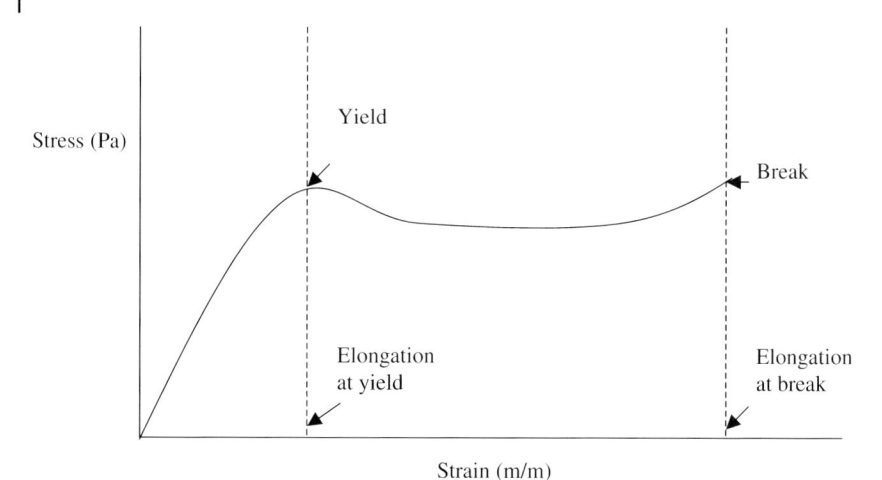

Fig. 4.1 A typical stress-strain curve.

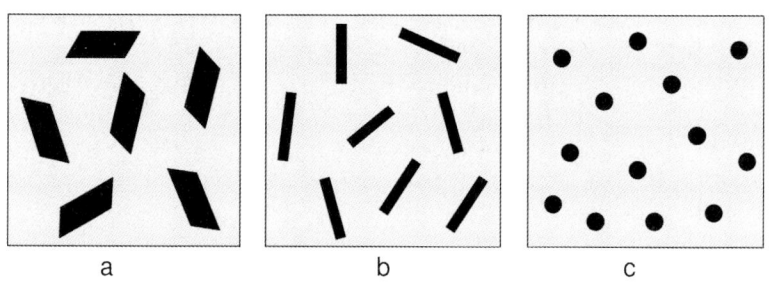

Fig. 4.2 Three types of polymer nanocomposites: a) polymer lamellar material nanocomposite; b) polymer nanotube/nanofiber nanocomposite; c) polymer spherical nanoparticle nanocomposite.

standard test bar (5/0.5/0.25 in.) deflects 0.010 in. under a stated load of either 0.45 or 1.82 MPa. The impact strength is a measure of toughness which is defined as the ability of the polymer to absorb applied energy.

From this beginning, we now arrive at the point where polymer-clay nanocomposites are one member of a group in which the dispersed particles have at least one dimension at the nanoscale. According to how many dimensions of the dispersed particles are at the nanoscale, the nanocomposites can be classified into three categories, which are shown in Fig. 4.2: polymer lamellar material nanocomposites, polymer nanotube/nanofiber nanocomposites, and polymer spherical nanoparticle nanocomposites. In the first type of nanocomposites, the dispersed particle has only one nanoscale dimension and polymer-clay nanocomposites fall into this class. The sheets of clay particles in the polymer matrix are in the form of one to a few nanometers thick, and hundreds to thousands of nanometers wide and long. In the second type, there are two dimensions of dispersed particles at the nanoscale; this classification includes carbon nanotube and nanofiber nanocomposites. The third type of nanocomposite contains three dimensions of dispersed particles at the nanoscale, which includes the spherical particle nanocomposites.

There have been numerous publications and patents appearing each year on polymer nanocomposites, which greatly piques the interest of both academic and industrial researchers. In this chapter, we will focus on the one-dimensional nanomaterials and their preparation, characterization, properties and potential applications will be discussed.

4.2
Polymer Lamellar Material Nanocomposites

4.2.1
Types of Lamellar Nano-additives

There are a large number of lamellar materials which can be used as nanofillers for polymer nanocomposites and these are shown in Table 4.1.

Amongst all of the potential nanocomposite lamellar fillers, clays have received the most attention and are the most widely investigated, due to their easy availability and well-established intercalation chemistry. As can be seen in Table 4.1, a large number of clays are used. The typical clay is either an alumniosilicate or a

Table 4.1 Layered host crystals susceptible to intercalation by a polymer.

Chemical nature	Examples
Element	Graphite[a]
Metal chalcogenides	$(PbS)_{1.8}(TiS_2)_2$[b], MoS_2[c]
Metal oxides	TiO_2[d]
Carbon oxides	Graphite oxide[e]
Metal phosphates	$Zr(HPO_4)$[f]
Clays and layered silicates	Montmorillonite, hectorite, saponite, talc, zeolite, kaolinite, magadiite, fluormica, vermiculite, ...
Layered double hydroxides	$M_6Al_2(OH)_{16}CO_3 \cdot nH_2O$; M=Mg[g], Zn[h]

a L. Chazeau, J.Y. Cavaille, G. Canova, B. Debdievel, *J. Appl. Polym. Sci.*, **1999**, 71, 1797–1808.
b L. Hernan, J. Morales, J. Santos. *J. Solid State Chern.*, **1998**, 141, 327–329.
c D.J. Harris, T.J. Bonagamba, K. Schmidt-Rohr *Macromolecules*, **1999**, 32, 6718–6724.
d A. Laachachi, M. Co chez, M. Ferriol, E. Leroy, J.M. Lopez-Cuesta, in *Fire & Polymers: Materials and Concepts for Hazard Prevention*, Eds., C.A. Wilkie and G.L. Nelson, American Chemical Society Symposium Series #922, Oxford University Press, Oxford, **2005**, pp. 36–47.
e Y. Matsuo, K. Tahara, Y. Sugie, *Carbon*, **1997**, 35, 113–120.
f Y. Ding, D.J. Jones, *Chem. Mater.*, **1995**, 7, 562–571.
g O.C. Wilson Jr., T. Olorunyolemi, A. Jaworski, L. Borum, D. Young, A. Siriwat, E. Dickens, C. Oriakhi, M. Lerner, *Appl. Clay Sci.*, **1999**, 15, 265–279.
h C.O. Oriakhi, LV. Farr, M.M. Lerner, *Clays and Clay Minerals*, **1997**, 45, 194–202.

silicate with a layered structure. In particular, the smectite clays, such as montmorillonite, saponite and hectorite, have been widely studied because of their excellent intercalation ability. They all contain two tetrahedral sheets with a central octahedral sheet, but they have different chemical composition. The general chemical formulae for these clays are $M_x(Al_{4-x}Mg_x)Si_8O_{20}(OH)_4$ (montmorillonite), $M_x(Mg_{6-x}Li_x)Si_8O_{20}(OH)_4$ (hectorite) and $M_xMg_6(Si_{8-x}Al_x)O_{20}(OH)_4$ (saponite). Both montmorillonite and hectorite have the same silica tetrahedral sheets, but montmorillonite has an alumina octahedral sheet and hectorite has a magnesium octahedral sheet. Saponite has a magnesium octahedral sheet with both silica and alumina tetrahedral sheets. The smectite clays are different from talc and mica because of the unit cell layer charge. The unit cell layer charge of smectite clays (–0.6 to –1.4 per O_{20} unit) is intermediate between that of talc (0.0 per O_{20} unit) and mica (–2.0 per O_{20}). The amount of the unit cell layer charge is dependent on the substitution of higher-valent elements, such as Al^{3+} and Si^{4+}, with lower-valent elements, such as Mg^{2+} and Li^+. For example, the unit cell layer charge of $(Al_{3.4}Mg_{0.6})Si_8O_{20}(OH)_4$ is calculated according to the charges of the elements, Al (+3), Si (+4), Mg (+2), oxygen (–2) and OH (–1). The net layer charge of this clay is $[3.4(+3)] + [0.6(+2)] + [8(+4)] + [20(-2)] + [4(-1)] = 43.4-44 = -0.6$. So the unit cell layer charge of this clay is –0.6 per O_{20} unit.

In the following discussion, the layered additives used to form the polymer nanocomposites will be montmorillonite, unless otherwise noted. There are two principle reasons for choosing montmorillonite. Firstly, it exhibits a very rich intercalation chemistry so that it can be easily organically modified. Secondly, it is ubiquitous in nature and can be obtained in high purity at low cost. A review has recently appeared covering all of the literature in the area of clay-containing polymer nanocomposites, including the patent literature.

4.2.2
Montmorillonite Layer Structure

Montmorillonite (MMT) is a hydrated alumina-silicate composed of units made up of two silica tetrahedral sheets with a central alumina octahedral sheet so that the oxygen ions of the octahedral sheet also belong to the tetrahedral sheet, the structure is shown in Fig. 4.3. From electron microscopy, it is composed of aggregates, whose size ranges from 0.1 to 10µ; these aggregates are comprised of a number of primary particles (structural units), which consist of a number of super-imposed lamellae. The height of the primary particle is 80–100 Å with a diameter of 300 Å and the thickness of the lamella is 9.6–10 Å. Thus, each primary particle contains approximately 8 lamellae or 16 planes.

4.2.3
Modification of Clay

Naturally occurring clays exhibit isomorphous substitution of one element for another in both the octahedral and tetrahedral layers, for instance aluminum or

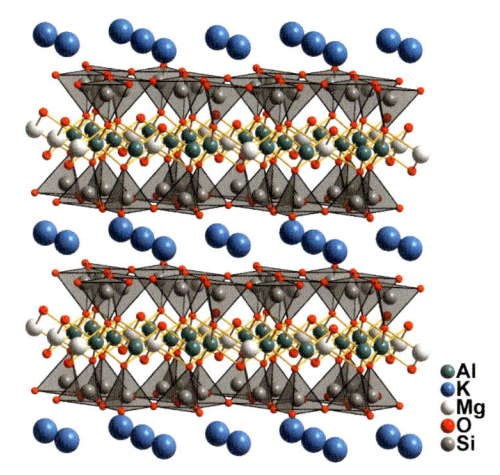

Fig. 4.3 Crystal structure of montmorillonite.

iron may substitute for silicon in the tetrahedral sheets and magnesium, or other divalent metals, may substitute for aluminum in the octahedral layer. These substitutions create a net negative charge on the clay layers, which must be balanced by a corresponding positive charge at some location. The charge in the naturally-occurring clay is balanced, typically, by a sodium cation which lies between the layers in what is called the gallery space. The ability of the clay to improve the properties of the polymer is primarily determined by the extent of its dispersion in the polymer matrix, which, in turn, depends on the clay particle size. However, the hydrophilic nature of the clay surfaces impedes homogeneous dispersion in the organic polymer phase. To overcome this problem, it is usually necessary to render the surface organophilic prior to its use. Three methods have been used to modify the sodium clay: a) exchange the gallery cation (Na^+) with quaternary organic cations, such as ammonium and phosphonium salts; b) directly modify the clay layers using organic coupling agents, such as silane coupling agents; and c) use crown ether to complex the clay cations. The role of the organic compound in the org-MMT is to reduce the surface energy of the MMT, thereby improving the wetting characteristics by the polymer or monomer. The successful formation of a hybrid is, to a large extent, determined by the miscibility of individual components of the system.

One should differentiate between water soluble and water-insoluble polymers; for water soluble polymers, such as poly (vinyl alcohol) (PVA), poly (vinylpyrrolidone) (PVP), poly (dimethyldiallylammonium) (PDDA), and so forth, no organic modification is required while for organophilic polymers, such as polyamide 6, polystyrene, poly(methyl methacrylate), etc., it is necessary to organically modify the clay before a nanocomposite can be produced.

Organically-modified clays are available from a variety of suppliers and the types of surfactants include both nonpolar, e.g. some combination of long chains and methyl groups to give four substituents on an ammonium ion, or polar, which

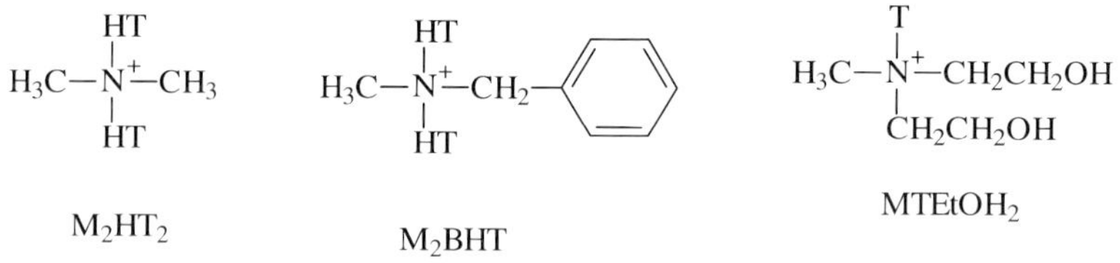

Fig. 4.4 Structures of some commonly used surfactants.

usually means that one or more of the substituents on the nitrogen contains a hydroxyl group. Figure 4.4 shows the structures of a few of the surfactants that have been used to modify clays. The preparation of an organically-modified clay simply involves ion exchange of the sodium cation for the ammonium cation and this is accomplished by dispersing the inorganic clay in water and adding the ammonium ion, also in water. Since the newly formed clay is now organically-modified, it does not disperse well in water and it will precipitate and can be filtered. Organically-modified clays have also been obtained by an *in situ* process of combining the inorganic clay, the surfactant and a polymer in some blending devise to obtain a nanocomposite.

4.3
Nanostructures and Characterization

One classification is based upon the registry of the silicate layers in the polymer matrix and two distinct nanostructures have been identified with layered silicates (Fig. 4.5). Delaminated, also known as exfoliated, hybrids retain little or no registry between the silicate layers, which are typically separated by as much as 5–15 nm. Intercalated hybrids, on the other hand, retain registry between the host layers, albeit with an expanded gallery height or distance between layers, which reflects the incorporation of polymer chains into the structure. A nanocomposite can contain both intercalated and exfoliated morphologies. Since exfoliated nanocomposites exhibit superior mechanical and barrier properties compared to intercalated systems, to obtain an exfoliated nanocomposite is the goal for every researcher who is interested in these properties. A third possible morphology is the immiscible nanocomposite, also known as a microcomposite, in which the clay is not well-dispersed and is behaving as a conventional filler. Another term which is commonly used is nano-dispersion. A system is said to be nano-dispersed if either an intercalated or exfoliated system is obtained. It is much easier to obtain good nano-dispersion using a polymerization technique than a blending technique because the smaller monomer can penetrate the gallery space more easily than the larger polymer molecules.

4.3.1
X-ray Diffraction and Transmission Electron Microscopy to Probe Morphology

There are two characterization methods in common usage to identify the nanostructures of polymer nanocomposites: X-ray diffraction (XRD) and transmission

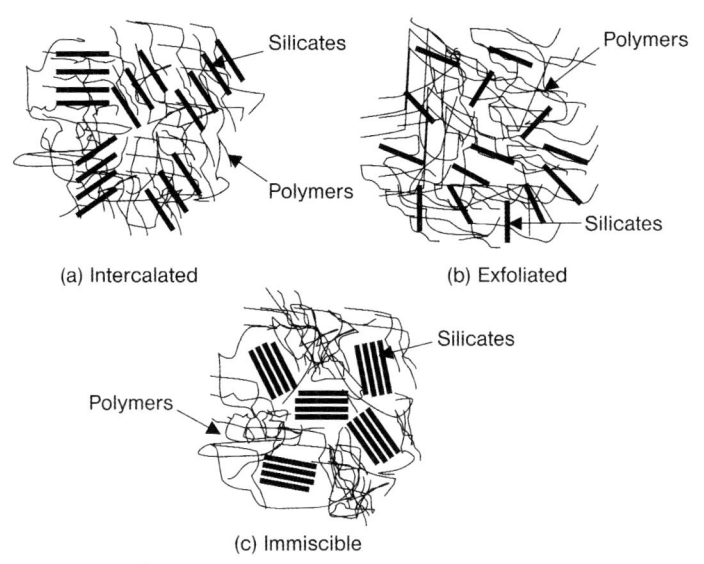

Fig. 4.5 Possible nanostructures.

electron microscopy (TEM). XRD measures the d-spacing of the gallery space of the clay, which tells if the size of the gallery space has changed upon insertion of the polymer chains. From the XRD traces, an intercalated nanocomposite will show a peak at a lower value of 2θ, which corresponds to an increase in the d-spacing, while there is no diffraction peak for an exfoliated nanocomposite. Figure 4.6 shows standard XRD traces for an intercalated and a delaminated nanocomposite. The d-spacing is calculated from Bragg's equation: $n\lambda = 2d\sin\theta$, where n is taken as unity, λ corresponds to the wavelength of the X-ray radiation, d is the spacing between silicate layers, and $\tilde{N}$ is the measured diffraction angle. If a strong, sharp peak is seen at a larger d-spacing, one can be somewhat confident that an intercalated nanocomposite has been obtained. Both the position at which the peak is found and its width are important. If the position is at a lower value of 2θ, this indicates that the d-spacing has increased and is indicative of nanocomposite formation. If the peak is broad, this may be interpreted as indi-

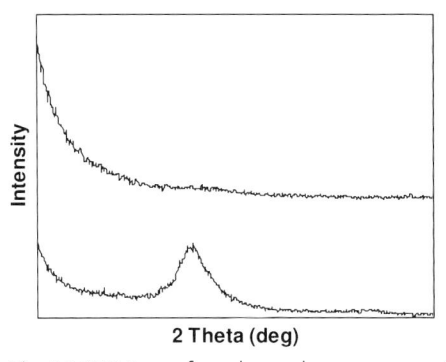

Fig. 4.6 XRD traces for polymer-clay nanocomposites.

cating disorder while if the peak is sharp intercalation is likely. On the other hand, the absence of a peak could mean either that a delaminated structure has been produced, which has lost registry and thus shows no peak, or that the clay layers have become disordered, meaning that no peak can be seen. While one can scan over any range of 2θ, the distance between the layers in pristine montmorillonite is about 1 nm and this can increase up to 4 or 5 nm in a nanocomposite so this is the region which is usually evaluated. If the source is copper, $\lambda = 1.54$ Å and 1 nm corresponds to a 2θ of about 7° so the region below 10° is typically scanned. In the typical XRD equipment, there is a minimum 2θ below which the instrument is not usable and this is somewhat smaller than 2°, so this defines the lower limit. Small angle X-ray scattering, SAXS, and small angle neutron scattering, SANS, are sometimes used to examine lower 2θ or larger d-spacings of the gallery space.

An additional measurement is required to elucidate the morphology and usually TEM is used. Both a low magnification image, to show that the clay is well-dispersed, and a high magnification image, to permit the imaging of the clay layers, are required. Figure 4.7 shows the TEM images of an intercalated nanocomposite and an exfoliated nanocomposite. One can see in the images on the left hand side that the clay is well-dispersed in both cases and there is little difference between these at this magnification. From the images on the right hand side, at high magnification, it is obvious that there is a significant difference between these two materials.

4.3.2
Other Techniques to Probe Morphology

Although TEM images together with XRD provides a very adequate description of the morphology, there are some limitations on these technologies. For example, for transmission electron microscopy only an extremely small portion of the sample is examined and the area is selected mainly relying on the skill and will of the researcher. Therefore, there is a need for some bulk measurement of morphology, which samples a much larger portion of the material, and three such techniques shall be mentioned here.

Nuclear magnetic resonance can be used as a technique to probe morphology. Scientists at the U.S. National Institute of Standards and Technology have pioneered the NMR technique to evaluate the type of dispersion of the clay in the polymer. The technique, which uses about one gram of sample and thus qualifies as a bulk technique, depends upon the fact that clays naturally contain iron and paramagnetic iron, in proximity to the protons of a polymer, will promote relaxation of those protons; the term that is evaluated is the proton longitudinal relaxation time, T_1^H. In a microcomposite the protons are rather far removed from the iron, since the clay is not well-dispersed. In an intercalated nanocomposite the protons are closer to the iron and thus more affected by it and the relaxation time is shorter. Finally, in the case of a delaminated nanocomposite the iron is, on average, quite close to the protons and the relaxation time is the fastest. Considering the nanocomposites for which the TEM images are shown in Fig. 4.6, the T_1^H

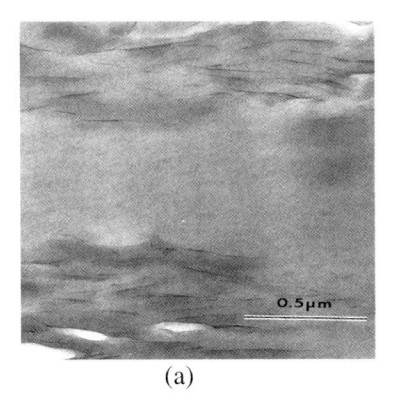

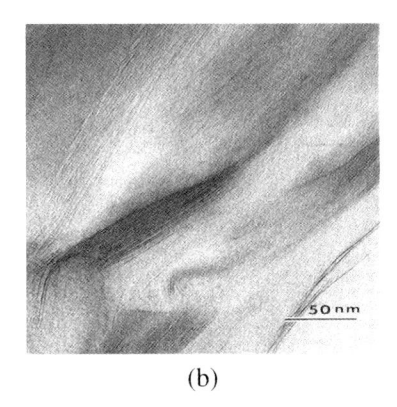

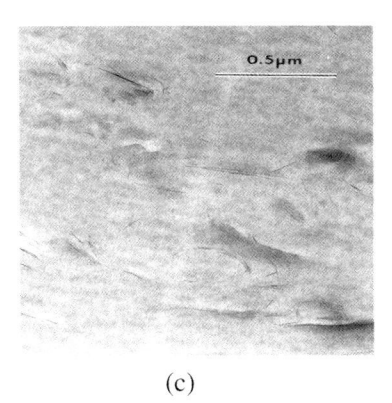

Fig. 4.7 TEM images of polymer nanocomposites: a) and b) an intercalated polystyrene nanocomposite; c) and d) a delaminated polystyrene nanocomposite. On the left is the low magnification image which shows if the clay is well-dispersed, while on the right is the higher magnification image which shows the actual sate of the clay dispersion. (Reprinted with permission from *Chem. Mater.* **2001**, 13, 3774–3780 © 2001 American Chemical Society.)

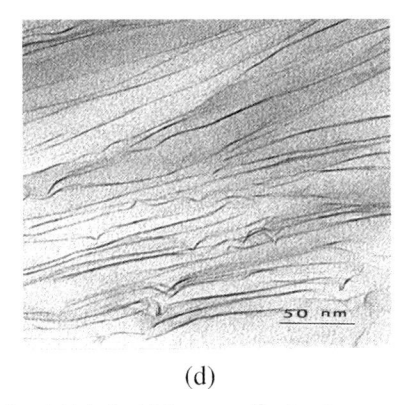

for virgin polystyrene is 38.58 sec while it is 14.21 sec for the intercalated system and 9.72 sec for the delaminated system. The authors have used this data to discuss the percent delamination in the systems.

Cone calorimetry can also be used as a technique to probe morphology. Cone calorimetry is a technique that is used to probe the fire properties of polymers; a picture of a cone calorimeter is shown in Fig. 4.8. This consists of a load cell on which the sample is placed and its mass can be followed as the sample burns. The name cone comes from the cone-shaped heater that is used to irradiate the sample. The concentration of oxygen in the effluent gases is measured and from this one may calculate the heat that is released, based on the assumption that almost all hydrocarbon fuels release 13.1 kJ per one gram of oxygen consumption if the fuel is completely combusted. Among the parameters that may be obtained are

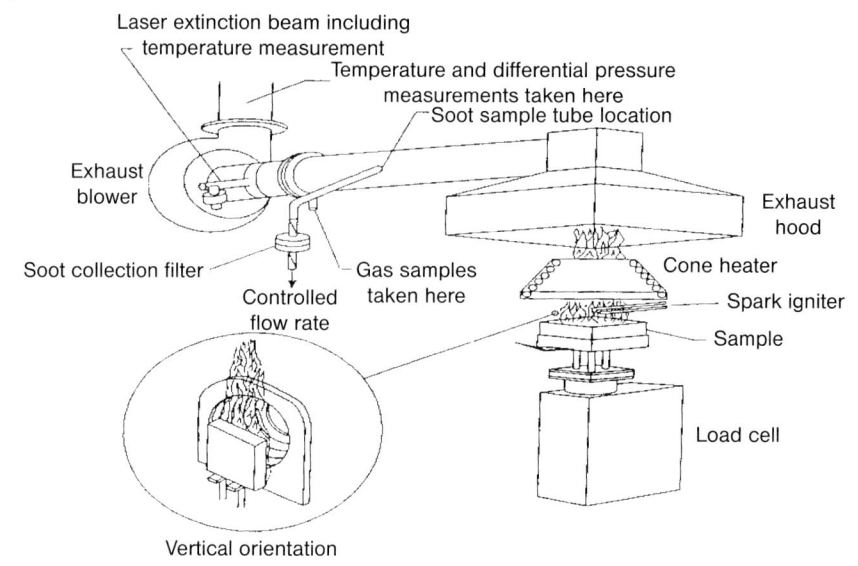

Fig. 4.8 The cone calorimeter.

the heat release rate and especially its peak value. It has been observed that microcomposites give a minimal reduction in the peak heat release rate while both intercalated and delaminated nanocomposites give much larger reductions, with the reduction depending upon the polymer under investigation. There are some cases in which the results from cone calorimetry agree very well with the combination of XRD and TEM but there are also cases where the XRD and TEM combination seems to imply good nano-dispersion and there is a minimal reduction in the peak heat release rate and the reverse, poor dispersion from XRD and TEM but a large reduction in the peak heat release rate, has also been seen. A single sample for cone calorimetry uses about 35 grams of material so this is again a bulk measurement.

Mechanical and rheological properties may be used as a probe of morphology. It is well-known that the mechanical properties are enhanced by nanocomposite formation. Neither mechanical nor rheological properties are commonly used at this time to evaluate the morphology of the nanocomposites but, as we move into the future, we may find that this becomes more widely used.

4.4
Preparation of Polymer-clay Nanocomposites

There are several ways to prepare polymer-clay nanocomposites (PCN). One promising approach is polymer melt-direct intercalation into layered silicate hosts by using a conventional polymer extrusion process. The formation of a PCN via melt intercalation depends upon the thermodynamic interaction between the polymer chains and the host silicates and the transport of the polymer chains from the bulk

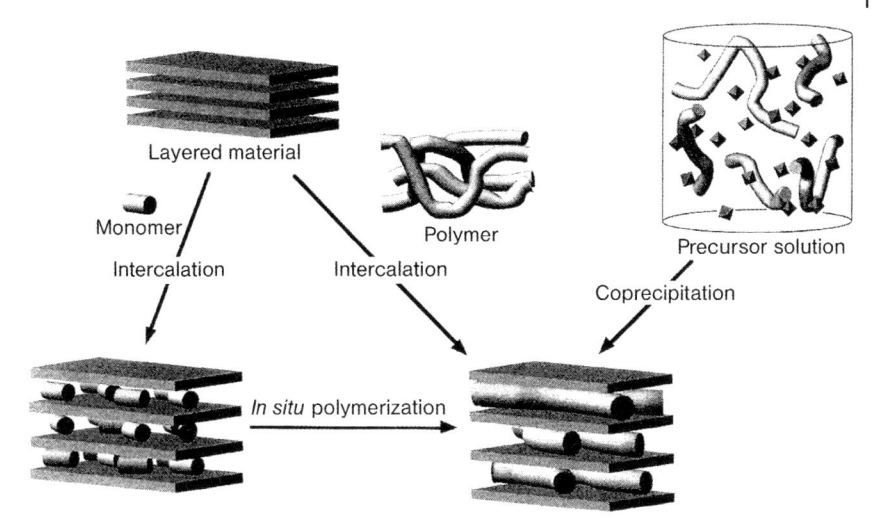

Fig. 4.9 Schematic view of preparation methods for polymer nanocomposites. (From Y. Komori and K. Kuroda in *Polymer-Clay Nanocomposites*, Eds T J Pinnavaia and G W Beall, © 2000, John Wiley and Sons Ltd. Reproduced with permission)

melt into the silicate layers. Another promising approach is *in situ* polymerization on or in the organically modified clay. In most cases, the synthesis involves intercalation of a suitable monomer and then exfoliating the layered host into nanoscale elements by subsequent polymerization. The nanocomposites can be also prepared by mixing the polymer with clay in a suitable solvent. The general process of nanocomposite formation is described in Fig. 4.9.

4.4.1
Solution Mixing

Solution mixing is a quite simple method to make polymer nanocomposites. Polymers are agitated in a suitable solvent with the desired amount of clay and the nanocomposites are obtained after evaporating the solvent. In general an organically-modified clay will disperse well in most organic solvents, so one need only to identify a suitable solvent for the polymer. The solution mixing method is used to make polymer nanocomposite coating, adhesives, films, and so forth. Solution blending is likely not industrially acceptable for preparing thermoplastic nanocomposites due to the presence of the solvent and the necessity to remove and either dispose or reuse it.

4.4.2
Polymerization

Polymer nanocomposites can be easily prepared by polymerization in the presence of an organically-modified clay. Any polymerization method can be used, including bulk (*in situ*), solution, emulsion, and suspension. Since the nanocomposites are made from monomers, a well-dispersed nanostructure is easily obtained, as long as a suitable organic-modification has been used. In order to improve the interaction between polymer chains and silicate layers, it may be necessary for the organically-modified clay to contain some reactive groups that can react with either monomer or polymer. To achieve good delamination, it is quite helpful to attach a monomer, a catalyst or an initiator on the clay for polymers such as polystyrene and poly (methyl methacrylate). It appears that if some of the functional groups can be attached to the clay cation that the chances of delamination are increased; this is shown schematically in Fig. 4.10. Another delaminated polystyrene nanocomposite was obtained by incorporating a living free radical polymerization initiator onto the clay, as shown in Fig. 4.11. While the bulk polymerization technique can relatively easily produce nicely delaminated nanocomposites of many polymers, it may not be suitable for large scale production, because it requires that a production line be available for only that material and this is most often not practical. It must be remembered that the product that results from a polymerization process may not be at thermodynamic equilibrium and, if it is subjected to melt blending, it may revert to an intercalated system.

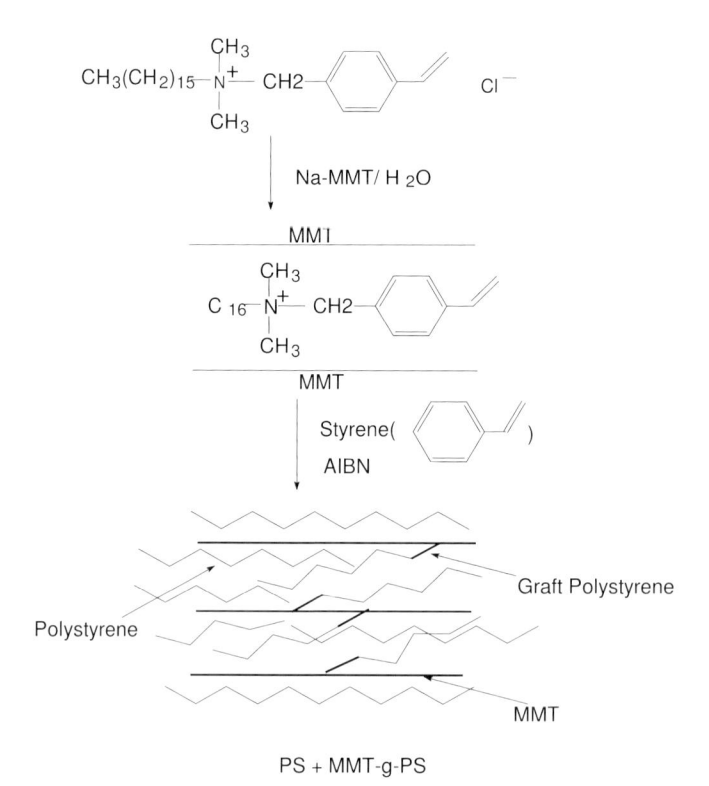

Fig. 4.10 Polystyrene nanocomposites that can be formed when the surfactant contains a reactive double bond.

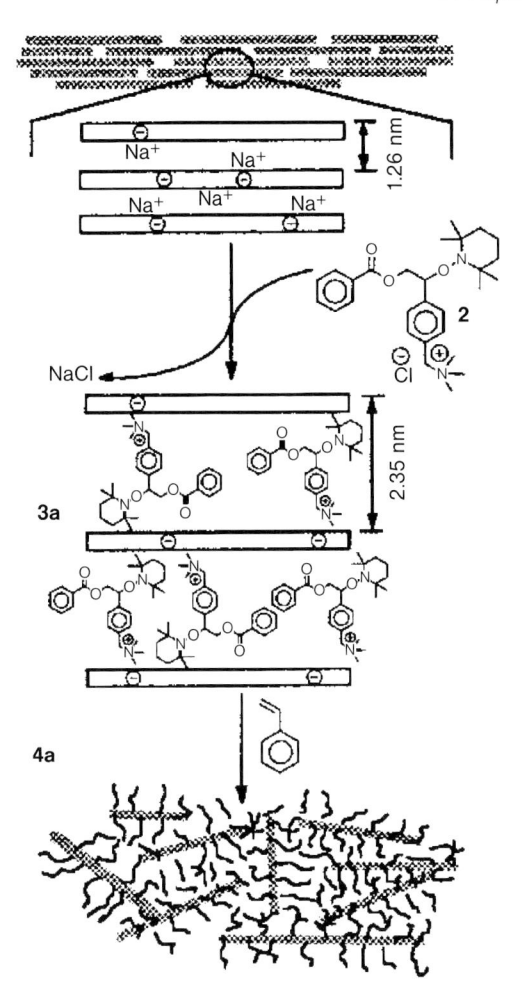

Fig. 4.11 Polystyrene nanocomposite from initiator bearing organoclay. (Reprinted with permission from *J. Amer. Chem. Soc.*, **1999**, 121, 1615–1616 ©1999 American Chemical Society.)

4.4.3
Melt Compounding

The melt blending process is the most promising and practical method for use in industry, but the clay exfoliation level achieved in the polymer matrix is normally lower than that obtained in *in situ* polymerization. Since the organically-modified clay does not react with the polymer, almost all nanocomposites made by melt blending are either immiscible, intercalated or, sometimes, partially exfoliated nanocomposites. The only possibility to achieve full exfoliation is when there is a much stronger interaction between the clay and polymers during the melt blending process, such as the presence of some functional groups which can react with the polymers. For example, an exfoliated polyamide 6-clay nanocomposite can be made through melt blending if the clay loading is no more than 10 wt.%.

Melt blending of nonpolar polymers is very difficult and almost always results in the formation of immiscible systems, unless a compatibilizer is used. For a polymer such as polypropylene or polyethylene, the usual compatibilizer is maleic anhydride. This is most commonly used by preparing a masterbatch of the graft copolymer, PP-g-MA, with the organically-modified clay and then combining this with the virgin polymer so that the maleic anhydride content is not too high. There have also been attempts to use maleic anhydride as an additive in the melt blending operation for polypropylene, polystyrene and polyethylene. The dispersion is significantly better when polymer-g-MA is used as the compatibilizer rather than maleic anhydride alone. Epoxy monomers as well as poly (dimethylsiloxane) have been used as swelling agents to enable the formation of nanocomposites. Of particular interest is that the addition of the epoxy does not exfoliate the clay but rather it was suggested that the intercalation of the epoxy facilitates the penetration of the polymer at the edges of the clay particle. These workers have treated the formation of polymer-clay nanocomposites in terms of the solubility parameter.

The small molecules typically used to modify clay have low thermal stability and organoclays with higher thermal stability must be developed for those polymers requiring higher processing temperatures. Oligomerically-modified clays have been developed for this purpose. The organic modifier is an oligomer with high thermal stability. The resulting organoclays seem to exhibit better thermal stability than is seen from the conventional materials. Another advantage of using oligomerically-modified clay is that the oligomer in the clay can act as a compatibilizer for the formation of nanocomposites. In this instance, the presence of the oligomer on the cation of the clay brings about spreading of the clay layers so that the polymer can enter. A large number of oligomerically-modified clays have been produced, including those based on styrene, acrylates, butadiene and capralactone.

4.5
Polymer-graphite and Polymer Layered Double Hydroxide Nanocomposites

While clay has been by far the most common nano-dimensional material that has been used to prepare nanocomposites, it is not the only material that has been used. In this section, we shall briefly cover the work that has been carried out using graphite as the nano-dimensional material and, in the next section, we will examine those materials that have been produced through the use of the so-called layered double hydroxides and salts.

Two strategies have been exploited to form polymer-graphite nanocomposites, the use of expanded graphites and the use of graphite oxide. In the arena of the expanded graphites, two systems have been used, potassium graphite and graphite sulfuric acid. Briefly, these materials are graphite intercalation compounds that are produced through the interaction of the indicated components. In the case of potassium graphite, several stages are know, which correspond to insertion of potassium between every graphite layer, that is alternating layers of potassium and graphite, stage 1, two layers of graphite between each layer of potassium, stage 2, and so forth. A representation of both stage 1 and stage 2 graphite intercalation compounds is shown in Fig. 4.12. Styrene has been polymerized using the potassium-graphite compound, a reaction which has been used on several occasions to polymerize styrene, and the nanocomposite that was produced was studied. The d-spacing has increased from the 5.54 Å to 15.1 Å in the nanocomposite. The reduction in the peak heat release rate suggests that good nano-dispersion has been obtained. Graphite-sulfuric acid has been used by several investigators. The foci have been on transport and electrical properties or nanocomposite characterization and fire retardancy. The reduction in the peak heat release rate is about the same for polymer-graphite and polymer-clay nanocomposites, so it can be assumed that the systems function by a similar process.

Graphite oxide, which is obtained by the oxidation of graphite, is another example of a layered material. The oxidation causes the formation of functional groups on the graphite, with the formation of a negative charge on the graphite layers which therefore requires the presence of some counter ion, which will occupy the gallery space between the layers. Since one may use the same type of surfactants that are used with clays, there is some similarity between these systems. Polystyrene nanocomposites have been prepared by emulsion polymerization in the presence of graphite oxide and a mixed intercalated/delaminated material was obtained. In another case, mixed intercalated/delaminated styrene-butyl acrylate nanocomposites were prepared and the fire properties of these systems have been studied. Well-dispersed styrene nanocomposites have been prepared and it has been shown that one must use a polymerization technique with graphite oxide because it decomposes at elevated temperatures to give back graphite; nanocomposites with graphite oxide cannot be prepared by melt blending at elevated temperature.

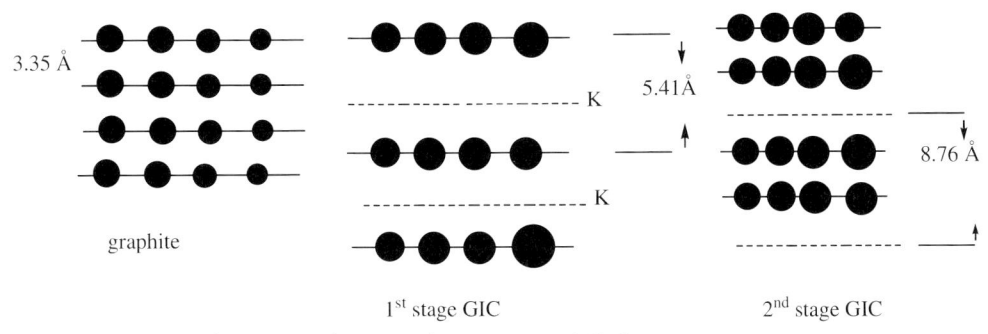

Fig. 4.12 A stage 1 and stage 2 graphite intercalation compound (GIC).

4.5.1
Nanocomposites Based on Layered Double Hydroxides and Salts

Layered double hydroxides (LDHs) are synthetic minerals with positively charged brucite-type layers of metal hydroxides. The structure of LDH is similar to that of montmorillonite, except that exchangeable anions located in the interlayer spaces compensate for the positive charge on the surface rather than the cation in the clay. Compared with the cation exchange capacity in montmorillonite, the anion exchange reaction in LDH is more difficult due to their high selectivity for carbonate anion and large anion exchange capacity. CO_2 must be carefully excluded during the ion-exchange reaction. Organic sulfates and sulfonates and carboxylates are widely used to modify LDH before polymer intercalation. A large number of polymers, such as polyaniline, polypyrrole, poly(vinyl alcohol), polyamides, and poly(vinyl sulfate) have been intercalated to form nanocomposites. The mechanical and thermal properties of polyimide were dramatically improved when LDH nanocomposites were formed. Biomolecules, such as DNA, can be also be intercalated into the LDH interlayers and these materials will be pH sensitive and can be considered as gene reservoirs.

The "normal" double layered hydroxide contains both magnesium and aluminum but it is quite possible to replace either of these metals and obtain polymer/LDH nanocomposites. The formation of an exfoliated linear low density polyethylene nanocomposites using a zinc-aluminum layered hydroxide has been reported; exfoliated nanocomposites using PE-g-MA and a magnesium-aluminum layered hydroxide have also been reported.

It has been shown that one may easily prepare epoxy nanocomposite using double layered hydroxides and good nano-dispersion was achieved with the reduction in the peak heat release rate better than what has been observed with clay as the nano-dimensional material. Likewise, double layered salts have also been used as the nanodimensional material with poly(methyl methacrylate). It is apparent that only microcomposites have been formed so far with this nano material but further work may lead to improved results.

Vermiculite has also been used as a layered silicate filler to prepare polymer nanocomposites. Vermiculite is a type of silicate that is more abundant and lower in cost than montmorillonite, hectorite, or saponite. Kaolinite and magadiite can also be used to make polymer nanocomposites. Kaolinite consists of SiO_4 tetrahedral sheets and $AlO_2(OH)_4$ octahedral sheets. Different from other layered fillers discussed above, kaolinite has neither cations nor anions present between layers and the layers are bounded to each other by hydrogen bonds between hydroxyl groups on the sheets. Magadiite is a layered polysilicate consisting mainly of silica tetrahedral sheets with some silanol groups and exchangeable cations in the interlayer region; the chemical formula of magadiite is $Na_2Si_{14}O_{29} \cdot nH_2O$. Magadiite can be hydrothermally synthesized from SiO_2, NaOH and water. Compared with that of the smectites, magadiite has higher cation exchange capacity; this clay contains very large platelets. The organically-modified clay shows good dispersion in ethylene-co-vinyl acetate as measured by XRD, TEM and cone calorimetry.

4.6
Properties of Polymer Nanocomposites

Owing to the nanoscale dispersion of clay, polymer-clay nanocomposites (PCN) exhibit greatly enhanced mechanical properties including strength, modulus and dimensional stability, thermal stability and heat distortion temperature, fire retardancy and reduced smoke emission, decreased gas, water and hydrocarbon permeability, ionic conductivity, chemical resistance, surface appearance, optical clarity, and so on, relative to the virgin polymers or conventional (micro-scale) counterparts.

The thermal stability of polymers is enhanced upon nanocomposite formation, but the extent of the increase in thermal stability is different for each polymer. This is due to the different mechanisms by which clay can act on the polymer degradation pathway. Clay can act as a barrier to hinder mass transport of the volatile degradation products and it can also prevent the polymer degradation. The clay in the nanocomposite also has a variable effect on the glass transition temperature, T_g, and the melting temperature, T_m, for different polymers. For example, the presence of clay increases T_g of epoxies, but it leads to a lower T_g for polycarbonate. Generally, the clay has strong heterophase nucleation effect on crystalline polymer. For instance, clay will increase the crystallization temperature of polyamide-6 and promote the γ crystalline form. Two T_m values are observed in the polyamide-6 nanocomposite, one for the γ crystalline form (214 °C) and the other for the α crystalline form (220 °C).

The rheological features are strongly related to the morphology of the polymer nanocomposites. The frequency dependence of the storage-moduli of polystyrene (PS) nanocomposites with various nanostructures is shown in Fig. 4.13. The

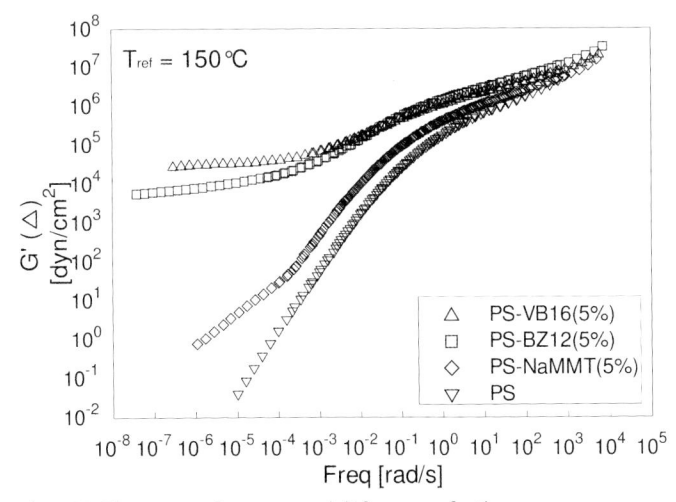

Fig. 4.13 The curves of storage moduli-frequency for the polystyrene and its layered-silicate nanocomposite. The reference temperature is 150 °C for all the samples.

storage modulus is related to the measurement of energy stored during deformation. It is measured by dynamic mechanical analysis (DMA), which measures the ability of materials to store and dissipate mechanical energy upon deformation. DMA measurement will determine which materials can be used in dynamic applications. At high frequency, the nanocomposites, either exfoliated PS-VB16 or intercalated PS-Bz12, have linear melt state viscoelastic properties which are similar to those of immiscible PS-NaMMT or pure polystyrene. At low frequency, there is dramatic difference with various morphologies of polymer composites. Nanocomposites show much higher moduli than immiscible systems or polystyrene and a pseudo-solid-like character is observed. The pseudo-solid-like behavior for the nanocomposite at low frequency suggests the percolation of a three-dimensional filler network.

4.7
Potential Applications

It has been said that polymer nanocomposites may prove to be useful in four areas: enhanced barrier properties, fire retardancy, increased heat distortion temperature and mechanical properties. The enhanced mechanical property improvements of polymer nanocomposites bring about numerous applications in the automotive area as well as for general applications. The automobile applications include the potential for mirror housings, door handles, engine covers, belt covers, and so forth. The general applications include the impellers and blades for vacuum cleaners, power tool housings, mower hoods and covers for mobile phones and pagers.

We feel that the major uses of nanocomposite technology will be in the areas of enhancing the barrier properties and improving the fire retardancy of materials. The advantage for enhanced barrier properties arises from the greatly reduced permeability of gases through polymer-clay nanocomposites compared to that in the virgin polymers. The explanation for the reduced permeability is that, in a delaminated nanocomposite, the clay layers are randomly dispersed throughout the polymer so the gases will have to follow a path of increased tortuousity in order to escape. This is depicted in Fig. 4.14. The tennis ball that is now used in Davis Cup matches contains a nanocomposite designed to reduce the permeability of air so that it is retained for a longer time within the tennis ball. Other potential applications include improving the air retention in tires by incorporating a nanocomposite on the inside of the tire, improved solvent, oil and flame resistant gloves with excellent dexterity and flexibility and improved protection in food packaging with thinner coatings.

Fire retardancy has been an active area of investigation for several years. The initial observation was that polymer-clay nanocomposites had a significantly reduced peak heat release rate, which may be approximated as the size of the fire. This indicates that the presence of the clay has a pronounced effect on combustion. The initial explanation for this was offered by scientists at NIST (National In-

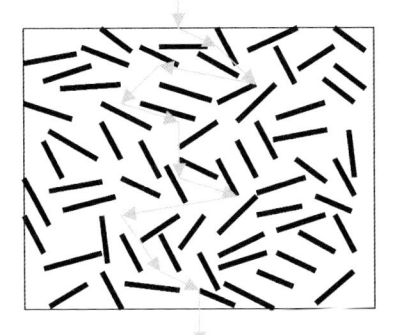

Fig. 4.14 The pathway for a molecule through a nanocomposite is more tortuous than through a simple polymer, which leads to enhanced barrier properties.

stitute of Standards and Technology, USA) who suggested that the clay could function as a barrier to mass transport and to insulate the fire from the underlying polymer. Work that has been carried out using X-ray photoelectron spectroscopy has confirmed that a barrier is formed. It was also found that very low amounts of clay, less than 1%, were also effective in lowering the peak heat release rate, which prompted the suggestion that the iron present in the clay could function as a radical trap and thereby limit the degradation. Very recently it has been shown that the products of degradation of a nanocomposite are different from those of a virgin polymer, and this has been used to suggest that the clay causes a retention of the degrading polymer radicals for a longer period of time which permits radical recombination reactions to occur.

It was initially felt that nanocomposite formation alone may be sufficient to solve the problem of fire and polymers. It is now recognized that nanocomposite formation is simply another tool that can be used is some cases and may not be of value in other cases. It is necessary to incorporate nanotechnology with conventional flame retardants to obtain the performance needed for fire applications.

4.8
Conclusion and Prospects for the Future

The future of polymer nanocomposites is a bright one. The initial exuberance that greeted their introduction has now been tempered by time and it is now realized that they will not be the solution to all problems but they can be a part of the solution. In particular, barrier properties are significantly enhanced and this, together with mechanical properties and fire retardancy, may be the areas in which nanocomposites make their major contribution. Although clay can bring enhanced stiffness and strength, the notched impact strength of nanocomposites is dramatically decreased. In order to maintain good impact property, some impact modifiers must be added to the polymer nanocomposites. In considering mechanical properties, we wish to make it clear that while mechanical properties on their own may be a target, the presence of a nano-dimensional material can potentially overcome any deleterious effects from other additives that are present to achieve some particular goal. This may be particularly true in fire retardancy, where the combination of a fire retardant may bring about plasticization or brittleness, depending upon the additive, and the nano-dimensional material may be able to counteract this effect.

The development of new surfactants must continue. The current common surfactants are thermally unstable at relatively low temperatures and it is hoped that new surfactants become available that will permit one to obtain delaminated systems by a melt blending process at temperatures as high as 300–350 °C.

The vast majority of the work that has been carried out to date has been with clays and one particular clay, montmorillonite. We believe that the future will see much more attention paid to other nano-dimensional materials, especially inorganic oxides and chalcogenides, such as titanium dioxide or molybdenum sulfide, and the double layered salts and hydroxides. This is not to suggest that clay has run its course and will be discarded, there will certainly be new innovations in the utilization of clays.

There is a definite need for new tools for the characterization of the nanocomposites. As was noted above, the standard combination that is now used is X-ray diffraction combined with transmission electron microscopy. Every person who examines a TEM image may not arrive at the same conclusion and some standardization of how to perform TEM, i.e. how many images of the sample are required to permit one to say that this is representative of the whole and how to interpret these images must develop and be agreed to by the practitioners in the field. Some bulk technique to characterize the morphology of the sample will be developed and will gradually become the standard method for this determination.

Further efforts are needed to identify the exfoliation mechanism of clay layers during polymer nanocomposite formation. The design of exfoliated nanocomposites can only be accomplished if the factors controlling exfoliation are well-understood.

Bibliography

Clay science

U. Hoffmann, K. Endell, D. Z. Wilm, *Krist*, **1933**, *A-86*, 340–348.

J. Mering, *Trans. Faraday Soc.*, **1946**, *42B*, 219–229.

A. Moet, A. Akehah, *Mater. Lett.*, **1993**, *18*, 97–102.

X. Kornmann, L. A. Berglund, J. Sterte, *Polym. Eng. Sci.* **1998**, *38*, 1351–1358.

E. Ruitz-Hitzky, B. Casal, *Nature*, **1978**, *276*, 596–597.

Y. H. Yao, J. Zhu, A. B. Morgan, C. A. Wilkie, *Polym. Eng .Sci.* **2002**, *42*, 1808–1814.

S. C. Tjong, Y. Z. Meng, Y. Xu, *J. Polym. Sci. Part B: Polym. Phys.*, **2002**, *40*, 2860–2870.

Y. Komori, K. Kuroda, in *Polymer-Clay Nanocomposites*, Eds. Pinnavaia, T. J., Beall,

G. W., John Wiley & Sons, Ltd., **2000**, pp. 3–18.

History of polymer nanocomposites

A. Blumstein, *J. Polym. Sci.: Part A*, **1965**, *3*, 2653–2664.

A. Blumstein, *J. Polym. Sci.: Part A*, **1965**, *3*, 2665–2672.

A. Blumstein, F. W. Billmeyer, *J. Polym. Sci.: Part A-2*, **1966**, *4*, 465–474.

A. Blumstein, R. Blumstein, T. H. Vandersppurt, *J. Colloid Interface Sci.*, **1969**, *31*, 236–247.

A. Blumstein, S. L. Malhotra, A. C. Watterson, *J. Polym. Sci.: Part A-2*, **1970**, *8*, 1599–1615.

A. Blumstein, K. K. Parikh, S. L. Malhotra, R. Blumstein, *J. Polym. Sci.: Part A-2*, **1971**, *9*, 1681–1691.

M. J. Kawasumi, *Polym. Sci. Part A: Polym. Chem.* **2004**, *42*, 819–824.

Y. Kojima, A. Usuki, M. Kawasumi, A. Okada, Y. Fukushima, T. Kurauchi, O. J. Kamigaito, *Mater. Res.*, **1993**, *8*, 1185–1189.

Mechanical property definitions

V. Shah, *Handbook of Plastics Testing Technology*, **1984**.

Reviews of polymer nanocomposites

M. Alexandre, P. Dubois, *Mater. Sci. Eng.*, **2000**, *28*, 1–63.

S. S. Ray, M. Okamoto, *Prog. Polym. Sci.*, **2003**, *28*, 1539–2641.

Z. Wang, J. Massam, T. J. Pinnavaia, in *Polymer-Clay Nanocomposites*, Eds. T. J. Pinnavaia, G. W. Beall, John Wiley & Sons, Ltd., **2000**, pp. 127–149.

L. A. Utracki, *Clay-Containing Polymeric Nanocomposites*, Rapra Technology Limited, Shropshire, UK, 2004. 2 volumes, 786 pages.

E. P. Giannelis, *Adv. Mater.*, **1996**, *8*, 29–35.

K. A. Carrado, L. Xu, S. Seifert, R. Csencsits, C. A. A. Bloomquist, in *Polymer-Clay Nanocomposites*, Ed. T. J. Pinnavaia, G. W. Beall, John Wiley & Sons, Ltd., **2000**, pp. 47–63.

Polymer nanocomposite preparation

M. Alexandre, G. Beyer, C. Henrist, R. Cloots, A. Rulmont, R. Jérôme, P. Dubois, *Chem. Mater.*, **2001**, *13*, 3830–3832.

D. Wang, J. Zhu, Q. Yao, C. A. Wilkie, *Chem. Mater.*, **2002**, *14*, 3837–3843.

J. Zhu, A. B. Morgan, F. J. Lamelas, C. A. Wilkie, *Chem. Mater.*, **2001**, *13*, 3774–3780.

S. Su, C. A. Wilkie, *J. Polym. Sci. Part A: Polym. Chem.*, **2003**, *41*, 1124–1135.

S. Su, D. D. Jiang, C. A. Wilkie, *Polym. Adv. Tech.* **2004**, *15*, 225–231.

M. W. Weimer, H. Chen, E. P. Giannelis, D. Y. Sogah, *J. Am. Chem. Soc.* **1999**, *121*, 1615–1616.

L. M. Liu, Z. N. Qi, X. G. Zhu, *J. Appl. Polym. Sci.* **1999**, *71*, 1133–1138.

D. Wang, C. A. Wilkie, *Polym. Degrad. Stab.*, **2003**, *80*, 171–182.

J. Zhang, C. A. Wilkie, *Polym. Degrad. Stab.*, **2003**, *80*, 163–169.

H. Ishida, S. Campbell, J. Blackwell, *Chem. Mater.*, **2000**, *12*, 1260–1267.

S. Su, D. D. Jiang, C. A. Wilkie, *Polym. Degrad. Stab.*, **2004**, *83*, 321–331.

S. Su, D. D. Jiang, C. A. Wilkie, *Polym. Degrad. Stab.*, **2004**, *83*, 333–346.

S. Su, D. D. Jiang, C. A. Wilkie, *Polym. Adv. Tech.* **2004**, *15*, 225–231.

J. Zhang, D. D. Jiang, C. A. Wilkie, *Thermochim. Acta*, **2005**, *430*, 107–113.

J. Zhang, D. D. Jiang, C. A. Wilkie, *Polym. Adv. Tech.*, **2005**, *16*, 549–553.

J. Zhang, D. D. Jiang, C. A. Wilkie, *Polym. Degrad. Stab.*, **2004**, *84*, 279–288.

J. Zhang, D. D. Jiang, C. A. Wilkie, *J. Vinyl Add. Tech.*, **2004**, *10*, 44–51.

X. Zheng, C. A. Wilkie, *Polym. Degrad. Stab.*, **2003**, *82*, 441–450.

Nanostructure characterization

D. L. VanderHart, A. Asano, J. W. Gilman, *Macromol.*, **2001**, *34*, 3819–3822.

D. L. VanderHart, A. Asano, J. W. Gilman, *Chem. Mater.*, **2001**, *13*, 3781–3795.

D. L. VanderHart, A. Asano, J. W. Gilman, *Chem. Mater.*, **2001**, *13*, 3796–3809.

R. D. Davis, J. W. Gilman, D. L. VanderHart, *Polym. Degrad. Stab.*, **2003**, *79*, 111–121.

S. Bourbigot, D. L. VanderHart, J. W. Gilman, W. H. Awad, R. D. Davis, A. B. Morgan, C. A. Wilkie, *J. Polym. Sci.: Part B: Polym. Phys.*, **2003**, *41*, 3188–3213.

S. Bourbigot, D. L. VanderHart, J. W. Gilman, S. Bellayer, H. Stretz, D. R. Paul, *Polymer*, **2004**, *45*, 7627–7638.

J. W. Gilman, T. Kashiwagi, M. Nyden, J. E. T. Brown, C. L. Jackson, S. Lomakin, E. P.

Gianellis, E. Manias, in *Chemistry and Technology* of Polymer Additives; S. Al-Maliaka, A. Golovoy, C. A. Wilkie, Eds.; Blackwell Scientific: London, **1998**; pp 249–265.

Flame retardancy of polymer nanocomposites

M. Zanetti, G. Camino, D. Canavese, A. B. Morgan, F. J. Lamelas, C. A. Wilkie, *Chem Mater* **2002**, *14*, 189–193

J. Zhu, P. Start, K. A. Mauritz, C. A. Wilkie, *Polym. Deg. Stab.* **2002**, *77*, 253–258

A. B. Morgan, P. D. Whaley, T. S. Lin, J. M. Cogen, in *Fire and Polymers IV: Materials and Concepts for Hazard Prevention*, C. A. Wilkie and G. L. Nelson, Eds., ACS Symposium Series #922, Oxford University Press, Oxford, **2005**, pp. 48–60.

J. Zhu, C. A. Wilkie, *Polym. Int.*, **2000**, *49*, 1185–1189.

J. W. T. Gilman, Kashiwagi, M. Nyden, J. E. T. Brown, C. L. Jackson, S. Lomakin, E. P. Gianellis, E. Manias, in *Chemistry and Technology* of Polymer Additives; S. Al-Maliaka, A. Golovoy, C. A. Wilkie, Eds. Blackwell Scientific: London, **1998**; pp 249–265.

J. W. Gilman, C. L. Jackson, A. B. Morgan, R. Harris, E. Manias, E. P. Giannelis, M. Wuthcnow, D. Hilton, S. H. Phillips, *Chem. Mater.*, **2000**, *12*, 1866–1873.

J. Wang, J. Hao, J. Zhu, C. A. Wilkie, *Polym. Degrad. Stab.*, **2002**, *77*, 249–252.

J. Du, J. Zhu, C. A. Wilkie, J Wang, *Polym. Degrad. Stab.*, **2002**, *77*, 377–381.

J. Du, D. Wang, C. A. Wilkie, J. Wang, *Polym. Degrad. Stab.*, **2003**, *79*, 319–324.

J. Du, J. Wang, S. Su, C. A. Wilkie, *Polym. Degrad. Stab.*, **2004**, *83*, 29–34.

J. Zhu, F. M. Uhl, A. B. Morgan, C. A. Wilkie, *Chem. Mater.*, **2001**, *13*(12), 4649–4654.

B. Jang, C. A. Wilkie, *Polymer*, **2005**, *46*, 2933–2942.

B. Jang, C. A. Wilkie, *Polymer*, **2005**, *46*, 3264–3274.

M. Costache, D. Wang, C. A. Wilkie, *Polymer*, **2005**, *46*, 6947–6958.

Polymer graphite nanocomposites

W. Rüdorff, *Adv. Inorg. Radiochem.*, **1959**, *1*, 223–266.

G. R. Hennig, *Prog. Inorg. Chem*, **1959**, *1*, 125–205.

F. M. Uhl, C. A. Wilkie, *Polym. Degrad. Stab.*, **2002**, *76*, 111–122.

W. Zheng, S. C. Wong, H. J. Sue, *Polymer*, **2003**, *73*, 6767–6773.

W. Zheng, S. C. Wong, *Composite Sci. Tech.*, **2003**, *63*, 225–235.

F. M. Uhl, Q. Yao, H. Nakajima, E. Manias, C. A. Wilkie, *Polym. Degrad. Stab.*, **2005**, *89*, 70–84.

F. M. Uhl, Q. Yao, C. A. Wilkie, *Polym. Adv. Tech.*, **2005**, *16*, 533–540.

R. Ding, Y. Hu, Z. Gui, R. Zong, Z. Chen, W. Fan, *Polym. Degrad. Stab.*, **2003**, *81*, 473–476.

R. Zhang, Y. Hu, J. Xu, W. Fan, Z. Chen, *Polym. Degrad. Stab.*, **2004**, *85*, 583–588.

F. M. Uhl, C. A. Wilkie, *Polym. Degrad. Stab.*, **2004**, *84*, 215–226.

Polymer–LDH nanocomposites

E. Ruiz-Hitzky, *Mol. Cryst. Liq. Cryst.*, **1988**, *161*, 433–452.

F. Leroux, J. P. Besse, *Chem. Ater.*, **2001**, *13*, 3507–3515.

H. B. Hsueh, C. Y. Chen, *Polymer*, **2003**, *44*, 1151–1161.

W. Chen, L. Feng, B. Qu, *Chem. Mater.*, **2004**, *16*, 368–370.

W. Chen, B. Qu, *Chem. Mater.*, **2003**, *15*, 3208–3213.

E. Kandare, D. Hall, D. D. Jiang, J. M. Hossenlopp, in *Fire and Polymers IV: Materials and Concepts for Hazard Prevention*, Eds, C. A. Wilkie and G. L. Nelson, ACS Symposium Series #922, Oxford University Press, Oxford, **2006**, pp. 131–143.

Properties of polymer nanocomposites

T. Lan, P. D. Kaviratna, T. J. Pinnavaia, *Chem. Mater.* **1994**, *6*, 573–575.

R. A. Vaia, S. Vasudevan, W. Krawiec, L. G. Scanlon, E. P. Giannelis, *Adv. Mater.* **1995**, *7*, 154–156.

K. Yasue, S. Katahira, M. Yoshikawa, K. Fujimoto, in *Polymer-Clay Nanocomposites*, Eds. T. J. Pinnavaia, G. W. Beall, John Wiley & Sons, Ltd., **2000**, pp. 111–126.

Y. M. Zhang J. Zhu, E. A. Verploegen, E. P. Giannelis, U. B. Wiesner, *Polymer Preprints* **2002**, *43*, 1049–1050

R. Krishnamoorti, S. Silva, in *Polymer-Clay Nanocomposites*, Eds. T. J. Pinnavaia, G. W. Beall, John Wiley & Sons, Ltd., **2000**, pp. 317–341.

Polymer nanocomposite applications
InMat corporation web site,
  http://www.inmat.com/

5
Porous Hybrid Materials
Nicola Hüsing

5.1
General Introduction and Historical Development

Porosity is ubiquitous to most known materials, with the exception of metals and ceramics that are fired at high temperatures. Even in nature, many materials are porous, including wood, cork, sponge, bone, or the skeleton structures of very simple organisms such as diatoms, radiolaria, etc.

Mankind has been using porous materials for a long time, certainly dating back to prehistoric times, e.g. as charcoal for drawings in ancient caves, or for purification of water or medical treatment. However, it was only in the first half of the 20th century that the deliberate design of porous materials, i.e. their composition, pore structure and connectivity, became possible. Early examples include materials such as aerogels with porosities above 95%, or the development of novel synthetic routes to crystalline zeolite lattices with defined pore size and structure. The most promising, namely template-based approaches towards porous materials, have been advancing rapidly since the end of the 20th century, a typical example being the M41S type of materials or inverse opal structures (see below). In addition, the range of compositions has been extended dramatically from purely inorganic, i.e. metals or metal oxides through carbons to porous organic materials, i.e. polymers such as poly(styrene)–poly(divinylbenzene) or organic foams such as poly(urethanes). A large variety of inorganic–organic hybrid porous materials are accessible today – one prominent recent example being three-dimensional (3-D) metal–organic frameworks, so-called MOFs.

For chemical routes to porous materials, deliberate control over the positioning of molecular network-forming building blocks within a material is crucial, since the arrangement of the different building blocks forming the solid framework determines not only the chemical composition, but also the size, shape and arrangement of the pores (Fig. 5.1).

Within this chapter, a selection of different hybrid inorganic–organic porous materials will be presented, focusing on porous inorganic matrices with organic functions. However, the reader is reminded that this covers the field only to some

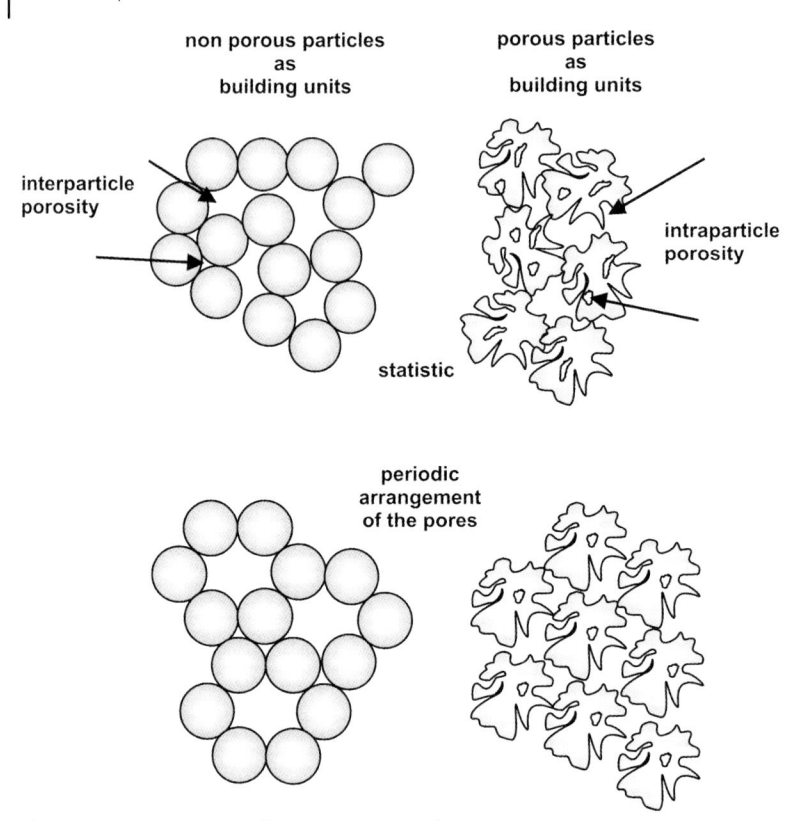

Fig. 5.1 Porosity: interparticle versus intraparticle porosity (top left and right) and statistic versus periodic arrangement (top and bottom) of the pores.

extent, since many organic materials are porous and can easily be modified with inorganic species. Since most textbooks do not even mention porous materials, despite their technological importance in many different areas, this chapter starts with a short introduction to porosity, followed by a brief overview of different types of porous solids. At the end, the reader is introduced to the different options for synthesizing hybrid porous materials, focusing on the problems and challenges for the given materials. The chosen examples are somewhat arbitrary, but to cover all types of porous hybrid materials would be far beyond the scope of this chapter.

This chapter gives an introduction to:
- Materials that are characterized by different types of porosity regarding the size and arrangement of the pores.
- Hybrid materials that differ in the way the organic part is incorporated; e.g. inclusion compounds, materials with covalently anchored organic functions, or even materials in

which the organic entity is an integral part of the porous network structure.
- Different synthetic strategies for hybrid materials, ranging from post treatment of a preformed matrix, *in situ* synthesis by mixing all the inorganic and organic components, or approaches in which the precursors already contain the inorganic and organic part within the same molecule.
- Some of the major applications of the different materials.

5.1.1
Definition of Terms

Porous materials A solid is called porous, when it contains pores, i.e. cavities, channels or interstices, which are deeper than they are wide. A porous material can be described in two ways: a) by the pores or b) by the pore walls. Some porous materials are based on agglomerated or aggregated powders in which the pores are formed by interparticle voids, while others are based on continuous solid networks.

For most applications the pore size is of major importance. However, pore sizes are not susceptible to precise measurement, because the pore shape is usually highly irregular, leading to a variety of pore sizes within one single material. Nevertheless, the use of three different pore size regimes was recommended by IUPAC and this terminology will also be used throughout this chapter:
- Micropores, with diameters smaller than 2 nm;
- Mesopores, with diameters between 2 and 50 nm;
- Macropores, with diameters larger than 50 nm.

This nomenclature is not arbitrarily chosen, but is associated with the different transport mechanisms occurring in the various types of pores, i.e. molecular diffusion and activated transport in micropores; while in mesopores Knudsen transport, surface diffusion and capillary condensation are the major mechanisms (Knudsen diffusion occurs when the mean free path is relatively long compared to the pore size, so the molecules collide frequently with the pore wall); and in macropores, bulk diffusion and viscous flow dominate.

As already mentioned, a wide variety of porous inorganic frameworks is known (Fig. 5.2). Today, zeolites or MOFs are the most prominent examples for microporous materials. Mesoporous solids with pore sizes between 2 and 50 nm can be found for example in aerogels, pillared clays and M41S materials, while macroporous solids are for example glasses, foams or inverse opal structures. In addition, these materials can be distinguished by the arrangement of the pores – periodic or random – and the pore radii distribution, which can range from either narrow with a rather uniform pore size distribution to quite a broad distribution.

In the discussion of porous materials, not only pore size distribution and pore diameters are of interest for later applications, but also the connectivity of the pore system or its dimension is of high interest. Porous channel systems, e.g. the ones

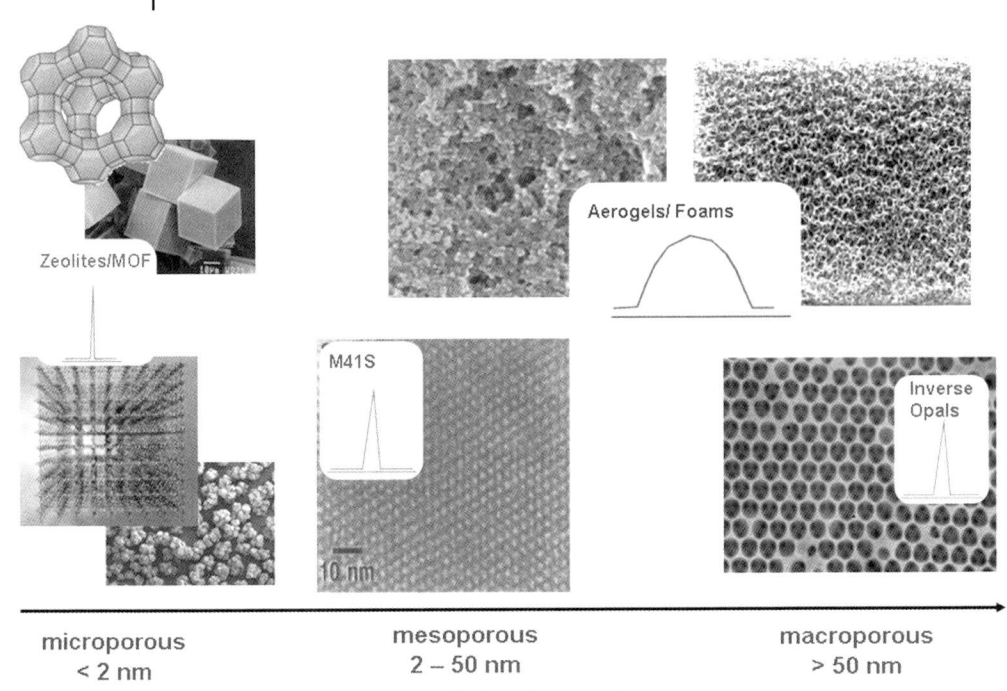

Fig. 5.2 Different porous materials classified according to their pore size and pore size distribution (insert).

in M41S phases, may be one-dimensional (1-D) as found for the hexagonally organized pore systems with their long channels, or 3-D as found for a cubic ordered pore structure. Two-dimensional (2-D) systems are layered materials which will not be discussed in this context (Fig. 5.3). In addition to the dimensionality of the pore system, two different surfaces must be distinguished in porous materials. The outer or exterior surface is an outward curving surface (convex) with a completely different reactivity as the inward curving surface (concave) that is typically found in the interior of the pores. This effect is of importance for functionalization reactions as discussed in the later sections of this chapter.

Nanocomposite Traditionally composites have been fabricated from preformed components in a process that organizes them in a matrix and with a particular arrangement. The integration of the different components is often a top-down approach and therefore, the structure and composition of the interfaces between the constituent parts is typically not under molecular scale control. In addition, the material may be divided into macroscopic domains with sizes of the order of milli- or micrometers (see also macroscopic phase separation). The bottom-up approach is an appealing solution to the interface problem, by co-assembling molecular inorganic and organic precursors into a *nanocomposite* material with molecular level command over interfaces, structure and morphology.

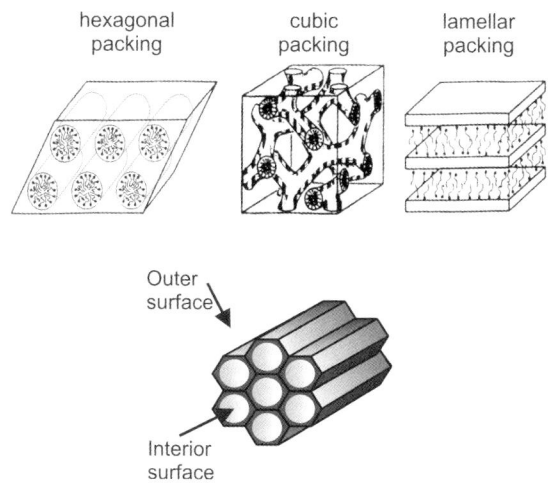

Fig. 5.3 Hexagonal, cubic and lamellar packings resulting in 1-D and 3-D pore dimensionalities and description of interior versus outer (exterior) surfaces.

Phase separation An inherent problem in the synthesis of inorganic–organic hybrid materials is the incompatibility of many organic moieties with the aqueous-based synthesis of inorganic matrices. This incompatibility may result in macroscopic phase separation during the synthesis if the organic and inorganic components are mixed together. Typically this type of phase separation is not desired and synthetic pathways are developed to circumvent this problem, e.g. by linking the inorganic and organic moiety within one molecule such as in organically-modified trialkoxysilanes as already discussed in Chapter 1.

Nevertheless, the processing of such inorganic–organic hybrid precursors may still result in microphase separation, where two phases are formed on the nanometer scale.

However, in some cases phase separation is deliberately induced, especially in the formation of porous materials, such as M41S materials (see Sections 5.1.2.1; 5.1.2.2). Here, an organized texture is formed via self-assembly of template molecules. This texture is based on microphase separation which divides the reaction space into a hydrophilic and a hydrophobic domain.

5.1.2
Porous (Hybrid) Matrices

With an organic modification of porous solids a wide field of porous hybrid materials can be obtained that, by the combination of inorganic and organic building blocks, benefit from the properties of both parts; an approach which already has been performed on a wide variety of different matrices. The organic groups can be placed selectively on the internal and/or the external pore surfaces or even within the pore walls. The organic modification in principle permits a fine tuning of materials properties, including surface properties such as hydrophilicity/hydrophobicity or potential interaction to guest molecules. In addition, the surface reactivity can be altered and the surface can be protected by organic groups with respect to chemical attack, but also bulk properties, e.g. mechanical or optical properties can be changed. This flexibility in choosing organic, inorganic or even hybrid building blocks allows one to control the materials properties to optimize them for each desired applications.

The selection of porous matrices that are discussed in the following sections of this chapter are chosen somewhat arbitrary, but can be taken as representative examples for the general reaction schemes and the problems associated to the different pathways in the synthesis of hybrid materials.

5.1.2.1 Microporous Materials: Zeolites

Zeolites are microporous crystalline oxides, typically composed of silicon, oxygen and aluminum with cavities that are interconnected by smaller windows. Since their first discovery in the middle of the 18th century, zeolites had been generally regarded as microporous crystalline aluminosilicates having ion-exchangeable cations and reversibly desorbable water molecules (analog to natural zeolites). Today this definition has been extended to quite some extent for several reasons, i.e. in 1978, a purely siliceous zeolitic material, silicalite, was synthesized, which does not have an ion-exchanging ability (it is an aluminum-free material) or in 1982, the first aluminophosphates as microporous crystalline molecular sieves, again with an electrostatically neutral framework, were prepared. The progress made can be related to some extent to the better understanding of the synthesis mechanism, which relies typically on host–guest reactions, with inorganic or organic cations as structure-directing agents. It would be far beyond the scope of this chapter to cover all aspects of zeolite chemistry and their microporous analogs. The reader is referred to the Bibliography at the end of this chapter.

As a definition, zeolites can be described as "open 4-connected 3-D nets which have the general (approximate) composition AB_2, where A is a tetrahedrally connected atom and B is any 2-connected atom, which may or may not be shared between two neighboring A atoms" (Fig. 5.4). For classical zeolites this means that A is either a SiO_4- or AlO_4-tetrahedron and two tetrahedra are linked by a corner-sharing oxygen atom.

In zeolites, the pores are formed as an inherent feature of the crystalline inorganic framework – thus they are also periodically arranged. When discussing pores in zeolites, the reader should be aware of the fact that one has to distinguish between a cage, in which molecules can be accommodated, and the windows to this cage that are typically smaller than the actual cage. Therefore, molecules that fit into the cage are not necessarily able to cross the windows, thus diffusion within the material can be drastically limited. The size of the cage (pore) must be spacious enough to accommodate at least one molecule. For the accommodation of water molecules the pore diameter must exceed 0.25 nm which is the lower limit for the pore size in zeolites. Today a wide variety of zeolitic structures either nat-

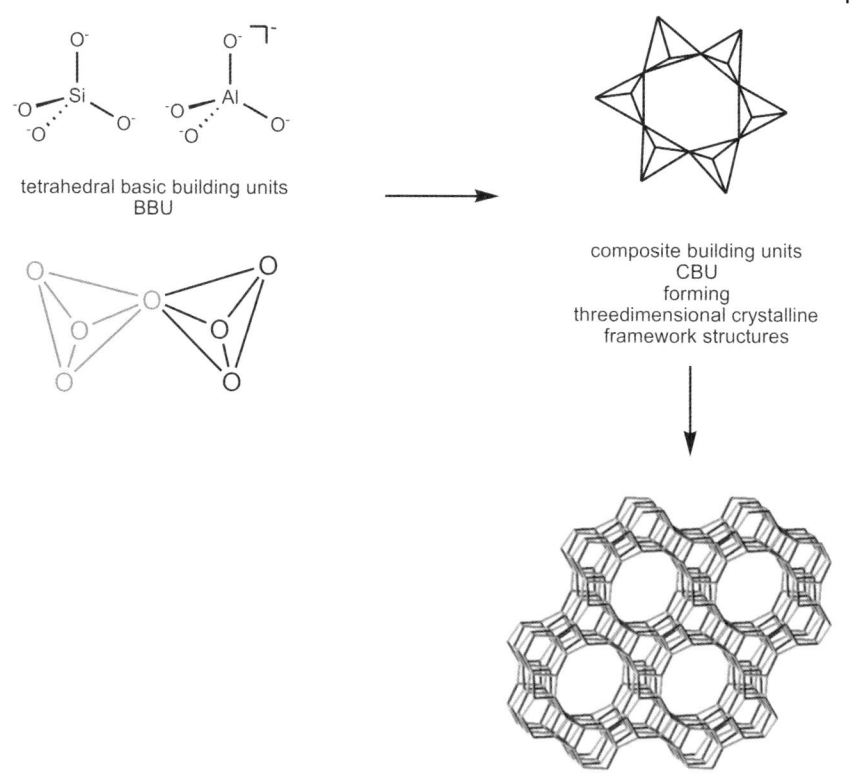

Fig. 5.4 Scheme of the zeolite synthesis forming a three-dimensional network.

ural or synthetic is well known, covering the pore size regime from 0.25 to 1.5 nm. In addition to the different pore sizes, zeolites can be classified as uni-, bi- and tridirectional zeolites, depending on whether the channel system is arranged along one, two, or the three Cartesian axes. This directionality is extremely important with respect to the ability of guest molecules to diffuse within the zeolite matrix.

Zeolites owe their importance not only to the presence of active moieties, e.g. acid centers, in the matrix, but to their general use as catalysts in gas-phase, large-scale petrochemical processes, such as catalytic cracking, Friedel–Crafts alkylation and alkylaromatic isomerization and disproportionation. In addition to their importance in heterogeneous catalysis, it is likely that these solids will also attract interest in the development of functional materials and in nanotechnology, for which zeolites provide an optimal rigid matrix which allows for inclusion of some active components.

Modification of zeolites can be performed by different approaches. A very common way is the substitution of framework atoms by heteroelements, e.g. Co, Mg, B, Ga, Ge, Fe, and many more, to add new properties to the microporous framework. Another possibility relies on the intrinsic ability of zeolites to exchange cations, which is due to the isomorphic substitution of silicon as a tetravalent framework cation by trivalent cations (typically Al) resulting in a net negative charge of the network.

With respect to the synthesis of inorganic–organic hybrid zeolites, five different approaches for the modification should be considered and they will be discussed in the following sections:
- Post-synthetic ion exchange reactions;
- Post-synthesis reactions with the surface groups;
- *In situ* modification during the synthesis via structure-directing agents;
- "Ship-in-the-bottle-synthesis";
- Application of hybrid inorganic–organic precursors.

The classification of zeolites is based on the symmetry of their unit cells, with each structure coded by three capital letters, e.g. MFI, LTA, *BEA, etc. More information on the different structures (especially for the zeolites mentioned in this chapter) can be found in the structural data base of the International Zeolite Association (see Bibliography).

5.1.2.2 Mesoporous Materials: M41S and FSM Materials

Zeolite chemists were always interested in extending the accessible pore sizes to above 1.5 nm. It was only in 1992, when hexagonally ordered mesoporous silicate structures were discovered by Mobil Corporation (M41S materials) and by Kuroda, Inagaki and their co-workers (FSM, folded sheet materials). These materials are mesoporous solids with a periodic and regular arrangement of well-defined pores, with tunable pore sizes between 2 and 50 nm, and (for silica-based materials) amorphous inorganic framework structures. They share characteristics of both gels and zeolites, and are typically characterized by a high specific surface area.

In addition to single molecules such as tetramethylammonium bromide used for the preparation of zeolites, supramolecular assemblies, as found in liquid crystals, can also be used for templating inorganic matrices, which opened a new pathway in host–guest reactions. This supramolecular templating relies on the ability of amphiphilic molecules, such as surfactants, e.g. soap in water, to self-assemble into micellar structures that, when concentrated in aqueous solutions, undergo a second stage of self-organization resulting in lyotropic liquid crystalline mesophases. Molecular inorganic species can cooperatively co-assemble with these structure-directing agents (templates) to eventually condense and form the mesoscopically ordered inorganic backbone of the final material (Fig. 5.5). The mesostructured nanocomposite is typically either calcined, ozonolyzed or solvent extracted to obtain a porous inorganic material.

With this discovery, research in the field of templating and patterning inorganic structures to get perfectly periodic, regularly sized and shaped channels, layers and cavities has expanded dramatically.

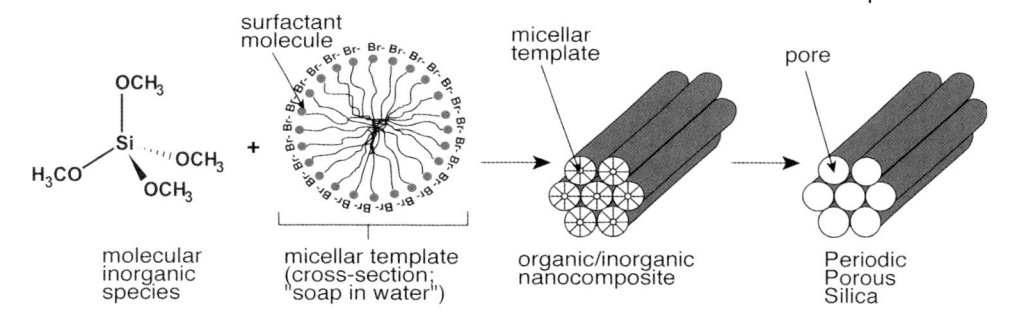

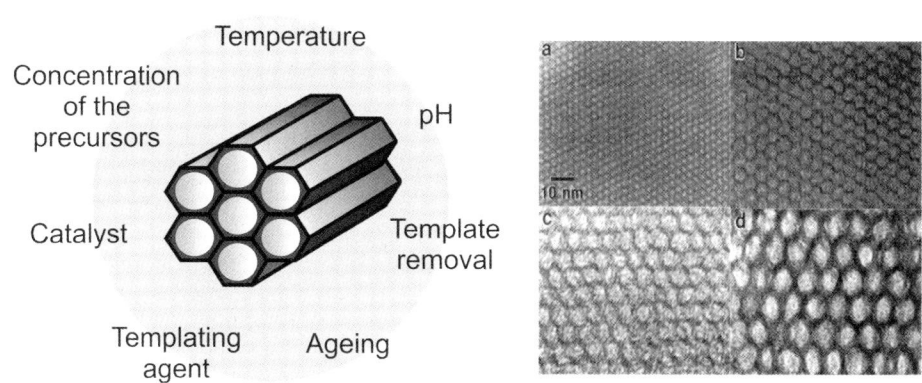

Fig. 5.5 Synthetic cooperative self-organization pathway to M41S materials (top), reactions condition variables influencing the final structure (bottom left) and a transmission electron micrograph showing M41S materials with a hexagonal arrangement of pores of different sizes (bottom right).

The pore dimension in the porous material is particularly related to the chain length of the hydrophobic tail of the template molecule and is typically in the range of 2–30 nm with a rather monomodal distribution of the size. Due to the different liquid-crystalline phases that are accessible – such as cubic, hexagonal and lamellar arrangements – different porous structures with regard to connectivity and dimension are obtained in the final mesoporous material (see Fig. 5.3). Templating of a lamellar lyotropic phase results in a nonporous inorganic matrix due to collapse of the lamellar structure upon removal of the templating agent. Hexagonal phases result in 1-D channel systems with a high aspect ratio – thus, the length of the channels is a manifold longer than the diameter of the pore, and cubic phases give a 3-D pore system with a cubic symmetry (Fig. 5.3).

A number of designations are used when describing M41S or FSM materials. The most relevant materials with respect to this chapter and the formation of inorganic–organic hybrid systems in general are listed in Table 5.1.

Table 5.1 Relevant mesoporous matrices, their abbreviations, pore connectivity and synthesis conditions (HMS = hexagonal molecular sieves, SBA = Santa Barbara Amorphous).

Designation	Pore system	Synthesis conditions
MCM-41	hexagonal	basic conditions, cationic template
MCM-48	cubic	basic conditions, cationic template
FSM-16	hexagonal	from Kanemite (layered clay), cationic surfactant
HMS	poorly ordered	neutral amine templates
SBA-15	hexagonal	acidic conditions, neutral block copolymer template

For both, zeolites and M41S materials, modifications of the inorganic backbone with organic groups are required to provide a certain specific surface chemistry or active sites in the pores or on the inner pore surface. This makes the material more viable for applications in catalysis, sensing or separation technologies, for example.

5.1.2.3 Metal–Organic Frameworks (MOFs)

The 3-D crystalline framework of zeolites (see Section 5.1.2.1), built from corner-sharing TO_4 tetrahedra (T = Si, Al), defines interconnected channels and cages. Constructing any zeolite structure with a molecular construction tool kit requires only two components: tetrahedral Si or Al atoms (the *connectors*) and linear or bent *linkers* (the oxygen atoms).

The construction of coordination polymers and 3-D metal–organic frameworks (MOFs) is based on the same principle, that is assembling connectors and linkers to networks. For clarity, the term coordination polymer is used to describe any extended structure based on metals and organic bridging ligands, whereas the term metal–organic framework is normally used for structures which exhibit porosity (Fig. 5.6).

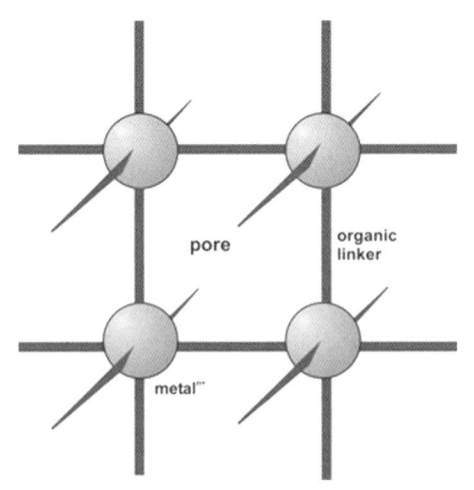

Fig. 5.6 Schematic representation of a MOF.

The basic building units of MOFs and coordination polymers, the connectors and linkers are characterized by the number and orientation of their binding sites.

- The connectors are mostly transition metal or lanthanoid ions or polynuclear clusters with various coordination numbers (up to 10) and coordination geometries.
- The linkers are organic or inorganic bi- or multidentate ligands with various linking directionalities.
- The interaction between connectors and linkers is based on coordinative or ionic interactions.

The greater variety of coordination geometries of metal ions and the possibility to design linker ligands with certain geometrical and chemical properties allows the construction of new and unusual 1-D-, 2-D- or 3-D topologies. In the simplest case a transition-metal ion (connector) is reacted with an organic ligand, which acts as a linear bridge (linker) to form an infinite 1-D-, 2-D- or 3-D framework. If the framework is one-dimensional, linear polymers are formed.

5.2
General Routes Towards Hybrid Materials

By modifying porous materials with organic groups the spectrum of properties is improved without deteriorating the existing positive characteristics. The favorable physical and derived materials properties of porous materials are typically a consequence of the highly porous structure. Therefore, any chemical modification of the materials must retain this structure. In the case of crystalline microporous materials, the crystallinity and stability of the material should remain unchanged, for mesostructured porous materials, the periodicity of the structure must also be retained. For example, unmodified M41S silica-based materials are rather hydrophilic, which is unwanted for many applications. The material can be rendered hydrophobic by the introduction of hydrophobic organic groups, e.g. methyl or phenyl groups (see below).

Chemical modification of porous materials in general, and covalent modification by organic entities in particular, can be achieved at various stages of the preparation process, as described in the following paragraphs:

5.2.1
Post-synthesis Modification of the Final Dried Porous Product by Gaseous, Liquid or Dissolved Organic or Organometallic Species

The post-synthesis modification is a well-studied option for the modification of porous hosts. Here, the porous matrix with the desired pore size, pore connectivity, surface area etc. is prepared prior to the modification step (Fig. 5.7).

The organic species can enter the porous network by simple adsorption of non-reactive (with respect to the pore wall surface) compounds from the gas or liquid

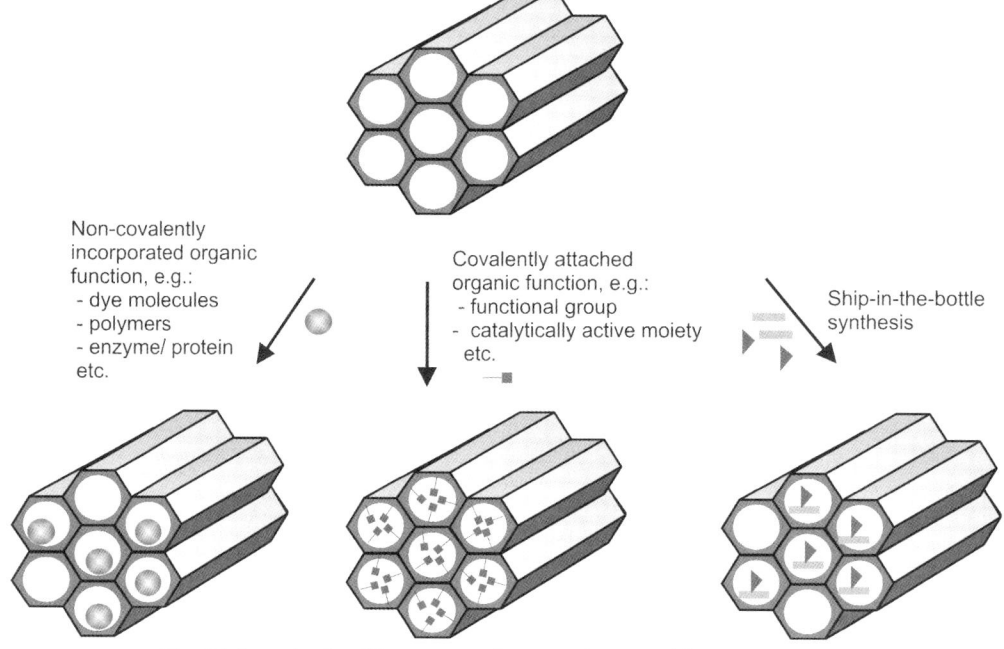

Fig. 5.7 Examples for different types of post-synthesis modification.

phase – here adsorption is based on noncovalent interactions. The advantage of this approach lies in the ease of processing; however loading might be a problem since many species tend to agglomerate at the pore entrance resulting in low loadings. As an additional advantage, the porous network typically retains its structural features.

Incorporation of the desired functionality is also possible by covalent bond formation between organic moiety and pore wall. In this case, reactive organic molecules are added to the preformed solid via the gas or liquid phase. As for the incorporation of organic entities without covalent attachment, the choice of organic molecules is large. Furthermore, the organic functions can be reacted in a second step via traditional organic reaction schemes to new functionalities. In most cases this approach also allows the porous host to retain its structural characteristics. However, depending on the loading, which again is difficult to control, and the size of the organic entity, the pore size is possibly reduced by twice the length of the incorporated species. The difficulties that are encountered are the same as for noncovalent anchoring, such as pore blocking, low loadings etc. A preferential reaction at the pore entrances hinders the diffusion of the reactive molecules into the pore interior, which might result in a very inhomogeneous distribution of the functional moieties and low loadings.

In another approach the organic molecule can be constructed step by step within the confinements of the porous structures – this is also termed *"ship-in-the-*

bottle-synthesis". This allows for inclusion of large compounds and molecules that otherwise could not pass the small pore windows into the zeolite cage.

5.2.2
Liquid-phase Modification in the Wet Nanocomposite Stage or – for Mesostructured Materials and Zeolites – Prior to Removal of the Template

The synthesis of a porous host typically starts with reactions of the network forming precursors to build the inorganic matrix, followed by the removal of the template if applicable, and the final drying step. Post-synthesis modification can be applied not only after drying as discussed in Section 5.2.1, but also in the so-called "wet" state, in which the template/solvent removal or drying still has to be performed.

Because of the special synthesis procedure when applying an organic template molecule, the as-synthesized material comprising the inorganic matrix, water and the templating agent, is already a nanocomposite or inorganic–organic hybrid material.

Sometimes this nanocomposite host matrix can even be modified by a simple ion-exchange process. This applies especially for samples, such as zeolites or M41S type of materials, which contain templating agents that can be exchanged by a new organic moiety as schematically shown in Fig. 5.8.

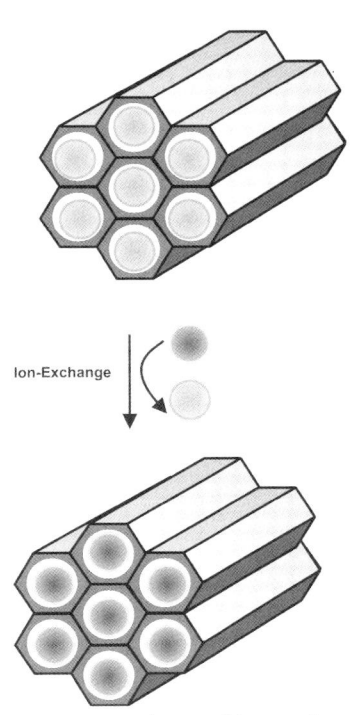

Fig. 5.8 Ion-exchange of the templating agent in MCM-41.

In addition to a simple exchange of the templating molecules to new entities, the template can also be used directly as the functional agent.

5.2.3
Addition of Molecular, but Nonreactive Compounds to the Precursor Solution

Modifications can be performed not only after the host matrix has been formed, but also during the synthesis. The same schemes as presented above apply: either a molecular, but nonreactive compound can be added to the precursor solution, or a reactive compound which interacts with the network-forming species to yield a covalently modified inorganic–organic hybrid material.

When an additional component that does not interact with the network forming precursors is added to a given synthetic mixture, several things must be considered. Here, the inorganic network is formed in the presence of a novel compound with different properties (polarity, charges, etc.). This might result in an unexpected network structure, loss of porosity or even in macroscopic phase separation during the synthesis. Therefore, the reaction conditions and the type of molecule have to be chosen very carefully.

5.2.4
Co-condensation Reactions by the use of Organically-substituted Co-precursors

The co-condensation method is an one-pot synthesis approach, in which e.g. tetraalkoxysilanes [$Si(OR)_4$ (tetraethoxysilane, TEOS or tetramethoxysilane, TMOS)] are condensed to form an inorganic network in the presence of organically-substituted trialkoxysilanes [$R'-Si(OR)_3$, see also Chapter 2 and Chapter 11.4] (Fig. 5.9). The hydrolysis and condensation (sol–gel) chemistry of these precursors and the variability of this approach is discussed in detail in Chapter 6; just for the sake of clarity: a pure *silsesquioxane*, is a network built from $R-SiO_{1.5}$ units only.

With respect to the formation of a porous network, several things must be considered: compared to the post-synthesis modification (see also Sections 5.2.1 and 5.2.2), pore blocking is no problem during co-condensation, since the organic moieties are part of the inorganic network structure. In addition, a better distribution of the organic groups within the matrix is achieved. However, the co-condensation approach also has some disadvantages. First, network formation can be disturbed to a high degree, e.g. for silica-based M41S materials, the degree of periodicity is strongly influenced by the amount of organosilanes – the higher this amount, the lower is the resulting degree of periodicity. Second, high loading with organic groups is in general rather limited – only few examples are known in which the network is built to 100% from an organically-substituted precursor.

Another inherent problem of this approach is the different condensation kinetics of the precursors. Homo-condensation is very often favored over co-

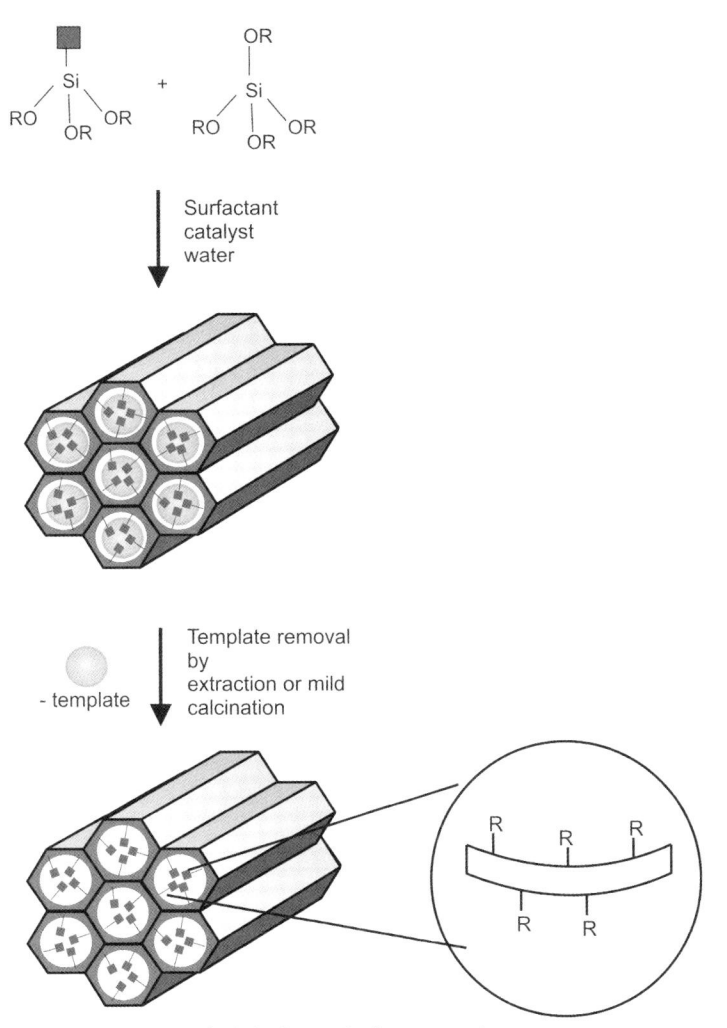

Fig. 5.9 Preparation of a hybrid periodically organized material by co-condensation of a tetra- and organically-substituted trialkoxysilane.

condensation which not only limits the degree of loading, but also influences the reaction time and the distribution of the organic groups in the network.

One more methodological disadvantage, which has to be considered in the synthesis of templated materials such as zeolites or M41S-type of materials, is that the removal of the templating agent must be performed very carefully. High temperature treatments, which are often used, would lead to the simultaneous destruction of the organic function, therefore, often time-consuming extraction processes have to be applied.

5.2.5
The Organic Entity as an Integral Part of the Porous Framework

In addition to materials in which the organic entities are either located in the pores or covalently attached to the pore walls, the organic "function" can also be an integral part of the porous framework itself. Two different strategies will be presented within this chapter, comprising first, the use of bridged bis(trialkoxysilyl) molecules in which the bridge can be composed of a wide variety of organic spacers including alkylene, arylene, but also even functional molecules, i.e. $([(RO)_3Si(CH_2)_3]_2NH)$, and second, microporous metal–organic frameworks based on coordination chemistry (MOF). Figure 5.10 shows schematically the build-up

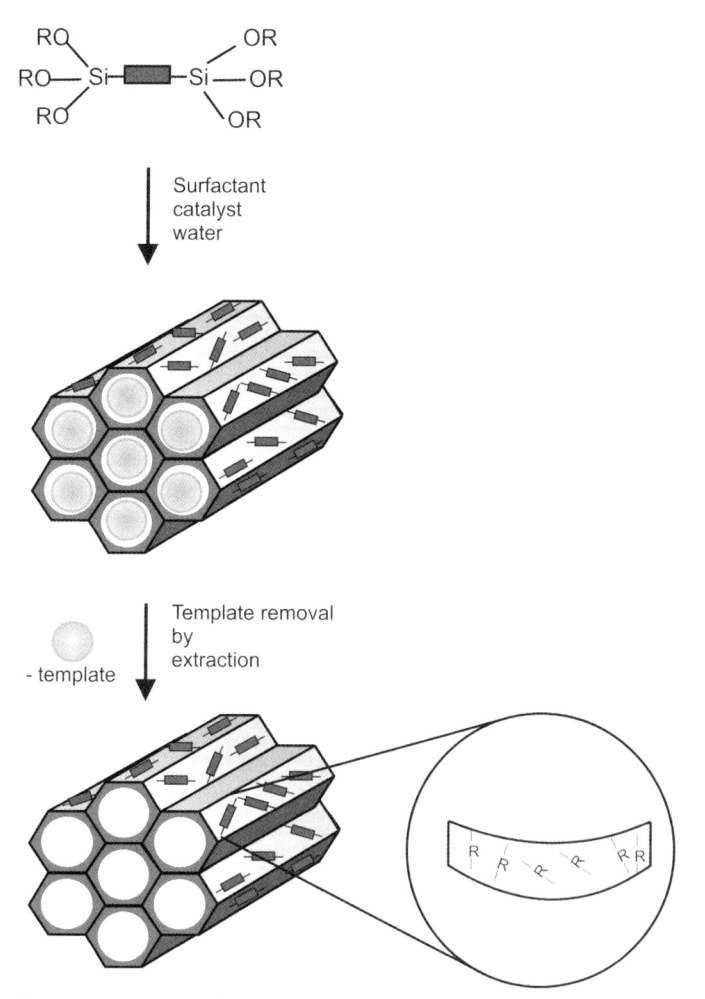

Fig. 5.10 Preparation of mesoporous hybrid frameworks with organic groups as an integral part of the network.

of an inorganic–organic hybrid material with the organic entities incorporated into the pore walls.

In principle, the general strategies for the modification of porous materials, as presented above, are the same for most of the different porous frameworks ranging from micro- to macroporous nets, but different problems are encountered for the various materials. Therefore the applicability of the approaches strongly depends on the desired material.

Examples for all presented approaches, and the relative importance of these methods will be discussed in the following sections. The discussion is focused on zeolites and M41S materials based on silica frameworks as examples for micro- and mesoporous materials. Both classes of material exhibit high surface areas and porosities. Zeolites are characterized by a crystalline framework and rather small but regularly arranged pores, while M41S materials are distinguished by an amorphous silica network with a narrow pore size distribution and a periodic and highly regular arrangement of the pores.

To summarize, the advantages and disadvantages of the different synthetic approaches are presented in Table 5.2.

Table 5.2 Advantages and disadvantages of the various modification procedures.

	Advantages	*Disadvantages*
Postsynthetic treatment	Matrix is preformed and not destroyed upon treatment Uniform surface coverage Good pore size control	Pore blocking Low degree of loading Diffusion limited Adsorption on outer surface
Ion Exchange	Easy processing Pore interior is modified	Limited to comparable substances with respect to charges Not quantitative
Co-condensation	Homogeneous starting mixture Homogeneous distribution in the matrix, when reaction rates are comparable	Polarity differences might induce phase separation (micro- and macroscopic) Limited in the degree of loading due to lower degree of connectivity Can change network formation completely Reaction rates of the precursors can be drastically different
Bifunctional precursors	Organic function is an integral part of the inorganic framework No pore blocking No problems with different reaction rates	Phase separation might occur Sometimes difficult precursor chemistry Cleavage of Si—C bond

The same procedures can be used for many other porous frameworks, such as porous glasses, aerogels, clays, also including macroporous skeletons, e.g. inverse opal structures, and in many cases even non-silica-based frameworks. Zeolites and M41S materials were chosen as examples for this chapter because the constraints given for the different materials with respect to pore sizes, framework build-up (crystalline versus amorphous phases) are representative examples for many kinds of porous matrices.

In addition to the organic modification of inorganic porous host materials based on the choice of modification procedure, another possible way to group the different hybrid materials is based on the way the organic functions are located in the network. Two different classes of hybrid materials can be distinguished:

a) doping with organic molecules, polymers or biological
   entities *without covalent* anchoring to the host materials or
b) *covalent interaction* between the organic function and the
   porous host, including coordination compounds.

5.3
Classification of Porous Hybrid Materials by the Type of Interaction

5.3.1
Incorporation of Organic Functions *Without* Covalent Attachment to the Porous Host

5.3.1.1 Doping with Small Molecules

Microporous solids Small molecules can be incorporated non-covalently bound into zeolitic structures either via ion exchange, via impregnation of the preformed matrix, via ship-in-the-bottle synthesis or as the (functional) template already during the synthesis procedure. When small molecules are incorporated into a zeolitic matrix, it is required that this guest becomes immobilized in the pores (cages). This situation can be regarded as mechanical immobilization, describing that the components of this host (zeolite)–guest (functional molecule) assembly cannot diffuse independently from the other and are held in place by physical forces.

When adding small molecules to a porous matrix it is also helpful to use the "@" notation that is used to indicate that a guest is assumed to be located inside the porous system and not exclusively on the exterior surface.

a) Impregnation
   For zeolitic systems different scenarios should be
   considered for an impregnation or post-synthesis loading of
   the porous matrix. Cationic species can be ion-exchanged
   into charged zeolitic frameworks, whereas neutral species
   can be inserted from the vapor or the liquid phase.

Typically, it is very difficult to introduce anionic species into the cavities of zeolites.

The chemical reaction of an ion-exchange process seems to be especially simple. A cation A^{n+} is exchanged against another cation B^{n+} (Scheme 5.1).

$$[A^{n+}(zeo)^-] + B^{n+} \longrightarrow [B^{n+}(zeo)^-] + A^{n+}$$

Scheme 5.1 Ion-exchange reactions in zeolites.

This simple reaction, however, can pose severe problems in some cases, e.g. upon exchange reaction, the crystallinity of the matrix can decrease, the pH is a critical parameter, especially if metal ions are involved resulting in problems due to precipitation of metal hydroxides, and ion exchange is an equilibrium process, which makes it difficult to get quantitative exchange. Nevertheless, a large variety of metal ions and even larger molecules can be incorporated into zeolitic frameworks by this approach.

b) Functional structure-directing agents

Typical molecular structure-directing agents for lab-based zeolite syntheses are aliphatic amines and tetraalkylammonium ions or, less prominent, oligoethers. These molecules, generally do not possess any specific properties nor do they exhibit specific reactivities e.g. act as a chromophore or have catalytic properties. Nevertheless, it is of high interest to identify functional molecules that can also act as structure-directing agents, because then the functionalities will be homogeneously distributed within the material, the loading will be very high and no post treatment step is required (saves time).

The use of functional templates in the synthesis of zeolite structures is similar to the next subsection, the *in situ* incorporation of organic moieties during the synthesis.

Today research is focused on finding new structure-directing molecules to access novel aluminosilicate crystalline frameworks. One example is the synthesis of the extra-large pore UTD-10 using a Co(II) complex of 1,8-bis(trimethylammonio)-3,6,10,13,16,19-hexaazabicyclo[6.6.6]icosane.

It was shown that in addition metal complexes are able to act as structure-directing agents in the hydrothermal synthesis of zeolite-type compounds (zeotypes, Fig. 5.11). This direct synthesis method, where the metal complexes become occluded by the crystallizing framework, leads to

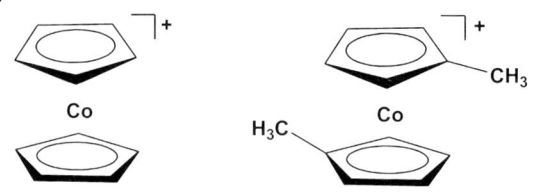

Fig. 5.11 Structure-directing agents based on metal complexes, here cobalticinium cations.

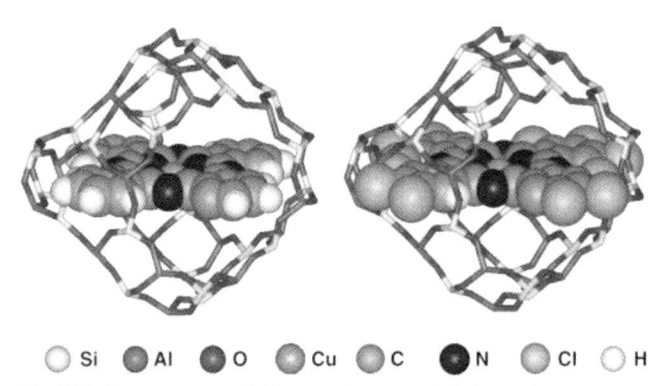

Fig. 5.12 Computer graphic images of copper phthalocyanine (left) and copper chlorophthalocyanine (right) encapsulated with the supercages of zeolite Y. (Taken from Thomas, Raja, J. Organomet. Chem. **2004**, *689*, 4110–4124).

stoichiometric and homogeneous compounds with an optimal loading of the metal complex, quite in contrast to the products of other post-synthesis modifications (ion-exchange, vapor-phase insertion, ship-in-the-bottle synthesis).

c) *In situ* encapsulation during the zeolite synthesis

Many molecules do not act as structure-directing agents, but can nevertheless be added to the synthesis mixture, thus are incorporated into the final zeolite structures. This approach could also be termed "build-the-bottle-around-the-ship". Two limitations must be considered here: the guest has to survive the relatively harsh synthesis conditions of zeolite formation in terms of pH and temperature for long periods, and the zeolite crystal structure should still be formed in the presence of the guest. For methylene blue and metal phthalocyanines it was shown that they can be introduced into the pore system of zeolites with structural and compositional integrity. Metal phthalocyanine complexes (MPc), e.g. CuPc, $CuCl_{14}Pc$, or FePc, have been encapsulated *in situ* via the zeolite synthesis approach (Fig. 5.12). Due to their large size, metal phthalocyanine complexes cannot be incorporated into zeolites by simple ion exchange.

The advantage of this co-inclusion method is the greater variability of the type of molecule that can be incorporated. However, a general disadvantage is the rather unspecific mode of incorporation and that inhomogeneous and nonstoichiometric materials are formed.

d) Ship-in-the-bottle-synthesis

If a molecule is too large to pass through the windows into a zeolitic cage, it can only be incorporated either via an *in situ* synthesis approach or via the piece by piece assembly after the formation of the host matrix just as it is done in a typical ship-in-the-bottle puzzle.

One archetypical reaction of a ship-in-the-bottle reaction is the synthesis of metallic phthalocyanines inside faujasite X by treating o-phthalodinitrile with a transition-metal exchanged faujasite at temperatures above 200 °C (Scheme 5.2). This synthesis allows species to be incorporated into the cavities of zeolites which would otherwise not be able to diffuse through the smaller windows.

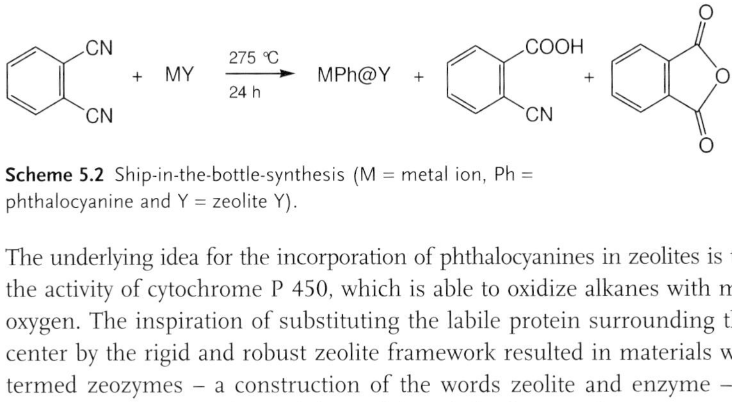

Scheme 5.2 Ship-in-the-bottle-synthesis (M = metal ion, Ph = phthalocyanine and Y = zeolite Y).

The underlying idea for the incorporation of phthalocyanines in zeolites is to mimic the activity of cytochrome P 450, which is able to oxidize alkanes with molecular oxygen. The inspiration of substituting the labile protein surrounding the metal center by the rigid and robust zeolite framework resulted in materials which are termed zeozymes – a construction of the words zeolite and enzyme –, already indicating the potential possibilites of these hybrid materials.

The ship-in-the-bottle-synthesis has been extended to the preparation of a large variety of molecules for various applications. Metal carbonyl clusters such as $Pd_{13}(CO)_x$ or $Rh_6(CO)_{16}$ or the heteropolyacid $H_3PW_{12}O_{40}$ have been successfully synthesized in zeolitic pores for catalytic applications. Purely organic molecules such as the 2,4,6-triphenylpyrylium ion have been synthesized in zeolitic channel systems for photocatalysis and many others for applications as ion photochromic systems, as sensors or even as molecular switches.

Mesoporous solids The same processes as just discussed above for microporous solids also apply for mesoporous matrices. However, leaching is one of the main problems, especially for these larger pore systems.

5.3.1.2 Doping with Polymeric Species

The combination of porous oxide materials as hosts and organic polymers as guests offers interesting opportunities for new types of hybrid materials, especially because the organic phase can be highly constrained. Similar materials are well known from the intercalation of polymer into layered structured, e.g. clays – as discussed in previous Chapter 1 and Chapter II.3.

One obvious application for organic polymers introduced into the pores of an inorganic matrix is the improvement of the mechanical stability, another, which is especially important for periodic micro- and mesoporous materials, is the option of forming aligned, mostly linear polymers which can be conducting, e.g. as molecular wire, or have interesting optical properties due to the space confinement. Zeolites and M41S materials are of special interest, because of their very well-defined channel or cavity systems.

The diffusion of polymers into preformed highly porous systems is very difficult, probably because of the associated loss of entropy of the polymer chain and steric and diffusional constraints. However, there are several other options how polymers can be incorporated into the porous host (Fig. 5.13).

1. The porous network can be formed around preformed polymers, that is, hydrolysis and condensation reactions can be performed in solutions of organic polymers.
2. Organic monomers can be added to the precursor mixture for the preparation of the inorganic host. Subsequent polymerization can either be performed simultaneously with the formation of the inorganic framework or after the inorganic network has been formed. A variation of the latter approach is the use of precursors with polymerizable organic groups such as 3-(methacryloxypropyl)trimethoxysilane. The silane part of these bifunctional molecules becomes part of the network structure during the hydrolytic polycondensation reaction and the unsaturated groups can be polymerized with the organic co-monomers. This results in a covalent tethering of the polymer to the inorganic framework.
3. The preformed porous material can be doped with organic monomers, which can then be polymerized inside the pore systems. This option is often used for M41S-type materials.
4. When structure-directing agents are involved in the synthesis another option exists: On account of the templating mechanism in the synthesis of the materials and the induced microphase separation, compartments of different polarity are formed, i.e. the hydrophobic core formed by a supramolecular arrangement of amphiphilic molecules as used in the synthesis of M41S type of materials. It is possible to use polymerizable template molecules which can be polymerized after self-organization

5.3 Classification of Porous Hybrid Materials by the Type of Interaction | 197

into supramolecular arrangements, thus the template is acting as the monomer.

Figure 5.13 displays options 1 to 3 in addition to the simple threading approach. One general problem for host–guest polymerizations is that the inner surface chemistry of the host compounds must interfere as little as possible, if at all, with the mechanism of the polymerization reactions.

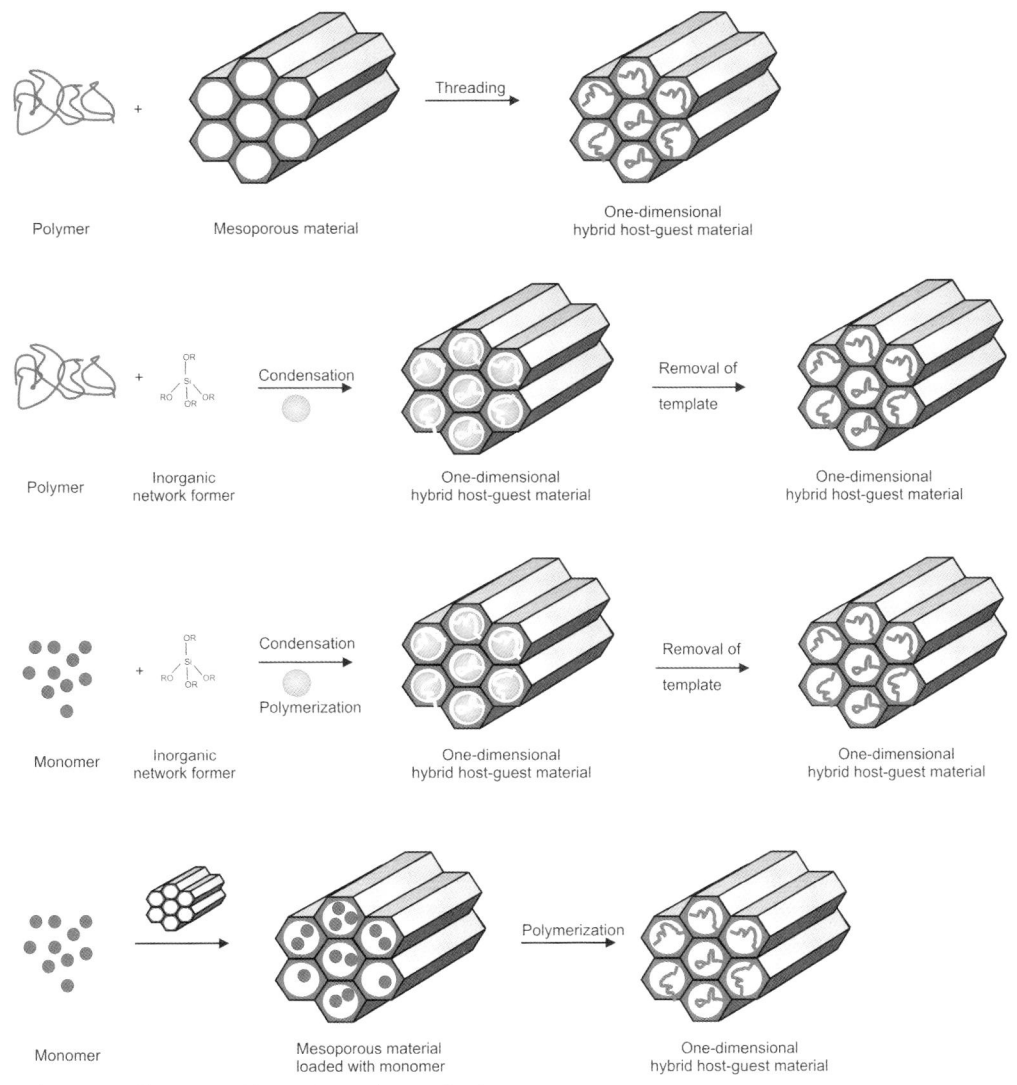

Fig. 5.13 Different possible reactions schemes for the incorporation of polymers into porous hosts.

Microporous solids Simple mixing of a polymer and a zeolite will result in a negligible penetration of the polymers into the zeolite pores (typical pore diameters are in the range of 3 to 13 Å), but is useful for an improvement of the mechanical properties of the polymer, thus as zeolitic filler reinforcement.

There are quite early reports (1978) on the polymerization of acetylene by zeolitic species, however, the authors were referring to polymer formation *on* the zeolite, not *in* the pores. From today's point of view, it seems almost certain that the polymer was formed inside of the zeolite (zeolite Y in this case). It is possible to polymerize monomers in the small cavities of zeolite crystals. Both, poly(styrene) and poly(ethyl acrylate) can be incorporated into zeolitic structures by adding the organic monomer to a preformed zeolite, followed by thermally induced polymerization reactions.

Conducting polymers, such as polypyrrole, polythiophene, polyaniline (Scheme 5.3), are interesting species for encapsulation in the cavity structure of zeolites, and it was clearly shown that *intra*-zeolite polymer formation took place. How could that be proven? The zeolites had been exchanged with Cu(II) and Fe(II) because the polymerization mechanism is believed to involve redox reactions at these metal centers. On copper exchanged zeolites with pore diameters (4 Å) definitely too small to accommodate pyrrole, the polymer was not formed.

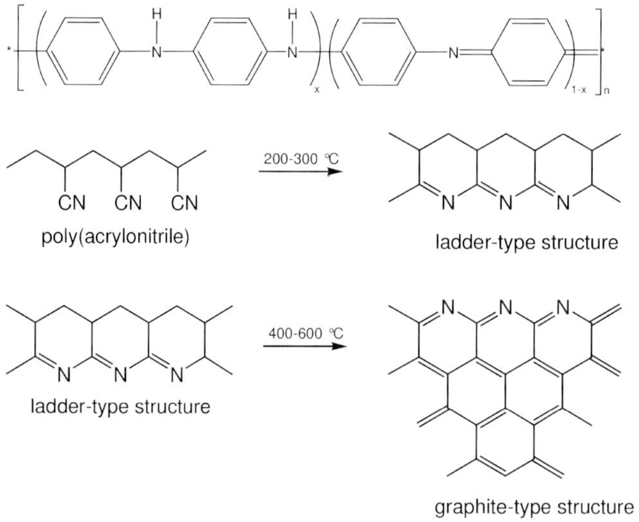

Scheme 5.3 Structure of polyaniline exhibiting the reduced and the oxidized form (top) and schematics of the formation of linear, graphite-like structure by pyrolysis of poly(acrylonitrile).

Mesoporous solids The feasibility of forming polymer networks within the pore channels of preformed M41S materials by polymerizing adsorbed monomers has been demonstrated by addition of styrene, vinyl acetate, and methyl methacrylate monomers into dehydrated and evacuated M41S hosts.

By this approach, nanocomposites that are not porous anymore with high polymer content can be synthesized. For polymerizations carried out in the restricted geometry of a mesoporous channel, confinement effects are observed in the final polymer. Only to mention two different examples: for poly(methylmethacrylate) (PMMA) an increase in chain length with decreasing pore size of the MCM-41 material was observed, probably due to less facile termination processes, and for polystyrene the glass transition temperature was affected.

Having a well-defined channel system, especially in MCM-41 with its straight hexagonally oriented channels, the incorporation of electrically conducting polymers is of special interest with regard to designing nanometer-scale electronic devices. The conducting polymer polyaniline was successfully stabilized as filaments in the channels of MCM-41 by adsorption of aniline vapor into the dehydrated host, followed by reaction with $(NH_4)_2S_2O_8$. In addition to polyaniline, other conducting species such as graphitic carbon wires generated from organic polymers were incorporated in MCM type materials. In this approach, the monomer acrylonitrile was introduced into the host through vapor or solution transfer, and polymerized in the channels. Pyrolysis of the long chains of poly(acrylonitrile) thus formed led to the formation of carbonized material in the channels of the host.

In another example, crystalline nanofibers of linear polyethylene with an ultrahigh molecular weight and a defined diameter of 30–50 nm were grown catalytically inside the channels of a MCM material with active transition metal complexes grafted to the silica walls. This concept was extended to a variety of metal complexes. An example in which the structure-directing agent forms the polymeric material inside the mesoporous channel is the use of an oligoethylene glycol functionalized diacetylenic surfactant as structure directing agent. It also acts as monomeric precursor for the conjugated polymer polydiacetylene because of its polymerizable groups.

5.3.1.3 Incorporation of Biomolecules

Why is the incorporation of biomolecules into porous matrices of interest? First of all, and probably, the most important factor is that it has been shown that the entrapment of biofunctional moieties into porous inorganic matrices allows for a stabilization of enzymes, proteins and cells even under severe conditions compared to the free biological entity in solution. In addition, for many applications immobilized enzymes on a solid support are necessary, e.g. as biosensors for the detection of chemicals and organisms within the environment and *in vivo* medical monitoring, or in biocatalysis where it is often advantageous to use immobilized enzymes because they are easier to separate from the reaction products.

Porous inorganic matrices, especially sol–gel derived silica-based materials, hold promise as biocompatible scaffolds for the immobilization of enzymes, proteins, cells etc. due to their biocompatibility, porosity, chemical inertness and mechanical stability. Here, especially mesoporous materials are of interest, since they fulfill many of the requirements for enzyme, protein or cell carriers such as sufficiently large pores, large surface areas, hydrophilic character, water insolubility, chemical and thermal stability, mechanical strength, suitable shape, regenerability, and toxicological safety. Inclusion of enzymes in the pores of microporous structures (i.e. zeolites) is a more or less impossible task since the pore size of these materials is too small (<20 Å).

Most importantly, the immobilization technique should allow the biomolecule to maintain its catalytic activity while diminishing other processes detrimental to the enzyme such as autolysis. One has to keep in mind that the interactions of these molecules with the inorganic matrix are typically not by covalent bonding but by hydrogen bonding or electrostatic interactions.

Different approaches can be taken for the immobilization of bioactive species within a porous network: the biomolecules can be incorporated either a) by a post-synthesis treatment or b) *in situ* during sol–gel processing.

Different problems arise with the two different approaches: When the entrapment process is performed via a post-synthesis treatment with a preformed inorganic matrix, it is unlikely that the proteins are evenly distributed in a given pore – they are more likely clustered at the pore entrance. This affects the amount of protein that can possibly be adsorbed significantly. If the entrapment of proteins is performed *in situ* during the synthesis, the synthesis conditions must be very carefully chosen, e.g. many biomolecules are very sensitive towards the presence of alcohol and in addition, severe pH conditions must be avoided. Therefore, the processing parameters must be adjusted in a way that enzymes and other bioactive species tolerate the gel synthesis conditions.

Many enzymes or proteins have a significant size, and to fit into a porous inorganic matrix, the pores must be sufficiently large. Conformational changes of the biomolecule due to size constraints inside the pore might lead to denaturation or loss of activity. The most widely used biological moiety in the studies of the incorporation of proteins into silica matrices is cytochrome *c*. Cytochrome *c* consists of a single polypeptide chain of 104 amino acid residues that are covalently attached to a porphyrin-based heme group. It has a molecular weight of 12 400 Dalton in solution and molecular dimensions of $2.6 \times 3.2 \times 3.3 \, nm^3$, which is already in the lower size range for proteins. Therefore, most studies concerning the entrapment of the proteins, enzymes and cells are performed with meso- or macroporous materials.

Besides cytochrome *c*, other examples are smaller biomolecules such as chlorophyll A, vitamin E or vitamin B12, proteins such as horse-radish peroxidase (HRP), lysozyme, penicillin acylase, trypsin, myoglobin, or even whole microbial cells such as *Arthobacter sp.*, *Bacillus subtilis*, and *Micrococcus luteus*.

Since the binding forces of the biological entity to the inorganic matrix are typically rather weak, leaching is a severe problem. One approach to prevent leaching can be silylation of the pore opening with organoalkoxysilanes. For many applications, e.g. biocatalysis, a stronger interaction is preferred. Here, crosslinking or covalent anchoring of the biological entity to the inorganic matrix can be performed via functional groups such as amino, carboxyl, hydroxyl, and sulfhydryl moieties (see also Section 5.3.2).

Biomolecules can be incorporated into mesostructured materials not only by simple impregnation techniques, but also as the template in the synthesis of periodic mesoporous silica. For example, vitamin E-TPGS (α-tocopheryl polyethylene glycol 1000 succinate) was found to be an efficient template for the synthesis of hexagonally mesostructured silica (DAM-1, Dallas amorphous material-1).

5.3.2
Incorporation of Organic Functions *with* Covalent Attachment to the Porous Host

5.3.2.1 Grafting Reactions

Grafting refers to post-synthesis modification of a pre-fabricated porous host by attachment of functional/ organic moieties to the inner surface. Grafting reactions can be performed before or after drying of the porous material. If template or structure-directing agents are employed in the synthesis of the material, grafting is typically performed after removal of the template, however, some examples show that grafting can also be used to catch two birds with one stone: template removal and surface functionalization e.g. for zeolites and M41S materials.

Microporous solids In the case of microporous zeolites, grafting of organosilanes that contain organic functional groups is not straightforward because a large fraction of the grafted functional groups become attached to the exterior surface of the zeolite crystals instead. Nevertheless, in principle the same reactions as discussed in the next section for mesoporous solids apply to zeolite materials. One has to keep in mind that the number of reactive groups at the interior surface of crystalline zeolitic matrices is not as high as for amorphous materials and steric constraints may influence grafting reactions.

Mesoporous solids In contrast to zeolitic materials, the number of surface silanol groups on the pore surface of M41S materials is very high just like in other amorphous silica-based materials. These silanol groups present ideal anchor groups. Besides esterification reactions, e.g. with ethanol, one very common reaction to modify the material with organic groups is a simple surface silylation reaction, e.g. to render the material hydrophobic (Scheme 5.4 and Fig. 5.14).

$$\equiv Si-OH + Cl-SiR_3 \xrightarrow{\text{base, 25°C}} \equiv Si-OSiR_3 + HCl / \text{base}$$

$$\equiv Si-OH + R'O-SiR_3 \xrightarrow{100°C} \equiv Si-OSiR_3 + HOR'$$

$$2\equiv Si-OH + HN-(SiR_3)_2 \xrightarrow{25°C} 2\equiv Si-OSiR_3 + NH_3$$

Scheme 5.4 Possible surface silylation reactions with chlorosilanes, alkoxysilanes and silazanes.

The silylation reactions shown in Scheme 5.4 occur on free ($\equiv$Si—OH) and geminal (= $Si(OH)_2$) groups. In general, the original structure of the porous host is retained upon these grafting reactions.

For an effective grafting via silylation reactions, it is desired to have a large number of surface silanols in the material. If the surfactant has been removed prior to

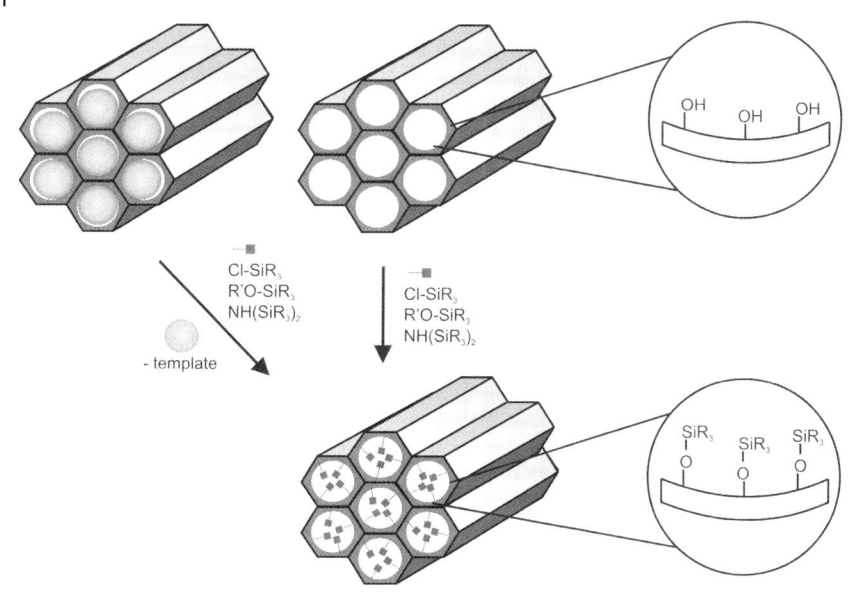

Fig. 5.14 Functionalization of M41S-materials by grafting reactions.

the grafting reaction by calcination, many of the silanol groups are lost due to condensation. In this case, the surface can be rehydrated by treatment in boiling water, or by acid hydrolysis or steam treatment. Removal of the surfactant by extraction processes minimizes the loss of silanols. However, without a heat treatment, many silanols show a lower reactivity due to intersilanol–hydrogen bonding.

Effective silylation reactions can also be performed on the composite material, which still contains the amphiphilic template molecule. Here, change in surface polarity – from a very hydrophilic medium to a hydrophobic pore interior after treatment with e.g. trimethylchlorosilane – results in an almost complete extraction of the template. Not only methyl groups, but a large variety of alkyl or aryl moieties can be incorporated into a M41S matrix via this approach and even silane-based coupling agents with functional organic groups can be used, e.g. carrying olefins, nitriles, alkylthiols, alkyl amines, alkyl halides, epoxides, etc. Incorporating functional groups into the porous host allows for a variety of further reactions, therefore, giving access to an almost unlimited choice of functionalities. Figure 5.15 shows a TEM image of a methacrylate-modified mesostructured material, prepared by post-synthetic grafting of a methacrylate-modified monoalkoxy-dimethylsilane to a preformed, as-synthesized (still containing the surfactant) silica matrix. The material carries 4.7 mmol functional methacrylate groups per gram.

Nitriles can be hydrolyzed to carboxylic acids, olefins (vinyl groups) can be modified by bromination or hydroboration, methycrylates can be used for polymerization reactions within the pores and even inorganic coordination chemistry is

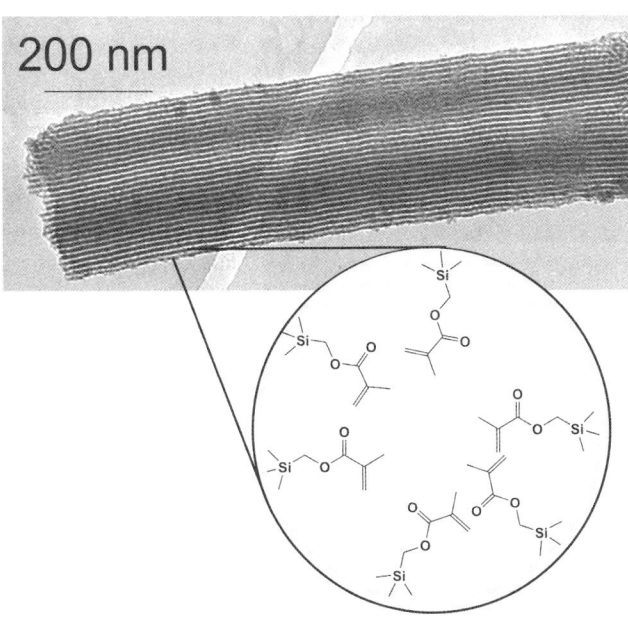

Fig. 5.15 Transmission electron micrograph of a hexagonally arranged (side view) region of a methacrylate-modified M41S material.

possible by grafting of ethylenediamine groups to the surface, to mention only a few of the numerous options.

When discussing porous materials one has to keep in mind that two different surfaces exist: the internal surface within the pore channels (concave) and the external surface on the outer part of the material (convex). In grafting reactions the two surfaces exhibit different reactivities. The external surface is more easily accessible and is functionalized predominantly. This can lead to pore blocking, resulting in an inaccessible inner surface of the material. Nevertheless, it has been shown that both surfaces can be modified in different ways. This controlled dual functionalization was shown by passivation of the external surface of a calcined MCM-41 material with Ph_2SiCl_2 and subsequent functionalization of the internal surface with an aminopropylsilane. This dual functionalization is also possible before removal of the surfactant (Fig. 5.16).

Besides these successes one has to be cautious about the selectivity of these approaches. Just remember that it is possible to modify all surfaces of the material with trimethylchlorosilane even prior to removal of the surfactant.

5.3.2.2 Co-condensation Reactions

For silica-based materials, many precursors are available that allow for an organic modification of the porous host via co-condensation reactions, e.g. as the most classical example co-condensation between a tetraalkoxysilane and an organically-

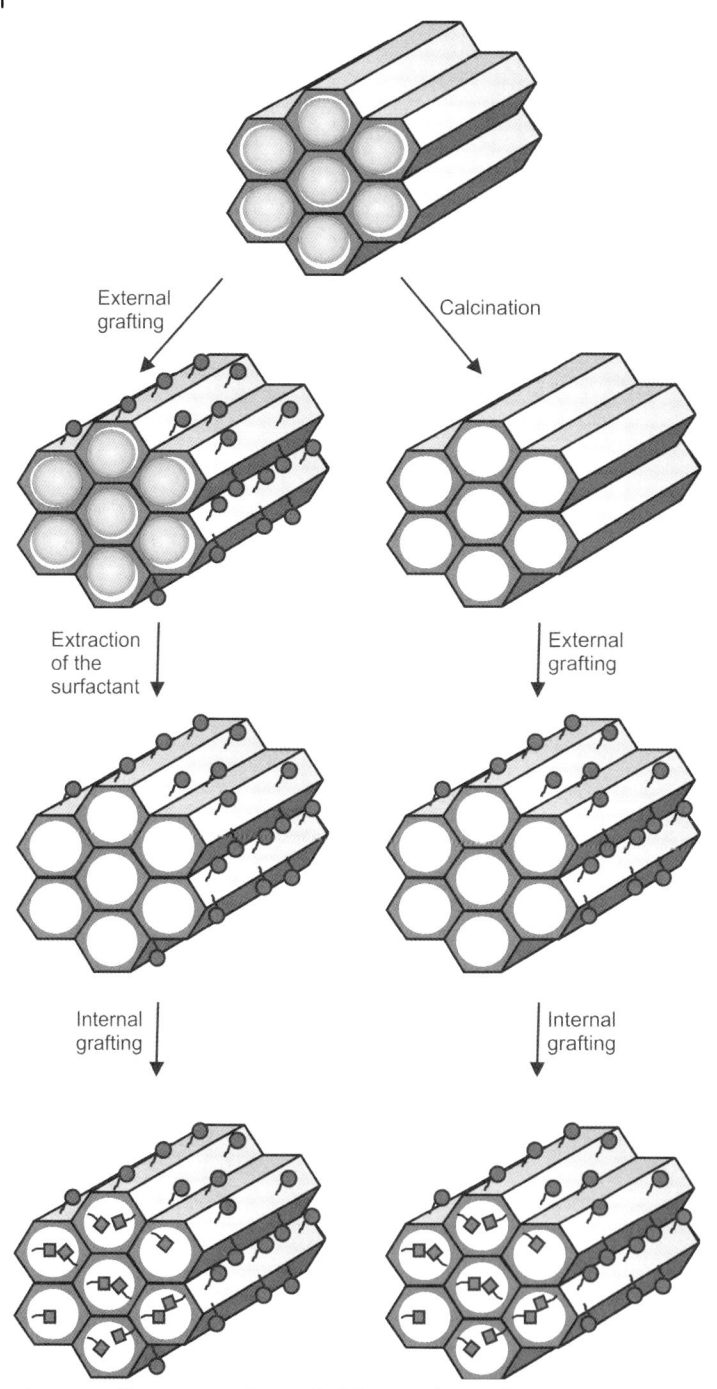

Fig. 5.16 Different approaches to dual functionalization by selective grafting of the internal and external surfaces in the material.

modified trialkoxysilane. Chapter 6 gives some selected examples of possible precursor molecules – the choices of which ranging from aliphatic, aromatic to even functional moieties. The Si—C(sp^3/sp^2)-linkage has proven to be stable under sol–gel processing conditions.

Different problems are encountered when this approach is applied to crystalline matrices such as zeolites or amorphous materials such as M41S phases.

Microporous solids   In principle, it is possible to synthesize zeolites with pendant organic functionalities. However, attempts at modifying zeolites by co-condensation with organically-modified precursors usually create structural defects (loss of crystallinity) or block the internal channels with large pendant groups. Further problems encountered include:

1. *Structural impacts*: The addition of small amounts of organic species to a zeolite synthesis mixture can have an extreme influence on the synthesis. Not only are mixtures of crystalline phases obtained, but in some cases crystallization does not occur at all.
2. *Phase separation*: In addition, phase separation of the organic phase from the aqueous zeolite synthesis gel is commonly observed.
3. *Porosity*: Not only structural defects, but also loss of porosity can be a problem. The organic groups are located in the micropores, thus spoiling the microporosity.
4. *Loading*: Only a very low amount of organic groups can be incorporated by co-condensation reactions with organically-modified silanes of about 1–3% of the Si atoms.
5. *Template removal*: How can one remove the template (typically calcination) without deteriorating the organic functional moieties on the inner surface?

These problems limit the scope of synthesizing molecular sieves with organic functionalities drastically. For zeolitic structures a major key to success in synthesizing organically-functionalized molecular sieves is the identification of a system that can be prepared in the absence of an organic structure-directing agent (SDA) such as zeolite NaY, from which the structure-directing agent can be removed by extraction.

By carefully choosing the system, some organically-modified zeolites have been prepared. For example, phenethyltrimethoxysilane has been used as modification agent for the preparation of zeolites with *BEA structure and intracrystalline phenethyl groups. Even polar groups have been successfully incorporated into zeolitic structures, e.g. aminopropyl-, mercaptopropyl-, 2-(4-chlorosulfonylphenyl)ethyl-, 3-bromopropyl-moieties, etc.

In addition, so-called follow-up reactions to convert functional groups have been successfully performed; see Fig. 5.17 for the sulfonation reaction of the phenethyl-moiety.

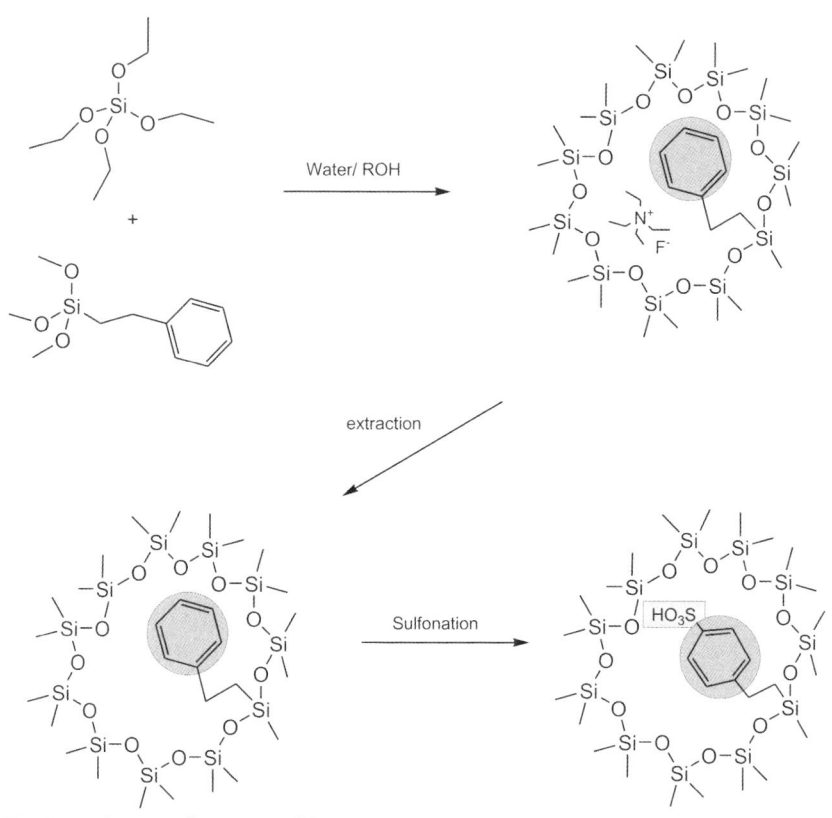

Fig. 5.17 Schematic illustration of the preparation procedure used to create a sulfonic acid-functionalized molecular sieve.

Mesoporous solids Similar co-condensation reactions – also called one-pot syntheses – have been applied to surfactant-templated systems. Hybrid mesoporous silicates have been prepared under a wide range of reaction conditions.

Again, related problems as already pointed out for the zeolitic systems must be considered. Especially, the avoidance of phase separation to achieve a uniform distribution of functional groups within the matrix and prevention of Si—C bond cleavage upon template removal are of major importance. Many examples show that co-condensation is one of the most successful pathways to mesoporous hybrid materials. From a mechanistic point of view, different routes can be distinguished. Following the original MCM-41 synthesis, co-condensation can be carried out with a cationic surfactant such as alkylammonium surfactants and anionic silica precursors, which are obtained under basic conditions. However, also other reaction pathways, e.g. with a neutral template such as a polyethylene oxide-based amphiphil under acidic conditions are possible. Not only hexagonal structures but also cubic structures have been accessed.

5.3 Classification of Porous Hybrid Materials by the Type of Interaction

The surfactant can be removed by either mild calcination at 350 °C (possible for example for temperature-stable phenyl groups) or by extraction processes with acidic ethanol.

The amount of organic groups that can be incorporated into a periodically arranged mesoporous material by a co-condensation approach is limited to about 20–40 mol% with regard to the tetrafunctional silane. The degree of mesoscopic ordering decreases continuously with increasing content of organoalkoxysilane. This is not only an effect of the change in polarity in the system, e.g. for hydrophobic phenyl groups incorporated into a hydrophilic matrix, but also due to the lower crosslinking density of the final network. The crosslinking density is defined as the possible points of crosslinking from one precursor: tetraalkoxysilanes are network forming agents having four branching possibilities, while trialkoxysilanes possess only a connectivity of three, typically resulting in a mechanically more unstable network.

Another problem that cannot be neglected for co-condensation reactions is the different reactivity of the precursors with respect to hydrolysis and condensation rates. It is well known for example, that compared to tetramethoxysilane, methyltrimethoxysilane reacts faster in an acidic medium and slower in basic environments. This can lead to an inhomogeneous distribution of the organic groups within the matrix – the extreme cases displayed in Fig. 5.18. In addition, this different reactivity might cause a lower concentration of organic groups in the

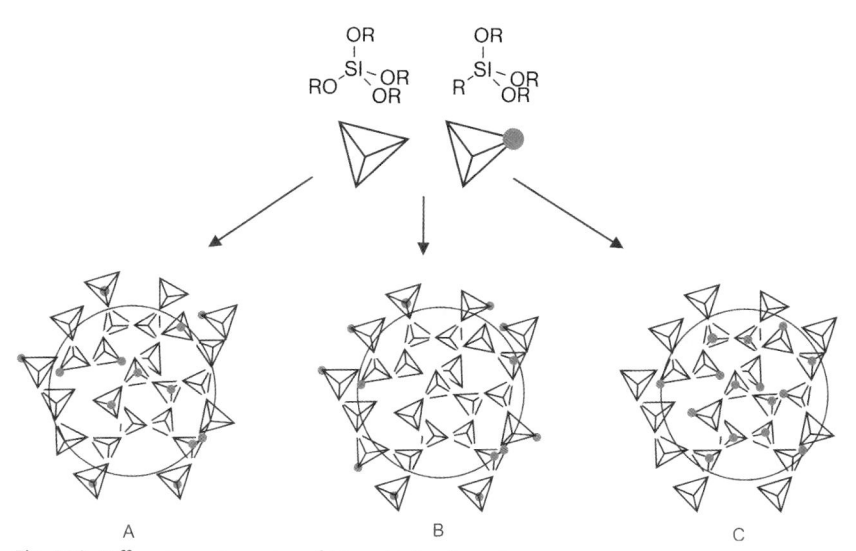

Fig. 5.18 Different reaction rates of tri- and tetraalkoxysilanes during co-condensation can lead to an inhomogeneous distribution within the framework. A) the reaction rates of both precursors are comparable; B) the tetraalkoxysilane reacts faster than the trialkoxysilane; C) the trialkoxysilane reacts faster, thus forming the particle core.

5 Porous Hybrid Materials

final material than in the initial solution. Thus, the reaction conditions have to be chosen carefully.

Multifunctional surfaces can also be prepared by incorporation of two or more functional groups in a one-pot synthesis. The same problems apply as just discussed, e.g. the location of the organic groups is not as controlled as by the grafting procedure – the groups are randomly distributed in the matrix.

This co-condensation approach can be extended to the preparation of a polymer–silica mesoporous nanocomposite as already briefly discussed in Section 5.3.1.2. Acid-catalyzed sol–gel reactions of TEOS with poly(stryrene-co-styrylethyltrimethoxysilane) have been performed in the presence of dibenzoyl-tartaric acid (DBTA) as a non-surfactant pore-forming agent. After removal of DBTA by extraction, a mesoporous material with a polymer covalently bound to the silica walls is obtained. Inorganic–polymer hybrid mesostructured materials with a polymer covalently attached to the inorganic framework can be prepared by co-condensation of TMOS with 3-methacryloxypropyltrimethoxysilane in the presence of a template. The template molecule has been extracted resulting in an ordered MCM hybrid material with covalently bonded methacrylate units. In a second step, methylmethacrylate was adsorbed into the material and polymerized resulting in a covalently coupled inorganic–organic composite material. A similar material was prepared in a one pot synthesis, using a complex mixture of a hydrolyzable and condensable inorganic precursor, water, hydrochloric acid, ethanol, surfactant, 3-methacryloxypropyl-trimethoxysilane as coupling agent, dodecylmethacrylate as monomer and hexanedioldimethacrylate as crosslinker. A layered polymer–silica nanocomposite, again with covalently bonded methacrylate units, was formed by a simple dip coating procedure relying on self-assembly driven by solvent evaporation (Fig. 5.19). The organic polymer was formed by UV curing in a second step. This example clearly shows that organic molecules can be placed in a very controlled fashion in mesostructured inorganic materials.

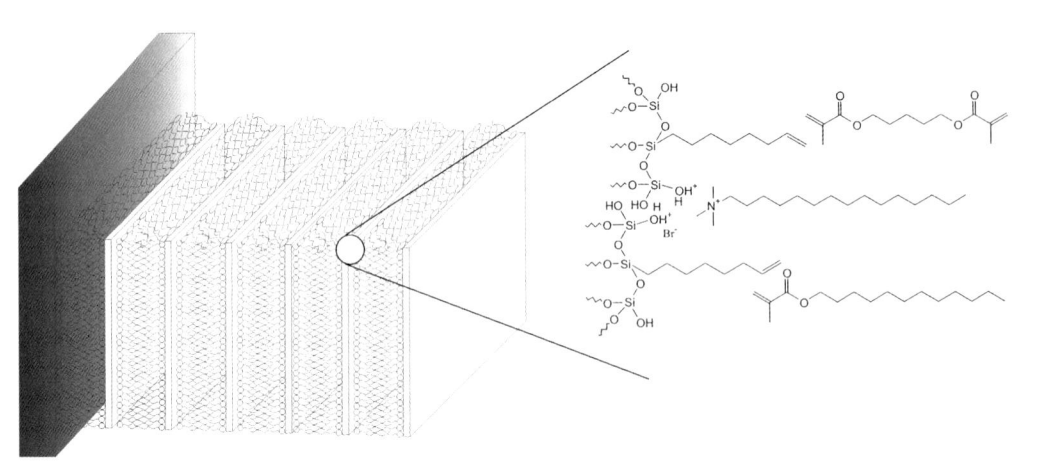

Fig. 5.19 Assembly mixture for a poly(dodecylmethacrylate) / silica nanocomposite.

5.3.3
The Organic Function as an Integral Part of the Porous Network Structure

5.3.3.1 ZOL and PMO: Zeolites with Organic Groups as Lattice and Periodically Mesostructured Organosilicas

Incorporation of organic groups as an integral part of the network results in materials that have a stoichiometric amount of organic groups in the inorganic matrix, thus resulting in higher loadings of the organic (functional) groups and more homogeneous distribution than by grafting methods or by direct co-condensation synthesis. In addition, the crosslinking density is the same as for the condensation of tetraalkoxysilanes due to the organic bridge. Because of the large variety of bridging organic functions that is available, tuning of the mechanical, surface chemical, electronic, optical and even magnetic properties of the hybrid composites by introducing suitable functional units into the walls can be envisioned.

Microporous materials As discussed above, modifying zeolites by adding organic groups is not trivial because of the steric constraints and the inherent structural defects imposed to the zeolite matrix. However, would it be possible to incorporate bridging organic groups?

ZOL materials are zeolite matrices with organic groups as an integral part of their lattice. In the synthesis of zeolites, these bridging moieties are limited to hybrid zeolites with methylene groups (—CH_2—) replacing a lattice oxygen atom by starting with bis(triethoxysilyl)methane having a bridging methylene group (Si—CH_2—Si) between two ethoxysilanes as the silicon source (Fig. 5.20). Different organic–inorganic hybrid zeolitic phases have been synthesized such as MFI or LTA structures. However, due to the reaction conditions, the synthesis is not as straightforward as anticipated. Although the Si—C bond is generally strong enough to be resistant to hydrolysis, Si—CH_2—Si is relatively easy to cleave by nucleophilic substitution via possible intermediate species, e.g. Si—CH_2^-. This carbanion can presumably be stabilized by the vacant d-orbital of the adjacent Si atom. Supposedly, thus formed purely inorganic Si species and organically-modified species co-crystallize to form the organically-modified material, which can be seen in the ^{29}Si-MAS NMR spectra of the final zeolite material that contains Q (SiO_4) and T (—CH_2—SiO_3 and CH_3—SiO_3) species (see also Chapter 1 and 6). The amount of organic groups can be quantified by measuring the amount of T units in the ^{29}Si-NMR spectra, which is about 30% of total Si. Compared to their purely inorganic counterparts very long crystallization times up to several weeks can be observed for the synthesis of these hybrid systems.

The use of other organosilanes, such as bis(triethoxysilyl)ethane, did not yield crystalline materials, presumably because two CH_2 groups are too large to replace a lattice oxygen atom.

Why is it so difficult to substitute the oxygen atom by a methylene group? Bond length and angles have to match rather closely to crystallize in a typical zeolite structure. The typical bond length of Si—C is 1.88 Å, which is longer than that of Si—O with 1.6 Å. In addition, the steric demand of a CH_2 unit is higher than a

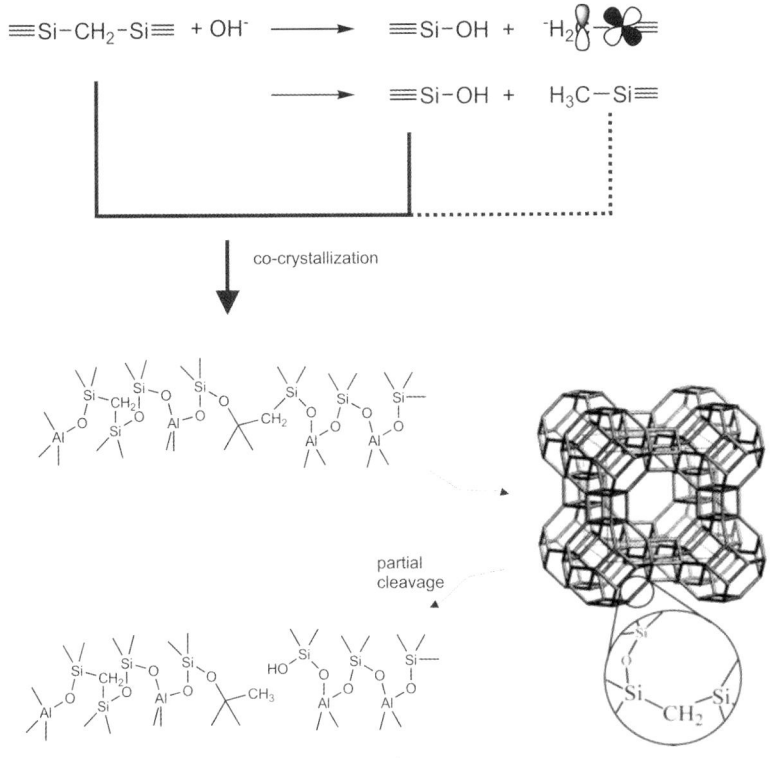

Fig. 5.20 Possible pathways to ZOL materials.

single oxygen. However, the Si—C—Si bond angle (~109°) is usually smaller than the Si—O—Si angle. This smaller bond angle compensates for the distance of two silicon atoms thus, enabling crystallization of this type of inorganic–organic hybrid zeolitic material, but the limits are already seen when the size of the organic spacer is increased.

It is not only for curiosity reasons that framework oxygen is replaced by organic spacers, but a substitution of bridging oxygen by organic groups (e.g. methylene) also provides zeolites with new functions as well as distinctively different lipophilic/hydrophobic surface properties.

Mesoporous materials Organically-bridged bis(trialkoxysilyl) compounds have been used in sol–gel processing for quite some time (see also Chapter II.3) and a wide variety of different precursor molecules is available (Scheme 5.5). These precursors have the advantage that the crosslinking density is not reduced as for monosubstituted poly(silsesquioxanes), $[R(SiO_3)_n]$, built from R—Si(OR)$_3$, with R being a nonbridging unit, where the 3-D network is formed via three siloxane bonds only. Many amorphous poly(silsesquioxanes) with bridging organic groups, e.g. with alkane, alkylene, aryl or even functional units have been synthesized, some still exhibiting porosity and high surface areas.

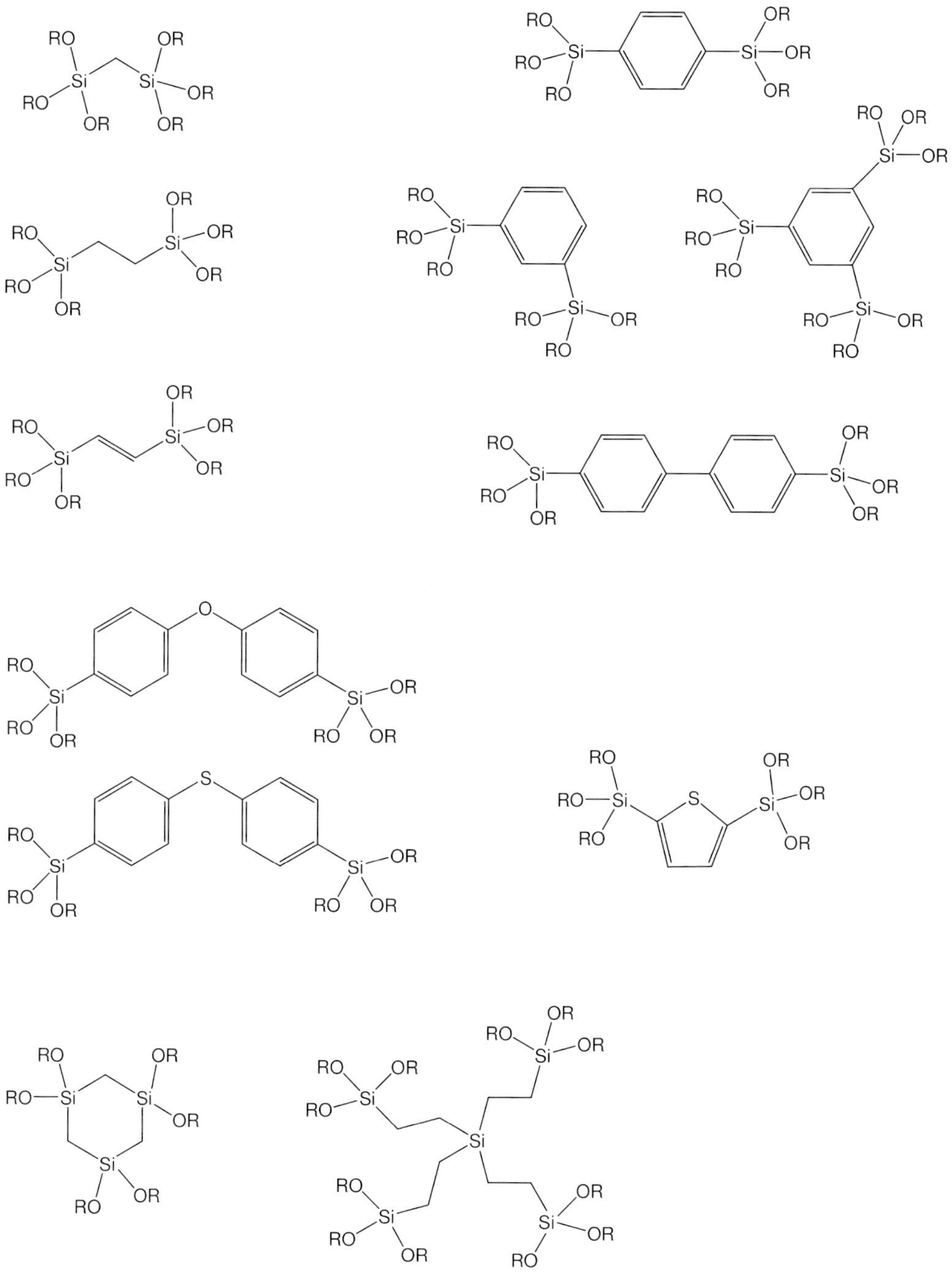

Scheme 5.5 Bridged organosilanes that have been used in the formation of micro/mesoporous frameworks.

Most of the available precursors can also be condensed in the presence of a structure-directing agent such as an LC phase, necessary for the preparation of periodically arranged mesoporous frameworks. Because the structure is amorphous and not crystalline as seen for zeolites, the templating process and structure formation is not governed by so many constraints and is more easily performed.

The first synthesis of a mesostructured silica-based material with organic functions as an integral part of the network (PMO) was reported in 1999, independently by three different groups.

PMOs exhibit several unique features built into their structure:
 a) high loading of organic groups,
 b) insignificant pore blocking,
 c) chemical reactive sites in the pore wall,
 d) homogeneously distributed groups,
 e) easily modified physical and chemical properties by flexible tuning of the organic bridge, and
 f) high surface area, uniform pore and channel size with nanoscale dimensions.

Furthermore, different periodic pore geometries (cubic – with a 3-D-channel system or hexagonal – with a 1-D channel system) are accessible.

The largest pores (cage-like pores of 10 nm) in a well-ordered mesostructured silica material with integrated organic groups were reported by using a block copolymer as surfactant and bis(triethoxysilyl)ethane as the framework-forming component. In contrast, super-microporous organic-integrated silica with periodic and uniform pore sizes of 1–2 nm were prepared from alkylamine surfactants and bis(triethoxysilyl)ethane. This already indicates that the synthesis of PMOs is very flexible with regard to the porous matrix, but also the chemical reactivity of the final material can be varied, e.g. by applying a functional bridging molecule already in the synthesis mixture, i.e. $[(CH_3CH_2O)_3Si(CH_2)_3]_2NH$, or by post-synthesis organic reactions such as sulfonation of phenylene moieties. In addition, different bridging units can be applied in the synthesis of a PMO, e.g. thiophene and phenylene bridges and the synthesis can be combined with co-condensation reactions of trialkoxysilanes. Scheme 5.5 also shows PMO precursors containing carbosilane bridges resulting in branched structural units able to form crosslinked robust mesoporous structures with high carbon content.

Typically, the framework of the MCM-type mesostructured materials is amorphous. However, an interesting feature was discovered when 1,4-bis(triethoxysilyl)benzene was used as network former. Reaction of a mixture of 1,4-bis(triethoxysilyl)benzene, octadecyltrimethylammonium salt, sodium hydroxide and water resulted in a material possessing a crystal-like pore-wall structure. This additional periodicity was attributed to a regular arrangement of $O_{1.5}Si—C_6H_4—SiO_{1.5}$ units in the pore walls due to noncovalent π–π-intermolecular and hydrophobic interactions between the phenylene groups und hydrogen bonding through C—SiOH groups (Fig. 5.21).

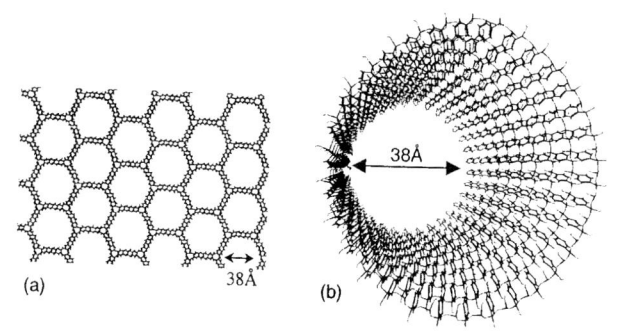

Fig. 5.21 Molecular scale periodicity in hybrid PMO materials by using 1,4-bis(triethoxysilyl)benzene.

5.3.3.2 Metal–Organic Frameworks

The self-assembly of metal ions, which act as coordination centers, linked together by a variety of polyatomic organic bridging ligands, can result in coordination polymers. However, the term coordination polymer is not very precise with regard to the structural features of the material. If a microporous solid is formed that displays attributes such as robustness due to strong bonding, linking units that allow for chemical modification and a geometrically well-defined structure, only then this type of material is labeled metal–organic framework (MOF). MOFs are very interesting inorganic–organic hybrid materials because they exhibit high porosities and world record surface areas.

The most prominent example of these MOF structures is MOF-5, a metal–organic framework built from an extension of the basic zinc acetate structure (an octahedral $Zn_4O(CO_2)_6$-cluster) by bridging carboxylate spacer ligands such as 1,4-benzenedicarboxylate. A typical synthesis of MOF-5 is performed by heating a solution of zinc nitrate (as the metal complex forming unit) in dimethylformamide (DMF) and 1,4-benzenedicarboxylate (BDC) (as the spacer unit) in a closed vessel at about 373 K (Fig. 5.22). A 3-D cubic array is obtained by combination of the $[Zn_4O]^{6+}$ cluster units (octahedral orientation of the binding sites) as connectors and the linear terephthalate ions ($^-OOC-C_6H_4-COO^-$) as linkers. The crystals have a very low density (0.59 g cm^{-3}); their framework is stable up to 300 °C. Their surface area and pore volume is higher than in most zeolites.

Assembly of the building blocks is achieved by using standard coordination chemistry methods that is the coordination of ligands to metal centers. Because the synthesis of the completely regular and highly porous solid materials occurs under mild mostly solvothermal conditions, the structural integrity of the building units is maintained throughout the reactions.

As mentioned above, the synthesis of MOF structures is regarded as a self-assembly approach. To obtain a crystalline product, e.g. the MOF-5 structure, the metal–ligand interaction has to be highly labile, meaning that bond formation is rapidly reversible, providing the initially formed products (typically the kinetic

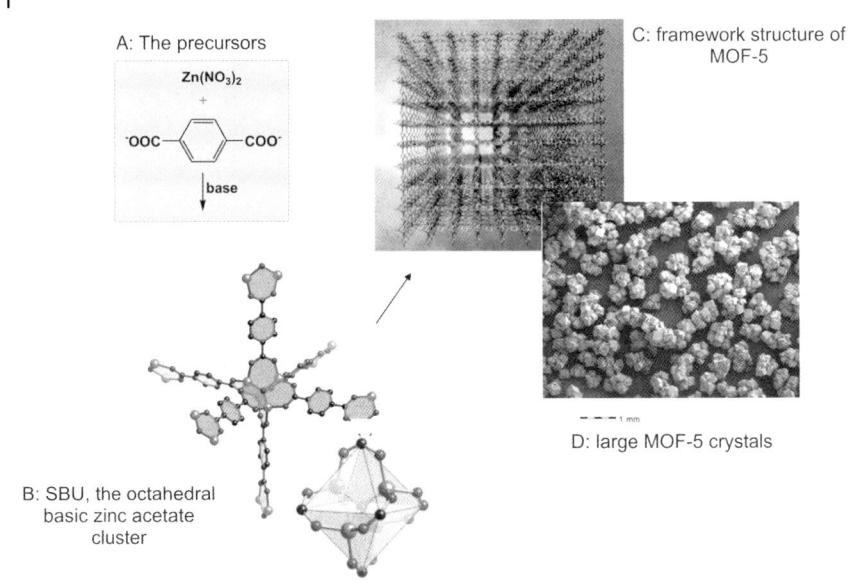

Fig. 5.22 Synthesis scheme and structural features of MOF-5.

A: The precursors
Zn(NO$_3$)$_2$ + $^-$OOC–C$_6$H$_4$–COO$^-$ | base

B: SBU, the octahedral basic zinc acetate cluster

C: framework structure of MOF-5

D: large MOF-5 crystals

product) with the opportunity to rearrange to give the thermodynamically favored material.

Some requirements for a successful MOF synthesis include: a) the metal center should have a preference for a certain coordination geometry, b) the bridging spacer should be rather rigid and c) the formation of 3-D structures by defined coordination environments must be possible. Keeping this in mind, the synthesis of MOF structures strongly resembles a modular or building block approach (Fig. 5.22). Considerations of the geometric requirements for a target framework and implementation of the design and synthesis of such a framework have been termed reticular synthesis (based upon identification of how building blocks come together to form a net, or reticulate). This process requires both an understanding of the local coordination patterns of the metal and organic units and foreknowledge of what topologies they will adopt.

The structural unit that self-assembles from a number of metal ions and ligands is referred to as a secondary building unit (SBU). Primary building units are therefore the ligand and the metal ions. A similar nomenclature is used in the synthesis of zeolites – here the primary building units are called basic building units (BBU, e.g., the SiO$_4$ and AlO$_4^-$ tetrahedra) and composite building units (CBU, larger structures formed by these tetrahedral units that can be found repeatedly in the given zeolite structure).

With respect to the ligand, problems arise when the ligand is too flexible. Clearly, if a ligand has a number of possible conformations, the framework geometry will be hard to predict and several products can be formed. Typically one general feature is common to almost all linkers used for the formation of MOFs: rigidity.

5.3 Classification of Porous Hybrid Materials by the Type of Interaction

Typical ligands can be cationic, anionic or neutral and very often include nitrogen or oxygen donor atoms, e.g., 4,4′-bipyridine, pyrazine, oxalate, benzene-1,4-dicarboxylates (terephthalate), benzene-1,3,5-tricarboxylate, etc., to name only a few (Scheme 5.6).

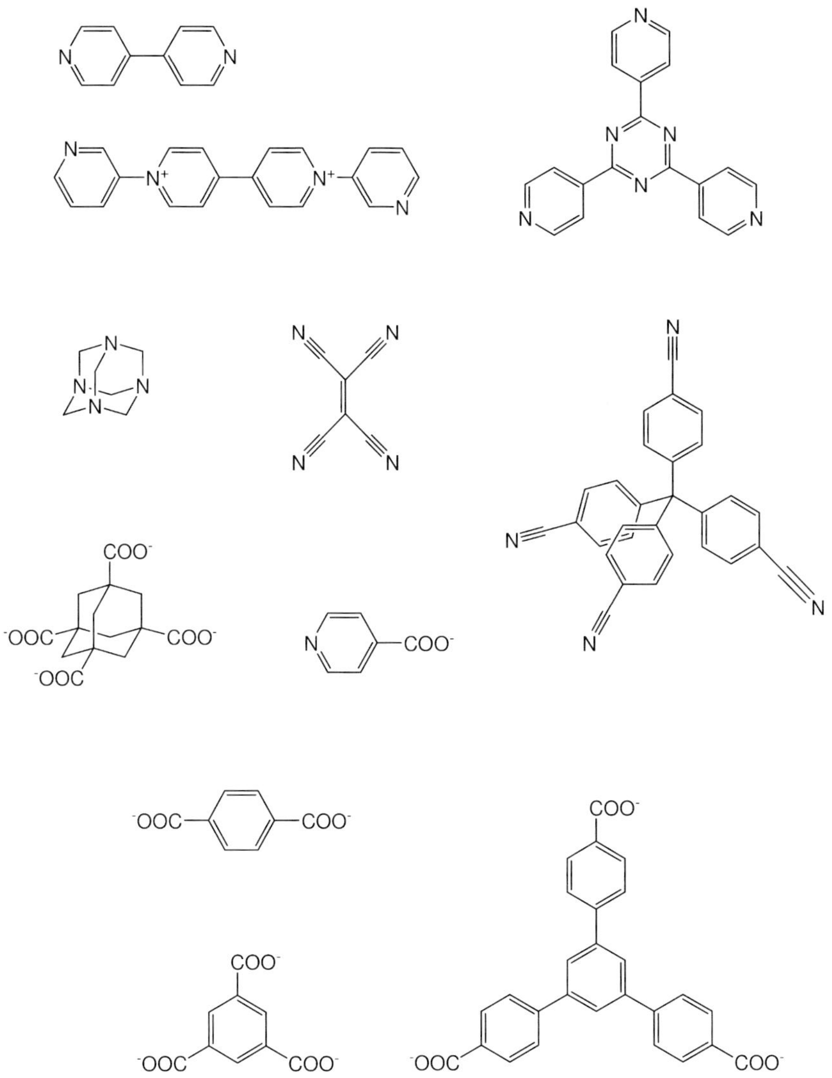

Scheme 5.6 Examples of linkers used in MOFs.

The cavities of MOFs are filled with solvent molecules in the as-synthesized form. These guest species can be removed not for all, but for some structures. The porosity can be investigated by measurements of gas adsorption/ desorption isotherms, typically using nitrogen or argon at liquid nitrogen temperature. All porous MOFs to date are microporous (after the IUPAC definition), featuring rather unique characteristics among crystalline porous materials with free pore diameters in the range of 0.4–1.9 nm and pore volumes of up to $2\,cm^3\,g^{-1}$. Changing the geometry and functionality of the organic spacer and/or the nature of the inorganic building unit allows to access a large variety of periodic structures – though not necessarily permanently open and thermally stable – with varying adjustable topologies, compositions and properties. MOFs exhibit the properties that are inherent to the building blocks themselves, e.g. magnetic exchange, non-linear optical properties, chirality or chemical functionality, but also to the 3-D pore structure formed, e.g. large channels available for the passage of molecules. Nowadays, a vast number of different MOF structures are known and is continuously increasing.

Three-dimensional metal–organic frameworks feature amongst the largest pores known for crystalline compounds. The open voids, cavities, and channels in these porous coordination polymers can make up more than half of the volume of the crystal. Because it is impossible to synthesize compounds with a large vacant space, the pores will always be occupied by guest or template molecules, such as solvent molecules or counteranions. Therefore, it is very important to select appropriate volatile or exchangeable guest molecules that can be removed after the synthesis. Up to now, all porous MOFs are microcrystalline by IUPAC definition of pore systems with pore diameters between 1–2 nm. The porosity and the specific surface area are typically determined from adsorption and desorption isotherms of gases such as nitrogen or argon at 77 K.

The series of MOFs sharing the formula $Zn_4O(L)$ with L being a rigid linear dicarboxylate spacer demonstrates impressively the possibilities in pore design given by a rational selection of the bridging entity (Fig. 5.23). Substituting 1,4-benzenedicarboxylate that is used in the synthesis of MOF-5 by 2,6-naphthalenedicarboxylate, 4,4′-biphenyldicarboxylate or 4,4′-terphenyldicarboxylate allows for the production of materials with larger pores. All these materials exhibit the same cubic morphology as the prototypical MOF-5 with octahedral $Zn_4O(CO_2)_6$ clusters as SBUs linked together along orthogonal axes by phenylene rings. This approach is called isoreticular (structures of the same net are formed) synthesis and the resulting structures are termed IRMOF-1 to IRMOF-16.

The rational design of MOF structures via the reticular synthesis approach is limited because not only can simple crystalline MOF lattices be formed, but framework catenation can occur, especially when large linkers are used. Catenation is the phenomenon of periodic entanglement of two or more frameworks at the expense of pore volume and can occur in two different ways (Fig. 5.24): 1) interpenetration, where the frameworks are maximally displaced from each other, or 2) interweaving, where they are minimally displaced and exhibit many close contacts.

The most immediate consequence of catenation is that the voids constructed by one framework are occupied by one or more independent frameworks. Such entangled structures can only be disentangled by destroying internal bonds.

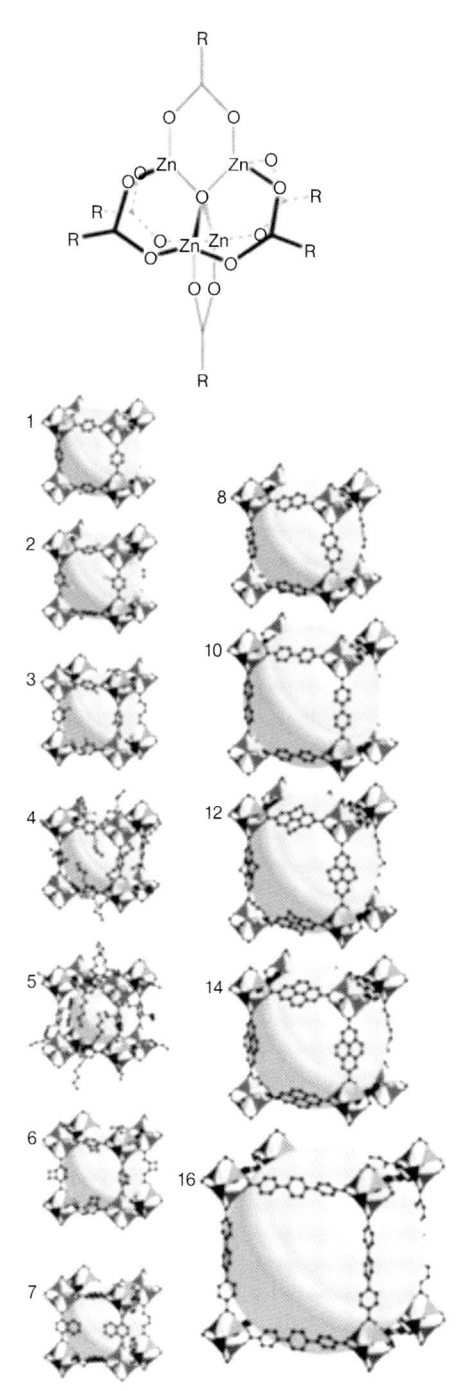

Fig. 5.23 Series of cubic frameworks based on dicarboxylate linkers and the octahedral $Zn_4O(CO_2)_6$-clusters as SBU. (Reproduced from S. L. James, *Chem. Soc. Rev.* **2003**, *32*, 276.)

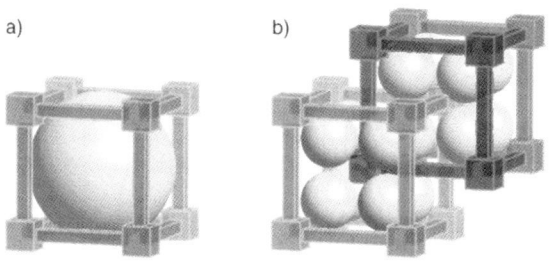

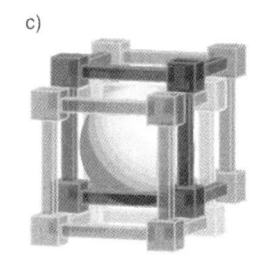

Fig. 5.24 Catenation: a) schematic representation of the repeat unit of a crystalline, single framework MOF with SBUs shown as cubes and linkers depicted as rods; b) interpenetration; c) interweaving. (Taken from Yaghi, *Angew. Chem. Int. Ed.* **2005**, *44*, 4670.)

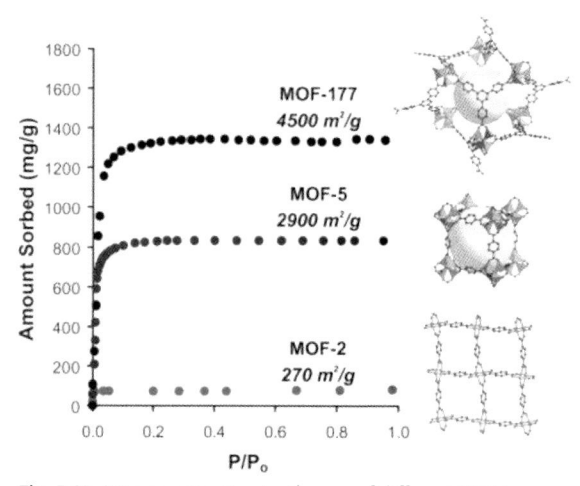

Fig. 5.25 Nitrogen sorption isotherms of different MOF structures with the world record specific surface area of MOF-177. (Taken from Yaghi, *Microporous Mesoporous Mater.* **2004**, *73*, 3.)

Why are MOFs of such interest for the scientific community and industry? MOFs have been proposed as attractive candidates for hydrogen storage. Despite the huge amount of research invested in this area, e.g. on solid metal hydrides, carbon nanotubes, and many more, hydrogen storage is still awaiting its breakthrough material. One outstanding property of MOFs that has prompted their study as gas storage medium is their extremely large surface areas, greater than $1000\,m^2g^{-1}$ for many cases. In particular MOF-177 shows an outstanding surface area with $4500\,m^2g^{-1}$ measured with nitrogen at 77 K (Fig. 5.25).

In addition to the large surface area, the main advantage of metal–organic frameworks compared with other porous materials is the easy adjustment of the pore and channel diameters (due to the easy variation of the linkers), the potential integration of chemical functionalities in the linkers as well as the metal centers, and the fact that the pore walls are constructed of organic entities, providing a "light material".

5.4
Applications and Properties of Porous Hybrid Materials

Hybrid inorganic–organic porous materials find applications in various fields, depending on their chemical composition, structure and porosity, surface area and pore size distributions.

Catalysis Catalysis plays a vital role in life today, such as in cleaning and production of energy and fuels, fine chemicals, pharmaceuticals, and commodity chemicals. Summing up, about 90% of the chemical manufacturing processes and more than 20% of all industrial processes involve catalytic steps.

Hybrid porous materials have been considered for a wide range of heterogeneous catalysis reactions. Heterogeneous catalysts consist, in most cases, of a catalytically active component carried on the surface of a solid support that is typically a porous material. Heterogenization of catalytically active species is of advantage for several reasons:
  a) the solid catalyst is easier to separate from the reaction
     solution (e.g. by filtration),
  b) the catalyst is easier to recycle, and
  c) shape- / and size selectivity can be imposed to the catalytic
     reactions by the constraints of the pore size dimensions.

In contrast to organic supports, inorganic supports do not swell or dissolve and leaching of the catalytically active species from the material can easily be prevented by covalent attachment. Two distinct approaches can be used, e.g. to bind chiral homogeneous catalysts to a solid support such as M41S matrices: the sequential and convergent approaches (Fig. 5.26).

Sometimes, superior regio- and stereoselective properties can be found for molecular catalysts confined in a porous environment. This is attributed to confinement effects and interactions between substrate, pore wall, chiral ligand and metal center, such as spatial restriction, electronic interactions, adsorption interactions and diffusion dynamics. These confinement effects are not visible when nonporous supports are used, because they have only external surfaces where no pore constraints are effective.

Nevertheless, the porous material must be chosen carefully for every catalytic reaction, especially with respect to its pore size and accessibility of the active species within the pores.

Catalytic reactions that have been performed include acid catalysis, base catalysis, oxidations, reductions, enantioselective catalysis, stereospecific polymerizations and a variety of reactions for the production of fine chemicals.

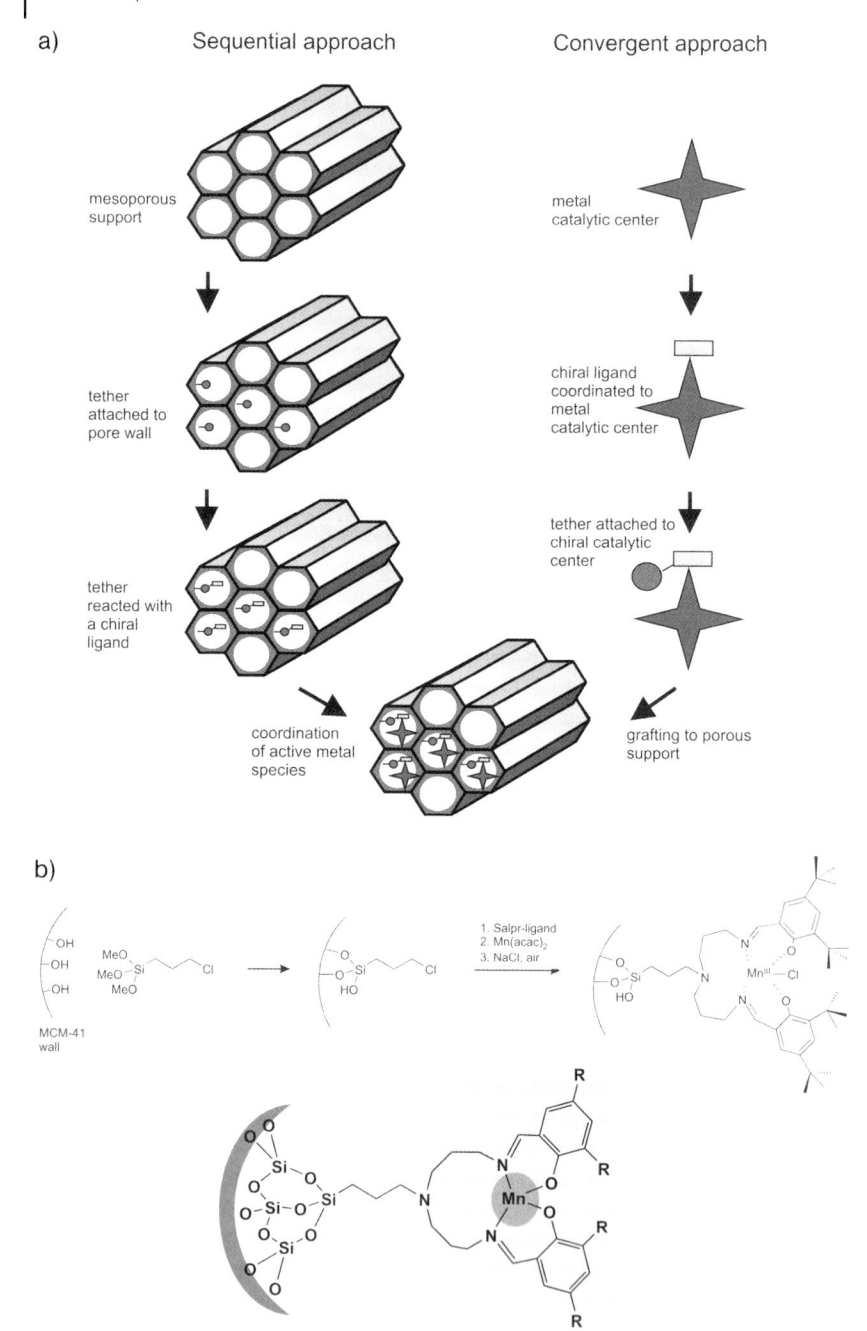

Fig. 5.26 a) sequential and convergent approaches to heterogeneous catalyst systems (via covalent attachment); b) a manganese salene complex derivatized on mesoporous silica.

Storage and adsorption of molecules Gas storage is of major importance especially with respect to the uptake of fuel gases such as methane or hydrogen, which are highly attractive candidates as replacements for fossil fuels. So far one of the major obstacles to their widespread use as fuel is safe and efficient storage (with regard to costs and volume of gas). This is of dramatic importance for mobile and portable fuel cell applications.

Porous media are of high interest as storage systems, because one of the fastest ways to charge and discharge a storage vessel with e.g. hydrogen is to maintain its molecular identity (this is not the case for the alternative storage media based on chemical hydrides). As a requirement for gas storage, the host material must provide a large gravimetric and volumetric uptake, facile release of the gases, reproducible cycling and the material must be economically produced.

The ideal pore size for maximal attraction of an adsorbate molecule to a porous host is the same as its diameter, because then an optimal interaction with all the surrounding adsorbent walls is provided. MOF systems are of special interest for gas storage applications due to their large surface areas and as advantage over nanostructured carbon materials, which are also tested for their hydrogen storage capacity, they provide isolated phenylene rings. It was proposed that for an effective storage medium the wall of the adsorbent should be as thin as possible and highly segmented, which is exactly what is found in MOFs. However, even in MOFs with small pore sizes, e.g. MOF-5 with an approximately spherical pore diameter of 1.5 nm, the pore is much larger than the 0.289 nm kinetic diameter of hydrogen (Fig. 5.27).

Nevertheless, the flexible synthesis protocol for MOF structures allows for improving this situation by several approaches. First, MOF structures can be impregnated with a nonvolatile guest, e.g. C_{60} was already successfully incorporated into MOF-177 from the solution phase. Second, catenation can be used to reduce the pore diameter as seen in Fig. 5.4. Third, open metal sites may be used for a stronger binding of hydrogen into the pore and last, the organic linker can

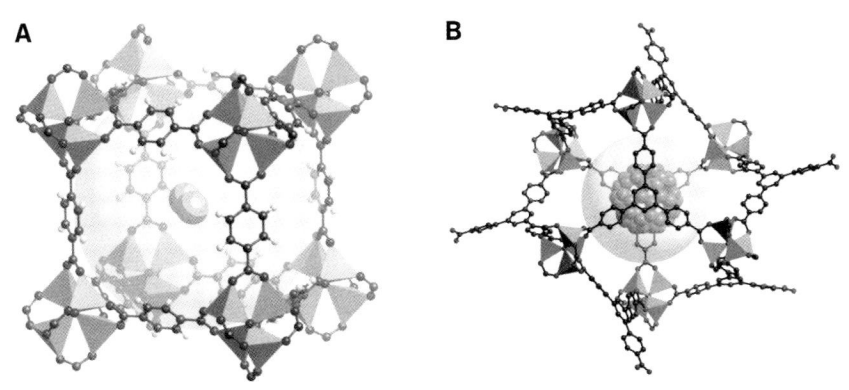

Fig. 5.27 Hydrogen in the pore of MOF-5 (A) and C_{60} in the pore of MOF-177 (B). (Taken from Yaghi in *Angew. Chem. Int. Ed.* **2005**, *30*, 4670).

be used to increase the interaction of hydrogen with the host matrix. Organic linkers with aromatic backbones such as phenylene, naphthalene, bipyridine, and biphenylene have been employed in MOFs to increase the rigidity, but at the same time, these units mimic the network structure of nanostructured carbon with sp^2-hybridized carbon atoms. Therefore, it is believed that adsorption energetics should be similar for both classes of materials.

Besides all these promising results, further research is necessary to identify the best materials for gas storage applications. The examples given above, can give only indications where to go in the future.

Membranes and monolithic chromatography columns   Hybrid inorganic–organic membranes offer consistent and unique opportunities to combine the specific transport properties of organic and inorganic materials in order to produce highly permselective membranes.

Organically-modified silica gel particles are extensively used as a packing material of columns for high performance liquid chromatography, HPLC. Recently, porous silica monoliths, exhibiting macro- and mesopores have been applied as the stationary phase in HPLC separation. These porous columns are typically modified in a post-synthesis grafting reaction by long aliphatic chains such as C18.

Many more applications that are not even mentioned in this chapter are under investigation, e.g. the alignment of dye molecules in microporous channel systems for laser applications, or the use of functional groups on the surface of porous matrices for heavy metal removal. In addition, these porous matrices can also be used as templates for new functional matrices.

The list of hybrid materials, the choice of functionalization agent and the potential applications are numerous and only a very small selection could be covered within this chapter. The field is rapidly expanding and it would be beyond the scope of this book to be comprehensive. In the future one can expect increasingly complex structures that combine not only hierarchical pore sizes with multiple organic and inorganic functional groups, but also a strategic placement of these groups on the internal and external surfaces. It is up to the creativity of the reader to imagine the many more possibilities available for this type of materials.

Bibliography

General

On porous materials in general, an excellent materials collection is given in the five volumes of:

Handbook of Porous Solids, Eds. F. Schüth, K. S. W. Sing, J. Weitkamp, Wiley-VCH, Weinheim, **2002**.

Zeolites

http://www.iza-online.org/ and
http://www.iza-structure.org/databases/

C. Baerlocher, W. M. Meier, D. H. Olson, *Atlas of Zeolite Framework Types*, 5th revised ed.; Elsevier, Amsterdam **2001**

H. L. Frisch, J. E. Mark, Nanocomposites Prepared by Threading Polymer Chains through Zeolites, Mesoporous Silica or Silica Nanotubes, *Chem. Mater.* **1996**, 8, 1735.

A. Corma, H. Garcia, Supramolecular Host-Guest Systems in Zeolites Prepared by

Ship-in-the-bottle Synthesis, *Eur. J. Inor. Chem.* **2004**, 1143–1164.

M41S materials

A. Stein, B. J. Melde, R. C. Schroden, Hybrid Inorganic-Organic Mesoporous Silicates-Nanoscopic Reactors Coming of Age, *Adv. Mater.* **2000**, *12*, 1403.

W. J. Hunks, G. A. Ozin, Challenges and advances in the chemistry of periodic mesoporous organosilicas (PMOs), *J. Mater. Chem.* **2005**, *15*, 3716.

K. Möller, T. Bein, Inclusion Chemistry in Periodic Mesoporous Hosts, *Chem. Mater.* **1998**, *10*, 2950.

A. Sayari, S. Hamoudi, Periodic Mesoporous Silica-Based Organic-Inorganic Nanocomposite Materials, *Chem. Mater.* **2001**, *13*, 3151.

S. Spange, A. Gräser, A. Huwe, F. Kremer, C. Tintemann, P. Behrens, Cationic Host-Guest Polymerization of *N*-Vinylcarbazole and Vinylethers in MCM-41, MCM-48, and Nanoporous Glasses, *Chem. Eur. J.* **2001**, *7*, 3722.

L. Nicole, C. Boissière, D. Grosso, A. Quach, C. Sanchez, Miso structured hybrid organic-inorganic thin films, *J. Mater. Chem.* **2005**, *15*, 3598.

Metal–organic frameworks

J. L. C. Rowsell, O. M. Yaghi, Metal-organic frameworks: a new class of porous material *Microporous Mesoporous Mater.* **2004**, *73*, 3.

S. Kitagawa, R. Kitaura, S.-I. Noro, Functional Porous Coordination Polymers, *Angew. Chem.* **2004**, *116*, 2388; *Angew. Chem. Int. Ed.* **2004**, *43*, 2334.

C. Janiak, Engineering coordination polymers towards applications, *Dalton Trans.* **2003**, 2781.

M. Eddaoudi, D. B. Moler, H. Li, B. Chen, T. M. Reineke, M. O'Keefe, O. M. Yaghi, Modular Chemistry: Secondary Building Units as a Basis for the Design of Highly Porous and Robust Metal–organic Carboxylate Frameworks, *Acc. Chem. Res.* **2001**, *34*, 319–330.

J. L. C. Rowsell, O. M. Yaghi, Strategies for Hydrogen storage in Mental-Organic Frameworks, *Angew. Chem.* **2005**, *117*, 4748; *Angew. Chem. Int. Ed.* **2005**, *44*, 4670.

U. Mueller, M. Schubert, F. Teich, H. Puetter, K. Schierle-Arndt, J. Pastré, Metal–organic Frameworks – Prospective Industrial Applicatons, *J. Mater. Chem.* **2006**, *16*, 626–636.

S. L. James, Metal–organic frameworks, *Chem. Soc. Rev.* **2003**, *32*, 276–288.

6
Sol–Gel Processing of Hybrid Organic–Inorganic Materials Based on Polysilsesquioxanes
Douglas A. Loy

6.1
Introduction

Hybrid organic–inorganic materials are materials with both organic and inorganic components combined in such a way that can lead to dramatically enhanced mechanical, thermal and chemical properties. This chapter focuses on a class of hybrid organic–inorganic materials called polysilsesquioxanes (Fig. 6.1) that have the basic repeating unit of $[RSiO1.5]_n$. These materials can assume many forms and serve manifold applications due to the marriage of the siloxane bond networks with organic constituents.

Polysilsesquioxanes, first reported in the 19th century, became the basis of the crosslinking agents in some room temperature vulcanizing silicones and the most common silating agent for modifying surfaces. More recently, polysilsesquioxanes have been used to protect stonework from the ravages of weathering, encapsulate fluorescent organometallic compounds in oxygen and carbon dioxide sensors and as nanoscale filler for advanced composites. Presently, these materials, prepared by the hydrolysis and condensation of organotrihalosilanes or organotrialkoxysilanes, represent the method of choice for many who wish to introduce organic functionalities into sol–gel processed materials.

The versatility of silsesquioxanes comes from the commercial availability of a diverse range of monomers as silating agents (Fig. 6.2) and the relative ease with which monomers with practically any organic groups can be prepared. This versatility gives consummate control over the sol–gel polymerization chemistry and the ultimate properties of the final hybrid organic–inorganic materials. While the power of this method for introducing organic functionalities into hybrid materials is clear, very few researchers realize the extent to which the nature of the organic group will determine the course of sol–gel chemistry being used (Fig. 6.1). The purpose of this chapter is to introduce the sol–gel chemistry and processing using organotrialkoxysilanes to researchers and workers using polysilsesquioxanes.

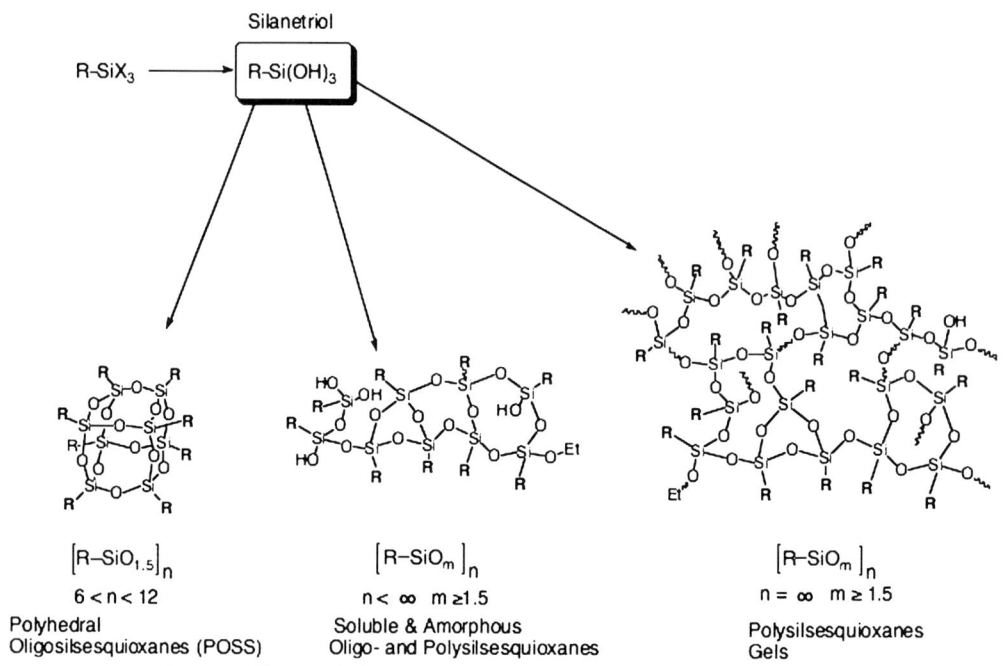

Fig. 6.1 Different polysilsesquioxanes form from the hydrolysis and condensation of organotrichlorosilane or organotrialkoxysilane monomers, $RSiX_3$. What product is obtained depends on the monomer, its concentration, the pH of the solution, the amount of water, type and amount of catalyst, the identity of the solvent, and reaction temperature.

Structures shown:
- $[R\text{-}SiO_{1.5}]_n$, $6 < n < 12$: Polyhedral Oligosilsesquioxanes (POSS)
- $[R\text{-}SiO_m]_n$, $n < \infty$, $m \geq 1.5$: Soluble & Amorphous Oligo- and Polysilsesquioxanes
- $[R\text{-}SiO_m]_n$, $n = \infty$, $m \geq 1.5$: Polysilsesquioxanes Gels

Starting materials: $R\text{-}SiX_3 \rightarrow R\text{-}Si(OH)_3$ (Silanetriol)

6.1.1
Definition of Terms

The first step will be to introduce some of the terminology commonly used with sol–gel processing, hybrid organic–inorganic materials and polysilsesquioxanes. The term hybrid organic–inorganic refers to any composite based on organic and inorganic components. The organic component can be composed of molecular (<500 Daltons), oligomeric (500–10 000 Daltons) and macromolecular (>10 000 Daltons) constituents based primarily of hydrogen, carbon with some oxygen, nitrogen and halogens. The inorganic component includes an even broader venue of metallic elements, generally in the form of some oxide. In polysilsesquioxanes, the inorganic component is an oligomeric or polymeric siloxane chain, branched structure or network to which the organic component is attached through a Si—C bond. The basic repeating unit or monomeric building block is a tetra-coordinate silicon with up to three siloxane bonds and an organic substituent. The term silsesquioxane describes the average stoichiometry of 1.5 or sesqui oxygens per silicon that arises in oligomers and polymers because each siloxane bond connects

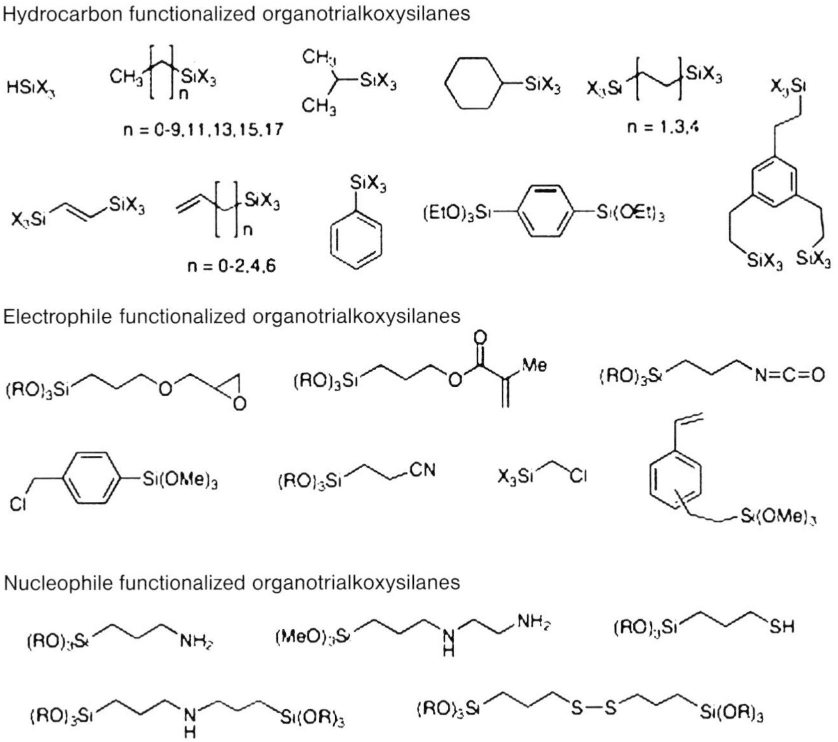

Fig. 6.2 Some commercially available organotrialkoxysilane monomers for preparing polysilsesquioxanes.

two silicon atoms. Polysilsesquioxanes are also called t-resins, a term with its roots in the alphabetic nomenclature developed for siloxane polymers during the 1950s. T-resins have three siloxane bonds to each silicon atom. This notation has survived as shorthand for describing polyhedral oligosilsesquioxanes (Fig. 6.3, Type I) and distinguishing between different silicon-29 NMR resonances. The silsesquioxane repeating unit can be incorporated into a number of architectural motifs that are divided here into groups: Type I–III (Fig. 6.3).

Type I materials (Fig. 6.1) are discrete polyhedral oligosilsesquioxanes (POSS) that often crystallize from solution. The most common size is the octameric POSS, T8, that is frequently drawn in the form of a distorted cube. Type II includes amorphous oligosilsesquioxanes, high molecular weight macromolecules and gels with pendant organic groups. Unlike the crystalline polyhedral oligomers, these amorphous silsesquioxanes are not single molecular species, but are characterized by a range of molecular weights, structures and even stereochemistry. Type II also includes copolymers of silsesquioxane monomers with other monomers such as tetraalkoxysilanes or even functionalized macromolecules. In these materials, there may be substantially less organic functionality mostly restricted to the surface of a macromolecule or gel. Type III materials are based on monomers in

TYPE I   **TYPE II**   **TYPE III**

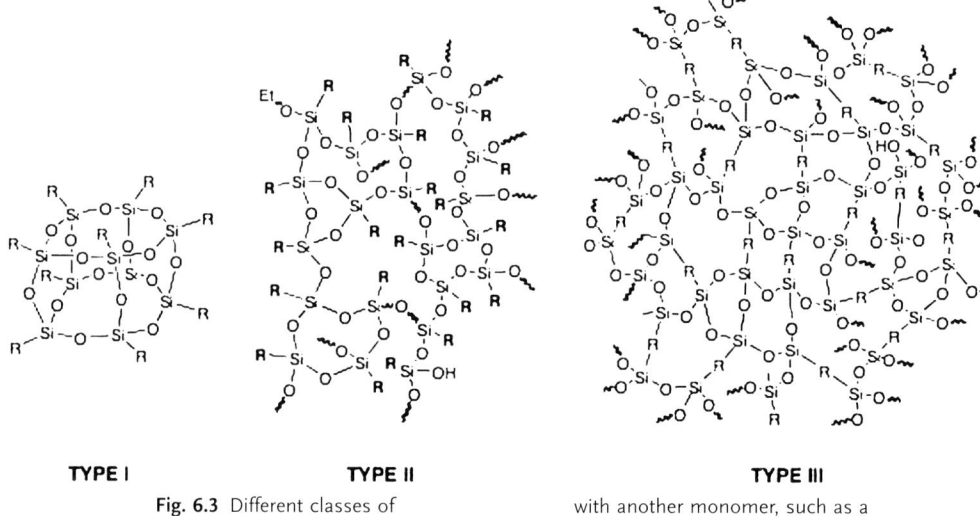

Fig. 6.3 Different classes of polysilsesquioxanes: crystalline, molecular polyhedral oligosilsesquioxanes composed of fewer than 20 monomers (Type I). Amorphous oligomers and polymers with pendant organic groups, R, prepared from the homopolymerization of an organotrialkoxysilane or its copolymerization with another monomer, such as a tetraalkoxysilane (Type II). Type III silsesquioxanes are bridged polysilsesquioxanes in which two or more silsesquioxane centers are bridged by an organic group. The organic group, R, may range from hydrogen (for Type I & II) to macromolecules.

which the organic group acts as a bridge between two or more trialkoxysilyl groups. Simplified to the most basic form, this class includes hexamethoxydisilane in which there is a Si—Si bond bridging the two silsesquioxane groups. The majority of the representatives of Type III silsesquioxanes have organic bridging groups that are molecular in size. Type III also includes examples where linear macromolecules and dendrimers have been modified to have trialkoxysilyl groups to yield nanocomposites with silsesquioxane crosslinks upon hydrolysis and condensation. Types II and III may form gels during their formation; Type I will form gels only through the reaction of the pendant organic groups. The following sections will describe the polymerization chemistries that are common to forming all three types of polysilsesquioxanes. Each of the three types will then be described and discussed in more detail.

6.2
Forming Polysilsesquioxanes

6.2.1
Hydrolysis and Condensation Chemistry

For the most part, the polysilsesquioxanes are prepared by sol–gel polymerizations of organosilane monomers to form silsesquioxane oligomers, polymers or networks. Organotrichlorosilanes, organotrimethoxysilanes and organotriethoxysilanes are the most commonly used monomers. Most of the early investigations of silsesquioxanes were based on materials prepared by the hydrolysis and condensation of organotrichlorosilanes (Fig. 6.4), very reactive monomers prepared by Grignard, organolithium, and radical chemistry from silicon tetrachloride.

Hydrolysis involves mixing an organotrichlorosilane with a minimum of 1.5 equivalents of water. This vigorous reaction will rapidly afford silanols that can co-condense with each other or with a chlorosilane to afford a siloxane bond. The production of up to three equivalents of hydrogen chloride for each equivalent of monomer does help make the reaction thermodynamically favorable, but limits its application to those systems that are tolerant to strong acid. The extreme reactivity of the chlorosilane monomer can also lead to the formation of precipitates and heterogeneous mixtures even before mixing is complete. Best results are often obtained by separately pre-mixing water and monomer in non-nucleophilic solvents such as ethers or acetone, then carefully mixing the solutions together to start the hydrolysis and condensation process. Organotrichlorosilanes are frequently used as the starting materials for preparing polyhedral oligosilsesquioxanes. They are

Hydrolysis reactions

$$n\ R\text{-}SiCl_3 + n\ H_2O \longrightarrow n\ R\text{-}Si(OH)Cl_2 + n\ HCl$$

$$n\ R\text{-}Si(OH)Cl_2 + n\ H_2O \longrightarrow n\ R\text{-}Si(OH)_2Cl + n\ HCl$$

$$n\ R\text{-}Si(OH)_2Cl + n\ H_2O \longrightarrow n\ R\text{-}Si(OH)_3 + n\ HCl$$

Condensation reactions

$$n/2\ R\text{-}Si(OH)_3 + n/2\ Cl\text{-}SiCl_2\text{-}R \longrightarrow n/2\ R\text{-}Si(OH)_2\text{-}O\text{-}SiCl_2\text{-}R + n/2\ HCl$$

$$n\ R\text{-}Si(OH)_3 \longrightarrow n/2\ R\text{-}Si(OH)_2\text{-}O\text{-}Si(OH)_2\text{-}R + n\ H_2O$$

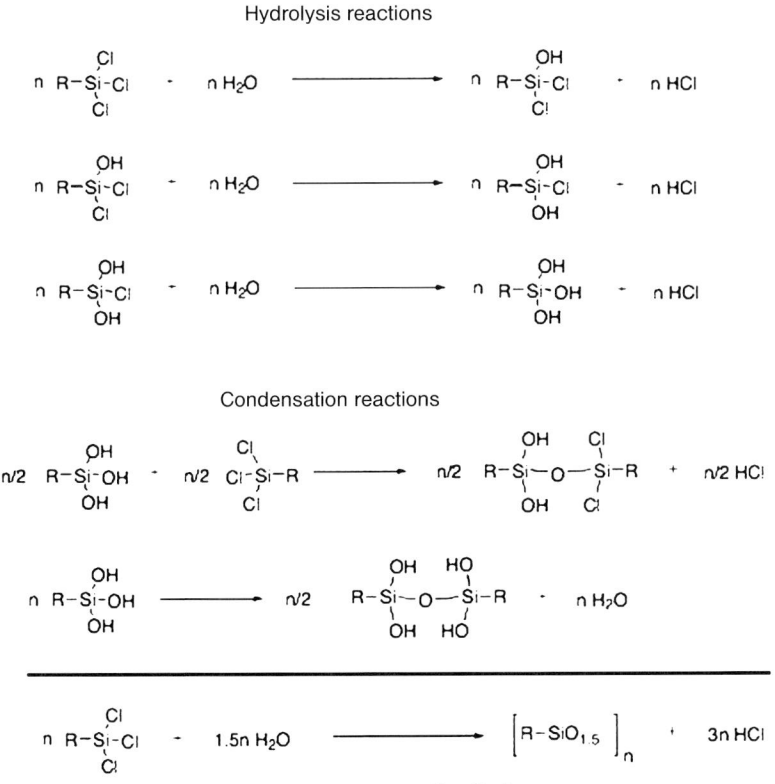

$$n\ R\text{-}SiCl_3 + 1.5n\ H_2O \longrightarrow [R\text{-}SiO_{1.5}]_n + 3n\ HCl$$

Fig. 6.4 Hydrolysis of organotrichlorosilanes rapidly affords silanols and HCl. The silanols can condense with other silanols or chlorosilanes to afford siloxane bonds.

also used as a co-monomer with an equivalent quantity of organotrialkoxysilane for Lewis acid catalyzed, nonaqueous sol–gel polymerizations.

Hydrolysis of alkoxide groups to afford silanols (Fig. 6.5) is orders of magnitude slower than chlorosilane hydrolysis. The rate of hydrolysis of alkoxysilanes decreases as the steric bulk of the alkoxide group increases. The relative rates for the first hydrolysis of alkoxysilanes reveal that MeO >> EtO > n-PrO > n-BuO. Most sol–gel chemistry is performed with organotrimethoxysilanes and organotriethoxysilanes with the later being favored for the lower toxicity of ethanol relative to methanol. As silanols are less bulky than alkoxides, hydrolysis rates have generally been shown to increase as alkoxide groups are lost. Conversely, siloxane bonds are bulky, restricting access to the silicon centers.

By varying the solution, pH, water content and solvent it is possible to exert greater control over the development of the polysilsesquioxane structure. It has been demonstrated with tetraalkoxysilanes that the choice of the alkoxide can affect the condensation rates and the proportion of cyclic siloxane rings that form during the sol–gel polymerization.

It is well known that the pH determines whether an acid or base-catalyzed mechanism of hydrolysis and condensation is followed by alkoxysilanes. The change over from acid to base-catalyzed mechanisms is marked by a minimum in the reaction rate (Fig. 6.6). For hydrolysis of alkoxysilanes this rate minimum is at pH 7.

Above pH 7 there is a greater abundance of hydroxide ions in solution, while below pH 7 there is a greater abundance of hydronium ions. At high pH the

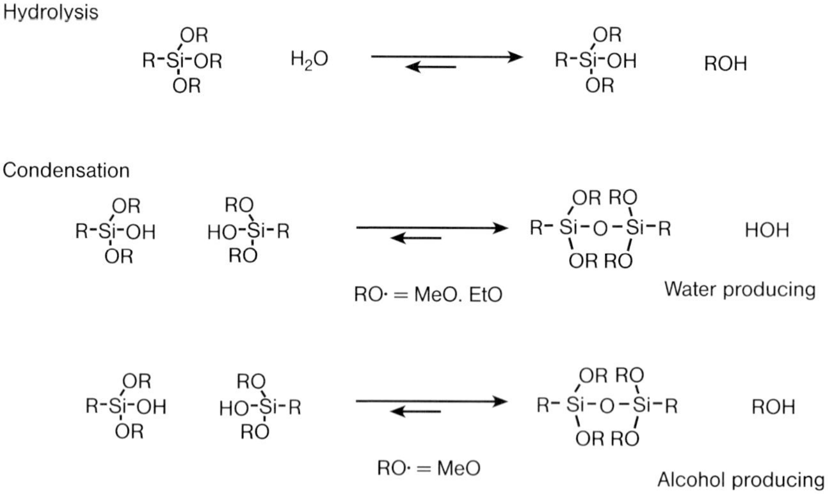

Fig. 6.5 Condensation reactions observed under sol–gel polymerization of organotrialkoxysilanes. Water producing condensation occurs when two silanols react. Alcohol producing condensation occurs when a silanol reacts with a methoxysilane.

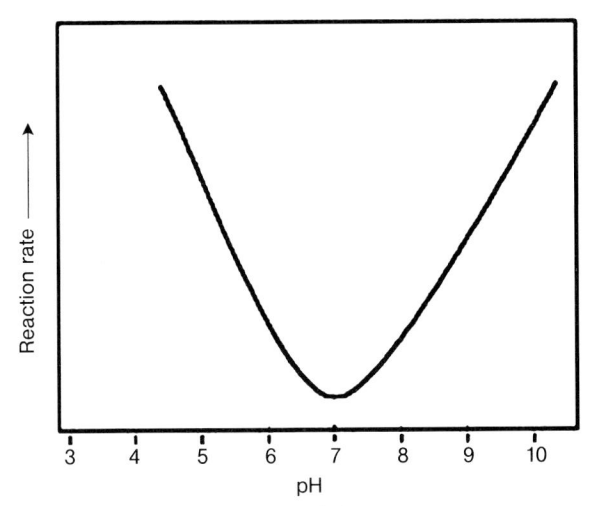

Fig. 6.6 Plot of silicon-alkoxide hydrolysis rate versus pH representing the minimum in hydrolysis kinetics near pH 7.

nucleophilic hydroxide groups replace the alkoxide in two steps. First, the hydroxide ion adds to the silicon to form a pentacoordinate siliconate intermediate. In the second step, this intermediate loses an alkoxide group to afford the silanol. Alternatively, the intermediate can lose the hydroxide to reform the product. Acids catalyze the hydrolysis by protonating the oxygen making the alkoxide into an oxonium leaving group that is more readily replaced with the attack of water.

Silanols are much more acidic than alcohols (pKa = 18) or water (pKa = 15.6) and are deprotonated under basic conditions to afford more nucleophilic silanolates (Fig. 6.7). Silanolates become the predominant nucleophiles under basic conditions. At pH 10 and above, the condensation rate decreases because of columbic repulsion between the negatively charged silanolates that predominate in the alkaline solution. Another consequence of the acidity of silanols is that the silicon-29 resonances can shift between 10–15 ppm upfield with deprotonation. The complex, pH dependent process of hydrolysis and condensation is made even more complicated by the fact that the acidity of silanols has been shown to be strongly dependent on hydrogen bonding. A number of acidities of silicate (Q) and silsesquioxane silanols have been determined. It is widely held within the sol–gel community that the acidity of silanols is dependent on the number of siloxane bonds attached to the silanol silicon. According to this model the electron-withdrawing, siloxane bonds stabilize silanolates formed by deprotonation. Thus, the greater the number of siloxane bonds, the more acidic the silanol. Yet, the evidence does not support this model. Instead, it appears that substituent effects are eclipsed by the effects of hydrogen bonding on the acidity (Fig. 6.7).

For example, "vicinal" hydrogen bonding results in acidic silanols with pKa value between 7 and 8. In absence of intramolecular hydrogen bonds in geminal triols and diols or isolated silanols are significantly less acidic with pKa between 8.9

Hydrogen bonding: pKa 7-8

Silica surface with "Vicinal" silanols

Soluble dimer with "Vicinal" silanols

POSS with hydrogen bonding triols

No hydrogen bonding: pKa 8.9-10

Silicic acid tetrahydroxysilane

Phenylsilanetriol A "geminal triol"

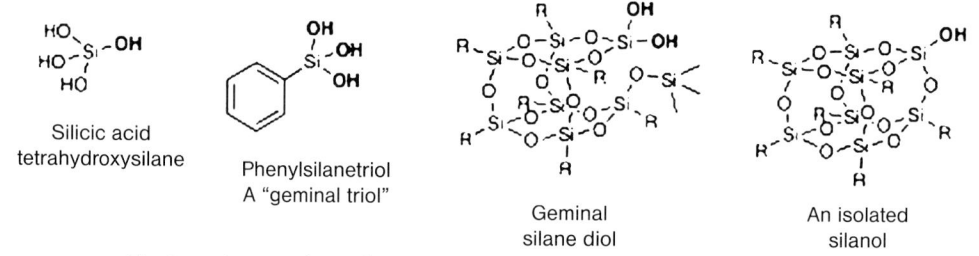

Geminal silane diol

An isolated silanol

Fig. 6.7 Relative acidities of silanols as a function of hydrogen bonding. The lower the pK_a, the more acidic the silanol. Silanols that can participate in intramolecular hydrogen bonding are more acidic than those unable to hydrogen bond.

and 10. This group includes silanols with three siloxane bonds that would be expected to be the most acidic. The pH rate profile for phenylsilanetriol (Fig. 6.8) reveals a minimum in the condensation rate at pH 4.2 that is consistent with the measured pKa of 9.5.

The nature of the organic group (R) has strong electronic and steric effects on the hydrolysis and condensation chemistry of alkoxysilanes and exerts a profound influence on the progress and ultimate product of the sol–gel polymerization. Under acidic conditions, electron-donating groups such as alkyl groups increase the rates of hydrolysis and condensation relative to tetraalkoxysilanes. The opposite has been shown with hydroxide-catalyzed polymerizations. Electron withdrawing groups, such as chloromethyl or cyanoethyl, slow hydrolysis relative to tetraalkoxysilanes under acidic conditions, but are thought to increase the rates under basic conditions.

Hydrolysis reactions of alkyltrialkoxysilanes and condensation of the resulting silanols are very sensitive to the steric bulk of the alkyl group with the rates of reaction decreasing as the alkyl substituent changes from methyl to ethyl, propyl, and butyl. Steric effects from longer alkyl groups essentially negate any rate enhancements over tetraalkoxysilanes due to the electronic effects of the alkyl

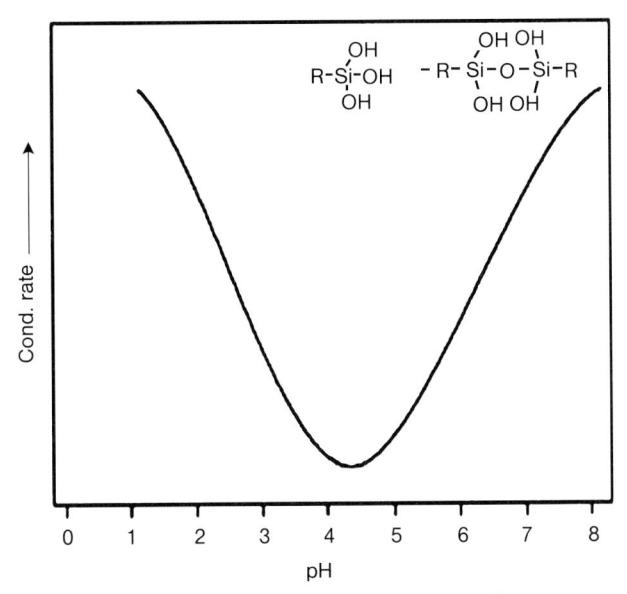

Fig. 6.8 Plot of condensation rate of triols reaction to form a dimer with pH. The reaction rate minimum is at pH 4.2.

groups. Branched groups, including iso-propyl, iso-butyl, sec-butyl, and tert-butyl groups, predictably retard hydrolysis and condensation rates even more. Silsesquioxanes with such sterically bulky groups exhibit a marked tendency to form cyclics such as polyhedral oligosilsesquioxanes. In the case of the tert-butyl- or phenyl-substituted monomers, silane triols are stable enough to condensation at ambient temperature to permit their isolation as crystalline solids. For surface modification applications such as modifying automobile glass windows to be non-wetting, this reduced tendency of silsesquioxane monomers to self-condense allows solutions of the silanes to have an acceptable shelf life, yet still react with the silanols on the surface of glass.

Hydrolysis reactions of the organotrialkoxysilanes are susceptible to exchange with nucleophilic solvents such as alcohols (Fig. 6.9). A common mistake is mixing monomers with different alkoxide groups or dissolving a monomer in an alcohol that is different from the alkoxide substituent. For example, carrying out the sol–gel polymerization of methyltrimethoxysilane in ethanol would increase the complexity of the kinetics through exchange of methoxide groups with ethanol from the solvent.

Fig. 6.9 Exchange of alcohol from solvent for methoxide groups.

6.2.2
Alternative Polymerization Chemistries

While the mildness and water-based chemistry of sol–gel processing permits the creation of new materials and modifiers without requiring stringent control over the environmental conditions, there are still a few drawbacks. The first is that the alcohols are still considered a volatile organic contaminant and methanol has significant health hazards from exposure. The second is losing three molecules of alcohol for every molecule of monomer results in a significant loss of mass and volume during the processing. This change in volume, worsened by the need for alcohol as a solvent to mix the hydrophobic monomer with water, can result in cracking of monoliths and thick films. One interesting strategy for circumventing these challenges is to use chemistries that do not use water to prepare the siloxane bonds.

There are a few nonaqueous alternatives to hydrolysis and condensation of alkoxysilanes that may be used to prepare polysilsesquioxanes. For the most part, these reactions require the use of strong Lewis or Bronsted acids as catalysts or reagents that preclude many of the applications currently being developed that are pH sensitive. For the most part, these reactions are nonaqueous chemistries. For example, organotrialkoxysilanes will react with organotrichlorosilanes in the presence of catalytic quantities of a Lewis acid to form the desired siloxane bond and eliminate alkyl chloride as a condensate (Fig. 6.10).

Polymerization leading to silsesquioxanes can also be accomplished by generating silanols under anhydrous conditions. One example is the reaction of organotrialkoxysilanes with anhydrous formic acid. Silanols are generated with the formation of alkylformate. The silanols can then condense to afford siloxane bonds. Another promising and unconventional technique has been applied to polymerizing organotrichlorosilanes with dimethylsulfoxide (DMSO) supplying the oxygen atoms in the siloxane bonds (Fig. 6.11).

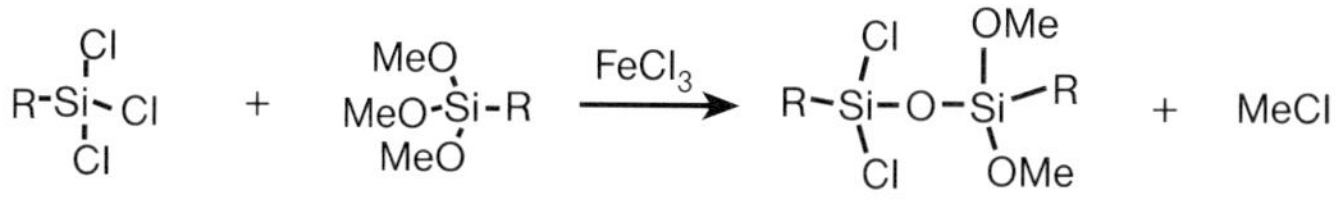

Fig. 6.10 Lewis acid-catalyzed condensation of chlorosilanes with alkoxysilanes to afford siloxane bonds.

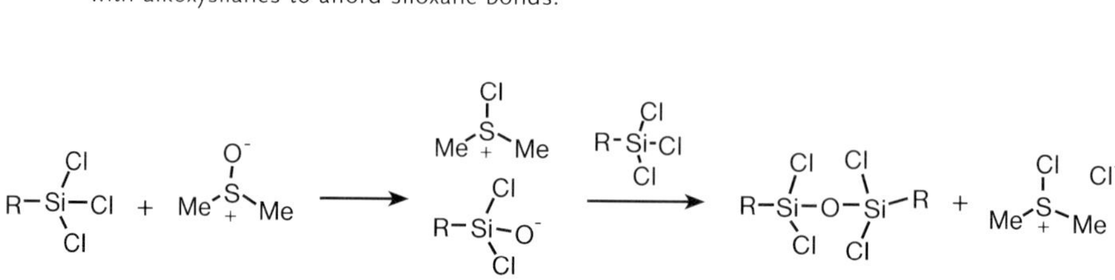

Fig. 6.11 Reaction of chlorosilanes with dimethylsulfoxide to form siliconates that condense with chlorosilanes to afford siloxane bonds.

The oxygen of the DMSO attacks the silicon, replacing a chloride group. The chloride group subsequently attacks the sulfur completing the transfer of the oxygen in the form of a siliconate that attacks another chlorosilane to create the siloxane bond. Transition metal catalyzed "hydrolysis" of organosilanes, [RSiHO]n, can also be used to generate silanols. With early transition metal catalysts or catalytic Bronsted bases and acids, oligomeric and polymeric alkylhydridosiloxanes undergo disproportionation to form silsesquioxane networks and an equal amount of the analogous organosilane. While this chemistry can be used to prepared polysilsesquioxane foams, one cannot avoid the inherent waste associated with losing 50 mol% of the starting material through the disproportionation chemistry.

6.2.3
Characterizing Silsesquioxane Sol–Gels with NMR

Much of what we know about the hydrolysis and condensation of alkoxysilanes has come from solution nuclear magnetic resonance spectroscopy (NMR). In particular, 29Si NMR provides an unparalleled window into the types of silicon species involved during sol–gel polymerizations. Organotrialkoxysilanes generally show a single, sharp 29Si resonance or peak in the NMR spectrum between −40 and −46 ppm for alkyl-substituted monomers, close to −56 to −59 ppm for aryl-, hydrido- and vinyl- substituted monomers (Table 6.1).

Hydrolysis of organotriethoxysilanes results in three new silicon resonances downfield of the monomer representing the products with one two and three silanols (Fig. 6.12). Furthermore, the peaks are found progressively farther downfield with greater degrees of hydrolysis and are well resolved by about 2 ppm. Unfortunately, there is no such corresponding trend observed with the hydrolysis of organotrimethoxysilanes. In this case, the resonances from the three silanol products are downfield from the parent monomer, but are in no particular order and are unevenly separated. In all cases, NMR experiments should be performed in a single solvent system as the silicon-29 resonances of silanols will lie as much as 5 ppm farther upfield in hydrogen bonding solvents such as alcohols, acetone or dimethylsulfoxide than in deuterochloroform.

Condensation of silanols to afford siloxane bonds results in new resonances upfield from the parent monomer (Fig. 6.13). The greater the number of siloxane

Table 6.1 Silicon-29 NMR chemical shifts for the liquid monomers and for their solid polymers.

Monomer	T0	T1	T2	T3
MeSi(OMe)3	−40.1	Not Observed	−57	−65
MeSi(OEt)3	−44.1	−51	−59	−65
PhSi(OEt)3	−57	−62	−71	−78
VinylSi(OEt)3	−59.1	Not Observed	−72	−80

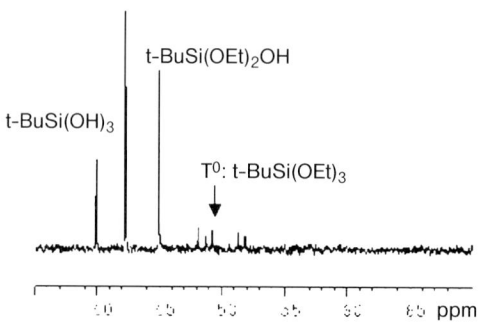

Fig. 6.12 Hydrolysis of ethoxide groups from organotriethoxysilanes, such as t-butyltriethoxysilane (above), can be monitored by the appearance of new peaks downfield of the parent monomer.

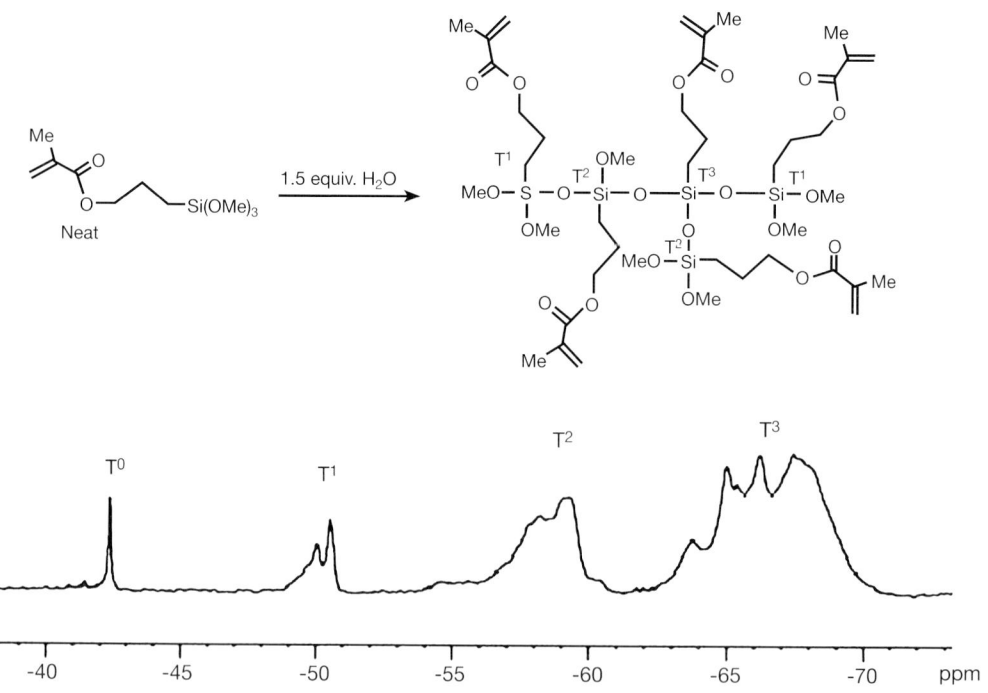

Fig. 6.13 The solution ^{29}Si NMR spectrum of the polymerization reaction pf methacryloxypropyltrimethoxysilane differentiates between silicons based on their extent of hydrolysis and condensation. The spectrum reveals monomer (T^0), singly (T^1), doubly (T^2) and fully (T^3) condensed silicons spaced 7–8 ppm increasingly upfield, respectively.

bonds, the farther upfield the corresponding resonance is found. The first group of peaks, denoted as T1 and lying approximately 8 ppm upfield from the monomer, would be silicons with a single siloxane bond and varying states of hydrolysis. A second cluster of peaks from silicons with two siloxane bonds would lie an additional 8 ppm down field. These are called T2 silicons. Resonances from fully condensed, T3 silicons are located an additional 8 ppm farther upfield but are generally broad and often indiscernible. Unlike the hydrolysis experiments described above, chemical shifts of the groups of T1, T2 and T3 silicon-29 peaks are relatively unaffected by starting with organotrimethoxysilanes.

The up-field shift in silicon resonances with increasing numbers of siloxane bonds is remarkable in its apparent contradiction with the common belief that siloxane bonds act as more electron deficient substituents than hydroxide or alkoxide groups. The more siloxane bonds to a silicon atom, the more electron deficient it should become. This generally results in a downfield shift in the position of a resonance. The upfield shift observed in silsesquioxanes with increasing extent of condensation suggests that the back-donation of electron density from the oxygen lone pairs into the σ^* orbitals and, to a lesser extent, d-orbitals on the silicon are more important than the inductive withdraw of electron density by the more electronegative oxygen atom.

A number of silsesquioxane researchers have used 29Si NMR to study the kinetics of hydrolysis and condensation. Solution NMR has also been used to characterize soluble polysilsesquioxanes and have provided some evidence that would argue against well-organized ladder polymer structures. Application of solid-state 29Si NMR to the characterization of silsesquioxane gels will be discussed in a later section.

6.2.4
Cyclization in Polysilsesquioxanes

Structures of polysilsesquioxanes are based on a finite number of siloxane bond configurations. After three monomers have formed single siloxane bonds to form a linear chain, the next step may be cyclization to afford a six-membered ring or addition of another monomer to give either the branched or the linear tetramer (Fig. 6.14). The four-membered siloxane ring cannot be made through hydrolysis and condensation chemistry. The six-membered ring is kinetically favorable enough to form from hydrolysis and condensation chemistry, but is not common because ring strain makes it less thermodynamically favorable. For example, six-membered rings are not observed in polysilsesquioxanes prepared under thermodynamic conditions such as in sol–gel polymerizations performed in nucleophilic solvents such as alcohols. However, it is possible to prepare fair yields of tetrahedral hexamers composed of six-membered siloxane rings from the hydrolysis and condensation of organotrichlorosilanes in non-nucleophilic solvents. Under these conditions, ring opening through nucleophilic attack is minimized.

Formation of cyclic structures during the sol–gel polymerization has been extensively studied by solution 29Si NMR spectroscopy and mass spectroscopy.

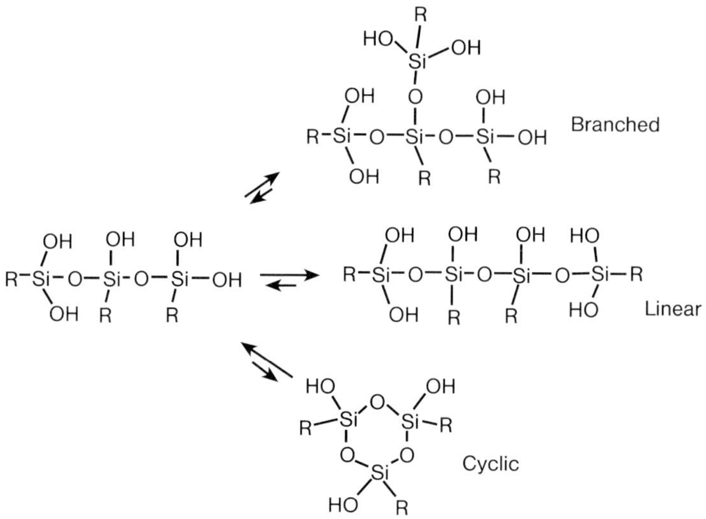

Fig. 6.14 Once the trimer forms, condensation reactions can lead to branched, linear and cyclic products.

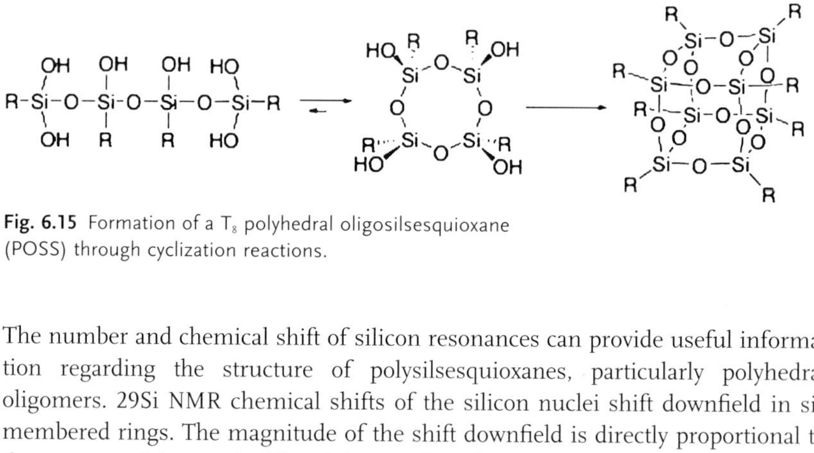

Fig. 6.15 Formation of a T$_8$ polyhedral oligosilsesquioxane (POSS) through cyclization reactions.

The number and chemical shift of silicon resonances can provide useful information regarding the structure of polysilsesquioxanes, particularly polyhedral oligomers. 29Si NMR chemical shifts of the silicon nuclei shift downfield in six membered rings. The magnitude of the shift downfield is directly proportional to the amount of ring strain. The eight-membered ring, resulting from the intramolecular condensation of a tetrameric sequence of monomers, is both kinetically and thermodynamically favorable. A consequence of the tendency to form cyclics is the widespread observation of polyhedral oligosilsesquioxanes (POSS) such as the octameric T8 "cube" that is composed of six eight-membered rings (Fig. 6.15). The formation of this POSS structure is particularly favored when the organic group is relatively bulky. Cyclic tetramers have been isolated for a number of silsesquioxanes. The phenyl-substituted tetramer with all of its hydroxide groups on the same side or "syn" to the ring is easily induced to dimerize forming the T8 product.

Another consequence of cyclization is its influence on the formation of gels. Despite having three reactive groups that could theoretically be converted to an infinite network of siloxane bonds, polymerization of organotrialkoxysilanes rarely affords gels. For example, macromolecules with molecular weights greater than 105 Daltons can be prepared from organotrialkoxysilanes without gelation. Determination of whether branching or linear chain growth occurs is not as simple as the kinetic and thermodynamic contributions favoring cyclization. Sites on the end of siloxane chains that would react to extend the linear chain are referred to as terminal silanols and those with two siloxane bonds to each silicon atom that would react to afford branches as interior silanols. One must consider electronic effects such the electrophilicity of a terminal silicon or interior silicon in a linear chain and steric differences at the sites. There will, depending on the system in question, be different numbers of terminal and interior sites. For example, for the trimer, there would be six terminal silanols and two interior silanols resulting in a ratio of 2:1 favoring the terminal sites. The longer the linear chain the smaller this ratio becomes. It is known for silica gels that a higher degree of branching results from sol–gel polymerizations at high pH than those conducted at low pH. This is confirmed in polysilsesquioxanes by measuring the degrees of condensation with solid-state MAS 29Si NMR (Fig. 6.16).

However, the origin of the difference is less certain. Many researchers hold that the branching is favored at high pH because siliconate intermediates are more stabilized by electron withdrawing siloxane bonds. Conversely, acid catalysis would favor terminal sites because of the electron poor transition state. However, the distribution of branched and linear products from the oligomerization of tetramethoxysilane at high pH mirrors what would be expected statistically based on the numbers of reactive sites. This means the only perturbation from expected probabilities occurs with low pH and that the siloxane bond may exert a significantly smaller electronic effect than has been attributed to it.

Besides controlling the steric bulk of the organic group, managing the amount of cyclization can be accomplished using the nature of the alkoxide group and the reaction conditions. It is well known that silica gels are far more readily formed from tetramethoxysilane than tetraethoxysilane due to the significantly higher population of cyclic structures in the silica formed from the latter.

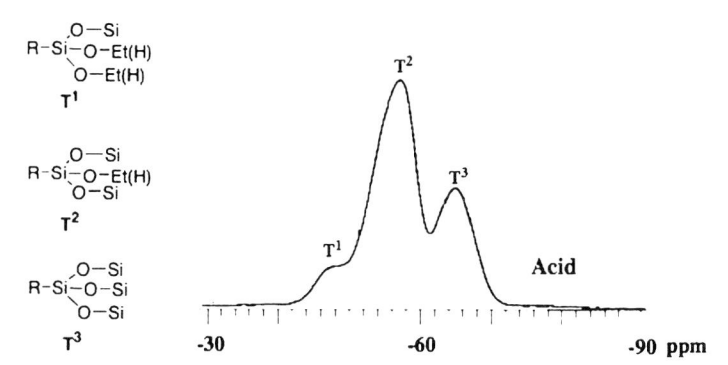

Fig. 6.16 Solid-state silicon-29 NMR spectrum of 1,6-hexylene-bridged polysilsesquioxane.

6.3
Type I Structures: Polyhedral Oligosilsesquioxanes (POSS)

6.3.1
Homogenously Functionalized POSS

Polyhedral oligosilsesquioxanes or POSS are crystalline solids based on six-, eight-, and ten-membered rings annulated into a 3-dimensional structure. POSS are widely investigated as nanoscale building blocks for materials, models for silica surfaces, and nanoscale "filler" for polymer based nanocomposites. While a number of POSS are commercially available, the first material incorporating POSS has reached the market as a durable dental composite. The most commonly observed form is the T8 cubical polyhedral octamer that has six faces, each composed of an eight-membered ring with eight identical organic groups. The organic substituents project outward from the corners of the cube shielding the siloxane core. Other POSS include the strained T6, T10 and T12 polyhedra (Fig. 6.17).

The POSS can be prepared from a number of organotrialkoxysilanes, but successful syntheses historically originated from the hydrolysis and condensation of the organotrichlorosilane in non-nucleophilic solvents. Colorless parallelipoid crystals (Fig. 6.18) are often obtained even from the sol–gel polymerization of

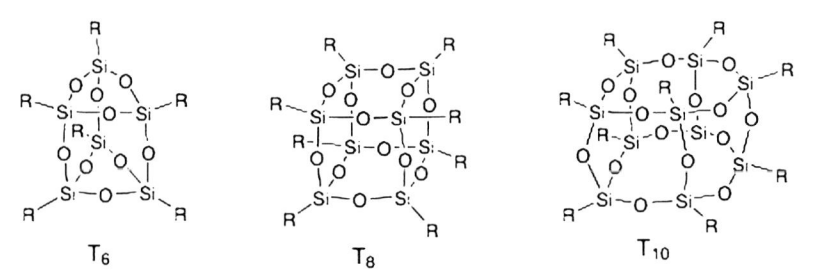

Fig. 6.17 Polyhedral oligosilsesquioxanes including the T_6 hexamer, T_8 octamer and T_{10} decamer. A number of POSS, including T_8 POSS with R = phenyl, cyclohexyl, cyclopentyl and T_{10} with R = phenyl, are commercially available.

Fig. 6.18 Crystals of T_8 octaethyl-POSS grown from the hydrochloric acid catalyzed hydrolysis and condensation of ethyltriethoxysilane in ethanol.

organotrialkoxysilanes. POSS formation is favored by bulky groups that hinder intermolecular condensations that might preclude full cyclization to the polyhedral forms. However, acid or base formulations would convert less than half of the monomer to POSS with amorphous oligomers making up the majority of the product mixture. By switching to fluoride catalysis in the polar aprotic solvent, tetrahydrofuran, T8 POSS in yields as high as 95% can be prepared from trialkoxysilanes and water in a single day.

When only one monomer is used, pure samples of various POSS can be isolated in pure form. Attempts to prepare POSS from several monomers with different organic groups leads to complex mixtures of isomers and different sized oligomers that cannot be separated. POSS formation is favored by bulky groups that sterically hinder intermolecular condensations that might preclude full cyclization to the polyhedral forms. However, the T6 oligomer is relatively unstable, requiring special reaction conditions to achieve high yields. Treatment of the T6 oligomer with acid or base in the presence of additional monomer leads to formation of the more stable T8 oligomer. Once formed, T8 POSS with hydrocarbon substitutents are stable to 500 °C.

The structures of POSS are often determined by X-ray crystallography, though mass spectrometry and NMR are becoming increasingly applied. Silicon-29 NMR with its sensitivity to symmetry and ring strain is a powerful tool for rapidly characterizing the oligomers. In addition to complete POSS, incompletely formed intermediates, such as cyclic tetramers, bicyclic hexamers, and T7 cubes with one corner missing, can occasionally be induced to crystallize in good yields (Fig. 6.19). Incomplete T7 cubes with three adjacent silanols oriented to act as a trifunctional chelating ligand were used to model silica-supported metal catalysts. Studies of the T7 POSS have also provided important insight into the forces, particularly hydrogen bonding, governing the acidity of silanols.

The incomplete condensed POSS, such as T7, have also provided the means for covalently attaching individual oligosilsesquioxanes to polymers to create composite materials. The T7 triol reacts with a second organotrichlorosilane to fill the last corner of the cube and permit the synthesis of cubes with seven unreactive, cyclohexyl substituents and a single reactive substituent (Fig. 6.20). Since there are a large number of functionalized organotrichlorosilanes available, it is possible to prepare new hybrid monomers with polymerizable pendant groups including vinyls, protected alcohols, amines, carboxylic acids, etc. that have enabled the

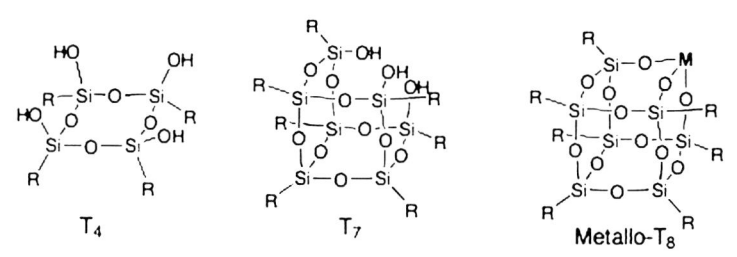

Fig. 6.19 POSS intermediates and derivatives. Heptaphenyl, cyclohexyl T_7 POSS are commercially available.

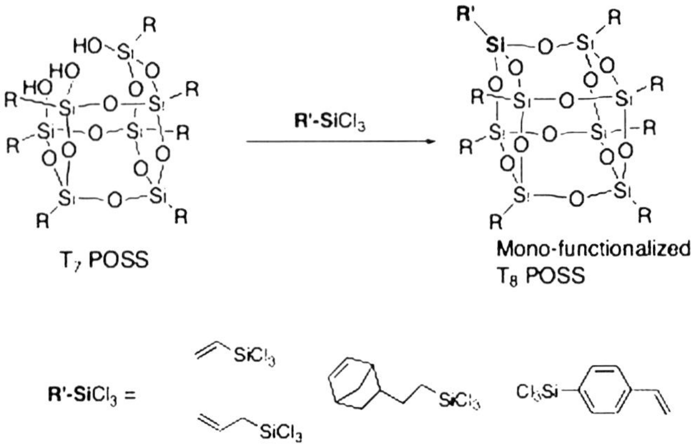

Fig. 6.20 Formation of POSS with a single polymerizable organic substituent from a T_7 POSS and an organotrichlorosilane.

preparation of a whole new class of POSS functionalized polymers. The POSS group has been shown to improve the thermal stability and mechanical properties of organic polymers while avoiding the difficulties of physically mixing in an inorganic filler phase.

Alternatively, cubes can be prepared with reactive functionalities on each organic group that can be chemically modified or used to build materials. Early efforts included the octavinyl T8 that could be hydrosilated, the octahydrido cube that is a convenient precursor to T8 spherosilicates, chloromethylphenyl-substituted cubes whose chloro substituents can be converted into a number of other reactive functionalities including phosphines for building organometallic networks, and phenyl groups that can be completely hydrogenated to cycloalkanes, nitrated then reduced to the amino-functionalized cube, or sulfonated to a strongly acidic POSS.

The ease with which the T8 cube can be prepared and functionalized, coupled with its well-defined dimensions, make it an excellent nanoscale building block for constructing materials. For example, chloromethylphenyl substituted cubes have been used as cores for star polymers built by atom transfer polymerizations. Aminopropyl- and glycidyloxypropyl-substituted POSS have been reacted with each other to afford silsesquioxane-epoxy composites. Amino-functionalized cubes have also been reacted with isocyanate co-monomers to form polyurea networks crosslinked at POSS. However, the high degree of functionality and irreversibility of most of these polymerizations prevents networks and gels with high degrees of order from being assembled.

6.3.2
Stability of Siloxane Bonds in Silsesquioxanes

While it is known that the condensation reaction is reversible, the equilibrium constant, Keq, favoring the siloxane bonds, is between 40 and 250 under sol–gel conditions. Increasing the water concentration shifts the equilibrium sufficiently to slow gelation but not enough to depolymerize the polymer. With strong acid or base, it is possible to open or hydrolyze the siloxane bond leading to opened cages or redistributed siloxane network polymers. As with silica gels, siloxane bonds in silsesquioxanes are particularly susceptible to opening with alkali base. For example, the mixtures of T4–T8 cyclic oligomers with toluenyl substituents polymerize to high molecule weight polysilsesquioxanes by heating with base to 100 °C. Heating with base to 300 °C can, in some instances, convert the polymers back to cyclic oligomers.

6.4
Type II Structures: Amorphous Oligo- and Polysilsesquioxanes

6.4.1
Gelation of Polysilsesquioxanes

Forming polysilsesquioxane gels depends on the nature and concentration of the monomer being used, the pH and to a lesser extent, the amount of water and catalyst. Since gelation occurs when macromolecules in the solution percolate throughout the solution, the molecular weight of the building block can have a pronounced effect on the relative ease of forming the required network. The organic groups in silsesquioxanes act as blocking agents to shield one side of the monomer from reactions that would lead to bond formation. The result is that a significant portion of siloxane bonds in polysilsesquioxanes is located in siloxane rings. Furthermore, as silanols are converted into siloxane bonds, the polysilsesquioxanes may become sufficiently hydrophobic to induce phase separation as a precipitate or resin rather than gelation. In many cases, the polymeric resins are tacky or even viscous liquids due to plasticization of the polymers with solvent.

Because it is well known that three functional groups per monomer would nominally result in highly crosslinked networks or gels, the solubility and tractability of many polysilsesquioxanes made it apparent that the structures must include a large number of siloxane rings. This hypothesis has been unequivocally supported by the isolation of cyclic tetramers, and numerous polyhedral oligosilsesquioxanes. It is possible to construct models for soluble polysilsesquioxanes in the form of linearly connected cyclics or "ladder polymers" that would also explain the solubility of many polysilsesquioxanes. However, careful examination of polyphenylsilsesquioxane revealed that a structure based on lightly interconnected rings more accurately represented the structure (Fig. 6.21) and more accurately portrayed the solution properties of the macromolecules.

Thus, a general lack of convincing X-ray diffraction data, coupled with NMR evidence, and a better understanding of the mechanism and thermodynamics of polysilsesquioxane formation appear to rule out ladder polymers save where additional templating or nonbonding effects may have been operating. However,

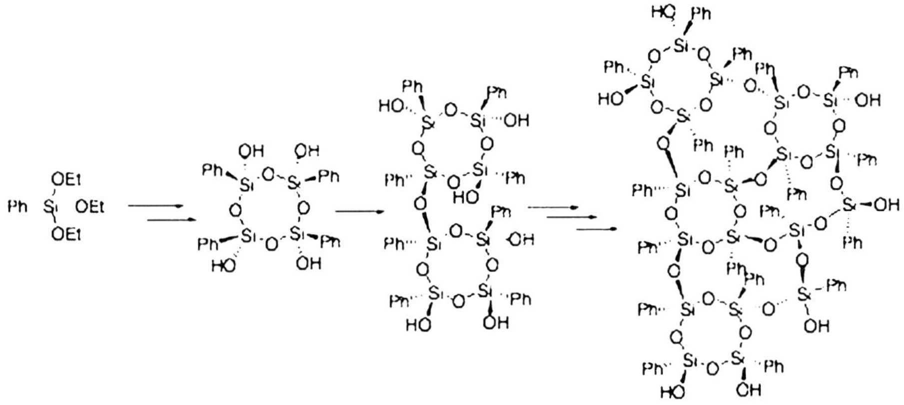

Fig. 6.21 Polymerization of phenyltriethoxysilane affords networks with sufficient cyclic structures to avoid gelation.

the hypothetical ladder polymer structure has been and is still frequently assigned to polysilsesquioxanes without the benefit of any definitive proof.

Silsesquioxane monomers that form gels can be divided into several classes (Fig. 6.22). The first class of monomers form products called "simple" silsesquioxanes

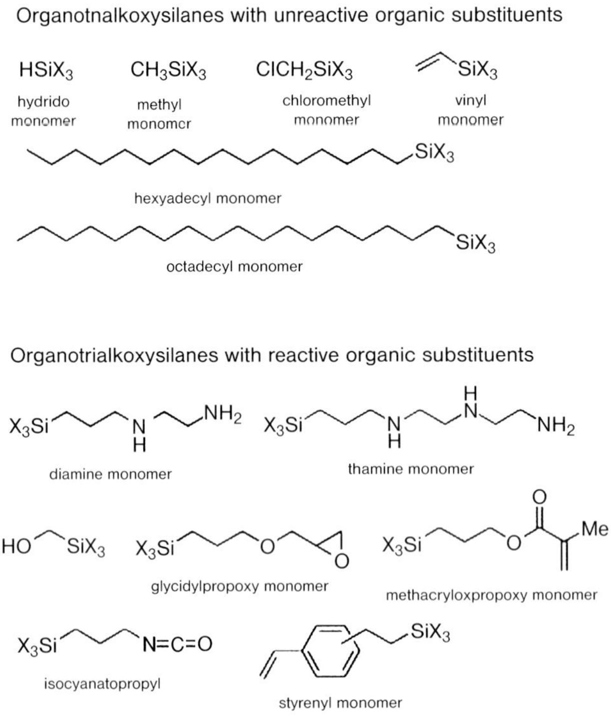

Organotrialkoxysilanes with unreactive organic substituents

$HSiX_3$ — hydrido monomer
CH_3SiX_3 — methyl monomer
$ClCH_2SiX_3$ — chloromethyl monomer
SiX_3 — vinyl monomer

hexadecyl monomer — SiX_3

octadecyl monomer — SiX_3

Organotrialkoxysilanes with reactive organic substituents

diamine monomer

thamine monomer

$HO\sim SiX_3$

glycidylpropoxy monomer

methacryloxpropoxy monomer

isocyanatopropyl — $X_3Si\sim N=C=O$

styrenyl monomer

Fig. 6.22 Gel forming silsesquioxane monomers, $RSiX_3$.

with pendant organic groups. This class can be further divided into monomers with hydrocarbon groups and those monomers that are more polar and possess functional groups that can participate in nonbonding interactions or are reactive enough to form new bonds. The first class is composed of the monomers with hydrogen, methyl, chloromethyl, vinyl, hexadecyl, and octadecyl substituents. The first four monomers have organic substituents that are too small to interfere with polymerization and gelation. Monomers with longer and branched hydrocarbon substituents generally do not form gels. The hexadecyl and octadecyltrialkoxysilanes are a special case in this class of monomers that form gels as oligomers due to the long chain alkyl groups organizing into nonbonding assemblies. These gels, which show signs of lamellar structures in micrographs, can be redissolved with gentle heating.

Monomers with polar functional and reactive groups that can form gels include those with diamine, and triamine groups, hydroxyalkyl, isocyanate, propoxymethacrylate, glycidyloxypropyl and styrenyl groups (Fig. 6.22). The diamine and triamine monomers react to afford oligomers that form gels through hydrogen bonding. These oligosilsesquioxanes are useful as thermoreversible gels for smart material applications. The hydroxyl methyl monomer probably forms gels through a contribution of factors including low sterics, hydrogen bonding, and bond formation from the condensation reaction between the hydroxyl group on the methyl and silanols. The isocyanate, propoxymethacrylate, glycidyloxypropyl, and styrenyl monomers probably form gels because they possess reactive groups capable of polymerizing that complement the sol–gel derived siloxanes in forming a gel forming network. The isocyanate group is so reactive to nucleophilic attack that excess water, base catalysts, and methanol and ethanol solvents should never be used for sol–gel polymerizations. The next most reactive functionality is the epoxy group in the glycidoxypropoxysilane monomers that ring opens, especially at higher pH.

6.4.2
Effects of pH on Gelation

Considering the influence of pH on hydrolysis and condensation, its not surprising that pH also effects the time it takes for a sol–gel polymerization to produce a gel. The plot of gelation time versus pH for silica gels (Fig. 6.23) prepared from water glass or tetraalkoxysilanes are nearly identical with a maximum in gelation time being observed near pH 2, a minimum near pH 5 and then a steady increase in gelation time with pH. In contrast, silsesquioxane silating agents dispersions in water are most stable to phase separation or precipitation at pH 4.5. Because it is relatively difficult to prepare gels from polysilsesquioxanes, studies of the effects of pH on gelation times were impractical until the development of bridged polysilsesquioxanes. Not surprisingly, the pH-gelation time curve revealed gels formed rapidly under acidic conditions (pH < 2) and basic conditions (pH > 11) and gels formed the slowest at pH 5.

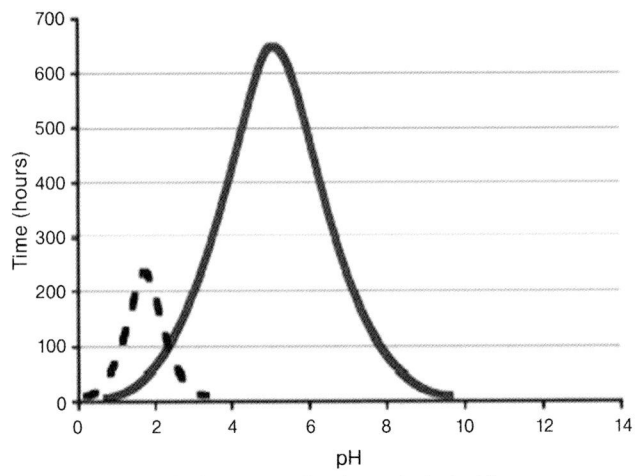

Fig. 6.23 Graph of gelation time for silica gel (dashed line) and an alkyl substituted silsesquioxane (solid line).

6.4.3
Polysilsesquioxane Gels

The gels themselves vary from transparent, brittle, colorless monoliths, which are nearly identical in appearance to silica gels, to fragile, opaque, white monoliths, which are more reminiscent of Styrofoam®. As a rule the gels that form the most readily, such as those from the hydrolysis and condensation of CH_3SiX_3, $ClCH_2SiX_3$, or bridged monomers, are generally the most glass-like. Those with more sterically bulky substituents, are usually white and opaque. Examination of the dried gels with electron microscopy reveals that the gels are networks of pores created by colloidal aggregates formed during the sol–gel polymerization (Fig. 6.24).

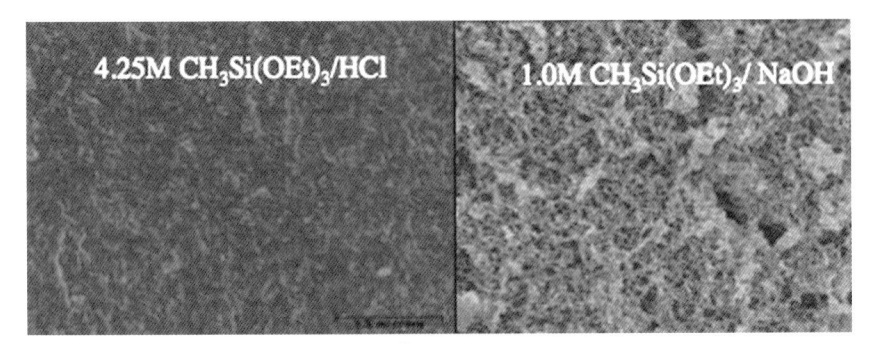

Fig. 6.24 Scanning electron micrographs of methylsilsesquioxane xerogels prepared by acid- (left) and base- (right) catalyzed polymerizations.

As with silica gels, the size of the particles making up these gels often is directly affected by the pH of the sol–gel polymerization. In many cases, acid-catalyzed polymerization affords finer grained structures (Fig. 6.24, left) than those formed under base-catalyzed conditions.

6.4.4
Polysilsesquioxane–Silica Copolymers

One of the most common approaches to hybrid organic materials is to copolymerize an organotrialkoxysilane with a tetraalkoxysilane (Fig. 6.25). In this manner, organic functionalities can be covalently incorporated into a silica network. Since many organotrialkoxysilanes will not polymerize to afford gels, silica gel precursors such as tetramethoxysilane are used to provide sufficient crosslinking to create a gel. This approach is particularly useful when only small quantities of the organic component are needed, such as in dyes for sensor or communications applications or organometallic catalysts. One must be careful to insure that the monomers are reacting with one another. Differences in reactivity, especially at pH ranges where the silica (pH 2) or the silsesquioxane (pH 5) precursors do not quickly form siloxane bonds, could lead to phase segregation and unpredictable properties. Performing the copolymerizations at near pH 1 or above pH 11 should minimize differences in reactivity between the monomers, though it is prudent to monitor the cross-reaction between the growing silica and silsesquioxane species using 29Si NMR.

The other result of copolymerizing organotrialkoxysilanes with tetraalkoxysilanes is that the steric retardation of gelation, common with many organotrialkoxysilanes, will be observed with higher ratios of the silsesquioxane to silica monomers. For example, with 75 mol% phenyltrimethoxysilane with tetramethoxysilane, no gel is obtained. Furthermore, even if gels are the result of a copolymerization, the resulting xerogel will be nonporous if too much of the silsesquioxane co-monomer is used. For example, 75 mol% methyltrimethoxysilane with tetramethoxysilane yields nonporous xerogels.

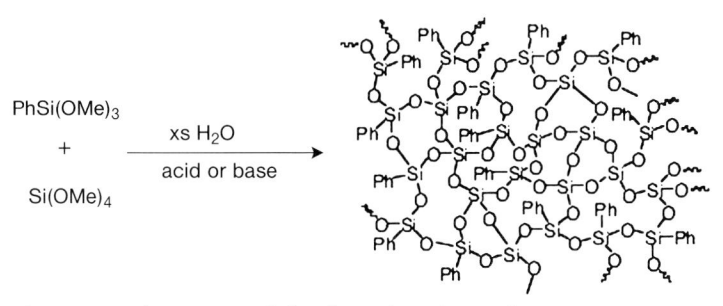

Fig. 6.25 Copolymerization of phenyltrimethoxysilane with tetramethoxysilane yields a silsesquioxane-silica hybrid.

$$n \ \begin{matrix} RO \\ RO-Si-R-Si-OR \\ RO \end{matrix} \begin{matrix} OR \\ \\ OR \end{matrix} + 3n\,H_2O \longrightarrow \left[O_{1.5}Si-R-SiO_{1.5} \right]_n + 6n\,HOR$$

Fig. 6.26 Bridged silsesquioxane are prepared from the hydrolysis and condensation of monomers with two or more trialkoxysilyl or trichlorosilyl groups covalently attached to an organic bridging group (R = arylene, alkylene, alkenylene, alkynylene).

6.5
Type III: Bridged Polysilsesquioxanes

It is possible to make as diverse a body of hybrid organic–inorganic materials based on a single type of silsesquioxane building block as is possible with the copolymerizations of organotrialkoxysilanes with tetraalkoxysilanes. This is accomplished by placing two or more trialkoxysilyl groups on a "bridging" organic group (Fig. 6.26). The larger number of polymerizable trialkoxysilyl groups allows the siloxane network to grow unimpeded with the organic group an integral part of the polymeric scaffolding. As a result, the monomers react quickly to form gels that can be air dried to afford xerogels or supercritically extracted to aerogels. Variations of the basic bridged polysilsesquioxane design include dendrimers with up to 32 separate trialkoxysilyl groups and oligomers and polymers with pendant trialkoxysilyl groups.

6.5.1
Molecular Bridges

A number of monomers with bridging groups are currently commercially available (Fig. 6.27) for modifying or protecting surfaces. The tetrasulfide-bridged monomer has been used as a silating agent for modifying the surface of silica filler particles destined for composites. Dipropylamine-bridged systems have been used for coatings, metal scavengers and "smart" materials that respond to their environment. Several alkylene-bridged monomers have also been used in various protective coating, membrane, and low κ dielectric formulations.

A far greater number of bridged systems have been prepared by hybrid organic–inorganic materials research groups investigating how the bridging configuration can be used to manipulate the physical and chemical properties of the hybrids. Rigid groups, such as phenylene- or acetylene-bridges, generally result in porous xerogels and aerogels. Flexible groups, such as alkylene bridges longer than six carbons (hexylene), can form gels whose pores collapse during drying (Fig. 6.28). This collapse is thought to occur when the network is too compliant to resist capillary stress during drying.

The compliance of a network is dependent on the degree of crosslinking in the network and the flexibility of its structural components. The degree of condensation, measured by solid state 29Si NMR, in a silsesquioxane gel is the measure of

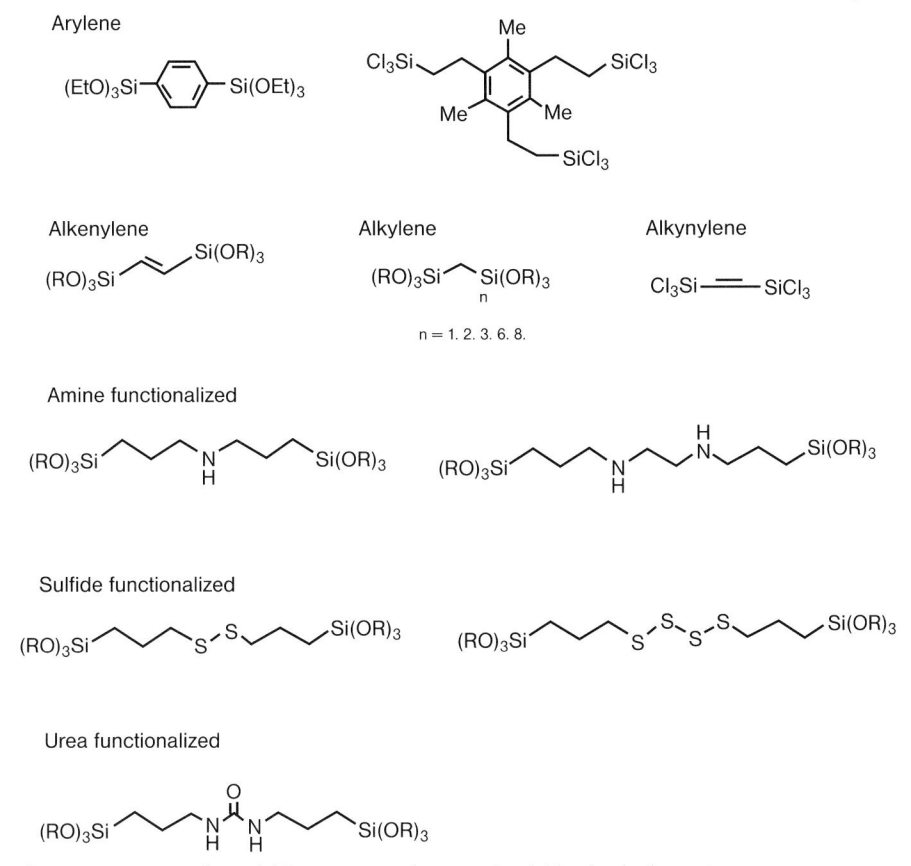

Fig. 6.27 Commercially available monomers for preparing bridged polysilsesquioxanes.

the number of siloxane bonds formed. Generally, the higher the degree of condensation, the less compliant a gel is and the more likely a porous xerogel will be obtained. Base-catalyzed sol–gel polymerization of alkylene-bridged monomers affords bridged polysilsesquioxanes that are more crosslinked than those prepared under acidic conditions and generally are porous. Gels prepared by acid catalyzed sol–gel polymerizations will be nonporous when the bridging group is six carbons in length or longer. One must be careful, however, because both acyclic (crosslinking) siloxanes and cyclic (noncrosslinking) siloxanes contribute to the experimentally measured degree of condensation.

The real value of bridged polysilsesquioxanes is in their utility for preparing highly functionalized gels. For example, high capacity adsorbents for heavy metals and chemical and biological weapons have been prepared by copolymerizing the phenylene-bridged monomer with high loadings of mercaptopropyltriethoxysilane. Porous xerogels of the thiourea-bridged polysilsesquioxanes can be used to recover transition metals from water. High surface area catalysts can be

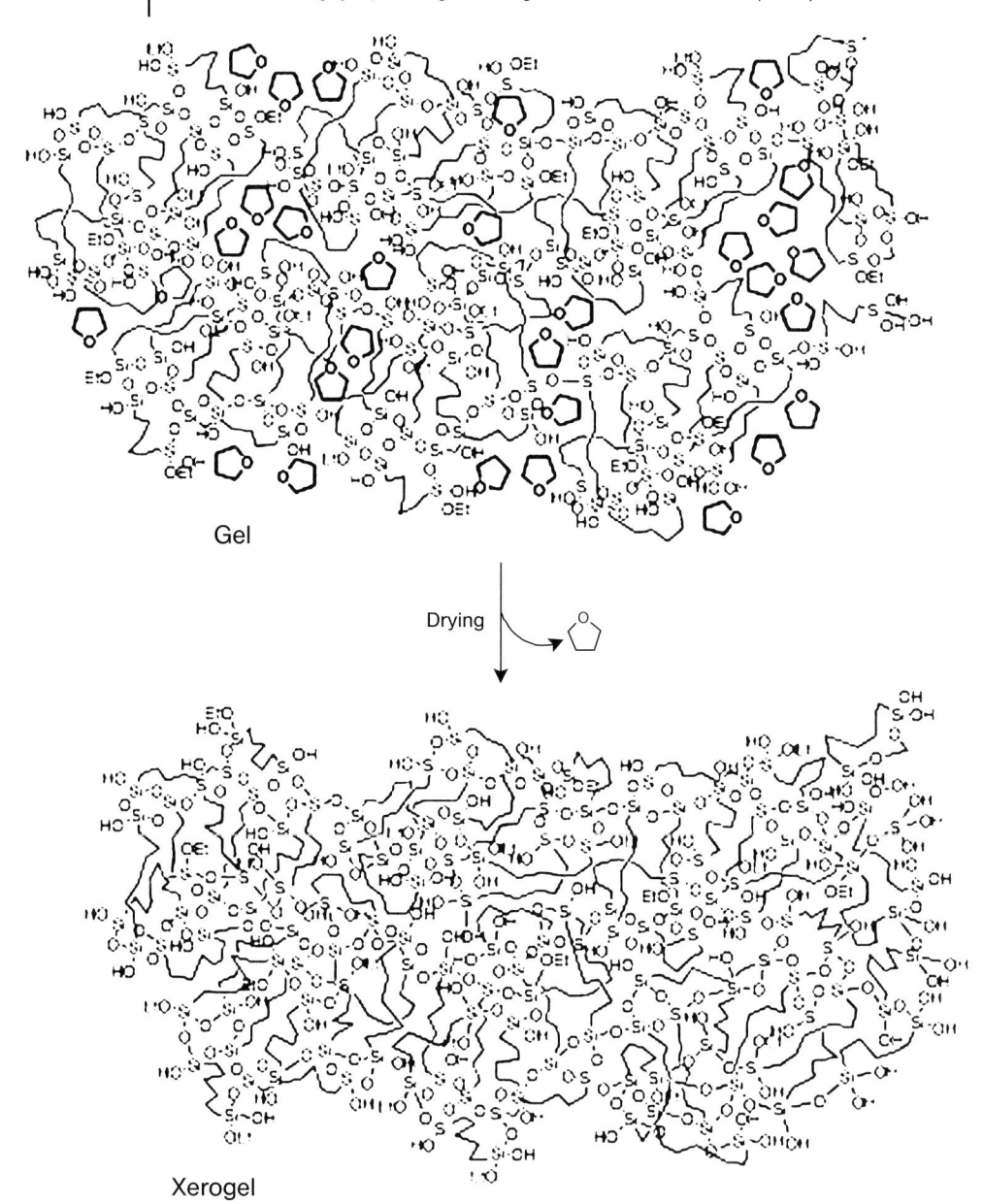

Fig. 6.28 Collapse of solvent occupied spaces during drying of the gel can afford nonporous xerogels. Collapse is believed to be facilitated by flexible structures.

prepared using organometallic compounds as the bridging group. In all of these cases, the ligands for binding metals can be incorporated into high surface area xerogels without causing the porosity to collapse. Similar levels of functionality in silsesquioxanes with pendant organic groups would result in sufficient loss of crosslinking to result in nonporous polymers.

The capacity of bridged polysilsesquioxanes to form as highly functionalized gels has been used for encapsulation. For example, organic and organometallic dyes or inorganic quantum dots can be dispersed in optically transparent, polysilsesquioxane xerogels for photonics applications. Alkylene-bridged polysilsesquioxanes were polymerized around enzymes to slow their denaturization. Construction of bridged monomers in which the dye is the bridging group extends the lifetime of nonlinear optical effects in electrical field oriented films. Covalent attachment also prevents dyes from leaching out of the matrices and retards some photochemical degradation processes. Encapsulation has been shown to slow denaturization of encapsulated proteins while allowing small molecules to be transported to and from the biological catalysts.

Sol–gel processed polysilsesquioxanes are generally amorphous solids with no long-range order such as that commonly observed in crystalline zeolites. However, the bridged structural motif and high degree of functionality lends itself to introducing order through nonbonding interactions established by the organic bridging group. There has been some evidence of order in the development of birefringence in thin films of bridged polysilsesquioxanes with mesogenic bridging groups (Fig. 6.29). More recently, bridging groups with urea groups that can hydrogen bond in the solution before gelation have been used to prepare crystalline polysilsesquioxanes. Lastly, bridged polysilsesquioxanes have been adapted to surfactant templating procedures to generate mesoporous materials in which the spatially restricted sol–gel polymerization creates almost crystalline-like order.

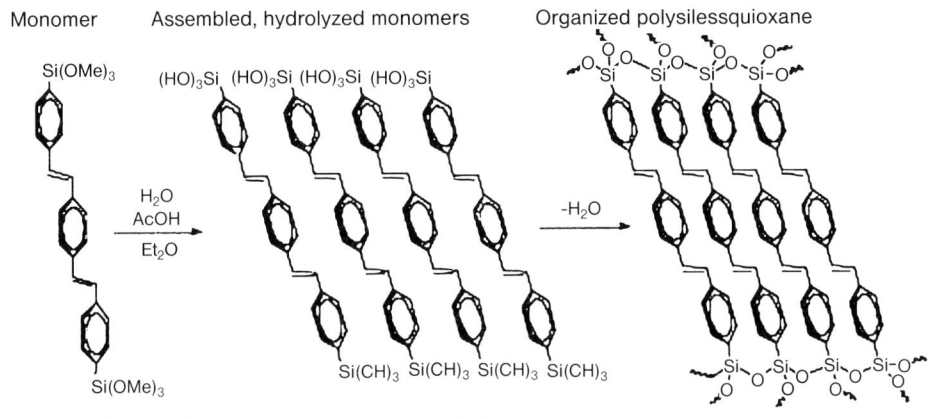

Fig. 6.29 Hydrolysis of monomers with mesogenic bridging groups can permit self-assembly into ordered structures during the sol–gel polymerization.

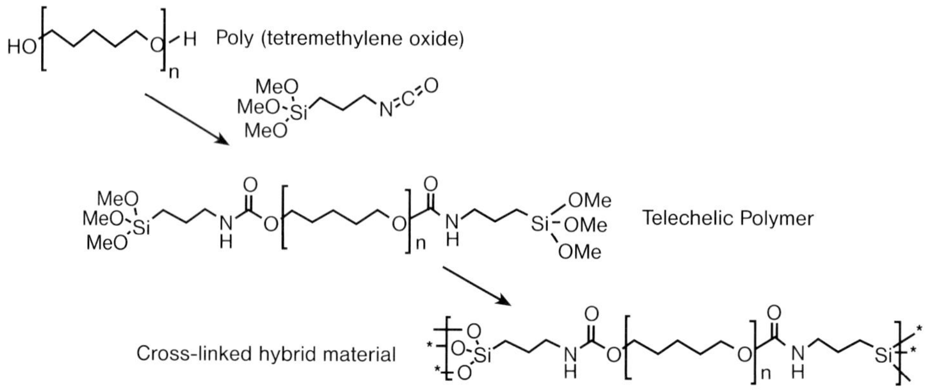

Fig. 6.30 Attachment of trialkoxysilyl groups to the ends of a polymer creates a macromolecule-macromolecular bridged polysilsesquioxane useful for preparing nanocomposites.

6.5.2
Macromolecule-bridged Polysilsesquioxanes

Macromolecular analogs of bridged polysilsesquioxanes were developed as an early form of ormosil nanocomposite with improved mechanical and adhesive properties (Fig. 6.30), with silsesquioxane groups at either end of the macromolecule.

These telechelic polymers permit the distance between crosslinks in the resulting composite to be exactly controlled. A common, less defined variant of this theme are silsesquioxane-functionalized polyolefins that have been commercialized as wire coating materials. There are a number of methods for preparing these materials. One method is to copolymerize vinyl monomers with trialkoxysilyl groups with olefin monomers. This works well with acrylates and styrene monomers that can polymerize by themselves or with other monomers under free radical polymerization conditions. These but not with simple olefins, such as vinyltrialkoxysilanes, which are relatively unreactive by themselves and poor monomers in copolymerizations. A more common approach to preparing silsesquioxane–polymer hybrids is to modify existing polymers with trialkoxysilylethyl groups using free radical chemistry. Regardless of the approach used, these silsesquioxane–polymer composites are used commercially for shrink-wrap coatings for electrical wiring.

6.6
Summary

The combination of organic and inorganic (siloxane) into a single building block makes silsesquioxanes a useful material for many commercial applications and for basic research into the creation of new polymers and composites. Three reactive groups in the monomer and three siloxane bonds to each repeat unit insures that silsesquioxanes can be formed in a wide range of architectures, ranging from polyhedral oligosilsesquioxanes (POSS) to macromolecular bridged silsesquioxanes composites. The potential for performing reactions at both silicon and organic components, coupled with the mild hydrolysis and condensation chemistry used to prepare siloxane bonds in the silsesquioxane component, extend the scope of these materials even further.

6.6.1
Properties of Polysilsesquioxanes

It is difficult to summarize the properties of polysilsesquioxanes because they are as diverse as the different forms that can be synthesized and processed. However, there are some generalizations that can be made. The thermal stability of polysilsesquioxanes can be as high as 500 °C, but varies considerably depending on the stability of the organic group. The most detrimental substitutents are nucleophilic groups, such as amines (Fig. 6.22), which aid in the breaking of siloxane bonds, particularly at higher temperatures. Physically, polysilsesquioxanes can be made as low modulus, weak resins (Type II) or as high modulus, brittle materials (Type III). These low modulus phases are not attractive for bulk applications, but allow polysilsesquioxanes to perform exceedingly well as coupling agents between organic polymers and inorganic fillers. The high modulus, Type III materials are generally amorphous, often porous glasses whose most desirable attribute is an organically tailored surface area. Bulk materials with intermediate properties are prepared by incorporating organic polymers into a composite structure or by combining POSS with organic polymers in compostes.

6.6.2
Existing and Potential Applications

Without question, the most important application for polysilsesquioxanes is for surface modification to strengthen composites or to provide protection to sensitive surfacs. Millions of kilograms of the disulfide-bridged silsesquioxane (Fig. 6.27) are used as a silica-rubber coupling agent. Large quantities of aminofunctionalized silsesquioxanes (Fig. 6.22) are used as coupling agents for glassfilled Nylon and in protective coatings for metal surfaces. Alkyl-silsesquioxanes solutions, kept at pH 4 to extend shelf-life, are used to make glass windows hydrophobic and easier to see through in rain. Methylsilsesquioxane formulations have been used to slow deterioration in architectural stonework and protect against weathering. Silsesquioxanes with long chain alkyl groups have been used to silylate and tailor the surface characteristics the surface of stationary phases in chromatographic columns. Recently, chromatographic column packing materials based on silica-methylsilsesquioxane and silica-ethylene-bridged polysilsesquioxanes particles have been shown to outperform many silica or surface modified silica columns in high resolution separations. Bulk applications of polysilsesquioxanes are relatively rare. An enormous amount of methyltrichlorosilane is burned to create fumed silica. Dental composites based on POSS modified polymers may be the first of this new class of materials to compete with engineering plastics and filled thermoplastics.

Potential applications of polysilisesquioxanes that have been demonstrated in the lab, but have yet to have broad impact on the market, are too numerous to mention. A few examples that highlight the benefits of polysilsesquioxanes to solving materials challenges include high capacity metal scavengers based on thiol-functionalized or thiourea, chemoselective sensor coatings, fuel cell membranes, encapsulants of biochemicals and living cells, and coupling agents for binding catalysts, enzymes or other biochemicals to surfaces.

Bibliography

Ronald H. Baney, Maki Itoh, Akihito Sakakibara and Toshio Suzuki, "Silsesquioxanes," *Chem. Rev.*, **1995**, 1409–1430.

E. P. Plueddeman, "Reminiscing on Silane Coupling Agents," *J. Adhesion Sci. Tech.*, **1991**, *5*, 261–277.

Philip G. Harrison, "Silicate Cages: Precursors to New Materials," *J. Organometallic Chem.*, **1997**, *542*, 141–183.

Alan R. Bassindale, Zhihua Liu, Iain A. MacKinnon, Peter G. Taylor, Yuxing Yang, Mark E. Light, Peter N. Horton, Michael B. Hursthouse, "A Higher Yielding Route for T8 Silsesquioxane Cages and X-ray Crystal Structures of Some Novel Spherosilicates," *Dalton Trans.*, **2003**, 2945–2949.

Kenneth J. Shea and Douglas A. Loy, "Bridged Polysilsesquioxanes. Molecular Engineered Hybrid organic–inorganic Materials," *Chem. Mater.*, **2001**, *13*, 3306–3319.

7
Natural and Artificial Hybrid Biomaterials

Heather A. Currie, Siddharth V. Patwardhan, Carole C. Perry, Paul Roach, Neil J. Shirtcliffe

7.1
Introduction

Materials that are implanted to repair, replace or augment existing tissues in the body are generally known as biomaterials. In the wider context covered in this chapter, biomaterials will also include all materials formed in biological systems, e.g. the specific products of biomineralization. Development of biomaterials, both as products and in understanding their *in vivo* behavior, has been driven largely by the desire to assist in the care for human patients. The materials forming processes occurring in living organisms require much milder reaction conditions than are currently used in the laboratory, such that a new area of materials chemistry, "biomimetics" has been established where scientists are taking ideas from biology to help generate "softer" routes to useful materials.

Biomaterials present in nature provide the necessary structure and architectures of all animal and plant species on earth and function to maintain the structure of organs as well as the organism itself. In nature the materials that are used are polymers, such as polysaccharides and proteins and a relatively small number of simple insoluble oxides and salts. These can be put together in a wide range of combinations to produce materials that are soft, materials that are hard, materials that are flexible, materials that are elastic, etc. In contrast, the range of available materials for biomedical applications is vast and includes metals, polymers, ceramics and composites thereof. In the design of medical devices materials are chosen to suit their intended use and the proposed implantation area. The materials that are used must have compatible properties with the location in which they are placed. Properties such as tensile strength, toughness, elasticity and hardness have to be considered and other factors, such as material transparency, may have to be thought about if, for instance, the device is to be used within the eye. Although the technologist has a wider array of materials at their disposal it is not a simple matter to come up with a material or series of materials that fulfils all the criteria required for successful implantation/biomedical use. This is due to the fact that as evolution has taken place over millions of years, the intricate natural materials that have developed are ideally suited to their function, whether it be support, sensing, use as an element store or as a deterrent. Nature still has a significant advantage over any bioengineer attempting to design materials to replace or mimic those in living organisms. In order to make progress it is imperative that a detailed understanding of these natural materials is gained before the full complexity of the problem can be appreciated and solutions proposed. On a positive note though, the future of biomaterials seems limitless. Various applications in this field that may have been thought of as radical last year are possible today. For example, metallic pins, wires and screws for skeletal fixation and repair were the first foreign materials implanted in the body. Such products are still used today but are forever changing in their design, to increase their use, lifetime and reduce patient trauma. Intelligent biomaterials are now being introduced, able to respond to the body's requirement rather than operating at a constant pre-determined rate, an example being the delivery of drugs.

This chapter will initially describe the building blocks (inorganic and organic) that are principally used to construct hybrid biomaterials before describing in some detail the process of biomineralization and the control mechanisms thought to operate in the generation of species specific materials that can be loosely described as natural hybrid biomaterials. Information on artificial hybrid biomaterials then follows including examples of paints from the ancient world as well as modern day solutions to controlled drug release. For any material to be used in the body, issues of biocompatibility arise and Section 7.5 describes the responses of the material and the host before conclusions are drawn.

7.2
Building Blocks

Natural bio-hybrids are produced by organisms for specific purposes including structural support, sensing, ion storage and toxic waste removal. The composite materials, such as bones, teeth, spines and shells (see below for more details), show a wide range of morphologies and yet are made from a relatively limited number of chemical components. It is the combination of these components assembled under genetic control that leads to composite materials with both intriguing structures and specific functions.

7.2.1
Inorganic Building Blocks

The most common metal ions used are calcium, iron and silicon and the most common non-metals are oxygen and oxygen in combination with carbon in carbonate and oxalate ions and oxygen in combination with phosphorous in phosphates. The compounds that are formed are "simple" salts and oxides that occur as both crystalline and amorphous phases, except for silica, which has only been

Table 7.1 Major biominerals.

Mineral	Forms	Functions
$CaCO_3$	calcite, aragonite, vaterite, amorphous	exoskeleton, eye lens, gravity device
$Ca_2(OH)PO_4$	apatite, brushite, octa calcium phosphate, amorphous	endoskeleton, calcium store
$CaC_2O_4(xH_2O)$	whewellite, whedellite,	calcium store, eterrent
Fe_3O_4	magnetite	magnet, teeth
$FeO(OH)$	goethite, lepidocrocite, ferrihydrite	iron store, teeth
SiO_2	amorphous	skeleton, deterrent

found in the amorphous form in living organisms. Table 7.1 gives the chemical formulae of the important biominerals together with their common names and functions.

For calcium carbonate, phosphate and oxalate there exist several forms (polymorphs) that differ in their arrangement of the cations and anions e.g. for calcium carbonate, calcite, aragonite, vaterite and an amorphous form are all known. These minerals all contain carbonate groups (CO_3^{2-}) that have a single carbon atom in the center of three oxygen atoms arranged in a triangle. How the mineral forms vary is in the coordination of the metal ion that varies between 6-fold coordination (calcite), 8-fold coordination (vaterite), 9-fold coordination (aragonite) and variable coordination (amorphous form). The structure of calcite is analogous to the cubic structure of sodium chloride with sodium and chloride ions being replaced by calcium and carbonate ions. The unit cell (the smallest structural component that represents the whole structure) is flattened to rhombohedral symmetry and the result is a structure that consists of alternating layers of cations and anions parallel to the main axis. The shape of a typical crystal is shown in Fig. 7.1a. The crystal form of aragonite is somewhat different, as seen in the Fig. 7.1b and arises from an essentially open hexagonal packing of metal ions with layers

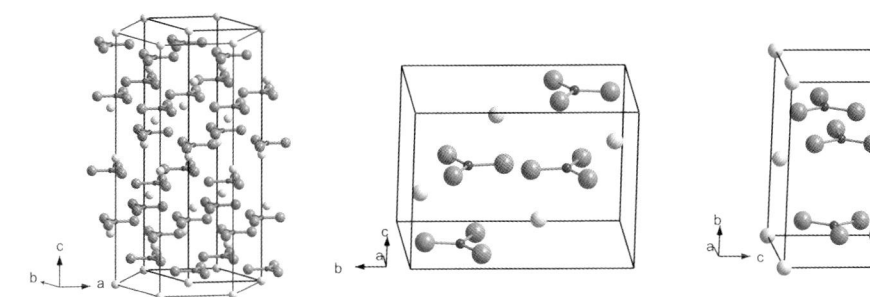

Fig. 7.1 Schematics of the crystal forms of three polymorphs of calcium carbonate – calcite, aragonite and vaterite (from left to right).

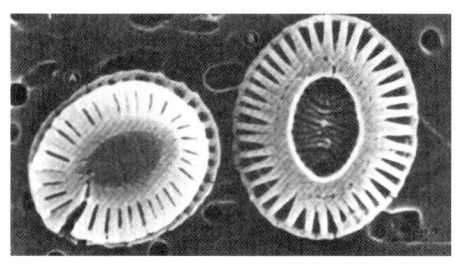

Fig. 7.2 Calcium carbonate (calcite) skeleton of a coccolith. (Taken from S. Mann, *Biomineralization: principles and concepts in bioinorganic materials chemistry*, Oxford University Press, New York, **2001**.)

of carbonate groups in-between, such that the cations coordinate with nine oxygen atoms. The symmetry of the mineral is reduced from hexagonal to orthorhombic as the corners of the triangular carbonate groups do not all point in the same direction, although crystals do sometimes show pseudohexagonal shapes. Vaterite is another form in which the calcium ions are eight coordinate. Crystallites of these different polymorphs of calcium carbonate are used to build up the beautiful species specific structures that we find in nature.

The minerals that are used by biology are simple but the structures that result are far from simple as a picture of a calcium carbonate "shell" around a single-celled organism show, Fig. 7.2, Section 7.3 on biomineralization will indicate some of the likely controls operating to generate such complex morphologies.

Only calcium carbonate and calcium oxalate approach a true stoichiometry (but note magnesium rich calcite is known) and for the others there are a considerable range of compositions that give rise to the individual phases. In particular, the apatitic phases that are used for bone and teeth can accommodate a range of other metal ions (Mg, Sr, Si) and anions (F, OH, CO_3) etc that lead to changes in strength/hardness of the material and its dissolution characteristics, e.g. carbonated apatite is found in bones and fluorapatite in teeth, with the latter being "stronger" than the former (see Section 7.3 for details).

As another example, although only amorphous (no ordering below *ca.* 1 nm), silica is known as a biomineral (for models of an amorphous and crystalline form of silica see Fig. 7.3). The properties of the material as isolated from different environments suggest that the molecular formula SiO_2 "hides" or "encompasses" many distinct forms of the mineral that differ in terms of water content, hydroxyl ion content and sizes of the fundamental units and their organization. The formula that more accurately describes the mineral is $SiO_n(OH)_{4-2n}$ ($n = 0$ to 2).

This ability to take "simple" materials and manipulate them to produce functional materials requires that all processes during their formation be controlled. For this to happen, both nucleation and growth of the mineral phases have to occur. The principles of crystal/particle nucleation and growth are described below. The controlling properties of organic components will be described in Section 7.3 on biomineralization, as it is by the juxtaposition of the inorganic mineral phase

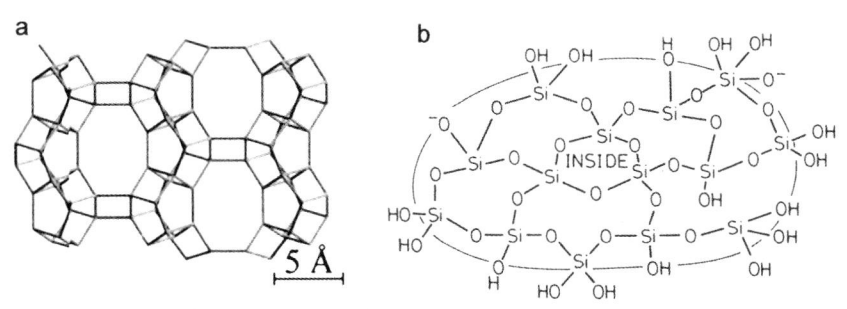

Fig. 7.3 Structures of (a) crystalline (zeolite) and (b) amorphous silica. Note in (b), the Si-O bond length and angle will be variable. (Image in (a) courtesy of Professor Geoffrey Ozin.)

with an organic "controlling" phase that composite materials showing properties of true hybrid materials are generated. It should be noted that artificial hybrid materials are not limited by availability of "raw materials" in water or toxicity, so the number of potential compounds is much greater.

7.2.1.1 Nucleation and Growth

Before we gain an understanding of biomineral formation, control strategies and the roles of organic phase in biomineralization, it is necessary to understand the fundamental aspects of mineral nucleation and growth. Biominerals are formed from an aqueous environment by:

- nucleation
- crystal growth or amorphous precipitation
- ripening.

In general, concentration of ions/molecules above the supersaturation level leads to the formation of tiny (a few angstroms in diameter) species or clusters that are called *nuclei*. In the case of crystalline mineral formation, the structure of these nuclei can control the polymorphs formed and also the ultimate shapes of the minerals. The levels of supersaturation/concentration of mineralizing precursors regulate mineralization (Lussac's law).

Supersaturation is when the concentration of the ions or molecular species in question is greater that the solubility product constant (a function of activities of ions in solution that are in equilibrium with pure solids). Supersaturation can be represented as follows (adapted from K. Simkiss, K. M. Wilbur, *Biomineralization*, Academic Press, San Diego, 1989):

$$K_{sp} = C_X^+ C_Y^- f_X^+ f_Y^- \tag{1}$$

where C_i is concentration and f_i is the square of the mean activity of a given ion. This equation implies that supersaturation can be achieved by regulating the concentration of one of the two ions involved in mineralization. Furthermore, different polymorphs typically have different K_{sp} values, which means that the concentration and activities of ions in a given solution can control the crystal structure.

In order for ions to form clusters to serve as nuclei, it is necessary for these molecular clusters to be stable in a given solution. The free energy for nucleation must be negative (implies a favorable process) and is comprised of two components, Equation 2. The formation of nuclei increases the surface energy. This increase in surface energy is counter-balanced by the formation of bonds in the "solid" or center of the nuclei (adapted from K. Simkiss, K. M. Wilbur, *Biomineralization*, Academic Press, San Diego, 1989).

$$G_N = G_{solid} + G_{surface} \quad (2)$$
$$\quad\quad (-ve) \quad (+ve)$$

As shown in Fig. 7.4 there is an optimum cluster size, known as the *critical nucleus radius*, r^*, at which ΔG reaches a maximum and beyond this point, the formation of stable nuclei is favored.

The number of nuclei formed is dependent on supersaturation levels as is the rate of their formation. Up to a certain supersaturation level the solution can

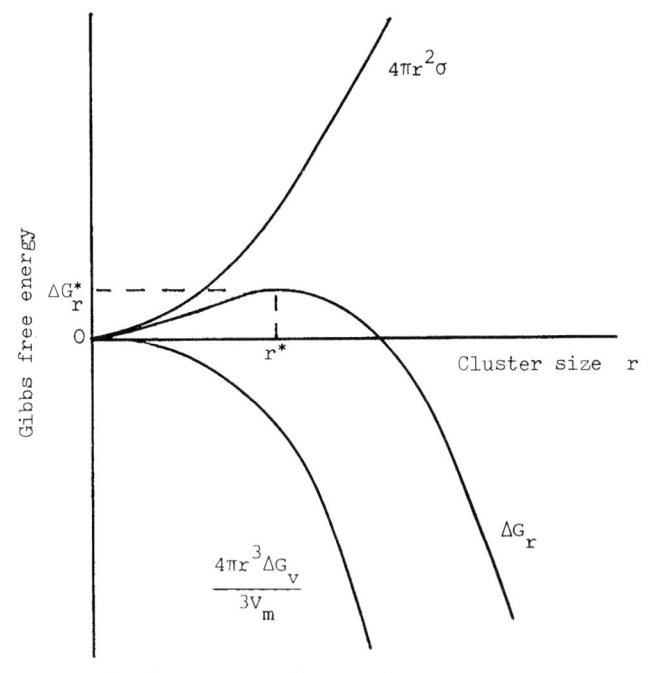

Fig. 7.4 Gibbs free energy as a function of cluster size indicating a critical nucleus size r^*. (Image taken from G. H. Nancollas, *Biological Mineralization and Demineralization*, Springer-Verlag, Heidelberg, **1982**.)

be regarded as *metastable* wherein no nucleation occurs. Once the supersaturation reaches its critical value S*, *homogeneous nucleation* takes place rapidly (see Fig. 7.5).

S* is attained in biological mineralization by several different reaction pathways that are discussed in Section 7.4 below. In the presence of an external surface, *heterogeneous nucleation* may occur. It should be noted that the energy barrier for heterogeneous nucleation is significantly lower than the barrier for homogeneous nucleation because less new surface is created and hence less surface energy cost.

The next stage in mineralization is either *crystal growth* (for crystalline minerals) or *amorphous growth and precipitation* (for amorphous minerals). The crystals can grow either by a slow *equilibrium process* whereby the crystal morphology and "type" is conserved throughout crystallization and the crystal size increases steadily. Or, faster *kinetic growth* may occur due to variations in local concentrations. In the case of kinetic growth, the growth of the crystal faces is non-uniform and can lead to changes in morphology. The role of other species (ions and organic molecules) in the control of morphology is described further below.

Amorphous growth and precipitation typically lacks a nucleation stage, although it is noted that the formation of silica nuclei has been regarded as an important step. Due to the lack of any fixed three-dimensional (3-D) structure, unlike that

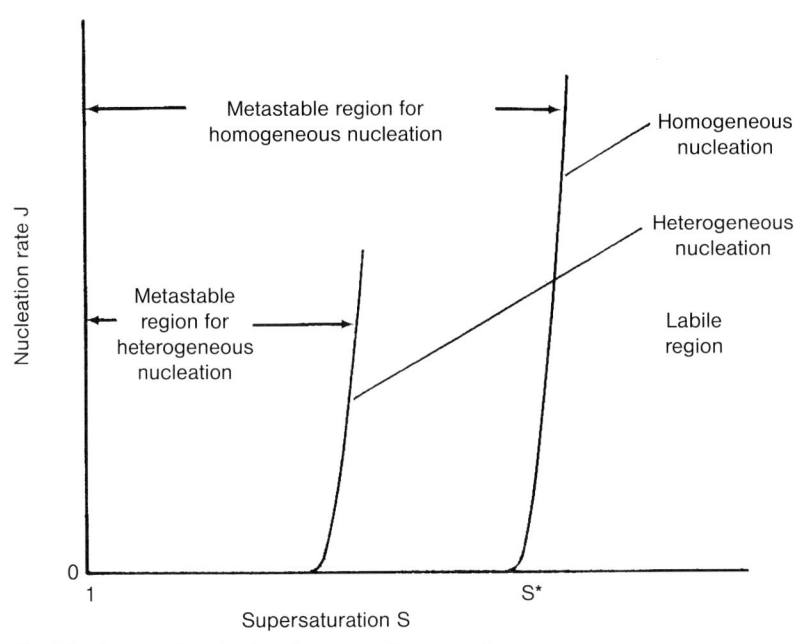

Fig. 7.5 The rate of nucleation shown as a function of supersaturation. (Image taken from G. H. Nancollas, *Biological Mineralization and Demineralization*, Springer-Verlag, Heidelberg, **1982**.)

observed in crystalline minerals, amorphous mineral formation usually occurs via localization of mineral precipitation. Such localization can be brought about using various strategies (see *spatial control* in Section 7.3).

Nucleation and growth is followed by what is termed *ripening or maturation*. In this stage of mineral formation, the growth of minerals continues at the expense of smaller, relatively unstable and more soluble species. The cessation of growth is associated with the cessation of supply/depletion of the required ions at the reaction site.

7.2.2
Organic Building Blocks

Hybrid materials are those that contain both organic and inorganic components and, as such, possess properties of each as well as properties that may be a consequence of intimate interactions at the molecular level between the different types of material. The organic biomolecules used in such natural "hybrid" materials are proteins, carbohydrates and lipids and/or combinations of these. The structural characteristics of each biomolecule type are described in some detail below.

7.2.2.1 Proteins and DNA
Proteins have diverse roles in nature and play important roles as structural components, in reactions as enzymes and in immunology.

The side chain functionality R can be acidic, basic, aliphatic, hydrophobic, nitrogen containing, sulfur containing etc, which join to form the protein primary sequence. Fig. 7.6 gives some illustrative examples of selected amino acids that are commonly found in structural proteins such as collagen. Under physiological conditions (often around pH 7.4) many amino acids have considerable polar character and are involved in hydrogen bonding. This is very important in the folding of the protein to generate regular recurring orientations, such as α helix or β sheet structures – the secondary protein structure, see Fig. 7.7a. Furthermore, weak electrostatic interactions between amino acid side chains gives rise to a 3-D conformational shape, which is known as the protein tertiary structure, Fig. 7.7b; and finally the quaternary protein structure is the interaction of two or more different polypeptides or subunits to give a unique spatial relationship of these components.

Although there are a limited number of naturally occurring amino acids that are used to generate the protein backbone these may be modified by a cell controlled process known as post-translational modification to produce naturally modified amino acids, such as the conversion of cysteine to cystine and the formation of hydroxyproline from proline as found in collagen (discussed later). These modifications may also include phosphorylation (the addition of a phosphate group to hydroxyl containing amino acids) and glycosylation (the attachment of one or more oligosaccharide chains to a polypeptide backbone). The glyco-components have a strong impact on the biological properties of the biomolecules in which they are found including effects on solubility, biosynthesis, stability, action and turnover of these macromolecules.

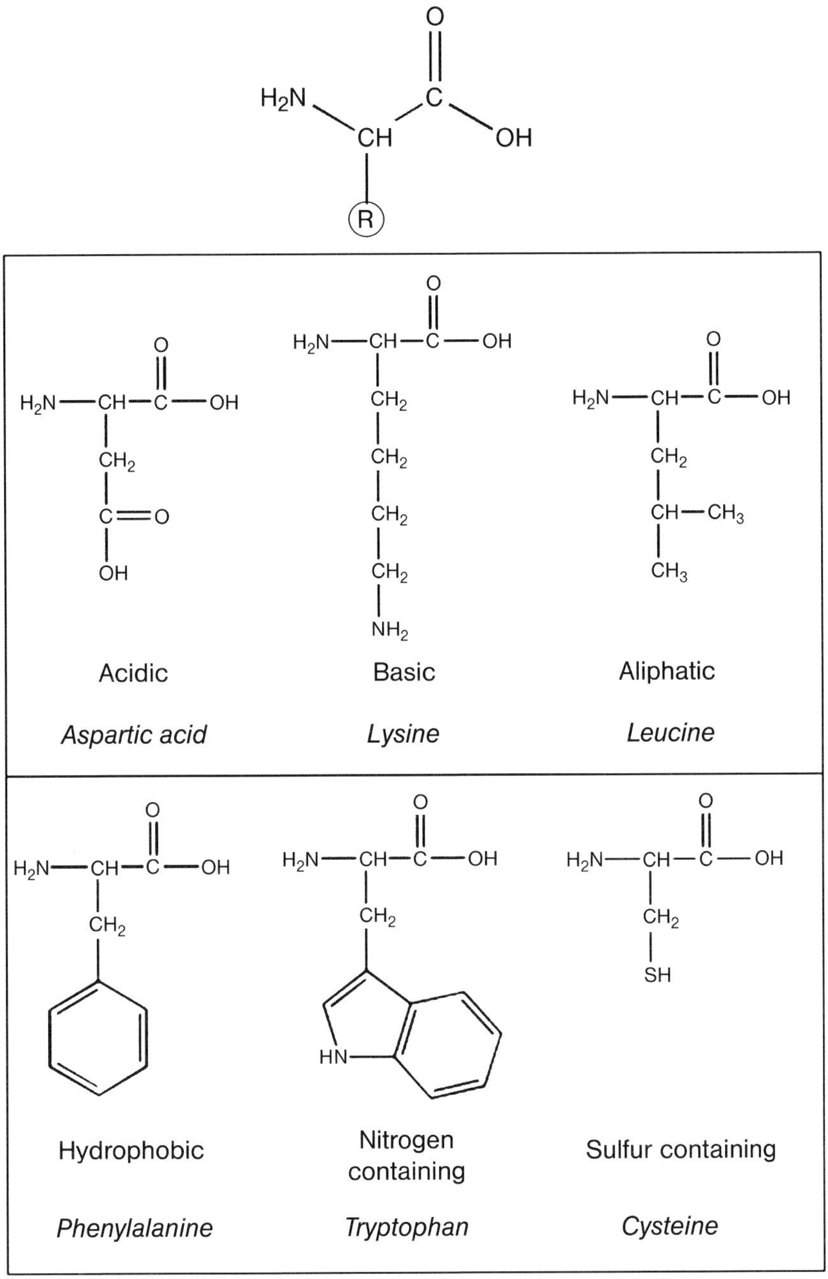

Fig. 7.6 The core structure of an amino acid (top), R representing where side chains of differing properties are attached. Representative amino acids showing a range of side-chain functionalities (bottom).

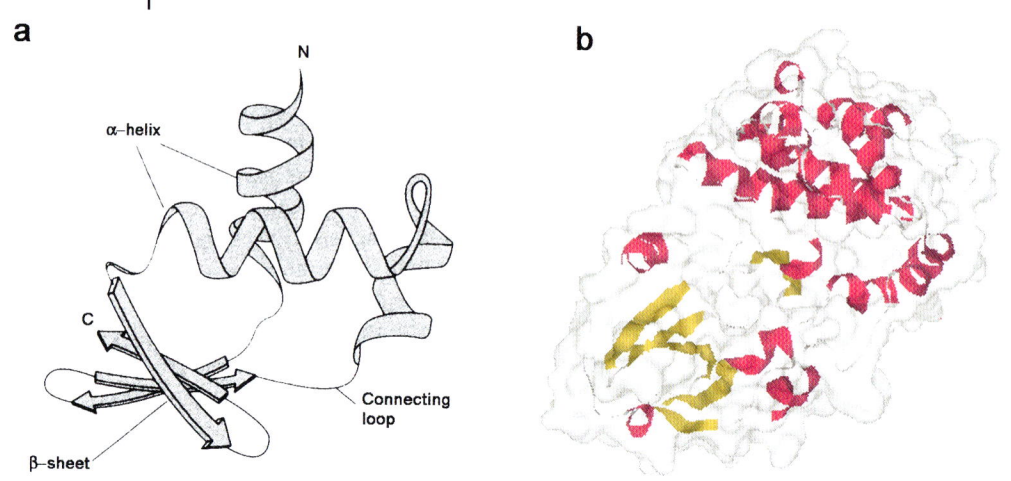

Fig. 7.7 (a) Schematic showing protein secondary structures present together in a tertiary structure. (Taken from P. C. Turner, A. G. McLennan, A. D. Bates, M. R. H. White, *Instant notes Molecular biology*, 2nd edn, Bios Scientific publishers, Oxford, **2000**.); (b) 3D representation of bovine serum albumin secondary structure highlighting α-helices and β-sheets.

The transfer of all genetic information is a result of nucleic acids in the form of deoxyribonucleic acid, more commonly known as DNA. Nucleic acids are linear polymers of purine (adenine or guanine) or pyridimine (cytosine or thymine) bases connected to a backbone of sugar units liked by phosphate groups. The individual polymeric chains which are formed are found as complimentary pairs with each adenine base perfectly matching a thymine base to link via hydrogen bonding and likewise guanine and cytosine also match permitting hydrogen bonding (Fig. 7.8). DNA is found as double helix, which is arranged into chromosomes found in the cell nucleus. An exception to this structure can be found in some viruses which may have genetic material in the form of single stranded DNA, however on infection of a cell the DNA is replicated as a double stranded molecule. The classification of DNA as one of nature's building blocks is as a result of it directing its own replication to RNA, which in turn directs its own translation to proteins. The sequences of purine or pyrimidine bases along the chain are transcribed by three different types of RNA and then directly translated with three adjacent bases (called a codon) corresponding to each amino acid.

7.2.2.2 Carbohydrates

The modern definition of carbohydrates is "organic compounds that can be hydrolysed to form monosaocharides, which vary with the number and orientation of carbon atoms and the functional groups, either aldehyde or ketone". The connection of one monosaccharide to another is brought about through the formation of a glycosidic bond or linkage. These bonds are formed when the

Adenine-thymine base pair

Guanine-cytosine base pair

Fig. 7.8 Complimentary pairs in DNA with each adenine base perfectly matching a thymine base to link (top) and guanine and cytosine match (bottom) permitting hydrogen bonding.

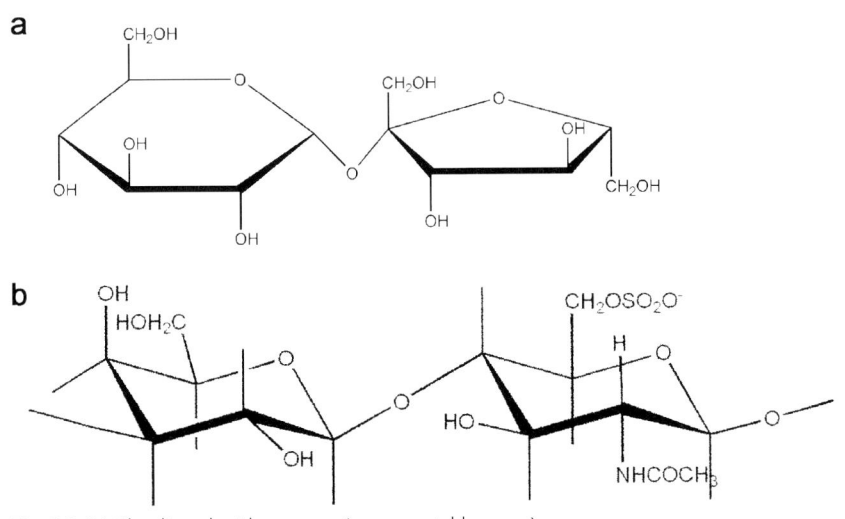

Fig. 7.9 (a) The disaccharide sucrose (common table sugar) which is formed from an α-1,2 glycosidic linkage between D-glucose and D-fructose. (b) Keratin sulfate, a glycosaminoglycan disaccharide.

hydroxyl group on an anomeric carbon of one monosaccharide reacts with the alcohol group of another monosaccharide, releasing water; the product is then classed as a disaccharide and a common example is shown in Fig. 7.9a. The glycosidic bond can theoretically form between the anomeric carbon and any available hydroxyl group of a second monosaccharide, giving a diverse range of structures. The glycosaminoglycan chains of proteoglycans are formed from repeating disaccharides containing at least one amino sugar, an example of which is shown in Fig. 7.9b.

Oligosaccharides are commonly only 5–15 monosaccharides long. The formation of an oligosaccharide chain can lead to a great number of possible structures. Three amino acids could form six variations of a tripepetide, however three hexoses could theoretically link to produce over 1000 different trisaccharides. The vast number of different glycoforms that can be produced on any one molecule is called heterogeneity. Polysaccharides are much larger chains of monosaccharides joined together by glycosidic linkages, such as starch, cellulose (e.g. wood, plant cell walls) and chitin (e.g. cockroach exoskeleton).

7.2.2.3 Lipids

Lipids are a complex and diverse class of biomolecules that are essential due to their many roles including the formation of membranes, transduction of cellular signals and as a source of energy in the form of triacylglycerols. The combination of the polar head group attached to one or more hydrophobic tails via a backbone unit leads to great deal of structural diversity and a vast array of different amphiphilic lipids can be found *in vivo*. The hydrophobic tail regions can be composed of saturated or unsaturated aliphatic chains or may also contain aromatic groups, while charged or uncharged polar moieties can be found at the head group. Some examples of the different lipid structures are shown on Fig. 7.10. As lipids are largely insoluble in polar environments, they are capable of self assembly in to more favorable structures when found in aqueous environments. The formation of a lipid bilayer, as found surrounding all cells, is brought about by self-organization of the hydrophobic regions to the inside of the bilayer, eliminating an unfavorable proximity to the polar environment. Another common structure of lipids is the formation of micelles, which again occurs in polar environments and has the polar heads surrounding a cluster of hydrophobic tails.

These organic building materials are, in nature, intimately associated with a vast array of mineral phases, utilizing the features of both to or materials which are ideally suited to their function. One clear example of this is the mineralization of collagens; glycosylated proteins, which are found in many different bio-logical materials from bone to cornea and are reliant on the individual properties of each component to generate these highly specific materials.

7.2.2.4 Collagen

In mammals, the group of proteins known as collagens accounts for nearly one third of an animal's protein content. These proteins are frequently glycosylated and are often found in conjunction with Ca in the form of hydroxyapatite

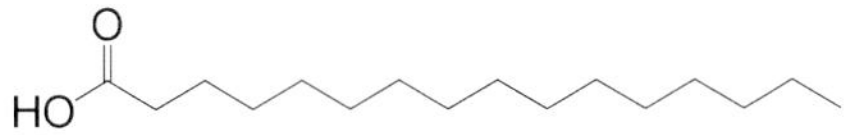

Palmitic acid, $CH_3(CH_2)_{14}COOH$

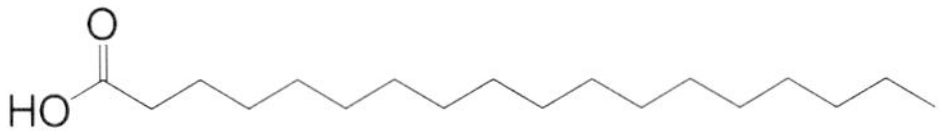

Stearic acid, $CH_3(CH_2)_{16}COOH$

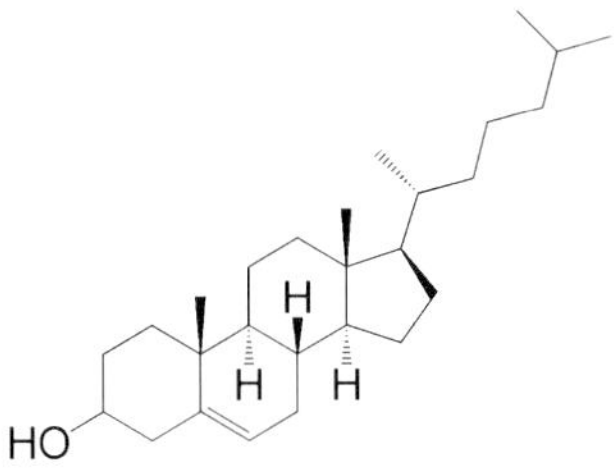

CHOLESTEROL

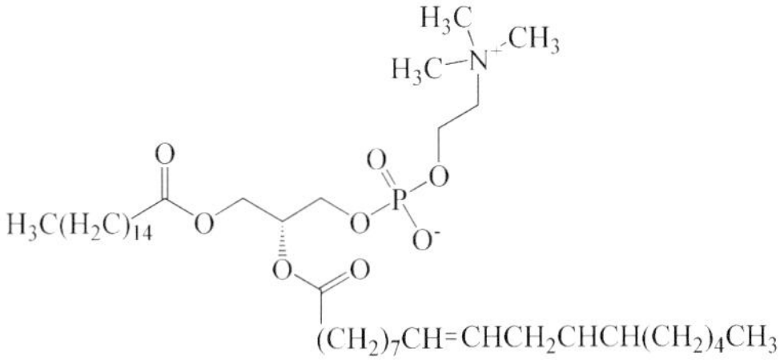

Phosphatidylcholine

Fig. 7.10 Some common fatty acids.

(discussed later). Collagen occurs in virtually all tissues in the body and is principally found as the major stress-bearing component of the connective tissues as a result of its immense tensile strength. Each collagen molecule consists of three polypeptide chains, which are individually coiled. The three chains then wind tightly like a rope into a right-handed triple helix, as shown in Fig. 7.11.

In order for a protein to be defined as a collagen it must conform to three characteristic traits. The protein must contain at least one triple helical domain; form super-molecular aggregates, known as collagen fibrils and maintain its structural integrity in the extracellular matrix where it is found.

The amino acid sequence of a collagen molecule is distinctive with a common repeating triplet of Glycine–X–Y where X and Y are frequently proline and its hydroxylated counterpart hydroxyproline. Collagen contains small amounts of hydroxylated amino acids and both positively and negatively charged

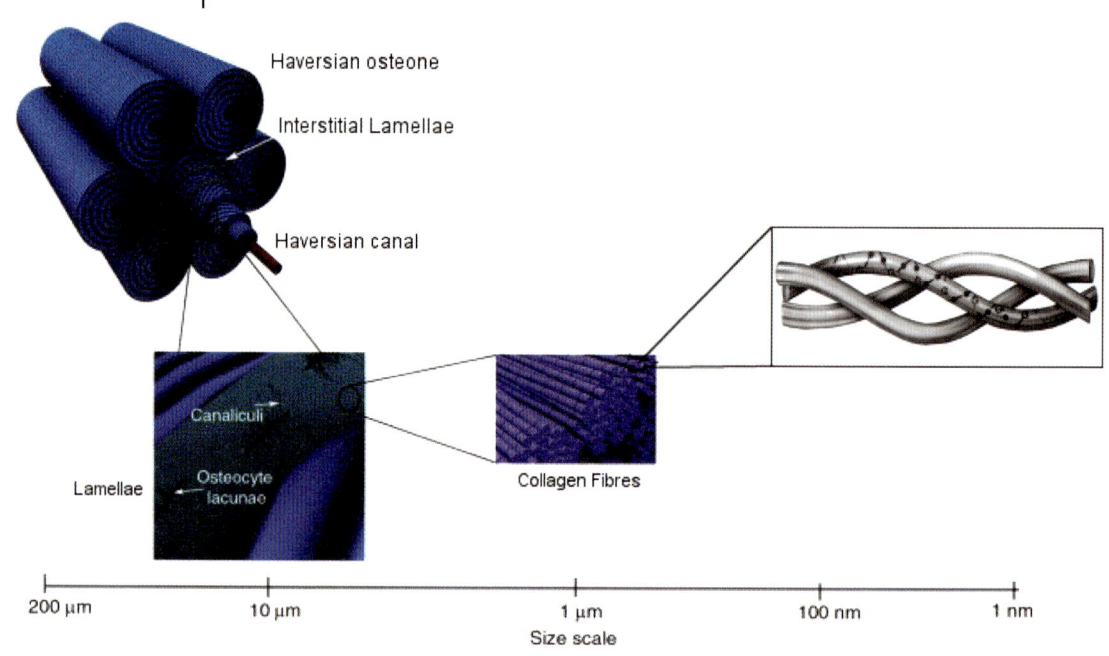

Fig. 7.11 Hierarchical assembly of collagen in bone formation. (Image adapted from M. Tirrell et al., *Surface Science* **2002**, *500*, 61–83, and cular E. et al., *Phys. Ther.* **1999** Mar, 79(3):308–319.)

amino acids distributed along the length of the individual protein chains. The hydroxylated amino acids are involved in hydrogen bonding both within an individual chain and between chains that serves to increase the stability of the molecule.

Once the triple helix has been formed the collagen molecules aggregate to form fibrils, filaments or networks. These structures can be of collagen alone or collagen in conjunction with components of the extracellular matrix. This assembly occurs spontaneously and involves the formation of electrostatic and hydrophobic interactions between adjacent collagen molecules. Also occurring in collagen fibrillogenesis is the loss of molecule-associated water. The fibrils formed by collagen proteins are tissue specific. In order to become fully active the aggregated collagen molecules are often found interacting with other components of the extracellular matrix, in particular proteoglycans.

To date 19 different types of collagen protein have been identified varying in the length of the non-helical fractions, the length of the helix itself and the number and nature of any carbohydrates attached to the polypeptide chain. Type I collagen is the most abundant, being present in a large variety of tissues including skin, bone, tendon, cornea and blood vessels, however these tissues all include other types of collagen molecules, often as minor components.

7.3 Biomineralization

7.3.1 Introduction

Biomineralization is a process by which biological systems produce inorganic minerals *in vivo*. Note: these materials are actually composites of biopolymers and inorganic salts or oxides and have physical and structural properties that may be somewhat different to those found for their individual components. In the process of biomineralization, organisms typically accumulate the precursors required to synthesize biominerals from their respective environments (water, soil, food). These precursors can be ions (e.g. ions of Ca, Fe and Mg), or small molecular complexes, the precise structures of which are currently unknown. Organisms are able to transport these precursors from the environment in which they are found, into the organism. Subsequently, the precursors may be stored, transported *in vivo* and converted into biominerals. The process of biomineral formation in many cases is regulated by genetic control and it is this control that produces species-specific, ornate biomineral structures with physical and mechanical properties fit for function.

Traditionally, biomineralization is broadly classified as follows:
- where minerals are synthesized for a specific biological function. Biomineralization in this class is strictly under biological (genetic) control (e.g. bone formation).
- where minerals are formed without any apparent specific function. These biominerals may be useful, detrimental or benign to the organisms producing them (e.g. kidney stones).

7.3.1.1 Biomineral Types and Occurrence

Biominerals are typically organic–inorganic hybrids that are hierarchically organized from the nano- to the macroscopic length scale. The organic components of biominerals include proteins, glycoproteins, polysaccharides and other small organic biomolecules. The organic phase occluded in biominerals may or may not be directly involved in biomineralization. The common ions involved in biominerals are Mg, Ca, Sr, Ba, Si and Fe as their carbonates, oxalates, sulfates, phosphates, hydroxides and oxides, see Table 7.1. The relatively rare ions are Mn, Au, Ag, Pt, Cu, Zn, Cd and Pb deposited largely in bacteria and often as sulfides. Over 60 different biominerals have been identified. The diversity in the occurrence of biominerals indicates the ability of biological organisms to manipulate and deposit inorganic compounds. Around 50% of biominerals are calcium-based minerals and of this, half are calcium phosphates of varying composition. In terms of structure, about 25% biominerals are amorphous in nature i.e. they do not show structural regularity at atomic scales (e.g. biosilica, amorphous hydrated iron phosphate, calcium carbonate).

7.3.1.2 Functions of Biominerals

In most organisms, biominerals are produced for specific functions. The functions can be classified as follows:
- mechanical/structural support
- protection
- motion
- as sensors
- cutting and grinding
- buoyancy.

These functions will be illustrated with selected examples. Marine sponges are known to form biosilica in the form of needle-like spicules that are a few tens of microns in diameter and can be as long as a few millimeters (Fig. 7.12). The primary role of biosilica spicules in sponges is to provide mechanical support to the animal and to protect from predators. Biominerals also act as gravity, optical or magnetic sensors thus providing useful functions to organisms. Some bacteria – *magnetotactic* bacteria – produce single crystals of magnetite (Fe_3O_4) which are called *magnetosomes* (Fig. 7.12c).

In a bacterium, these magnetite crystals are typically identical to each other and are arranged in a linear chain. Magnetosomes help bacteria navigate using the Earth's magnetic field. Similarly, gypsum deposited in jellyfish is used for gravity sensing. Limpet teeth, used for grinding, are biomineralized geothite (α-FeOOH), while chiton teeth are lepidocrocite (γ-FeOOH) or ferrihydrite ($5Fe_2O_3 \cdot 9H_2O$); (see Fig. 7.14b). Whewellite ($CaC_2O_4 \cdot H_2O$) and weddellite ($CaC_2O_4 \cdot 2H_2O$) are found in plants or fungi and are used to store calcium. Some marine molluscs and cephalopods use aragonite ($CaCO_3$) shells as buoyancy devices.

7.3.1.3 Properties of Biominerals

Biominerals are intriguing due to their unique characteristics in comparison with the mineral alone, such as:
- chemical composition
- structure and morphology
- mechanical properties.

Biological organisms are known to control the chemical composition of the biominerals that they produce. Very rarely are biominerals "pure" with respect to stoichiometric chemical composition and often are hybrids containing "major" components and "minor" components (e.g. additives or dopants). Doping biominerals with other ions and molecules is precisely carried out so as to regulate the physical and mechanical properties of the biominerals. For example, teeth (enamel) contain F^- ions that reduce the solubility of enamel. The F^- ion content of human enamel is 0.02 wt% while that of shark enamel is about 3.65 wt%, thus making shark teeth relatively more stable than human teeth to dissolution. Often, the presence of Mg^{2+} ions has been reported in calcite biominerals and strontium

Fig. 7.12 Examples of some biominerals – (a) sponge spicules, insets show high magnification images of a spicule tip and head. (b) limpet teeth and (c) magnetite crystals (magnetosomes) from bacteria. Bar = (a) 100 μm (1 μm for inset), (b) 200 μm and (c) 50 nm. (Image in (a) taken from S. V. Patwardhan, S. J. Clarson, C. C. Perry, *Chem. Commun.* **2005**, *9*, 1113 and reproduced by permission of the Royal Society of Chemistry. Image in (c) taken from S. Mann, J. Webb, R. J. P. Williams, *Biomineralization*, VCH, Weinheim, **1989**.)

has been detected in aragonite biominerals but the reasons for these variations appear complex.

Biosilica ($SiO_2 \cdot nH_2O$) occurs as an amorphous biomineral in many organisms ranging from single-celled diatoms to higher plants and animals. Although non-crystalline, biosilica in single celled diatoms exhibits repeating structural features of *ca.* 10–40 nm. In addition to organization of matter at different length scales, some biominerals occur as polymorphs i.e. identical chemical composition but different crystalline structures. Calcium carbonate, for example, occurs in several

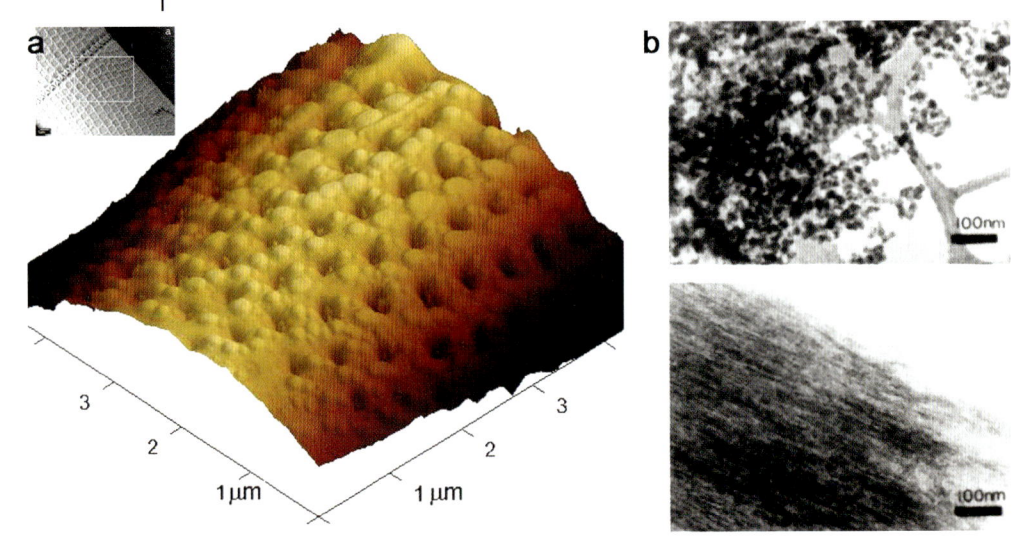

Fig. 7.13 Images of morphology of biosilica from (a) diatoms and (b) plants. (Image in (a) taken from S. V. Patwardhan, S. J. Clarson, C. C. Perry, *Chem. Commun.* **2005**, 9, 1113 and reproduced by permission of the Royal Society of Chemistry.)

crystalline and amorphous forms. $CaCO_3$ plates in nacre (see Fig. 7.18, next section) are made from aragonite while structures in coccolithophores (Fig. 7.2) are made from calcite crystals. Similarly, morphology of biominerals is also strictly controlled and is species-specific. An example of biosilica morphologies is given to illustrate this point; while the biosilica in the diatom cell-wall is a typically porous network (N.B. porosity only relates to exernal pores), biosilica in sponges takes a needle-like form and in higher plants several different arrangements of silica particles may be observed even within one cell, (Fig. 7.13).

Many organisms deposit biominerals for their protection. In order for biominerals to provide such protection, their mechanical properties need to be regulated with precision. In the case of bones, the strength and stiffness can both be increased with higher mineral content. However, the toughness optimising these and dictates the optimum composition (Fig. 7.14). It is thus not surprising that the mineral content in bone differs bone type to bone type and from species to species for the combination of stresses for the particular bone.

The above section has highlighted how sophisticated biominerals are in terms of their chemistry, structure and morphology, and other specific properties making them "fit for function" within a given organism.

7.3.2
Control Strategies in Biomineralization

One of the most intriguing features of biological mineral formation is the regulation of the entire process of biomineralization from the intake of ions and

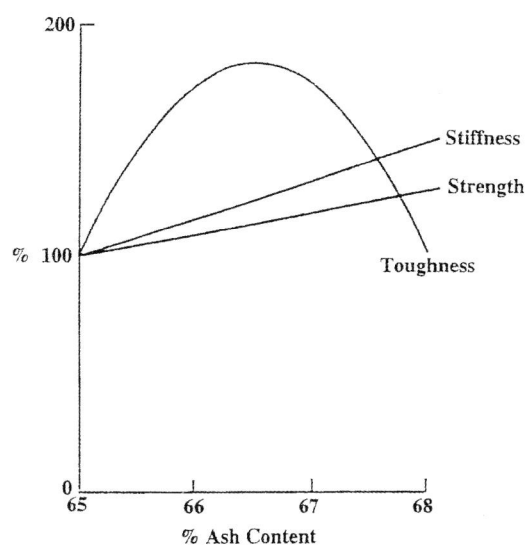

Fig. 7.14 Graph showing bone stiffness, strength and toughness as a function of ash content. (Redrawn from S. A. Wainwright, W. D. Biggs, J. D. Currey, J. M. Gosline, *The Mechanical Design of Organisms*, Edward Arnold, London, **1976**.)

molecules from the surroundings to the deposition of stunningly beautiful and organized structures as we have seen in the preceding sections. It is this control that separates the *in vitro* synthetic capabilities of mineralization, that are presently far removed from the sophistication observed *in vivo*. R. J. P. Williams clearly states – "The chemical character is clearly then a genetically controlled feature due to the deliberate movement of elements into specialized parts of biologi-cal space by ion pumps. However, inside this space, the chemistry of the surface and/or of growth inhibiting compounds will control the precise compound which is precipitated". This briefly suggests different control strategies involved in biomineralization and these are listed below (also see Fig. 7.15).

The aforesaid strategies will be considered individually, their roles explained and illustrative examples provided.

Chemical control in biomineralization constitutes one of the most important regulators. The basic requirement for the formation of any mineral is the attainment of supersaturation with respect to the mineral precursors as described earlier. Biological organisms control supersaturation by governing various chemical aspects. *pH* for example, dictates the ionic strength and activities of ions in a given solution. This in turn, according to solubility product (equation 1), may initiate or inhibit mineralization. In some cases even the slightest changes in pH can cause drastic effects on biomineral deposition. For example, iron reduction in some bacteria is triggered by changes in pH. The influx and efflux of certain ions to and from the site of biomineralization leads to changes in the *concentration of*

7 Natural and Artificial Hybrid Biomaterials

BIOMINERALIZATION
- **A. Controlled**
  - 1. Physical control
    - a. Spatial control / compartmentalization
    - b. Site specific deposition
  - 2. Chemical control
    - a. pH
    - b. Ionic and precursor Concentrations and solubilites
    - c. Surface and interface interaction
    - d. Organic phase
    - e. Remodelling / Phase transformation
  - 3. Temporal control
    - a. Order of events
  - 4. Biological control
    - a. Cellular and genetic
    - b. Biomolecule synthesis
    - c. Transport
- **B. Uncontrolled**

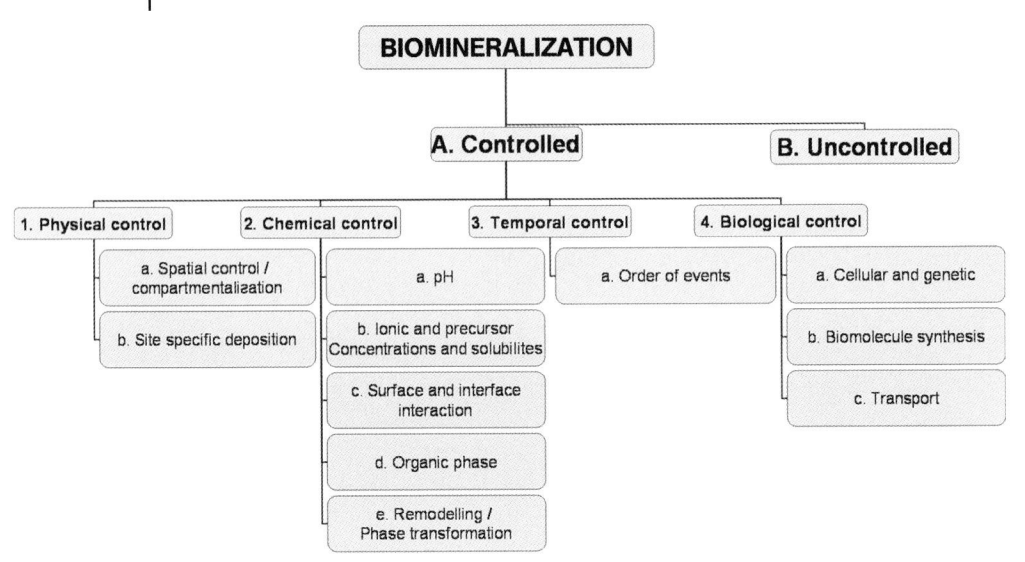

Fig. 7.15 List of various factors controlling biomineralization. See text for details.

ions and precursors thus altering their relative solubilities, activities and compositions. This effect, again, can regulate biomineral deposition. The control of the rate of availability and diffusion of individual species to the biomineralization site, their adsorption onto growing surfaces, their incorporation into biominerals and their inhibitory effects are all known to directly affect biomineralization. In the earliest stages of formation of hydroxyapatite (bone), Ca^{2+} ions are adsorbed onto glycoproteins thus altering local precursor concentrations.

Biomineral formation on *surfaces* (i.e. heterogeneous nucleation) is known to exhibit lower energy barriers than the energy barriers to be overcome in homogeneous nucleation. Hence the formation of biominerals on surfaces is favored, e.g. the formation of mollusc shells where layers of organic material including proteins with high acidic functionality provide nucleation sites for calcium carbonate formation. Furthermore, *interfacial interactions* between biomineral-liquid and biomineral-surfaces are also important and are precisely controlled by organisms. When a solid mineral phase forms, it gives rise to a solid–liquid interface and in turn to a solid–liquid interfacial energy (σ_{SL}). Higher surface energy leads to instability and thus to produce stable minerals, biology adopts different ways of optimizing σ_{SL}. One simple example to stabilize surface would be the adsorption of inorganic ions or *organic biomolecules* onto unstable mineral surfaces. This example leads us to a discussion on the roles of the organic phase in biomineralization. In passing we note that the organic phase is known to affect almost all aspects of biomineralization – from transport of ions, storage of precursors, regulating solubilities, catalysis and structure direction of biominerals to biomineral stability. Last but not least of the chemical controls is *remodeling and/or phase transformation* of biominerals. This is the "final touch" given to biominerals before they are actually functional. In crystalline biomineralization, the formation of one phase may dominate under given conditions, but that phase may not be suitable for its function and hence phase transformation is performed *in vivo*. For example, spicule formation in sea urchins occurs via the initial deposition of amorphous calcium carbonate that is "moulded" into the final required morphology before being converted into the crystalline form – calcite.

In addition to chemical regulatory effects, biomineralization also exhibits *physical controls*. These include compartmentalization or *spatial constraints* and localization or *site-specific deposition*. Organisms produce biominerals typically in compartments that are also termed biomineral deposition *vesicles*. Biomineralization within vesicles offers greater control over ion transport and supersaturation levels. Furthermore, biomineral structure – from the atomic to the macroscopic scale – can be precisely controlled in such compartments. Typically, vesicles are formed from assembled biomolecules such as lipids, proteins and polysaccharides. The pre-organization of phospholipid vesicles in the formation of magnetic crystals in magnetotactic bacteria is a well established example. The phospholipid vesicles regulate precursor transport inside the vesicles, the concentration of individual species and the crystalline and chemical nature of the biomineral. *Site-specific* biomineralization is the formation of biominerals at a given functional site that possesses activity to facilitate biomineralization. Diatom cell walls are proposed to be functionalized with peptides and biomolecules (silaffins and polyamines respectively). These functional biomolecules, according to *in vitro* studies, deposit biosilica due to their specific silica precipitating abilities.

In biomineralization, *temporal control* is very important as it dictates the order of all biological events and activities taking place prior to, during and after biomineralization. For example, the deposition of biominerals can only occur if the precursors have already been taken in by the organism. We note that temporal control is tightly related to *biological control* strategies. Biomineralization is typically under strict biological control as explained earlier. This includes both *cellular* and *genetic* regulation. Biological control is usually achieved by the cellular machinery – proteins. Organisms, through the genetic code, regulate the secretion of the "right" molecules at the "right" time and at the "right" place. In biomineralization, biomolecules, proteins in particular, are involved in activities such as ion uptake, storage, transport of molecules and organelles, etc. All regulation mechanisms described above are ultimately controlled biologically and this is why such a large difference is observed between minerals produced *in vivo* and *in vitro*.

7.3.3
The Role of the Organic Phase in Biomineralization

It is known that organic biomolecules are involved in most, if not all, stages of biomineral formation. Biomolecules may be in the form of peptides and proteins, lipids, polysaccharides, proteoglycans, etc. Traditionally, they are termed the "organic matrix"; however, herein we choose to call them the *organic phase* because matrix may imply a particular structure, and in many cases this may not be present. The organic phase can be defined as one that acts as a "mediator" of

Table 7.2 Selective examples of biomacromolecules involved in biomineralization. (Adapted from S. Mann, *Biomineralization: principles and concepts in bioinorganic materials chemistry*, Oxford University Press, New York, **2001**.)

System	Framework	Functional
Bone and dentin	collagen	glycoproteins osteopontin, osteonectin proteoglycans keratin sulfate, chondroitin sulfate Gla-containing proteins
Tooth enamel	amelogenin	osteocalcin glycoproteins enamelins
Mollusc shells (nacre)	β-chitin silk-like proteins lustrin A	glycoproteins nacrein
Crab cuticle	α-chitin	
Magnetic bacteria		proteins
Diatom shells		glycoproteins silaffins propylamines
Sponge silica	silicatein	proteins silicatein α, β and γ
Plant silica		glycoproteins carbohydrates

biomineralization. It can be divided into two categories – functional and framework. The former is directly involved in regulating biomineralization (see below and Table 7.2), while the latter, typically, only acts as support for the former. We will discuss the modes of interaction between the organic phase and inorganic minerals and then describe the various roles that the organic phase plays in biomineralization.

7.3.4
Mineral or Precursor – Organic Phase Interactions

It has been proposed by Mann that *molecular recognition* between organic and inorganic species is essential and important in biomineralization. This recognition can arise due to various factors, including electrostatic potential and stereochemistry. In particular, these organic–inorganic interactions can be viewed as *non-bonded chemical* and/or *physical interactions*. Non-bonded chemical interactions between the organic phase and inorganic biominerals arise from electrostatic forces, hydrophilic and/or hydrophobic effects, hydrogen bonding and van der Waals forces. Non-bonded interactions operate between atoms that are not linked together by covalent bonds. Non-bonded interactions vary in strength from 0.1 kcal mol^{-1} to several hundreds of kcal mol^{-1} depending on the environment in which the interaction occurs (vacuum through to water) and the nature of the specific interaction. In solution reactions, especially those taking place within a living organism, the medium for the reaction is water, except for reactions that occur in specific membrane-like compartments. However, reactions in life do not occur in "deionized water" free from additives but rather in a medium that contains ions and molecules and all of these are available for interaction, in principle, with other species that may be present. The same may be true for laboratory based reactions although here the number of components and the amounts of each component can be more easily regulated and/or modified. Non-bonded interactions include electrostatic interactions, hydrogen bonding, van der Waals interactions and the hydrophobic effect and they vary considerably in strength. As a comparison, covalent bond energies are of the order of 60–250 kcal mol^{-1}. Electrostatic interactions in water are of the order of 1–15 kcal mol^{-1}, hydrogen bonds are of the order of 2–5 kcal mol^{-1}, van der Waals interactions are of the order of 0.5–1 kcal mol^{-1} and the hydrophobic effect is of the order of 1 kcal mol^{-1}.

The *effect of charge*, distance and environment for reaction on the strength of non-bonded interactions can be understood by consideration of Coulomb's law where attraction between species of opposite charges (generally considered for positively charged nuclei and negatively charged electrons but can be more generally applied to any species for which there is a separation of charge) is represented as:

$$V = q_i q_j / 4\pi\varepsilon_0 \varepsilon_r r_{ij} \tag{3}$$

where q_i, q_j are charges, r_{ij} is their separation, ε_0 is the permittivity of free space and ε_r the relative dielectric constant of the medium (for water *ca.* 80, for methanol *ca.* 35 and for a lipid bilayer *ca.* 2). The charges may be taken as formal charges but partial charges may also need to be considered for some reacting species. The dielectric constant is dimensionless and accounts for solvation and charge shielding due to the presence of the solvent. Note that the effect of charge is felt much more strongly in a lipid bilayer or in a non-aqueous solvent such as methanol than in water. Part of this effect has to do with the ability of particular solvents to hydrate ions (by attraction between the solvent molecule and the solute) such that they effectively enlarge the ions. This enlargement shields the charges from each other thus diminishing the strength of interaction between such species.

Hydrogen bonds arise from electronegativity differences between an electronegative element and the hydrogen atom it is attached leading to a redistribution of charge within the covalent bond joining the two. For a hydrogen bond to form there needs to be one molecule with an electronegative element attached to hydrogen and another molecule containing an electronegative element. Clearly, for reactions in water, hydrogen bonding between solutes must compete with hydrogen bonding to the water molecules. This happens when protein folding occurs to give the secondary building blocks, alpha helices and beta sheets that further fold to generate specific conformations of active proteins.

Van der Waals interactions are most important for essentially uncharged atoms as they come close together in space. The effect of one atom on another is to deform the electron cloud due to electronic repulsion. This sets up transient dipoles on both atoms, resulting in a weak attractive interaction between them. Van der Waals interactions are weak but it is a result of these interactions that geometric specificity is achieved in biological systems.

The *hydrophobic effect* operates when there is a mixing of polar (e.g. water) and non-polar (e.g. oil) molecules. The effect is not due to the hydrophobic groups themselves rather due to a reorganization of the solvent to minimize the amount of water, for example, that is not ordered in its normal fashion.

Although non-bonded interactions are individually much weaker than covalent interactions, when they act in concert, as in a protein or a drug–receptor complex or indeed for a mineralization reaction in the presence of biomolecules, their effect is cumulative and all non-bonding interactions will collectively operate to reduce the free energy of the reaction system, whatever it is.

Such interactions can control biomineralization, for example by regulating the available concentrations of ions, molecules and/or biomineral precursors. These chemical effects arising from the organic phase may also alter the solubilities of specific species in a given solution thus affecting the nucleation and growth of biominerals. In the case of interactions between proteins and biominerals, the primary, secondary, tertiary and quaternary structures of proteins play a major role. The primary structure, i.e. the amino acid sequence, determines the chemical "nature" of the proteins present while the other – secondary to quaternary – structures define the "shape and topography" of proteins. Due to the advanced nature of this topic, readers are advised to refer to further reading (see Evans, 2003, Sarikaya 2003 and Shiba, 2003 in bibliography). *Physical interactions*, for example, refer to the availability of surfaces for adsorption and nucleation of ions and/or biominerals. We have seen previously that heterogeneous nucleation has a low energy barrier and requires the presence of suitable surfaces, such as the organic phase.

7.3.5
Examples of Non-bonded Interactions in Bioinspired Silicification

7.3.5.1 Effect of Electrostatic Interactions

Oligomers that are formed during silica polymerization are negatively charged at neutral pH and also possess free hydroxyl groups. It is thus likely that the presence of an additive may alter the stability of such intermediates and/or the silicic acid polymerization process due to electrostatic interactions. The presence of cationic species in solution affects silica formation. Effects on the kinetics of the early stages of oligomer formation, particle growth, aggregation and the nature of the materials formed are observed. Nitrogen-containing amino acids, particularly L-arginine and L-lysine (positively charged side chains) promote silicic acid condensation. Aggregation of the condensing silica in the presence of amino acids showed pI dependent behavior with the most significant increases being observed for L-arginine and L-lysine. This behavior can be explained by consideration of the increasing negative charge on silica particles as they grow and the neutralization of these charges in the presence of the positively charged amino acids, thereby allowing the silica species to interact with one another to produce larger aggregates. Experiments with homopeptides of lysine showed similar behavior. In addition, the presence of several charges within one molecule (increase in number of lysines per molecule) magnified the observed behavior with increased rate of condensation and aggregation being observed. The attraction between positively charged additives (e.g. metal ions, small organic molecules and macromolecules) and negatively charged silica species alters the stabilities of silica oligomers and particles, the activities of silica species and solvent, thereby influencing the solution chemistry and structure formation of silica.

7.3.5.2 Effect of Hydrogen Bonding Interactions

Hydrogen bond formation between silanol groups of silicic acid, oligomers and/or negatively charged silica particles and proton donors, such as free amines or hydroxyl groups present in the additives, will usually occur. The effect of additions of alkanediols to a model silicifying system was investigated. The diols investigated did not significantly affect any of the early stages of silica oligomer formation. However, the presence of the diols led to higher levels of silicic acid being present in solution and silicas with lower surface areas and increased porosity were produced. The observations can be explained by considering all of the species capable of forming hydrogen bonds. There are three species cable of constructing a hydrogen bond: silicic acid, water and the alkanediol species. Silicic acid is expected to hydrogen bond with the diol as well as with water. However, for all the experiments conducted, water was in excess and hence the effect of the diols would be insignificant compared with the effect of the solvent, as is indeed observed. As particle growth continues the presence of the additive promotes reorganization of the siliceous phase, but the diol is not incorporated into the structures that form.

7.3.5.3 Effect of the Hydrophobic Effect

The hydrophobic effect is expected to play an important role in silicic acid polymerization and silica – additive interactions. A series of organic additives possessing cationically charged end-groups and increasingly larger hydrophobic domains (from C_2 to C_{10}) have been used to investigate their interactions with silicas. Effects on silicic acid condensation, aggregation and materials properties were observed that could not be explained by consideration of the electrostatic effects alone. Increased rates of condensation and aggregation were observed and materials with lower surface areas were produced in the presence of molecules having progressively larger hydrophobic structural components. The increase in silicification rates was explained by the formation of a clathrate-cage-like water structure around the non-polar surfaces of the alkyldiamines. The cage-like structure may tie up some of the free water molecules (i.e. those not associated with ion hydration shells) resulting in higher reactant (silicic acid) concentrations in the bulk aqueous environment and also a possible reduction in the hydration shells around the anionic silica species. Under these conditions, reactions involving anions, such as the condensation of a silicate anion with a neutral silicate species, would be expected to show an increase in rate, as is observed. For shorter chain diamines an initial increase in observed aggregation rates was attributed to surface-charge neutralization of the negatively charged primary silica particles by the cationic diamines – the electrostatic effect. However, as the diamine chain length increased, the diamines were found to additionally bridge the particle double layer, resulting in accelerated growth. The coacervation of diamine coated silica particles (the hydrophobic effect), which increased with increasing chain length, manifested itself as the continued rate and size increase that was observed. In other words, the addition of nonpolar organic species was found to alter the chemical potential and structure of water (solvent) due to the hydrophobic effect and electrostatic interactions.

7.3.6
Roles of the Organic Phase in Biomineralization

Organic phases are able to regulate all aspects of biomineralization – from transport to deposition to biomineral stabilization. Their roles and effects can be categorized as follows:
- chemical
- spatial
- structural and morphological
- mechanical support.

The organic phase can exist in various forms. These include vesicles, networks, membranes, surfaces and at interfaces, self-assembled structures and in a soluble form. The organic phase is an effective tool used by organisms for controlling biomineralization.

Chemical effects include maintaining the local supersaturation of precursors. As an example, in bone formation, acidic glycoproteins bind Ca^{2+} ions thus increasing their local concentration thereby promoting bone formation. In some cases, the organic phase, and functional biomolecules in particular, exhibit enzymatic activity that can catalyze the formation of biominerals. As an example, it has been shown that the controlled formation of magnetite crystals in bacteria takes place only in the presence of proteins that are tightly associated with the crystals. In addition, the organic phase can also be involved in the inhibition of biomineral formation as well as in the stabilization of biominerals, with avoidance of mineral dissolution.

The *spatial effects* arise from localization of biominerals and their precursors. The organic phase can present itself in a range of forms and assembled structures. This organization exerts spatial constraints on biomineral formation. As an example, the protein ferritin forms a hollow cage, the size of which imposes a limit on the amount of ferrihydrite that can be synthesized within the protein cage. Organic biomolecules are not only known to control the final shapes and morphologies of biominerals, but they are also able to determine the atomic and molecular structures. In case of calcium based biominerals it has been proposed that stereochemical interactions occur between the organic phase and the growing mineral and that this interaction dictates the nature of crystals produced. An example is the regulation of coccolith formation. It has also been suggested that the organic phase in various biominerals acts as a mechanical support imparting desirable strength, toughness and/or flexibility for a given system. Bone is a classic example where the organic phase provides mechanical properties to the ultimate hybrid biocomposite. Variation in mechanical properties is achieved by the amount and type of collagen that is present as well as the amount of mineral associated with the collagenous matrix.

7.4
Bioinspired Hybrid Materials

An important question needs to be addressed now: how biological routes towards fabricating hybrid biomaterials can be transferred to artificial materials design. We try to address this aspect in this section. Sources for inspiration will be described followed by selected examples of bioinspired artificial materials synthesis.

As discussed above, the novelty about biomineralization is not only the structural control but also the use of biomolecules in synthesis and organization of inorganic minerals. Specific roles of functional organic biomolecules are being elucidated in order to reveal the *"active"* components. This information can be utilized in designing synthetic analogues that would facilitate *in vitro* mineralization. The example of silica synthesis is given in this case. It has been found that the protein containing extracts isolated from diatom biosilica possess unusually modified amino acids. The modifications, which are polyamines of 6–10 repeat units, are shown to be important for silica formation *in vivo*. A number of synthetic amines with a variety of sizes and architecture were designed and were found not only to facilitate the synthesis of silica under ambient conditions, which would otherwise not be possible, but also to produce tailored silicas in terms of their structure, morphology and porosities, for example (see a recent review for details: Patwardhan, Clarson and Perry, *Chem. Commun.*, **2005**).

In the case of some biomineral formation, organic molecules act as *templates or scaffolds* for inorganic biomineral deposition. The *self-assembly* of the organic molecules becomes an important prerequisite in determining the final biomaterials properties. This principle has been exploited for designing novel hybrid materials wherein organic molecules, e.g. polymers, peptides, surfactants, etc., are organized prior to the mineral growth. Before giving an example, the principles of self-assembly are briefly outlined. Self-assembly or organization means attaining an ordered state from a disordered one and results in net reduction in entropy of the system. It is well known that soft matter can be organized into various assemblies via weak molecular interactions, such as van der Waals forces, hydrogen bonding, hydrophobic forces, and ionic interactions as discussed above. Typically a combination of long-range repulsion and short-range attractive forces is a prerequisite for self-assembled structures (Table 7.3). A variety of molecules and macromolecules ranging from simple structures such as lipids and surfactants through to peptides and DNA, can participate in the formation of assembled structures resulting in a wide range of hierarchical structures, such are micelles, lyotropic phases, lamellae, cylinders, spheres, etc. These self-assembly guidelines, when applied to functional molecules, result in the fabrication of organized inorganic–organic hybrid materials as exemplified later. Inspired by ferritin – the iron storage protein whose outer structure is built up from self-assembled peptide units in such a way that they present a hollow core for iron storage, scientists have used various proteins and peptides to construct cages that are used for controlled precipitation of various minerals. This strategy allows the composition, shape and size of the inorganic materials synthesized to be controlled precisely.

The next important lesson learnt from biology in designing synthetic hybrid materials is that of *programmed assembly and recognition*. Biological processes are

Table 7.3 Examples of pairs of forces which can lead to self-organization. (Taken from Forster and Plantenberg, *Angew. Chem. Int. Ed.* **2002**.)

Long-range repulsion	Short-range attraction	Examples
Hydrophilic/hydrophobic incompatibility	covalent binding	micells, lyotropic liquid crystals
	covalent binding	block copolymers
Coulombic repulsion	electroneutrality	ionic crystals
excluded volume	minimum space required	thermotropic liquid crystals
electric dipole filed	electric dipole interaction	ferroelectric domains
magnetic filed	magnetic dipole interaction	magnetic domains

under strict control and are inherently programmed, e.g. protein expression, protein folding, etc. In addition, molecular recognition is a key aspect behind the success of biological systems. These ideas can be used in material design in order to fabricate organized and structured materials. An example of the assembly of nanoparticles using DNA hybridization is given in this case. Nanoparticles can be functionalized by single stranded DNA molecules. These functionalized nanoparticles spontaneously assemble onto surfaces that are that are pre-patterned with complementary single stranded DNA molecules, thus creating patterned surfaces using programmed assembly. This can be used in sensor technology.

This description of biomineralization and bioinspired materials naturally leads us to consider specific natural and non-natural composite-hybrid materials including bone and nacre as examples of naturally occurring hybrid materials.

7.4.1
Natural Hybrid Materials

7.4.1.1 Bone

An example of a naturally occurring hybrid material is bone. Bone can be considered as a three-phase composite of organic collagen fibers, inorganic crystalline hydroxyapatite crystals and bone matrix. This combination of structural components provides versatility and, when present as a skeleton, functions to protect vital organs, provides support and sites for muscle attachment, generates new blood cells for the protection and oxygenation of other tissues and lastly acts as a reservoir for calcium, phosphate and other important ions. The unique properties of bone are a result of the atomic and molecular interactions occurring within this hybrid material. Bone is composed of type I collagen (36%) which is intimately associated with hydroxyapatite crystals ($Ca_5(PO_4)_3OH$ (43%). Bone matrix is composed of more than 200 components including many non-collagenous proteins such as proteoglycans, glycoproteins and Gla (γ-carboxyglutamic acid) proteins. The cells responsible for the development of new bones during repair and growth are the osteoblasts.

At the highest level of organization two bone types exist; compact or cortical bone has a porosity of approximately 10% with the pores containing osteocytes, cells that maintain bone, and blood vessels. This is the predominant bone type in long, weight bearing bones such as those of the legs and arms. In contrast, cancellous bone (also known as trabecular bone) has a much higher degree of porosity with a honeycomb like structure filled with osteocytes and marrow; Fig. 7.16. Cancellous bone is predominant in short bones such as those of the ankles and wrists. The structural differences of these two bone types result in compact bone having a much greater compressive strength and a higher modulus of elasticity both around 10% greater. Flat bones such as the skull and ribs are composed of spongy bone sandwiched between layers of compact bone.

In the formation of new bone, osteoblasts first produce a matrix of organic material, predominantly small fibrils of type I collagen randomly orientated and include water within the structure instead of the mineral phase. This material that

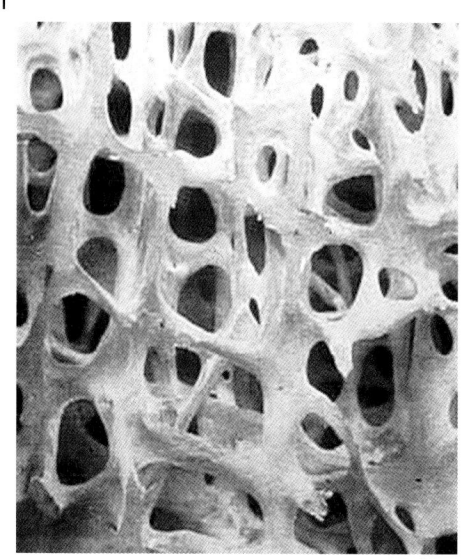

Fig. 7.16 An image demonstrating the porous nature of bone. (Taken from http://www.gcrweb.com/OsteoDSS/clinical/scope/pages/scope-definition.html)

forms initially is called osteoid, and it is then mineralized over a few weeks. The mineralization of the immature woven bone occurs in two stages, first nucleation of calcium phosphate crystals occurs by two different pathways. Homogenous nucleation is a result of super-saturation of the local environment with the appropriate ions then nucleating on the inner surface of small membrane bound matrix vesicles. Heterogeneous nucleation occurs at lower levels of supersaturation in the presence of surfaces that have a high affinity for calcium ions such as the surfaces of the proteins osteonectin and phosphoproteins that are often found in close association with collagen. Once nucleation has occurred the growth of crystals can proceed with the precipitation of amorphous calcium phosphate converting to octacalcium phosphate and then finally hydroxyapatite. Mineralization then continues with the newly formed crystals disrupting the vesicle membranes and fusing. The small crystals grow quickly and are aligned along the collagen fiber axis. The most developed bone, lamellar bone, contains the thickest collagen fibers which are aligned either in a linear or concentric (around a central blood vessel) orientation. The breakdown of bone in small packets to repair the structure is caused by cells called osteoclasts which resorb the bone.

The interactions between inorganic and organic phases in the process of biomineralization are still greatly debated throughout the scientific community. One such example is the possible interaction of aragonite plates of red abalone nacre with the organic matrix. Examination of Lustrin proteins – matrix proteins from the nacreous layer –a 24 amino acid polyelectrolyte domain within the protein Lusturin A, was identified. This domain, termed D4, is rich in aspartic acid and also

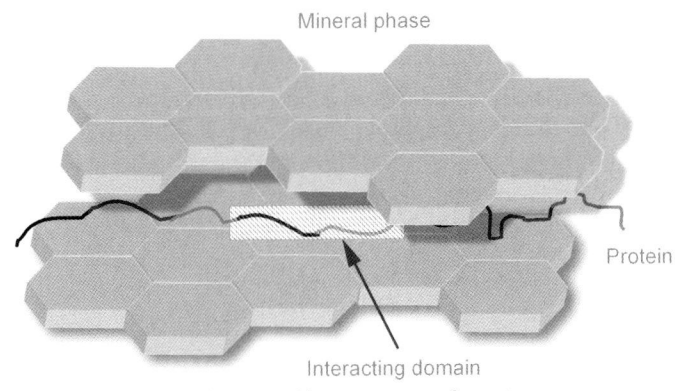

Fig. 7.17 Schematic of the possible interactions of proteins and minerals with an example of Lustrin interactions with calcium carbonate in nacre.

contains hydrogen donor/acceptor amino acids – asparagine, glutamine, arginine, threonine, serine and tyrosine. This structure was shown by NMR spectroscopy to adopt an open chain conformation in solution allowing side chain access for charged residues to the inorganic surface (Fig. 7.17). Further examination, using a modeled polypeptide of D4 revealed a 9 amino acid Ca(II) interaction sequence capable of binding Ca(II) ions in a 2 : 1 (Ca: peptide) stoichiometry *in vitro*. These studies also found the modeled D4 sequence affected the morphology of the $CaCO_3$ crystals grown *in vitro*, and the crystals retrieved form this assay also displayed the polypeptide as a bound species on the crystal surface.

These findings are put into question by a recent report which examined nacre of a different genus of red abalone. The aragonite crystals examined were found to have a continuous coating of amorphous $CaCO_3$ therefore not contributing to any epitaxial interactions with the organic matrix. This finding leads to uncertainty as to the involvement of the organic components and their possible interaction with the inorganic matrix.

Together the organic and inorganic components of bone are combined, creating a material which is both resilient and versatile (Table 7.4). Through variations of each component it is able to ideally form itself to the required function at each precise location. Artificial materials still need to be developed which have, in combination, the necessary strength to withstand load bearing and reasonable force. In addition to this, any novel material should be the source of minimal corrosion in the biological environment.

7.4.1.2 Dentin

Dentin, the most abundant mineralized tissue in the human tooth is another naturally occurring hybrid material containing apatite and fibrils of type I collagen. Although dentin contains the same basic materials as bone, the mechanical properties are considerably different in response to the functional requirements of dentin that is found beneath the enamel layer in teeth. This hybrid material

Table 7.4 A comparison of the mechanical properties of some naturally occurring biological hybrid materials and some metal and ceramic materials used as artificial biomaterials.

	Elastic Modulus (MPa)	Tensile Strength (MPa)
Compact bone	20 000	200
Dentin	17 400	97.8[a]
Nacre	64 900	130
Wood	10 000	100
Ti alloys	110 000–116 000	760–1100
Co—Cr alloys	210 000–253 000	655–1795
Dense hydroxyapatite ceramics	70 000–20 000	40–100
Bioglass (40%) HDPE[b] composite	2380–2700	9.44–10.86

a At strongest point, the dentino–enamel interface.
b High Density Polyethylene.

must be capable of enduring dental caries, ageing and disease. Dentin is a fiber reinforced composite material containing approximately 50% carbonated apatite and 30% organic matter which is predominantly cross linked type I collagen. Important features of dentin are the fiber-like tubules that provide reinforcement to the surrounding matrix. The tubules represent the tracks taken by the odontoblasts (dentin producing cells) from the dentino-enamel interface to the central tooth pulp, within which the new dentin continues to form throughout the life of the tooth. The lining of these tubules is composed of a highly mineralized cuff of intertubular dentin containing mostly small apatite crystals. The crystals are needle like near the pulp becoming more plate like closer to the enamel and there is very little organic material found in the tubules. Mineralization also occurs in the collagen fibrils that are randomly orientated perpendicularly to the dentinal tubules. The collagen fibrils of dentin are visible in Fig. 7.18. The elastic proper-

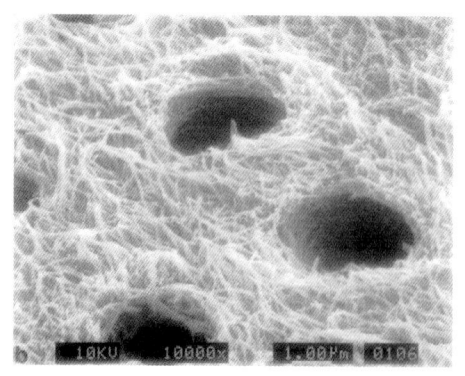

Fig. 7.18 Demineralized collagen showing the collagen fibrils perpendicular to the dentinal tubules. (Taken from Marshall et al., *J. of Dentistry*, **1997**, *25*, 441–458)

ties of dentin are a direct result of the perpendicular arrangement of the collagen within the dentinal tubules. As a material, the hardness of dentin is determined by the extent of mineralization and this may differ depending on the location of the dentin within the tooth, decreasing with proximity to the central pulp.

7.4.1.3 Nacre

Nacre, also known as mother of pearl, is a strong and resilient hybrid material formed by many families of mollusc including gastropods, cephalopods, and bivalves. This material, in spite of being based on a weak ceramic, calcium carbonate has a significantly greater bending strength ($220\,MN\,m^{-2}$) than natural ceramics such as slate and granite (100 and $150\,MN\,m^{-2}$ respectively). The properties of nacre are a direct result of its unique architecture. The inorganic component, calcium carbonate, is in the form of aragonite that results in an arrangement of hexagonal plates, layered as sheets <0.5 µm thick. Between the plates and sheets of calcium carbonate is an organic matrix which accounts for 1–4% of the total composition of the composite. The organic matrix is composed of chitin (a polymer of the monosaccharide *N*-acetyl glucosamine), layered between proteins rich in glycine and alanine (both hydrophobic). The surfaces of the organic matrix in contact with the calcium carbonate are hydrophilic acidic macromolecules rich in glutamic and aspartic acids, and these acidic molecules have been proposed to dictate the crystalline structure of the inorganic phase. The structure of this material relates to its strength in a number of ways. Firstly, the thin sheets of calcium carbonate (0.5 µm) are only one third of the critical crack thickness. This means that if one plate contains a flaw it is not sufficient to start a crack running through the material. Secondly, as shown in Fig. 7.19, the plates are staggered into a brick and mortar type arrangement. This means that failure of the material in the matrix will result in the crack continually meeting a plate and it will then be deflected around the edge of the plate. Thirdly, the laying down of nacre to form a shell occurs so the sheets are parallel to the shell surface so any crack travels through the nacre in the direction of its greatest strength.

7.4.1.4 Wood

Perhaps this well known natural material can be considered as a fibrous composite composed of parallel columns of cells which are supported by cellulose (a polymer of glucose) wound spirally in one direction (Fig. 7.20). These cells are embedded in a matrix of a substance called lignin, a complex poly-phenolic resin. Typically wood has three distinct layers that differ in the angles at which fibers lie relative to the longitudinal cell axis. The fibers of the outermost primary layer are at 50–70′, the central secondary layer at 10–30′ and finally the core tertiary layer at 60–90′. The thickest of the three layers is the secondary layer that is responsible for 80% of the total wood thickness, as a result, it is this layer which provides the greatest strength and acts as the major load bearing component.

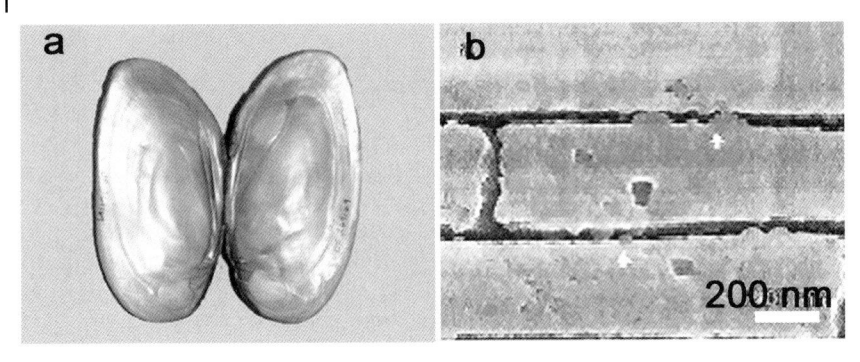

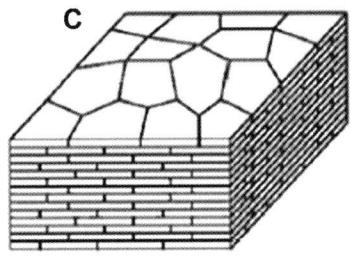

Fig. 7.19 (a) Nacre also known as mother of pearl formed by molluscs. (b) SEM of the structure of nacre showing its distinctive architecture, scale bar equal to 200 nm. (c) Diagrammatic representation of the brick and mortar structure of nacre with hexagonal aragonite plates forming the bricks surrounded by an organic matrix. (Image (a) taken from http://research.amnh.org/biodiversity/mussel/elliptiogenustext.html. SEM in (b) taken from Wang et al, J. Mater. Res., **2001**, *16*, 2485. Image in (c) taken from P. K. V. V Nukala and S. Simunivic, *Biomaterials*, **2005**, *26*, 6087–6098.)

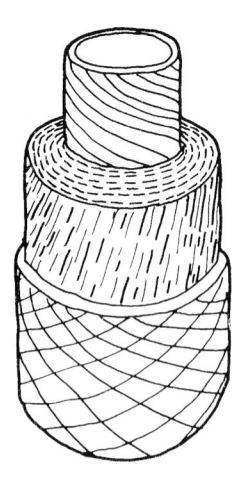

Fig. 7.20 A diagrammatic representation of the primary, secondary and tertiary layers of wood. (Taken from J. F. V. Vincent, *Structural Biomaterials*, Macmillan Press Ltd, London, **1982**.)

7.4.2
Artificial Hybrid Biomaterials

Synthetic hybrid materials can be made using components generated by biological organisms. These can be separated into those where the organic components are natural and those where the inorganic components are natural.

7.4.2.1 Ancient materials

Some of these materials have been used for many years. The only available materials were organic or minerals. This resulted in many hybrid materials being created to mix the properties of the constituents.

Dyes   Early dyes were often derived from plants, some were fixed using inorganic solutions, such as alum. This process created a hybrid material with the organic dye bound through an inorganic bridge to the organic fibers of the textile that was dyed. A well-known example of this is the red dye derived from madder, alizarin. The Maya civilization created a particularly resilient dye, Maya blue, which was a hybrid of the blue dye indigo (often used in the manufacture of denim jeans) and clay.

Paints   Early paints used a protein base, commonly egg white (albumin) or milk protein (casein) or an oil base, such as linseed oil and a powdered mineral as coloring. These paints are surprisingly resilient and are starting to be used once more, due to the hazards of organic solvents used in synthetic paints. The minerals used varied with their availability in each area and were chosen by their ease of powdering and lack of solubility in the protein/water base. There is some evidence that even early cave art was produced by mixing some kind of organic fluid, such as blood, with pigments. Pigments used included ochre (iron oxide), charcoal, lapis lazuli and malachite, along with many others.

The ancient Egyptians used tools to extract and process minerals that brightened the palette, including azurite blue, malachite green, yellow and orange from orpiment and realgar and vermilion red. They synthesized blue from cobalt ore and also from silica known as Egyptian blue frit or smalt. They made white from lead. Indigo blue and madder red were manufactured as dyes and pigment.

In Ancient Greece and Rome, tyrian purple from *Murex* was used as dye and pigment. It took an enormous amount of *Murex* to produce a tiny quantity of color. Verdigris green was made from copper acetate. In the Middle Ages, the development of the dyeing industry increased the number of organic pigments available. Dyes were extracted from flowers, roots, berries and insects to make lakes, which are translucent colors.

Others   Many other hybrid materials were used by ancient civilizations, but their particle size was usually quite large. Clays were, however, available as very small particles and were used in combination with animal and plant organic material, often dung, as building materials. Such materials are still in use in many parts of the world.

7.4.2.2 Structural Materials

When only the inorganic component is natural the hybrid materials are similar to fully synthetic hybrid materials. Extracted natural inorganic materials are often similar to their synthetic counterparts; natural systems derive their unique properties from the polymers that they use and from organization of the inorganic and organic components of the hybrid. Synthetic materials of this type include polymers filled with chalk or diatomaceous earth and bone cements containing hydroxyapatite from bone.

Organic polymers derived from nature can also be used to produce hybrid materials. One of the oldest is natural rubber, which can be combined with nanoscale carbon or clay powders to produce black and white rubbers as used in tyres. These materials are tough, resembling natural hybrids with high organic fractions. As with most synthetic hybrids the inorganic and organic domains are not organized and the final mechanical properties of the material rely upon an homogeneous mixing of the various phases on all length scales.

Ongoing work using collagen to mimic nacre and to attempt to generate structures of similar strength and toughness require growth of the inorganic material inside an organic matrix, as occurs inside organisms. This growth process has been partially copied synthetically, but no products are currently available. Ongoing work often uses proteins extracted from organisms or recombinant proteins, designed to act similarly to the natural ones. The systems used include calcium carbonate as found in nacre, calcium phosphate as found in bone and silica as plant surfaces or diatom skeletons.

Hybrid materials containing large particles are more common and have been produced in various forms for the building and composites industries. These often use aligned fibers and biological binders but are not nano-hybrids so fall out of the scope of this book.

7.4.2.3 Non-structural Materials

Non-structural materials have also been synthesized using biological materials as a source. These are usually core-shell particles, small particles of one material coated with another. These materials have wide uses, and some have been covered in previous chapters. Clay particles, titania and other materials can be coated with organic materials and used in makeup.

The reverse type of particle, with organic centers and inorganic surfaces can also be constructed and is often considered as a drug delivery method, where the inorganic material protects the drug and allows it to escape slowly. This is useful for drugs that would be broken down in the stomach or for a slow release, to allow the concentration of the drug to remain constant without having to inject it too frequently.

Current research involves many of the biological systems, researchers use extracts, recombinant proteins and other biological organic molecules to promote the growth of inorganic materials. Examples include the use of silica forming proteins from sponges and mares' tail ferns and the use of carbonate building proteins from corals and shellfish. This research has mostly resulted in porous particulate material to date, but much of it is targeted at the production of material of definable internal structure, size and shape.

7.4.3
Construction of Artificial Hybrid Biomaterials

In the search for new materials with useful properties, either as materials for use in the human body or as conjugates with, for example, novel electrical, catalytic or optical properties a range of approaches have been utilized, including the use of organic templates for the generation of inorganic materials and the semi-programmed or building block approach to materials (in its widest sense) assembly. The principles of both will be described and an illustrative example given.

7.4.3.1 Organic Templates to Dictate Shape and Form
The natural world makes materials (substance unspecified) that have inherent beauty because there is always control over shape and form. Organic chemistry and more recently supramolecular chemistry has been very successful in creating structures with spectacular morphologies but the synthesis of materials with shape and form using most of the elements of the periodic table (non carbon-based chemistry) has lagged behind. It is in conjunction with organic molecules such as small organic molecules, surfactants, synthetic polymers, organogels, proteins, viruses, etc. that shape can be transferred from the organic phase (with shape) to the inorganic phase (usually without shape) to generate a hybrid material with shape. The transcribing process can be divided into several stages:

(a) Firstly, the organic template, which may either be preformed or consist of self-assembled units, is brought into contact with the inorganic precursor, which may consist of ions in solution, vapor phase species or small particles of the required material that have been prepared previously. This stage often, but not exclusively, takes place in solution.
(b) Next, deposition of the inorganic material occurs on the inside and/or outside of the template, with the formation of an organic–inorganic hybrid material. The process could stop at this stage or;
(c) The organic template may be removed by washing with a solvent in which only the template is soluble or heat treatment (conventional thermal methods or use of microwave) that may lead to further structural changes in the material.

Materials that have routinely been formed using such a templated route include common oxides, such as silica and titania, but the methodology has recently been expanded to cover some nitrides and sulfides as well.

Biologically relevant molecules that have been used to template the formation of hybrid materials and inorganic materials with defined shape and form include ammonium DL-tartrate; proteins such as collagen that have positively charged surfaces that attract negatively charged species such as silica to themselves; DNA, tobacco mosaic virus; yeast cells, decalcified cuttlebone from the cuttlefish and wood.

In principle there are an infinite range of materials that could be combined to make novel hybrid materials if the chemistry at the interfaces between the two phases can be made compatible. One of the principle challenges in this area is in finding novel ways to obtain transcribed structures with a high degree of order over all three dimensions of the object being used as the template.

7.4.3.2 Integrated Nanoparticle–Biomolecule Hybrid Systems

Nanomaterials including metal and semi-conductor nanoparticles (having a range of shapes including spheres and rods) have dimensions on the same length scale (2–20 nm) as biomolecules such as proteins and DNA. The hybrid materials generated from the chemical or physical juxtaposition of these two types of materials generates novel bionanohybrids with a range of properties and functions. Currently available nanoparticles including metals (e.g. Au, Ag, Pt and Cu) and semiconductors (e.g. PbS, Ag_2S, CdS, CdSe and TiO_2) exhibit unique electronic, photonic and catalytic properties. Biomolecules, such as proteins, naturally play roles as enzymes, antigens and antibodies and these attributes can be harnessed within a composite hybrid material. There is much interest in generating 2D and 3D ordered structures of these biomolecules–nanoparticle hybrids both in solution and on surfaces and the materials generated have vast potential in the development of novel biosensors for use in the health-care industry.

Fig. 7.21 below shows schematically how nanoparticle–biomolecule devices can be generated. A range of functions for such materials is envisaged in areas as diverse as sensing, catalysis and electronics/optoelectronics. As many of these materials may eventually be intended for use in the human body the term "nanobiotechnology" has been coined to describe this research area.

7.4.3.3 Routes to Bio-nano Hybrid Systems

The main routes to producing nanoparticles are through solution-based approaches with clusters of the chemical components forming in the presence of surface-capping ligands that prevent aggregation and limit the final dimensions of the nanoparticles. Capping systems available include the use of hydrophobic or water-hating monolayers, positively or negatively charged hydrophilic or water-loving monolayers and polymer layers. The very presence of these chemical functionalities on the outside of the nanoparticles (initially present only to limit growth) is enabling the development of some truly novel hybrid materials as the surface functionality can be used to direct organization of the nanoparticles into 2D and 3D arrays. If this inbuilt functionality is combined with biomolecules that themselves have specific patterns for binding then site-specific binding of the nanoparticles to one another by self-assembly or to another material can be accomplished. Examples of biomolecular recognition include antigen-antibody interactions,

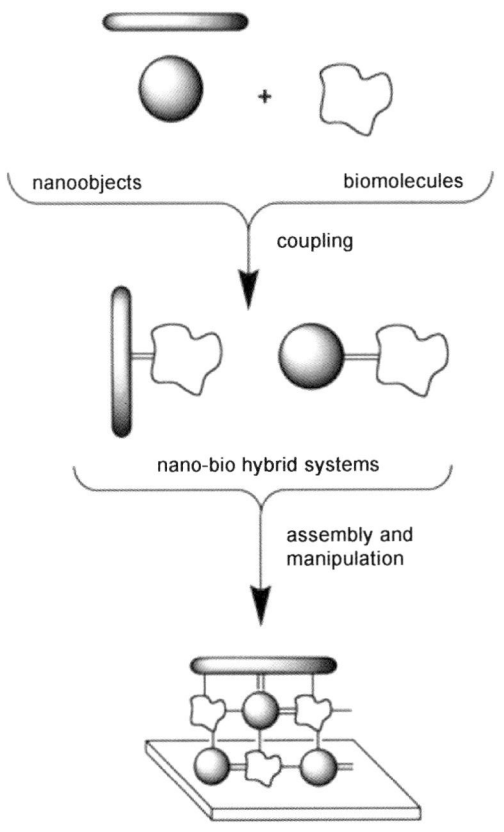

Fig. 7.21 A schematic of the interactions occurring between nano-objects and biomolecules. (Taken from Willner and Katz, *Angew. Chem. Int. Ed.* **2004**.)

nucleic acid-DNA interactions and hormone-receptor interactions. Some biomolecules exhibit two binding sites. Proteins can also be engineered to produce molecules with specific anchoring groups at specific locations on the surface thus increasing the options further. Hence it is possible to organize nanoparticles and even more than one type of nanoparticle into extended arrays by judicious choice of the chemistry linking the particles together.

Biomolecular functionalized nanoparticles required for such devices can be formed through:
 (a) electrostatic interactions, e.g. the binding of immunoglobulin G (IgG) to gold and silver nanoparticles functionalized with negatively charged citrate ions;
 (b) chemisorption and covalent binding through bifunctional linkers, e.g. oligopeptide binding to gold nanoparticles

previously functionalized with L-cysteine through thiol groups on the surfaces of the particles;

(c) programmed assembly using specific affinity interactions, e.g. streptavidin (Sav)-functionalized gold nanoparticles have been used for the affinity binding of biotinylated proteins and oligonucleotides as well as to other functionalized nanoparticles. This type of interaction is also feasible for nanoparticle bound antibody–antigen interactions.

A combination of the unique properties of nanoparticles and biomolecules provides innumerable opportunities for scientists and engineers to generate novel hybrid materials using a bottom-up approach to materials and device design based on the same principles of chemical recognition that leads to the structure and order that we see in nature.

7.5
Responses

Whether the hybrid biomaterial is generated from natural or artificial sources, if it is to be used in the human body then it must be compatible with the materials already present. In this section we consider the body's response to an implant and the fate of an implant once in the body.

Biocompatibility is closely related to cell behavior in contact with surfaces and can be described by two main factors; response of the host to implantation and the response of the biomaterial to the host. At each stage of implantation a foreign material is conditioned with biological fluid components. This interface can change with the response of the host organism and degradation of the implant material. Often living tissue rejects artificial materials. As part of an immunological response the host will act to destroy any potential pathogens and foreign bodies however this may also result in a reaction against the biomaterial. It is clear that many processes occur at the tissue–implant interface. Here, we attempt to give an overview of these processes to give the reader an insight into this environment and to suggest some of the complications that surround such terms as *biocompatibility*.

7.5.1
Biological Performance

Surface properties of implants are extremely important in obtaining the required host responses and possibly even more importantly, to prevent unwanted responses. The successful engineering of any product can only be achieved if the correct material is chosen.

Research has led to the development of biomaterial surfaces. Past focus has been on the bulk materials used for implantation, although it has now been shown that the surface of an implant is key in terms of its compatibility. For some time it has been understood that cells react to differing surface properties. Cell growth, proliferation and movement can be affected by changing the surface on which they reside. By changing the physical and chemical properties of biomaterials they can be fine-tuned to interact with the implant region in a specific way. By managing the implant interface, rapid and controlled healing could be possible, which may decrease health care costs and, ultimately, patient morbidity.

The role of protein adsorption and subsequent cell adhesion to biomaterials is important in terms of the performance of an implant; from the time of implantation and throughout its lifetime. Primarily it is the soluble proteins that are involved with adsorption onto implanted materials due to their availability.

7.5.2
Protein Adsorption

The adsorption of proteins to surfaces can be defined as the accumulation of protein at the solid–liquid interface. The surface concentration increases to a maximum, dependant on the bulk concentration and specific protein–surface interactions.

Protein–surface interactions are very important factors when considering the adsorbed state. A protein with a higher affinity towards the surface will adsorb at a faster rate, reaching an equilibrium after a shorter time. If the interaction is very high it is possible for the protein to deform, either on adsorption or some time after due to surface rearrangement (Fig. 7.22).

7.5.3
Cell Adhesion

Hybrid material properties arise from their individual components. Hybrid biomaterials benefit from differing properties in that they can be produced having

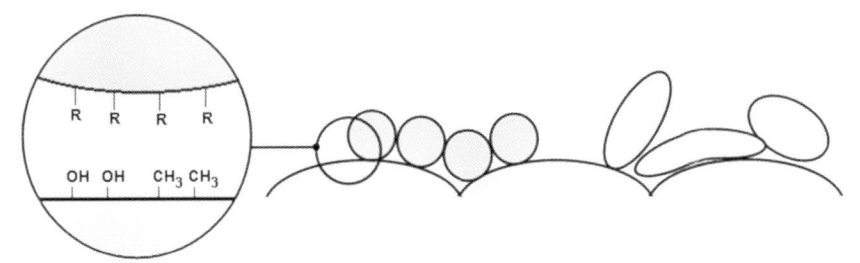

Fig. 7.22 Protein-surface interactions altered due to the chemical and topographical nature of the surface.

sufficient metal ion content to control (usually activate) cell adhesion and proliferation, and so increase the rate of biomaterials integration. More recent advances in materials technology, wherein nano-hybrid materials are produced using proteins and particle building blocks, utilize the effects of protein adsorption to govern the spacing between particles and aggregates thereof. Because these materials contain proteins, it is hoped that cell adhesion may be controlled at the molecular level, leading to the evolution of a whole new generation of materials. This is a good example of how the properties of individual hybrid components can be harnessed.

The orientation of adsorbed proteins is very important when considering the biological compatibility of a surface. As discussed earlier, the surface of a protein is crucial in detailing its interactions; both with the surface and with other proteins and cells. Specific regions on the outermost edge of the protein may have amino acid or oligosaccharide residues that have particular chemical functional characteristics, i.e. regions of charge or hydrophobicity. Such regions may become exposed once bound to the surface and may have particular functions in relation to cell attachment, e.g. if a surface-bound fibronectin molecule has the RGD (arginine–glycine–aspartic acid) tripeptide pointing upward from the surface, it may be available to bind to cell membrane receptors, allowing cells to attach, Fig. 7.23.

In a physiological environment cell adhesion always follows protein adsorption. Adhesion points can vary in strength depending on whether pre-adsorbed proteins from the extracellular matrix or proteins produced by the cell are utilized. Initial cell coverage occurs via cell–protein–surface attachments, followed by cell–cell connections as biomatter accumulates on the biomaterial surface.

The binding strength and spreading of cells is associated with the number of protein mediated surface attachments. These processes involve the continuous adhesion and release from the surface.

7.5.4
Evaluation of Biomaterials

There are many ways in which to test biomaterial properties to elucidate how they may be incorporated on implantation or react with the implant region. Biomaterial properties are important and must be closely examined, not only in standard physiological conditions but also in more harsh surroundings, which may be

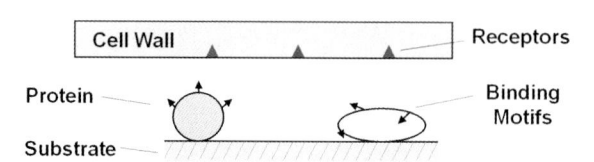

Fig. 7.23 Cell attachment to surface via a pre-adsorbed protein layer.

caused *in vivo* by numerous factors, not least of which is the initial breakdown products of implant materials.

There is no singular method which encompasses all biomaterial parameters with respect to response to integration. *In vivo* studies are usually more reliable because the complex physiochemical environmental parameters are tested at the same time. However, extrapolating absolute data from such experiments is usually very difficult. On this basis and for ethical reasons, an array of *in vitro* studies are usually conducted, each covering specific biomaterial reaction parameters.

The success of a biomaterial depends on its interaction with and acceptance by the host. Currently, the assessment of the host response is somewhat subjective and often limited to general histopathology and the identification of inflammatory responses. Determination of the implanted characteristics of materials is key in terms of biocompatibility – not only the absence of toxicity, but positive effects in the sense of biofunctionality are also required. Here we give a brief overview of some tests which can be performed on materials to assess how they will function on implantation, or more to the point, how the host will respond to the implantation.

Biomaterial testing requires that it be placed in the implant region and its performance be evaluated. Such *in vivo* (in a natural environment) investigations are expensive and often require long observation times in a number of test subjects. Due to the complexity of the biological system as described above and the diversity of each subject, results from such studies are often difficult to interpret. A more efficient method to assess the general performance of implant materials is to carry out a number of *in vitro* (in an artificial environment) model investigations. The pressure to reduce the number of tests on animals has also been a driving force behind *in vitro* methodology.

Evaluation of the processes associated with implant interfaces can be analyzed *in vitro* in a number of ways, each allowing specific results to be gathered about differing properties a material may have, e.g. toxicity, protein adsorption and cell culture. In addition, studies of material degradation are important so that the behavior of leached components or those produced by abrasive wear and fracture can be identified and understood. Although such tests minimize the number of animals used in research it must be appreciated that the results may not directly reflect how materials will react in the implant environment.

Once implanted, a material may undergo mechanical loading or stresses due to the environment. These forces may cause variation in material properties, e.g. a polymer having excellent osteoconductive properties may be used for bone repair, but it would be useless if it were to deteriorate under stress. Properties such as flexibility, elasticity, strength, rigidity, toughness, transparency, etc, may be of the utmost importance when considering a material for implantation. The physical bulk properties of materials often oppose those required for implantation so it is becoming more common to use one material for the bulk of an implant and another to give a biocompatible surface coating.

In brief summary, once a material is implanted into a biological system a number of steps are immediately underway. Proteins will adsorb from the surrounding serum and a protein adlayer will be formed within minutes. The composition of this layer will depend on the characteristics of the surface and the mixture of proteins surrounding the implant site. Initially those proteins in high abundance will adsorb and be replaced sequentially by proteins having stronger affinity to the surface, e.g. fibrinogen will replace albumin. Cells will then adsorb with their interaction mainly being mediated by pre-adsorbed proteins, although cells can specifically deposit proteins themselves. Adsorbed cells move around on the surface, spreading and proliferating. In this way the foreign material can be incorporated into the biological system through the progressive build-up of biomatter. Materials which do not permit normal biological function may not be suitable for use as implant surfaces. Biointegration therefore depends on the host response to the implanted material. A biocompatible material by definition is one which does not induce an acute or inflammatory response and does not prevent normal growth of tissues surrounding the implant.

7.6
Summary

Biomaterials, produced by biological organisms or designed as implants, are a technologically important class of materials. In this chapter, we have focused on hybrid biomaterials. Biological organisms produce inorganic–organic composite materials via a process termed biomineralization. We have shown how biology is a master at controlling and regulating biomineral formation by utilization of relatively simple building blocks – inorganic ions and organic biomolecules. The formation and properties of biominerals has been described and illustrated with selected examples. The mechanisms and control strategies underpinning biomineralization have been discussed where information is known. The principles underpinning the formation of biomaterials are useful to us both in terms of understanding existing materials and in our search for novel materials and processing technologies suitable for a variety of applications, some which are described in other chapters in this publication. From our new found knowledge, it is possible that new implantable hybrid biomaterials will be generated. These will be more readily integrated into the body on account of cell–protein–material interactions being considered at the design stage as well as the mechanical and physical characteristics of the materials being used to generate the composite phase.

Bibliography

Building blocks
W. D. Nesse, *Introduction to Optical Mineralogy*, Oxford University Press, New York, **1986**.
S. Mann, J. Webb, R. J. P. Williams, *Biomineralization*, VCH, Weinheim, **1989**.

E. J. Kucharz, *The collagens: biochemistry and pathophysiology*, Springer-Verlag, Berlin, **1992**.
D. Voet, J. G. Voet, *Biochemistry*, 3rd edn, John Wiley & Sons, **2004**.

Biomineralization

G. H. Nancollas, *Biological Mineralization and Demineralization*, Springer-Verlag, Heidelberg, **1982**.

S. Mann, *Struct. Bond.* **1983**, *54*, 127.

K. Simkiss, K. M. Wilbur, *Biomineralization*, Academic Press, San Diego, **1989**.

H. A. Lowenstam, S. Weiner, *On biomineralization*, Oxford University Press, New York, **1989**.

S. Mann, J. Webb, R. J. P. Williams, *Biomineralization*, VCH, Weinheim, **1989**.

S. Mann, *Biomineralization: principles and concepts in bioinorganic materials chemistry*, Oxford University Press, New York, **2001**.

J. H. Collier, P. B. Messersmith, *Annu. Rev. Mater. Res.* **2001**, *31*, 237.

B. A. Wustman, J. C. Weaver, D. E. Morse and J. S. Evans, *Connect. Tissue Res.* **2003**, *44*, Suppl 1:10–15.

B. Zhang, B. A. Wustman, D. E. Morse and J. S. Evans, *Biopolymers* **2002**, *63*, 358–369.

J. S. Evans, *Curr. Opin. Colloid Int. Sci.* **2003**, *8*, 48.

K. Shiba, T. Honma, T. Minamisawa, K. Nishiguchi, T. Noda, *Embo Reports* **2003**, *4*, 148.

M. Sarikaya, C. Tamerler, A. K. Y. Jen, K. Schulten, F. Baneyx, *Nature Materials* **2003**, *2*, 577.

S. Weiner, P. M. Dove, *Rev. Minerology Geochem.* **2003**, *54*, 1.

W. E. G. Muller, *Silicon Biomineralization*, Springer, Berlin, **2003**.

Bioinspired materials

P. G. de Gennes, **1991**, Nobel Lecture.

F. C. Meldrum, V. J. Wade, D. L. Nimmo, B. R. Heywood, S. Mann, *Nature* **1991**, *349*, 684.

S. Mann, *Biomineralization: principles and concepts in bioinorganic materials chemistry*, Oxford University Press, New York, **2001**.

C. M. Niemeyer, *Angew. Chem. Int. Ed.* **2001**, *40*, 4128.

S. Forster, T. Plantenberg, *Angew. Chem. Int. Ed.* **2002**, *41*, 688.

C. F. J. Faul, M. Antonietti, *Adv. Mater.* **2003**, *15*, 673.

S. V. Patwardhan, S. J. Clarson, C. C. Perry, *Chem. Commun.* **2005**, *9*, 1113.

D. Belton, G. Paine, S. V. Patwardhan, C. C. Perry, *J. Mater. Chem.* **2004**, *14*, 2231.

D. Belton, S. V. Patwardhan, C. C. Perry, *Chem. Commun.* **2005**, 3475.

Natural and artificial hybrid materials

H. Yamada, F. G. Evans, *Mechanical properties of locomotor organs and tissues*, Williams and Wilkins, Baltimore, **1970**.

J. F. V. Vincent, *Structural Biomaterials*, acmillan Press Ltd, London, **1982**.

G. Buxbaum, *Industrial Inorganic Pigments*, VCH, Weinheim, **1993**.

R. B. Martin, in *Introduction to Bioengineering* (Eds.: S. A. Berger, W. Goldsmith, E. R. Lewis), Oxford University Press, New York, **1996**, pp. 339.

C. C. Perry, in *Chemistry of advanced materials: An overview* (Eds.: L. V. Interrante, M. J. Hampden-Smith), Wiley-VCH, New York, **1998**.

D. A. Puleo, A. Nanci, *Biomaterials* **1999**, *20*, 2311. L. A. Polette, N. Ugarte, M. J. Yacaman, R. R. Chianelli, *Discovering Archeology* **2000**, 46.

F. Caruso, *Adv. Mater.* **2001**, *13*, 11.

S. H. Cypes, W. M. Saltzman, R. A. Gemeinhart, E. P. Giannelis, *Abstr. Pap. Am. Chem. Soc.* **2001**, 221: 657 CHED.

C. C. Perry, in *Encyclopaedia of physical science and technology, Vol. 2*, 3rd edn (Ed.: R. Meyers), Academic Press, San Diego, **2002**.

K. J. C van Bommel, A. Friggeri, S. Shinkai, *Angew. Chem. Int. Ed.* **2003**, *42*, 980.

E. Katz, I. Willner, *Angew. Chem. Int. Ed.* **2004**, *43*, 6042.

Responses

A. V. Recum, *Handbook of Biomaterials Evaluation: Scientific, Technical, and Clinical Testing of Implant Materials*, 2nd edn, Taylor & Francis, Philadelphia, **1998**.

J. E. Ellingsen, S. P. Lyngstadaas, *Bio-Implant Interface: Improving Biomaterials and Tissue Reactions*, CRC Press, Boca Raton, **2003**.

J. H. Kinney, S. J. Marshall, G.W. Marshall, *Crit. Rev. Oral. Biol. Med.* **2003**, *14*, 13.

B. D. Ratner, A. S. Hoffman, F. J. Schoen, J. E. Lemons, *Biomaterials Science*, 2nd edn, Academic Press, San Diego, **2004**.

J. J. Gray, *Curr. Opin. Struct. Biol.* **2004**, *14*, 110.

8
Medical Applications of Hybrid Materials

Kanji Tsuru, Satoshi Hayakawa, and Akiyoshi Osaka

8.1
Introduction

When artificial substances are introduced into the human body and come in contact with plasma (body fluid), many complex reactions and interactions take place. It is essential to recognize those issues before understanding the design concept of and behavior of inorganic–organic hybrids. Moreover, implant materials that are substituted for wounded or damaged parts of organs should be compatible in mechanical properties as well as in biological properties. It seems adequate to touch upon issues of fundamental significance for biomaterials and hybrids for biomedical applications.

8.1.1
Composites, Solutions, and Hybrids

Composites, solutions, and hybrids: those three all represent mixtures in which two or more components are joined together. Table 8.1 gives a rough idea of the differences among them. Carbon fiber-reinforced plastics (CFRP) are made of carbon fibers and a synthetic polymer matrix. The fibers are wound in lamellae, mimicking collagen fibrils in cortical bone (hard bone). Glass fibers are also employed for reinforcement. Common examples are golf club shafts, fishing rods, and tennis racket frames. The steel-belted automobile tire is another example of a composite. Composites between ceramic particles and metallic particles are possible; they are cermets (cer + met). Porcelains consist of a number of crystalline particles (grains) of various oxides, bonded together by firing at high temperature (sintering). It is not rare that such ceramic particles are integrated with each other to form a solid solution. When a mixture of MgO and NiO powders are heated, such particles are yielded which might be denoted chemically as (Mg, Ni)O. In the case of MgO and Al_2O_3, thorough heating yields a new compound, spinel. From the crystal chemistry viewpoint, the spinel structure is interpreted as magnesium and aluminum ions which are dissolved in the interstitial sites of the ordered

Hybrid Materials. Synthesis, Characterization, and Applications. Edited by Guido Kickelbick
Copyright © 2007 Wiley-VCH Verlag GmbH & Co. KGaA, Weinheim
ISBN: 978-3-527-31299-3

Table 8.1 Mixtures under different names.

States	Component size	Character of the components	Examples
Composite	~μm (at least one of the components)	similar dissimilar	porcelain, Pyroceram® CFRP, bone cements
Solution(*)	atomic and molecular level	similar dissimilar	gasoline, brass (CuZn), air, glass saline
Hybrids	atomic and molecular level	dissimilar	under development, involving molecular brush on substrate, oxide layer on metal

* Solid solution is an extreme form of hybridization.

packing of oxide ions. In contrast, moderate heating results in a sintered body (composite).

The crystal lattices of common ceramic materials are so flexible that the cations and anions can be replaced by foreign ions at their lattice sites: solid solutions are obtained. Such lattice modification or tuning modifies the properties of the original materials, and, is therefore a common technique for controlling the magnetic properties of ferrites or dielectric properties of barium titanate and its derivatives. Sometimes a vacancy (a lattice site not occupied by any ion) is important to stabilize lattices. The hydroxyapatite (HAp; $Ca_5(PO_4)_3OH$) lattice is one of the most flexible ones and it accepts a wide range of substituting ions. In the apatites involved in bone tissues, the hydroxy and phosphate ion sites are partly replaced by fluoride and carbonate ions while the calcium ion sites are frequently left vacant. When any other cations and anions replace the corresponding ions of HAp, novel properties are expected. Considering the effects of ion substitution, a solid solution is an extreme form of hybridization.

In the biomedical field, resin cements that comprise filler (ceramic particles smaller than ~100 μm in size) and synthetic polymer matrices derived from monomers having carboxyl, acryl, or epoxy groups are very common. Recently, dental resin cements are commonly employed for repairing caries teeth which involve fine particles (0.01~1 μm in size) of silica or β-$LiAlSiO_4$ (spodumene) dispersed in polymers, derivatives of methacrylate (resins) like bis-GMA (2,2-bis(p-2'-hydroxy-3'-methacryloxypropoxyphenyl)propane. Ba-containing silicate glass fillers are used for X-ray contrast purposes. Several silane coupling agents are used for securing filler-polymer bonding. Glass particles involving fluoride and aluminum ions are combined with such matrices to provide glass-ionomer cements. They are all composites.

Polymers consist of an infinite number of units of one kind. When two or more kinds of polymers are mixed together to yield polymeric solids, they are called polymer blends. Sometimes they are denoted as polymer alloys, but are rarely called hybrids since the components are similar in both structure (almost infinite repetition of a monomer unit) and property (soft, tough, deforming well). When chemical affinity of the components is low, they do not form a homogeneous mixture but an inhomogeneous one where the molecules of each component get together but are also segregated (phase separation). The mixture looks like mayonnaise, which consists of tiny isolated droplets of oil and water. A block copolymer material comprises repetition of oligomers of different kinds: [—(A—)$_m$—(B—)$_n$—]$_x$ where m and n are arbitrary, and x is practically infinite. Thus, a block copolymer is then probably one of the hybrid materials. Metallic alloys, in contrast, are actually solid solutions like (Mg, Ni)O mentioned above. For example, brass is obtained when a part of the copper atom sites of metallic copper is replaced by zinc. It must be noted here that ordinary metallic materials are polycrystalline, i.e. they consist of small crystallites, or grains. Solutions of two or more liquids or gases are also possible: glass is a homogeneous amorphous solid made of many oxide ingredients. A few oxides such as silica, phosphorus pentoxide, and boron oxide construct glass skeletons or networks due to, e.g. —Si—O—Si— bridging bonds. Alkali oxides (Li$_2$O, Na$_2$O, or K$_2$O) and alkaline earth oxides (MgO, CaO, or BaO) form ionic bonds like Si—O$^-$ •• M^{n+}, which are derived from breaking the bridging bonds. When glass is placed under aqueous conditions (in plasma), the Si—O$^-$ •• M^{n+} bonds are susceptible to hydrolysis.

$$Si-O^- \cdot\cdot M^{n+} + H_2O \rightarrow Si-OH(aq) + M^{n+}(aq) + OH^-(aq) \qquad (1)$$

The alkali and alkaline earth ions are leached and raise the pH of the surrounding medium. Glass is thermodynamically unstable, but its structure is frozen because structural relaxation at lower temperatures, say, room temperature, is too sluggish to be practical. When it is heated at moderate tem-peratures (400 °C~700 °C), structural relaxation is likely to occur, and glass precipitates crystallites which turn into glass-ceramics. Such a phenomenon is devitrification. This material is called glass-ceramic: Pyroceram® is a typical example. Ceravital®, Cerabone A-W®, and Bioverit® are glass ceramics commercially developed as bone-replacement materials. When only small amounts of ingredients are dissolved in a host medium, the former are solutes and the host is the solvent. Glass is a very good solvent for any oxide, but scarcely good for organic substances because glass melts only at higher temperatures that organic substances cannot tolerate, and because organic substances and inorganic moieties are incompatible with each other. However, those two components can be well mixed to yield hybrids via sol–gel routes; properly selected organic and inorganic molecules or those having both organic and inorganic groups are dissolved in a solvent and are subjected to hydrolysis and condensation to yield gels and solids.

Thus, hybrids stand between composites and solutions. Hybrids are not only homogeneous solids but also atomic- or molecular-level mixtures of components. The components have different chemical characters, but they are integrated, hence, the hybrid may exhibit novel properties, or at least those of each component at the same time. When a hybrid is composed of organic and inorganic components, it may behave as either an inorganic solid or organic solid, exhibiting properties characteristic of either. Table 8.2 gives a few examples among a huge number of molecules applicable to hybridization. All of them involve two reactive sites and organic/inorganic parts, except dimethylsiloxane, tetraalkylorthosilicate, and tetraalkylorthotitanate. Molecules with the same kind at both ends are also practically useful. Note that almost all parts of our tissues are constructed with proteins, even bones, that contain such chemically active sites as amide (—HN—CO)—, amino- (—NH$_2$), thiol (—SH), hydroxy- (—OH), carboxyl (—COOH), and sulfide (—S—S)— groups. The molecules having those at an end are likely to exhibit an affinity for living cells. That is, their molecular brush layers are likely to capture cells or proteins, and thoroughly improve the cell- or tissue-compatibility of the substrates on which the layers are established. Derivatives of Si are mainly selected here as examples because:

1. alkoxysilane groups (Si(OR)$_3$: R—OH is an alcohol) are susceptible to hydrolysis and yield silanol groups (—Si(OH)), which play an essential role in achieving material–tissue bonds;

Table 8.2 Typical molecules for preparing organic–inorganic hybrids for biomedical applications.

Molecule	Chemical formula	Reactive sites
tetraalkoxysilane	Si(OR)$_4$[a]	M—OR, M—OH (hydrolysis of MOR; M: Si, Ti)
tetraalkylorthotitanate	Ti(OR)$_4$	
dimethyldihydroxysilane oligomers (polydimethylsiloxane: PDMS)	monomer: Si (CH$_3$)$_2$(OH)$_2$ —(Si (CH$_3$)$_2$—O)$_m$ (m: ~200)[b]	—Si—OH
vinyltrimethoxysilane (VTMS) and its analogues	H$_2$C=CHSi(OCH$_3$)$_3$ (vinyl)—[ ]—SiX$_3$ [ ]: any organic skeleton; X: Cl, OR	vinyl group (H$_2$C=CH—) —Si(OCH$_3$)$_3$ —SiX$_3$
γ-methacryloxypropyl trimethoxysilane (γ-MPS)	CH$_2$=C(CH$_3$)COO(CH$_2$)$_3$Si(OCH$_3$)$_3$	methacryloxy group —Si(OCH$_3$)$_3$
γ-glycidoxypropyltrimethox-ysilane (GPTMS)	H$_2$C—CHCH$_2$O(CH$_2$)$_3$Si(OCH$_3$)$_3$ (with epoxide O)	—Si(OCH$_3$)$_3$ ethyleneoxide-like group
γ-aminopropyl triethoxysilane (γ-APS)	H$_2$NCH$_2$CH$_2$CH$_2$Si(OC$_2$H$_5$)$_3$	—Si(OC$_2$H$_5$)$_3$ —NH$_2$
titanium methacrylate triisopropoxide (TMT)	H$_2$C=CHC(CH$_3$)COO— Ti(OCH(CH$_3$)$_2$)$_3$	vinyl group —Ti(OCH(CH$_3$)$_2$)$_3$

a R: methyl, ethyl, 2-propyl, butyl.
b In the course of sol formation reactions, some Si—O—Si bonds are broken to give Si—OH.

2. silanol groups condense to give a siloxane skeleton
 (—Si—O—Si), from which the hybrids involve two skeletal
 bonds, the organic part presented as —[ ]— in Table 8.2
 and the inorganic siloxane part.

Points (1) and (2) also apply to the halosilane groups (—SiX$_3$; X: a halogen, usually Cl). The Si—X bond (X: Cl) is so reactive that it attacks a hydroxyl —OH group on some molecules to directly yield a Si—O—(molecule) bridging bond. Ti—O-hybridized materials are favorable when blood compatibility is of importance, and they might exhibit different properties from those of hybrids having Si—O bonds. In addition, hydrated silica and titania are compatible with hard tissues, as indicated later, and play a special role when embedded in the body. Oxygen tetrahedra with a Si atom in each center to form four Si—O bonds construct many kinds of silicate structures, networks, sheets, chains, and rings. Some of the silicate-based ceramics give such hydrated silica layers, exhibit significant tissue-compatibility, and have attracted much attention during the past three decades.

In the sol–gel routes of preparing oxides such as silica and titania, their alkoxides or halides are hydrolyzed to form sols consisting of colloidal particles or even smaller oligomers: those precursors yield gels due to further condensation of those sol particles at ambient temperature or ~150°C at the highest. When a sol is kept in a closed container with a cap or lid (a closed system) to gelation, hydrolysis and condensation proceed well while the solvent remains in the gel; the gel usually involves a larger volume fraction of micropore. In contrast, when a sol is kept in an open container (an open system), the solvent, if volatile, and even water will rapidly be evaporated to gelation. The product is xerogel. In this case, the gelation reactions are mostly incomplete, whereas the xerogel has a smaller volume fraction of micropore than the gel from the closed system. Taking advantage of being a fluid, the sols with controlled viscosity are shaped into fibers, coating, films, monoliths, and porous gels, employing drawing, dip-coating, spin-coating, and casting techniques.

The flexibility of sol–gel processing is crucial in fabricating organic–inorganic hybrids for biomedical applications. In addition to shaping flexibility, particular hybrids can be impregnated with active entities, like drugs, cells, or proteins, via the sol–gel route, with which they provide various kinds of biomedical functions such as:

- Surface modification of bulk materials by coating. The layer may be impregnated with physiologically active entities, which will be released into the body environment due to diffusion and sometimes accompanied by biodegradation of the hybrids themselves.
- The modification is also adequate for porous bulk materials with macro-pores of ~mm in size.
- Upon coating the pores with some hybrids, specific kinds of cells will be fixed on the surface and the system works as a cell-assisted bioreactor, or even as a "hybrid artificial organ".

- The proper components in hybrids also provide tissue-bonding ability, which will be described in Section 8.1.4, as well as machinability, described below.

It might be emphasized that macropores or macroporous hybrids can be fabricated due to a few techniques such as freeze-drying or the use of porogens, and that hybrids themselves are applicable to bioreactors. Fibers, films, and coatings are shaped in open conditions, and they are mostly xerogels. A ceramic structure is then obtained when the gels are calcined at appropriate temperatures. In the case of organic–inorganic hybrids, further treatment is seldom necessary for the gels. Those hybrids (hybrid gels) are sometimes able to be machined (machinable) with ordinary knives, scissors, and drill bites as they can be not so brittle or ceramic in character. Most of the gels, regardless of the oxide or being organic–inorganic, have microporosity, i.e. they have micropores of ~0.5 µm: the oligomers and colloidal particles in the sols are agglomerated and interconnected under gelation to leave particle stacking spaces. They appear white because of light scattering. When a transparent material is the final target of sol–gel processing, those micropores must be collapsed.

Metallic substrates coated with biologically or biochemically active molecular layers exhibit the properties of both components, and are involved in organic–inorganic hybrid materials because of their design concept. When the molecules are fixed at either of their end groups perpendicularly onto the substrate, the system looks like a brush, hence, it is called a "molecular brush". Oligomers are sometimes employed for the brush. Diverse functions will be exhibited dependent on the brush molecules. Metal–thin oxide-layer systems deserve to be taken as hybrids, too, in as much as the layer exhibits interesting biological activity. However, this topic is beyond the scope of this Chapter.

In biomedical fields, hybrid artificial organs have been proposed: living cells are fixed on a substrate to perform the functions of the original organs from which the cells are taken. In this case, the meaning of hybrid slightly differs from that meant in the present article (materials to immobilize such cells are organic–inorganic hybrids with proper porous characteristics). An attempt to explain the difference will be presented later (Section 8.4.1).

From the terminology point of view, biomaterials involve a wide range of applications, including beds and knives. As attention is focused on those applications associated with the cells and tissues of the body, they may be categorized into three groups as in Table 8.3: (1) structural materials, (2) therapeutic (clinical) materials, and (3) tissue engineering materials.

Organic–inorganic hybrids find quite a few applications.

8.1.2
Artificial Materials for Repairing Damaged Tissues and Organs

In a Hollywood motion picture "Robocop" released in 1987 the hero is not a robot but a cyborg. All of his movements are motor-driven, and he recognizes every item

Table 8.3 Biomaterials in three groups.

Applications	Location or property	Substance
(1) Biomedical materials (structural)		
Hard tissue substitutes	bone, joint, tooth & tooth root, casing of artificial heart (implant type), stent	metals, polymers, ceramics, hybrid coating (molecular brush)
Soft tissue substitutes	heart, kidney, lung, ligament, tendon	metals, polymers, ceramics, organic-inorganic hybrids, hybrid membranes
(2) Clinical materials		
Radioactive therapy	carrier of radioactive atoms and ions	glass and ceramics
Thermotherapy	materials susceptible to magnetic waves	glass and ceramics
Operations	suture (biodegradable)	synthetic and natural polymers
(3) Tissue engineering materials	carriers for biocatalysts (growth factors, cells, and enzymes) and scaffolds	polymers, ceramics, gels, organic-inorganic hybrids

of the environment in which he is placed through his peripheral sensing system: for example, his eyesight is dependent on a CCD camera. Yet, his brain, the only residual part of his human life, processes all information he gets through his sensing system, and gives electrical instructions to control all parts of his body as we do with our nerve system. When any part of the body suffers from malfunction, it is repaired in one of the following ways

1. Transplants and allografts:
   The whole or a part of a normally functioning organ or tissue of another person is transferred to replace the corresponding part of a patient.
2. Autografts:
   A part of a normal living tissue of a patient is transferred to replace his/her malfunctioning part.
3. Artificial materials and devices for implants:
   Materials (biomaterials) are implanted into the malfunctioning tissue or organ. An assembled system (artificial organ) replaces or accomplishes a complementary function of the damaged organ.

Materials derived from animals, mostly cattle, are frequently used for allografts. The most excellent tissue or organ replacements are what Mother Nature has created. Thus, transplanting and autografting are the most promising. Yet, the patient's immune system naturally responds severely and unfavorably against the transplanting and allografting of tissues and organs because they are foreign to the patient's body. In autografting, a limitation to the amount of the normal tissue to be taken must be encountered. It seems reasonable, therefore, to expect that a small part of any tissue could be proliferated industrially into a bigger piece and returned to the original tissue site. This is the concept of tissue engineering. Because of its promising applicability, those engaged in tissue engineering have carried out intensive and extensive research in recent years. It is one of the application fields of organic–inorganic hybrids. This therapy requires porous materials or scaffolds that have interconnected pores to accommodate the cells to be proliferated, to transfer oxygen and nutrients necessary for cell activity and to remove excretory waste. The porous structure should be so designed as to ensure the proliferation of coaxing cells and blood vessel formation. Once blood vessels appear in the scaffold, the transfer/removal processes become easier. Mainstream porous scaffolds are currently fabricated with a few synthetic polymers, such as poly(L-lactic acid) (PLA), poly(glycolic acid) (PGA), or their copolymer, poly(lactic-co-glycolic acid) (PLGA). While the scaffolds should be degraded at a proper rate, synthetic polymers are somehow unfavorable, though PLA is presumed to be degradable. In this respect, natural polymers such as gelatin and chitosan are preferable. In particular, bone tissue engineering scaffolds primarily stimulate bone cell proliferation and bone tissue generation; they must be osteoinductive and osteoconductive. Therefore, impregnating bone growth factors and mixing in hydroxyapatite or TCP has been attempted in clinical practice. As presented later, some chitosan-based organic–inorganic hybrids are promising materials for bone tissue engineering.

Still, tissue or cell proliferation takes a longer period of time. The patient probably cannot wait so long for his own tissue to be sufficiently proliferated. A highly sterilized environment for proliferation is essential at a tissue-engineering factory and the proliferated tissue must also be sterilized before transplanting: Any contamination (pollution with another microorganism, dust, or other undesirable things) might destroy the tissue under proliferation and causes fatal effects on the patient.

Man-made materials or devices are imperfect and inferior to the original tissues or organs in function, but at this moment are strongly in demand to replace damaged or malfunctioning tissues or organs. They are embedded or implanted in those places to perform the given function. Complicated larger systems also are employed at the patient's bedside that are supposed to maintain the life of the patients. They supplement and substitute the functions of damaged organs: artificial hearts, lungs, or kidneys. Among them, the artificial kidney or dialyzer is the most commonly applied system: dialysis is the indispensable medical treatment for diabetics (people suffering from diabetes). Artificial implants and devices will surely be used even when tissue engineering is improved in the future and becomes more practical, provided with excellent materials and processes by the research in progress nowadays. Fig. 8.1 is a famous drawing by Leonard da Vinci. It is a pleasant exercise to meditate upon what he had in mind as he drew this.

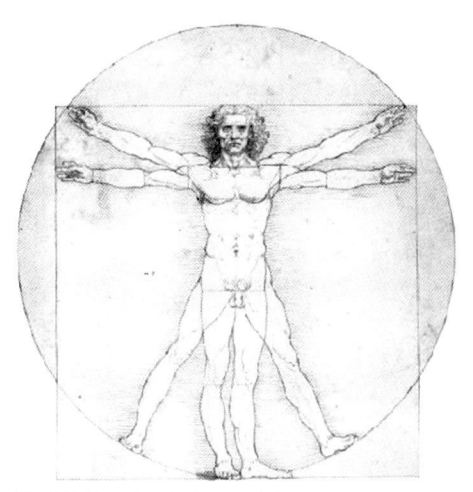

Fig. 8.1 Drawing by Leonard da Vinci.

Fig. 8.2 shows how near we have come to developing a cyborg or the Robocop. Most of the body consists of soft tissues, and the biomaterials presented here are mostly composed of many kinds of synthetic "polymers" including poly(tetrafluoroethylene) (TEFLON®), poly(methylmethacrylate), polyethylene, polyurethane, polyamide (Nylon®), to name a few. In contrast, metals and ceramics are

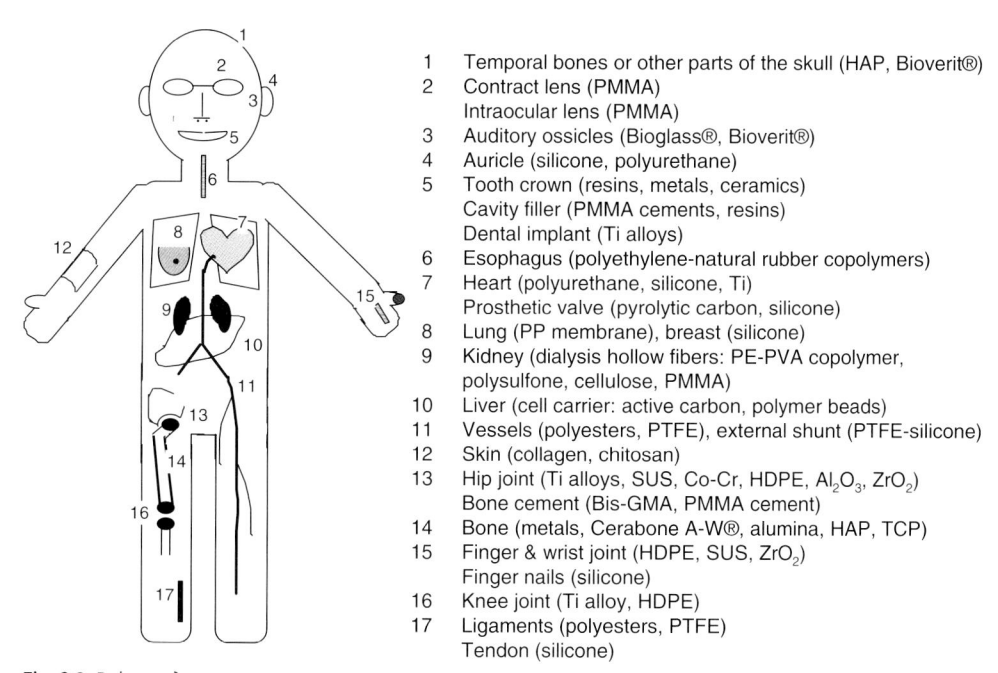

1	Temporal bones or other parts of the skull (HAP, Bioverit®)
2	Contract lens (PMMA)
	Intraocular lens (PMMA)
3	Auditory ossicles (Bioglass®, Bioverit®)
4	Auricle (silicone, polyurethane)
5	Tooth crown (resins, metals, ceramics)
	Cavity filler (PMMA cements, resins)
	Dental implant (Ti alloys)
6	Esophagus (polyethylene-natural rubber copolymers)
7	Heart (polyurethane, silicone, Ti)
	Prosthetic valve (pyrolytic carbon, silicone)
8	Lung (PP membrane), breast (silicone)
9	Kidney (dialysis hollow fibers: PE-PVA copolymer, polysulfone, cellulose, PMMA)
10	Liver (cell carrier: active carbon, polymer beads)
11	Vessels (polyesters, PTFE), external shunt (PTFE-silicone)
12	Skin (collagen, chitosan)
13	Hip joint (Ti alloys, SUS, Co-Cr, HDPE, Al_2O_3, ZrO_2)
	Bone cement (Bis-GMA, PMMA cement)
14	Bone (metals, Cerabone A-W®, alumina, HAP, TCP)
15	Finger & wrist joint (HDPE, SUS, ZrO_2)
	Finger nails (silicone)
16	Knee joint (Ti alloy, HDPE)
17	Ligaments (polyesters, PTFE)
	Tendon (silicone)

Fig. 8.2 Robocop?

employed to replace the hard tissues of the body, i.e. bone and joints, or to support fixation of the bone replacement materials to the bone, such as wires, pins, or small pieces of plate. Highly densified polyethylene with huge molecular weight (ultra-high-molecular-weight-polyethylene, (UHMWPE)) is used for the component of hip and knee joint replacement systems. No organic–inorganic hybrids appear here, because they are still in the research stage, though approaching the level of clinical and practical application, and the first artificial hybrids intended for medical application are calcium-containing ones derived from poly(dimethylsilane) and tetraethylorthosilicate via sol–gel rout (§ 2.3). It should be pointed out here that medical application of artificial materials is not necessarily limited to implants or implant-related issues; especially, one of the most significant roles of organic–inorganic hybrids is to provide a scaffold for cell culture with an optimum porous structure specific to the relevant cells.

8.1.3
Tissue–Material Interactions

All kinds of materials are possible components for hybridization. Those materials mentioned above are biocompatible, and not decomposed by plasma (body fluid). They are supposed to perform their functions properly at the site of replacement or substitution. In contrast, some biomaterials will gradually be decomposed into plasma, and new tissues are induced and reconstructed; a suture is a typical example. They are denoted as bioresorbable or biodegradable materials. Table 8.4

Table 8.4 Typical interactions between materials and tissues.

Influence of materials on tissues	Typical responses of materials in a body environment (chemical stability)	Typical responses of living tissues	Compatibility
Active	Stable accompanying surface reactions: Leaching and deposition of the components, changes in composition and surface structure	Blood clotting, denatured proteins, tissue-malfunction, severe inflammation, allergy (e.g., Ni hypersensitivity)	Toxic
	Chemically unstable: Bioresorption, biodegradation	Stimulated tissue generation and cell proliferation; Cell attachment, tissue-material bond formation, depositing apatite	Biologically active Biocompatible (positive biocompatibility)
Passive	No degradation: Inert	Recognition as foreign substance, fibrous tissue formation (encapsulation)	Non-toxic Bioinert, biotolerant. (passive biocompatibility)

summarizes examples of the interaction that takes place at the interface of materials and tissues. Note that any substance outside the body or artificially produced is foreign to the body, but the human living system should respond properly to such foreign components if they are covered with proteins or other suitable things in blood plasma. When such a layer is produced, the system feels relieved because the materials constructing the layer are its own. A layer of fibrous tissue covers materials only when they are bioinert. Our life system thus distinguishes its own things that it creates from foreign things. Any of the interactions described in Table 8.4 are expected after the fundamental response. This again confirms that biocompatibility control is by far significant. For example, however properly the rate of degradation or cell-attaching surface structure is designed, the material could never perform its function when a stable protecting layer, fibrous tissue, covers the material surface.

Biodegradation proceeds basically with continuous hydrolysis due to a group of hydrolyzing enzymes, hydrolases, that include, e.g. esterase, phosphatase, nuclease, peptidase, and lysozyme. It stands to reason, therefore, that they cannot hydrolyze synthetic polymers: common synthetic polymers in daily life such as Nylon, polyethylene, polypropylene, or polyurethane, are good examples of bioinert materials, while polylactic acid is supposed to be biodegraded, as well as chondroitin sulfate and hyaluronic acid, whose structures are presented in Fig. 8.3.

Here, bioinert stands for being biologically or biochemically inert in a body environment (36.5 °C, pH 7.4, 1 atm). Among numerous natural polymers, cellulose (cotton, wood cellulose) and silk fibrils are not biodegradable because the human body has no hydrolases for them, while gelatin, chitosan or chitin are more or less biodegradable. In particular, chondroitin sulfate and hyaluronic acid are produced in the body, and are therefore biodegradable. Those degradable ones are good components for hybridization. Lysozyme is sometimes used in the laboratory to hydrolyze gelatin and chitosan while it is also a hydrolase for glycoproteins. Weight loss of materials in a lysozyme solution is a good measure of biodegradation.

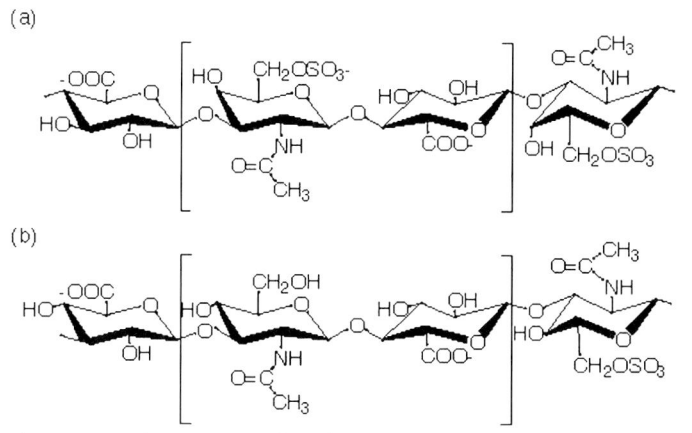

Fig. 8.3 Secondary structure of (a) chondroitin sulfate and (b) hyaluronic acid.

Biodegradability of organic–inorganic hybrids can be controlled by the components involved and their mixing ratio (composition), how they are hybridized (bonded to each other to form a rigid structure, or just mixed to be solidified: gels), etc. That is, they can either be fully biodegradable, partly dissolved only at the surface layer, or not degradable. Note that, in any level of biodegradability, the biodegradation reaction products are more or less released into plasma, hence, hybridizing components should be carefully selected and the toxicity of such products should also be taken into consideration.

Metals are mostly bioinert unless they are rigorously dissolved due to corrosion into the body fluid. Ca^{2+}, Mg^{2+}, Na^+, and K^+ are the major metal ions in plasma, and are essential for maintaining our life. Fe^{2+}, Fe^{3+}, Cr^{3+}, Zn^{2+}, or Co^{2+} are also essential but are involved in smaller amounts as a part of cells or enzymes. Fe and Co play important roles in cell functions and synthesizing Vitamin B_{12}, respectively. The ion concentration in plasma is equilibrated (balanced to be constant) by homeostasis. When ion concentration exceeds the controllable level for any reason, malfunctions like allergies will appear. Moreover, Ni^{2+} ions frequently cause heavy allergies (hypersensitivity). Therefore, a limited number of metallic materials appropriate for implants and bone substitutes are permitted by a national agency in each country (the FDA in the USA). A few of them are listed in Table 8.5.

Table 8.5 Metals in implant use.

Metals (Examples of commercially available items)	Components (mass %) (+: additional trace elements)	Applications	Comments
Stainless steel 316L (SS316L)	Fe(~70), Ni(~13), Cr(~18), Mo(~3), +	Temporary devices: bone plates, screws, pins	Less C content than SS316, giving better chemical corrosion resistance. Inferior in wear resistance under shear stress to Co—Cr alloy
Co—Cr alloys (Vitallium®)	Co(~60), Cr(~20), Mo(~5), +	Femur heads, bone plates, dentures	Heavier than Ti alloys Higher shear stress resistance
Ti6Al4V	Ti(90), Al(6), V(4)	Femurs, knee joints, tooth roots	Stronger than pure Ti, lighter than Co—Cr alloys
Ni—Ti alloys (Nitinol®)	Ni(50), Ti(50)	Dental arch wires	Shape memory effects (super elasticity), allergy-inductive (Ni hypersensitivity) due to corroded nickel ions

Titanium is one of the inert metals but it becomes slightly corroded and leaches titanium ions into plasma. The dissolved ions sometimes are reduced to a metallic state which are precipitated as very small particles (colloids) and accumulated in nearby tissue. Under shear stress and loads like those at joints, wear debris is yielded and those metallic colloid particles also cause inflammation.

Only a few kinds of ceramics are commonly employed for medical and dental purposes: aluminum oxide (alumina; Al_2O_3), zirconium oxide (zirconia; ZrO_2), hydroxyapatite (HAp; $Ca_5(PO_4)_3OH$), and tricalcium phosphate (TCP; $Ca_3(PO_4)_2$). Silica colloids and fluorine-containing glass particles are used as cement filler. When compacts of µm-size alumina and zirconia particles are heated at elevated temperatures, >1500 °C, and sometimes under pressure, the particles will be bonded strongly (sintered) to yield polycrystalline alumina and zirconia ceramics. Taking advantage of their mechanical strength and toughness, those ceramics are used for cup/head hip prosthesis. Sapphire (single crystalline alumina) was once frequently employed for artificial tooth roots. Alumina and zirconia are inert because they are highly chemically stable under ordinary conditions, and are passively biocompatible.

The crystal lattice of common ceramic materials is so flexible that its cations and anions will be substituted by foreign ions at the lattice sites when the size of the guest ions is comparable with that of the host ions. Such lattice modification or lattice tuning varies the properties of the original materials, hence, is a common technique to control the magnetic properties of ferrites or dielectric properties of barium titanate and its derivatives. The HAp lattice is one of the most flexible ones and it accepts a wide range of substituting ions, including vacancies, i.e. it provides a good base lattice for solid solution: the extreme form of inorganic–inorganic hybrids.

8.1.4
Material–Tissue Bonding; Bioactivity

Whatever biomaterials are implanted, they should remain for a while at the site and perform their functions. Take, for example, a surgical operation for total hip joint prosthesis to substitute the damaged hipbone and hip joint. The Ti6Al4V shell with a high-density polyethylene cup and bone implant will be fixed to the remaining femoral bone tissues with bone cement as well as pins, screws and wires made from Ti6Al4V. Since the alloy is bioinert, the pins and screws will be loosened in the long term. Current cement is based on methylmethacrylate and a filler, where the methylmethacrylate monomers are polymerized to yield poly(methylmethacrylate) (PMMA). In the course of polymerization, much heat is generated, which might damage the surrounding tissues. Sometimes unpolymerized monomers get into blood and cause fatal effects. Moreover, in the longer run, the cement will be degraded due to repetitive stress in the patient's daily life activity. When the implant is able to bond to living bone tissue, a cement-free implant operation is enabled. Polymers or hybrids that are expected to stay at the implanted sites to perform their functions present similar issues. Such tissue–bond ability widens their applicability.

All materials were bioinert (Table 8.3) except HAp and TCP before the invention of Bioglass® by Hench, consisting of SiO_2, CaO, and Na_2O as the major ingredients, with a small amount of P_2O_5. Any point within the composition diagram presented in Fig. 8.4 represents the composition of a glass: the mixing ratio of each oxide is proportional to the distance from the point to the edge (0%) opposing the apex (100%), while all of the glasses contain a constant 6% P_2O_5. Compositions in region A can be chemically bonded to bone tissue. The most frequently used is denoted as 45S5, containing 45% SiO_2, 24.5% CaO, 24.5% Na_2O, and 6.0% P_2O_5, (in weight) whose composition is located in the center of region A of Fig. 8.4. It forms a bond so strongly with bone that when a bone–glass–bone specimen is extracted after a certain period of implant and subjected to torsional stress, the bone part breaks instead of the bone–glass interface. After the invention of Bioglass®, a few other glass ceramics were developed, such as Bioverit®, Ceravital®, and Cerabone A-W®. Unfortunately, commercial production of the latter two has been terminated, though they are still used for research. Such achievements of tissue–material bond-ing are epoch-making, because no man-made substance has ever been bonded to tissue in the history of mankind for millions of years. (Bone cement provides only a physical fixation.) Moreover, even the collagenous constituent of soft tissue can strongly adhere to the glass within the range, at whose center 45S5 glass is present. The principle of the bonding has since then been extensively studied, and it was found that the essential event is spontaneous deposition of apatite on the glass surface when it is in contact with plasma. In the field of ceramics for biomedical applications, this specific property is called bioactivity.

A thorough understanding of the apatite deposition process in plasma is essential for designing novel bioactive materials, including organic–inorganic

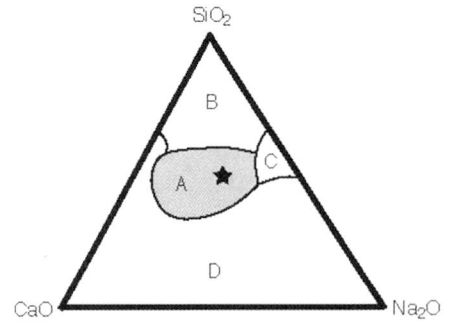

Fig. 8.4 Compositional dependence of bioactivity (bone bonding) of the glasses consisting of SiO_2, CaO, Na_2O, and 6mass% P_2O_5. Region A: bioactive glasses, Region B: inert, as window glasses, Region C: resorbable. Region D forms no glass, and has no practical significance. ★: 45S5 glass, bonding to soft-tissue. (Modified from L. Hench, *J. Am. Ceram. Soc.*, 74, 1487–1510, **1991**.)

Table 8.6 Inorganic ion concentration in human blood plasma, and ion product (IP) in terms of hydroxyapatite formation. The ion concentration of Kokubo's simulated body fluid (SBF) is also presented.

Ions	Na^+	K^+	Ca^{2+}	Mg^{2+}	Cl^-	HCO_3^-	HPO_4^{2-}	SO_4^{2-}
Human plasma	142.0	5.0	2.5	1.5	103.0	27.0	1.0	0.5
SBF	142.0	5.0	2.5	1.5	147.8	4.2	1.0	0.5

Log (IP: eq. (1)): −117; Log (IP for plasma according to eq. 1): −96.

hybrids. Table 8.6 shows the inorganic ion concentration of plasma as well as the ion–product (IP) of hydroxyapatite according to Eq. (2)

$$10Ca^{2+} + 6PO_4^{3-} + 2OH^- \rightarrow Ca_{10}(PO_4)_6(OH)_2 \qquad (2)$$

for an apatite formation reaction in an aqueous system. An increase in the concentration of ions on the left-hand side of the equation for any reason favors the formation of apatite. When equilibrium is established with respect to Eq. (2), IP is equivalent to solubility product (Ksp). Surprisingly, the value of log (IP) for hydroxyapatite in the literature is 10 orders of magnitude smaller than that calculated from the ion concentration data in Table 8.6. That is, human plasma is intensively supersaturated in terms of precipitation of apatite. Still, apatite would not be precipitated either in blood or on the wall of blood vessels, with one exception: bone tissue. A few people suffer from malfunctions due to the precipitation of apatite or stones (calculus) in organs such as liver and kidney. Bone t issue is a hybrid in which thin apatite platelets (20 ~ 40 nm in size) are precipitated on collagen fibrils constructing a wound lamellar structure. The constituents of bone, i.e. apatite crystallites and collagen fibrils, are organized into three-dimensional (3-D) structures. This apatite precipitation is analogous to the formation of frost. Frosting takes place near 0 °C. In contrast, another product due to vapor-to-solid transformation, diamond dust, will only be observed far below 0 °C, about −15 °C. Physical chemistry interprets the difference in temperature in terms of the nucleation mechanism: inhomogeneous nucleation for frost and a homogeneous one for diamond dust. The former nucleation is thermodynamically more favorable than the latter and needs less supersaturation. Therefore, this suggests that the body has some trick to suppress homogeneous nucleation of apatite in plasma and to prohibit an inhomogeneous one at the interface of plasma and soft tissues. Bone is living tissue, and bone formation involves osteoblast and osteoclast cells; an osteoblast is a bone-forming cell while an osteoclast absorbs them, and bone tissue is continuously renewed: this process is called remodeling. It may be inadequate, then, to presume that the hybridization of apatite and collagen deriving from bone tissue is so simple as to be understood by the nucleation scheme. As seen above, however, the apatite layer is located between the Bioglass® and the newly generated bone tissue. This indicates that osteoblasts operate after the apatite deposition. Thus, it is safe to conclude that the apatite layer is purely a result of physico-chemical reactions in which the inhomogeneous nucleation mechanism works. Ease and promotion of experiments of the laboratory level to minimize sacrificed animals as much as possible has been the stimulus to develop an artificial plasma that may simulate and mimic the plasma–material interaction in good reproducibility: a simulated body fluid (SBF), whose ion concentration is seen in Table 8.6. Owing to the metabolism in the body, plasma is higher in hydrogen carbonate content than SBF.

It seems accepted now that apatite crystallites are deposited on CaO—SiO$_2$—based glasses in the processes schematically demonstrated in Fig. 8.5.

1. The glass surface is corroded by plasma and leaching sodium and calcium ions.

$$\equiv Si-O^- \cdot Ca^{2+} + H_2O \rightarrow \; \equiv Si-OH(aq) + Ca^{2+}(aq) + OH^-(aq) \tag{3}$$

Si—O·Ca represents the ionic interaction between the oxide and calcium ions, and (aq) indicates a hydrated state of the species. The calcium ions and hydroxyl ions increase the supersaturation of plasma for apatite formation.

2. A hydrated silica gel layer is produced due to condensation of Si—OH:

$$\equiv Si-OH + HO-Si \equiv \; \rightarrow \; \equiv Si-O-Si \equiv \; + H_2O \tag{4}$$

Some fraction of the Si—OH groups changes into Si—O$^-$ by the hydroxyl groups.

$$\equiv Si-OH(aq) + OH^-(aq) \rightarrow \; \equiv Si-O^-(aq) + H_2O \tag{5}$$

Eqs. (3) and (4) yield a hydrated silica gel layer involving Si—OH(aq) and $\equiv$Si—O$^-$(aq).

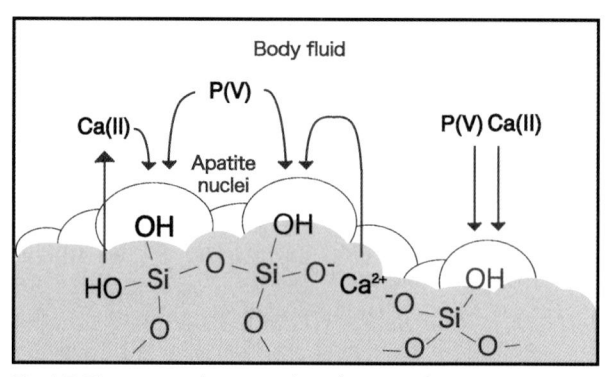

Fig. 8.5 The commonly accepted mechanism of apatite formation on CaO—SiO$_2$-based glasses in the body.

3. The hydrated silica gel layer induces inhomogeneous apatite nucleation. Once nuclei are produced, they grow, taking the component ions present in the plasma.

Bioglass® is prepared by melting component oxides at high temperatures (1200 ~1600 °C): if oxides are not easily available for any reason, like being too hygroscopic, corresponding carbonates or hydroxides that will be decomposed in the course of melting give corresponding oxides. The melts are cast into any shape, or crushed to granules. Fine powders are combined with polyethylene to produce a composite, HAPEX®, which is also applied to middle ear devices, as is Bioglass. Bioglass® granules are sometimes used to fill bone defects to stimulate bone induction and conduction. Glass ceramics having the same ability to be directly bonded to bone have been developed: Ceravital®, Bioverit®, and Cerabone A-W®. Their mechanical properties are compared with those of bone in Table 8.7, where K_{IC}, a measure of toughness, indicates that those glass ceramics are too brittle (less tough) but are too hard. Cerabone A-W® is comparable in bending strength with cortical (hard) bone.

Besides Bioglass® or other multicomponent ceramic materials, pure hydrated silica and titania gels derived from hydrolysis of tetraethylorthosilicate (= tetraethoxysilane) and tetraisoproylorthotitanate ($Ti(OC_2H_5)_4$), respectively, also deposit apatite in SBF. Those results indicate that an essential factor is the substrate; specific hydrated oxides serve nucleation sites. In addition, incorporation of calcium ions into the material enhances the ability to induce apatite nucleation if it has less ability for any reason. This applies to organic–inorganic hybrids.

Table 8.7 Mechanical properties of cortical bone and Bioglass® or relevant bone-substitute ceramics.

Materials	Bioglass®	Ceravital®	Bioverit II®	Cerabone A-W®	Cortical bone
major components	Na_2O, CaO, SiO_2, P_2O_5	Na_2O, K_2O, MgO, CaO, SiO_2, P_2O_5	Na_2O, K_2O, MgO, CaO, SiO_2, P_2O_5, F	Na_2O, K_2O, MgO, CaO, SiO_2, P_2O_5	
crystals precipitated	None	oxygenapatite	florogite, oxygenapatite	β-wallastonite, oxygenapatite	hydroxyapatite
Bending strength/MPa	85	100–150	90–140	215	150
Young's modulus/GPa	79	–	70	118	30
K_{IC}/MPa·$m^{1/2}$	0.5	3.5	–	7.4	6.3

8.1.5
Blood-compatible Materials

Blood compatibility is another important aspect of tissue–material interactions. When a material is placed in a body environment, inorganic and organic ions are attached on the surface, then proteins, and finally cells, or tissues come to the surface. Blood involves several key substances (factors), which are activated to initiate blood clotting cascade reactions in two routes, schematically indicated in Fig. 8.6. The left route leads to the formation of insoluble fibrin via the cascade reactions of coagulation factors, while the right route leads to the formation of white thrombus via activation and aggregation of platelets. Finally, insoluble fibrins (thrombi) and activated platelets are simultaneously aggregated together on the foreign surface. In clinical examinations, blood clotting times, represented as PTT and PT, are important factors to monitor the healthiness of blood. They represent the times for blood to yield thrombi after the clotting factors are activated, as indicated in Fig. 8.6. Note, however, that the blood components are condensed in the thrombi or clots, or some kinds of bone growth factors are present there. Thus, the blood clotting is not always unfavorable but is favorable in some cases. It has empirically been pointed out that not only titania gel layers formed on Ti substrates due to chemical treatment with hydrogen peroxide solution but also titania gels derived from tetraisopropylorthotitanate through sol–gel processing have excellent blood compatibility in terms of blood clotting time and platelet adhesion. Though the reason has not yet been clarified, hybrids incorporated with Ti—O bonds might be blood compatible. Those hybrids are applicable to anticlotting coating.

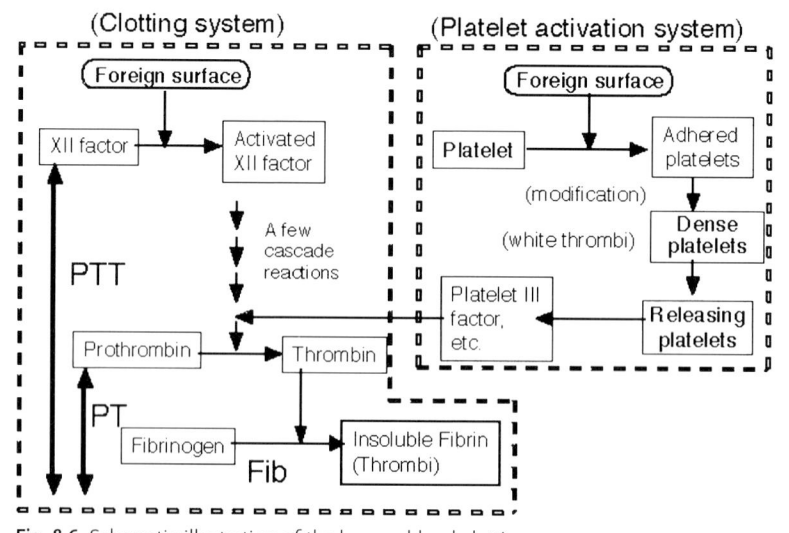

Fig. 8.6 Schematic illustration of the human blood clotting processes.

8.2
Bioactive Inorganic–Organic Hybrids

In the next few sections, the basic concepts of designing hybrid materials, plus the structural aspects and apatite-forming ability of individual hybrids are discussed. Before considering their practical application as scaffolds and so on, such characteristics must be understood.

8.2.1
Concepts of Designing Hybrids

Table 8.8 summarizes the general properties of metals, ceramics, and polymers in terms of biomedical applicability. Any material is basically employable as

Table 8.8 Mechanical, chemical, and biological characteristics.

Materials	Property modification	On-site machinability	Cell affinity and compatibility
Metals	Alloying atoms Oxide or molecular layer coating Not biodegradable	Possible with difficulty	Encapsulation with fibrous tissue without bioactive or cell-compatible oxide layers or molecular brush layers
Ceramics Glasses	Composition Crystallization Not biodegradable with exceptions: Bioglass®, TCP	Impractical	Encapsulation with exceptions (Bioglass®, Bioverit®)
Polymers — Synthetic	Monomers, co-polymerization, degree of polymerization (molecular weight) Forming composites with ceramic particles (HAp, glass ceramics A-W, titania)	Possible	Encapsulation without bioactive or cell-compatible layers
Polymers — Natural	Forming composites with HAp, hybridization with inorganic components		
Organic-inorganic hybrids	Hybridizing components, their ratio, doping salts	Controllable	Controllable

biomaterial as long as it is harmless. Usage is dependent on mechanical, chemical, and biological properties. Their properties are modifiable by coating, including molecular brush coating. It is advantageous that the coatings do not change the properties of the substrate that are determined by the chemical bonds comprising them. Yet, organic–inorganic hybridization drastically changes the properties, e.g. from ceramic to rubber. Metals are tough and hard, and are thus suitable for load-bearing materials such as bone substitutes, bone plates, pins and wires, or tooth roots, whereas no one would want to employ them for soft tissue substitutes. Wear resistivity is one of the key factors for materials on which larger shear stress is loaded. Corrosion in plasma must properly be taken care of: a protective oxide layer is naturally yielded (denoted as passivation) on such metals as stainless steel and the Co—Cr alloys in Table 8.8, but local electric cells are occasionally formed and enhance corrosive reactions. Though ceramics are mostly stable in a body environment, their reactivity with plasma depends on the composition. Materials chemically unstable in plasma are adequate for bioresorbable biomedical materials when the rate of degradation is controllable. Note that the addition of an oxide to a mother glass causes opposite effects on chemical stability depending on the glass composition and amount of the additive oxide. Zirconium oxide ceramic including a trace of yttrium oxide, denoted as partially stabilized zirconia (PSZ), is one of the toughest ceramics. The toughness is degraded due to stress under aqueous conditions because its crystalline form changes gradually and the tough aspect disappears. Alumina and PSZ are employed for artificial femoral heads in hip joint prosthesis, taking advantage of their hardness, toughness, and wear resistance. It follows that those items are hardly adjustable in shape at an operation site.

Organic–inorganic hybrids are naturally by far mechanically inferior to metals, but it is certain that they have potential in the following ways:

- forming bonds with soft and hard tissue since they involve Si—O or Ti—O bonds,
- such bioactivity is enhanced by incorporating calcium ions
- they are derived at low temperatures (room temperature ~100 °C) via sol–gel routes.
- their pore characteristics, porosity and pore size, are controllable.
- their selected components (active groups and length of the molecule skeleton), and mixing ratio of the components are the major factors on which the properties depend.
- they are good solvents for many kinds of chemical substances, with which their properties are modifiable.
- they are mostly machinable, but can be as brittle as ceramics and as flexible as rubber.

It should be pointed out that, as mentioned above in section 8.1.4, spontaneous apatite formation is crucial for tissue–material bonding. No synthetic polymers or organic materials have ever been found to provide apatite nucleation and

8.2 Bioactive Inorganic–Organic Hybrids

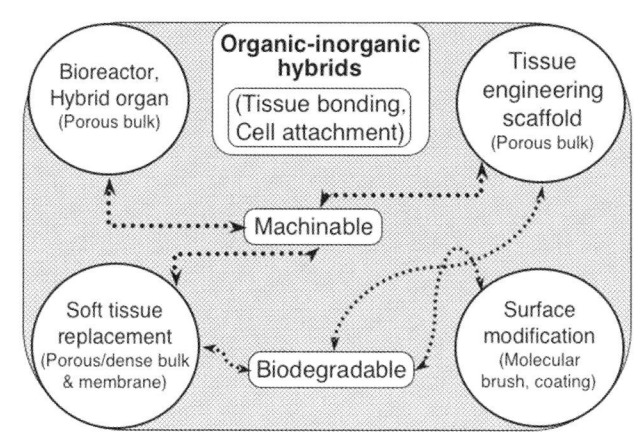

Fig. 8.7 Applications and properties of organic-inorganic hybrids.

deposition in a body environment. Thus, the incorporation of Si—O or Ti—O is essential. Figure 8.7 schematically demonstrates a rough idea of such organic–inorganic hybrid characteristics. They may have excellent tissue compatibility and secure cell attachment, while they often must be biodegradable and machinable. They may appear in bulk, membranes, coatings (molecular brush), and porous bodies with properly controlled pore size and porosity. Their application ranges from scaffolds for tissue engineering, and carriers of artificial organs where the active cells are fixed on the pore walls, to soft tissue replacement. In the following sections, examples of such organic–inorganic hybrids will be presented.

8.2.2
Concepts of Organic–Inorganic Hybrid Scaffolds and Membranes

As briefly discussed in section 8.1.2, tissue engineering denotes an industrial procedure to proliferate tissues by the use of porous materials (scaffolds). Suppose a scaffold has even excellent cell attaching and cell proliferating ability. Then, the scaffold works well to assist tissue regeneration, regardless of impregnation of a corresponding cell, when it is directly implanted in a defective part of an organ. The scaffold has a few roles:
- providing a cell-attaching surface
- supporting cell proliferation by being impregnated with additive agents – e.g. growth factors.
- diminishing the space volume that should be filled with regenerated tissues or cells; this helps earlier formation of blood vessels.

The scaffold should be so bioactive as to be fixed at the implanted site, and is hopefully biodegraded so that thorough replacement by the newly generated tissue is ensured. In addition, bone tissue scaffolds are required to be osteoinductive or osteoconductive. Osteoinduction is the process or ability of material to induce stem cells to differentiate into bone cells. Bone growth factors are involved in this process; interleukin is one of the bone cell differentiation and growth factors. Bone morphogenic proteins (BMP) and demineralized bone matrices are common osteoinductive materials. In contrast, osteoconduction is the process whereby osteogenic cells in porous material (grafts also) differentiate into bone cells or form bone tissue. The most important issue here is that new blood vessels are generated (neovasculation). In many cases, bone cell attachment and ingrowths are observed without such blood vessels. Osteoconductive materials include hydroxyapatite and tricalciumphosphate (TCP) ceramics as well as human- or animal-originated ones: autografts and allografts, and collagen. Figure 8.8 illustrates, for example, bone regeneration procedure using a bone scaffold that is bioactive and biodegradable. A scaffold with or without cells or bone growth factor proteins is implanted in a tissue defect. It is firmly fixed due to its bioactivity, and the newly grown cells differentiate into bone tissue. At the same time, biodegradation proceeds until the scaffold hopefully disappears when the defect fully recovers by the newly grown tissue. Most clinicians favor employing porous hydroxyapatite ceramics solidified at high temperatures for bone scaffolds as well as bone defect fillers. Some are against this because of sluggish biodegradation of such apatite ceramics: they prefer TCP. Yet, some post operational inspections have indicated that the bone tissue generated on those ceramics precipitates only apatite crystallites and involve no collagenous components or blood vessels, i.e. it is not real bone tissue; such bone tissue generation has been confirmed with Bioglass® granules. Despite those somehow contradicting results, clinical observation points out at least the importance of silica components that might be leached into plasma near the implant.

Membrane also plays an important role. When a bone tissue defect is filled with granular scaffold materials, the defect should be covered with a piece of membrane that holds the granules in the defect, and keeps fibroblast from coming into the defect. In the dental field, bioactive scaffolds are applicable for artificial periodontal membranes (fibrous tissue at the interface of tooth root and alveolar bone) for tooth root implant. The membrane will relax mastication stress that might

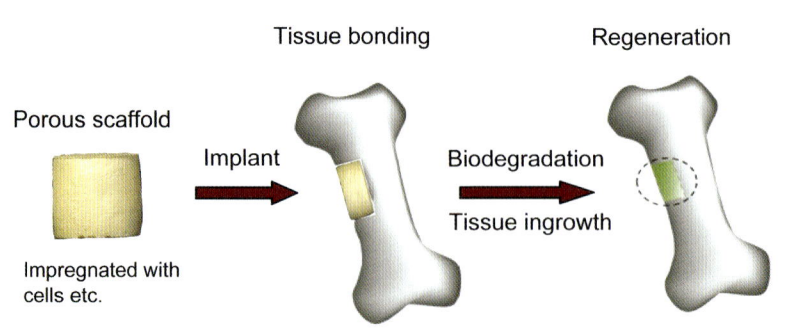

Fig. 8.8 A concept of how bioactive and biodegradable porous hybrids work when implanted in tissue.

8.2.3
PDMS–Silica Hybrids

Figure 8.9 illustrates the structure of poly(dimethylsiloxane) (PDMS)–silica hybrids, denoted as Ormosils (organically modified silicates). They are derived from dimethylsiloxane oligomers and tetraethylorthosilicate (TEOS) via sol–gel processing with HCl as the catalyst in a closed system. They consist of nm-sized silica clusters and PDMS chains and include microporosity, by which they appear opaque. Structural analysis based on the ^{29}Si MAS NMR technique confirms that the PDMS chains were grafted onto the silica particles. Strangely, the grafting was observed even if the methyl-terminated PDMS oligomers were employed for preparation. That is, some parts of the Si—O—Si skeleton in the chains are broken to yield $Si(CH_3)_2$—OH under HCl catalysis:

$$-Si(CH_3)_2-O-Si(CH_3)_2-\ +H_2O \rightarrow 2(-Si(CH_3)_2-OH) \tag{6}$$

and those —Si—OH groups react with the —Si—OH groups on the silica particles to eliminate H_2O.

$$-Si(CH_3)_2-OH + HO-Si-(silica) \rightarrow -Si(CH_3)_2-O-Si-(silica) + H_2O \tag{7}$$

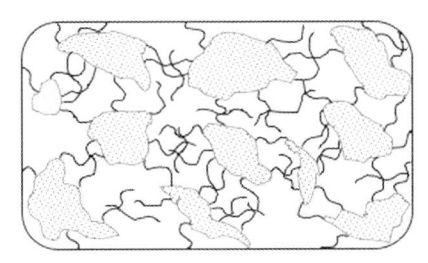

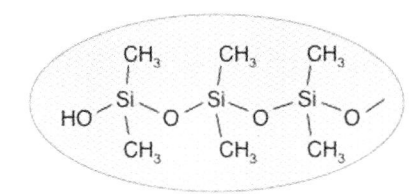

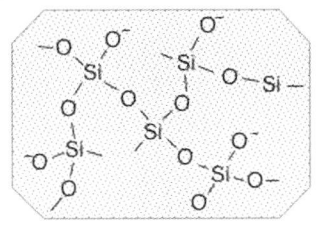

Fig. 8.9 Microstructure of an organically modified silicate (Ormosil) Poly(dimethylsiloxane) chains combine silica blocks derived from hydrolysis of an alkoxysilane.

Ormosils are basically bioinert, or they cannot deposit apatite in a body environment or in SBF (see Table 8.6). In contrast, Ormosils incorporated with calcium ions are bioactive: the induction period (the time necessary for apatite deposition) is about 1 day or longer, depending on the calcium ion content. Calcium ions are naturally more compatible with the silica particles than the PDMS matrix. It is observed that the more HCl is employed as the catalyst, the more calcium ions are incorporated, and hence, apatite-forming ability is enhanced. That is, admitting that the solubility of calcium ions in the silica matrix is greater than in the PDMS matrix, it is speculated that, with the help of HCl to yield —Si(CH$_3$)$_2$—OH, a larger fraction of calcium ions are stabilized in the silica matrix, and those calcium ions stimulate the apatite formation.

Ormosil analogues in which the Si—O component is partly substituted by Ti—O are derived in the system CaO—SiO$_2$—TiO$_2$—PDMS, employing tetraisopropylorthotitanate as the source of Ti. Their yield strength is as large as human sponge bone (cancellous bone) though they have rather weak bioactivity, taking 3 days in SBF before depositing apatite. Some attempt is made to improve the mechanical properties of the hybrids employing poly(tetramethyleneoxide) terminated with 3-isocyanatopropyltriethoxysilyl groups, instead of PDMS.

Taking advantage of the higher refractive index contribution of Ti—O compared to that of Si—O, Ti—O-containing hybrids are suitable materials for contact lens application. Figure 8.10 schematically shows the hybrid structure.

8.2.4
Organoalkoxysilane Hybrids

Organoalkoxysilanes with active groups on one end such as vinyl, cyano, amino and epoxy have been employed as precursors to synthesize inorganic–organic hybrids. For biomedical applications, the group of vinylalkoxysilane compounds listed in Table 8.2 has been well examined in terms of apatite deposition. Besides they are hybridized with other polymerizing molecules, they are polymerized to their own hybrids. They comprise polyethylene-type C—C skeletons and Si—O—

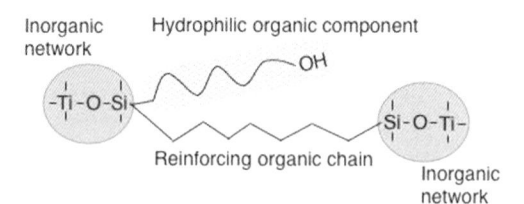

Fig. 8.10 A schematic structure model for Ti—O involving organic-inorganic hybrids for contact lens application, consisting of hydrophilic inorganic blocks and hydrophobic organic chains. The Ti—O species contributes to the refractive index.

8.2 Bioactive Inorganic–Organic Hybrids

Si networks. Figure 8.11 is a schematic illustration of the structure of a hybrid derived from organoalkoxysilane. Since a major fraction of the silanol (—Si—OH) groups, one of the key factors for bioactivity, is consumed for networking, the hybrids are almost barren unless calcium ions are introduced as in the case of Ormosils. Bioactive hybrids are obtained in bulk or membranes for vinyltrimethoxysilane and γ-MPS. After a ^{29}Si CP-MAS NMR analysis, calcium ions are associated with the silicate network as expected. The leaching of the calcium ions and formation of —Si—OH again stimulate apatite deposition in a body environment.

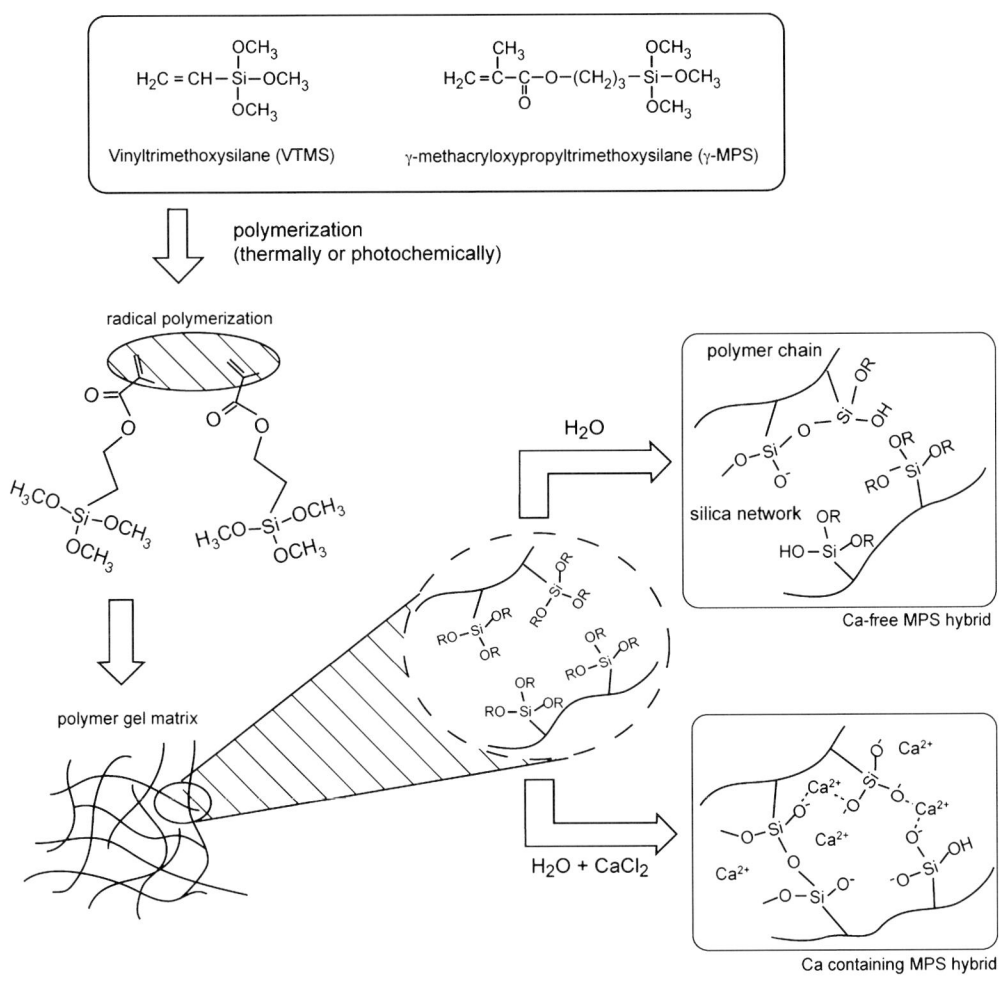

Fig. 8.11 Schematic illustration of inorganic-organic hybrid structure derived from organoalkoxysilane.

8.2.5
Gelatin–Silicate Hybrids

It is true that organic–inorganic hybrids are mechanically inferior to ceramics, but it is certain that they are applicable to small bone defects or soft tissue. With their advantages of bioactivity and controllable biodegradation, a few hybrids with natural polymers and silanes (listed in Table 8.2) have been prepared.

Gelatin is a good component for such hybridization since it has free amino groups (—NH_2) that are reacted with epoxy groups to form bridging bonds:

$$—NH_2 + HC(O)—CH— \rightarrow —NH—C(OH)—CH— \quad (8)$$

According to this concept, bioactive and biodegradable gelatin–GPTMS hybrids have been prepared with and without calcium ions. Figure 8.12 illustrates a schematic structure model on the basis of a ^{29}Si CP-MAS NMR analysis. The trimethoxysilane groups provide a silicate network, while the bridging bonds presented in Eq. (7) crosslink the two gelatin chains via the silicate network. Calcium ions or other modifying ions will be added to the starting solutions or to the sols for the hybrids, if necessary. A larger fraction of GPTMS yields a greater number of crosslink bonds, and constructs a rigid network. Indeed, the glass transition temperature (Tg) of the hybrids, derived from viscoelasticity measurements, increases with the fraction of GPTMS. In the gelatin-GPTMS system, Tg is in

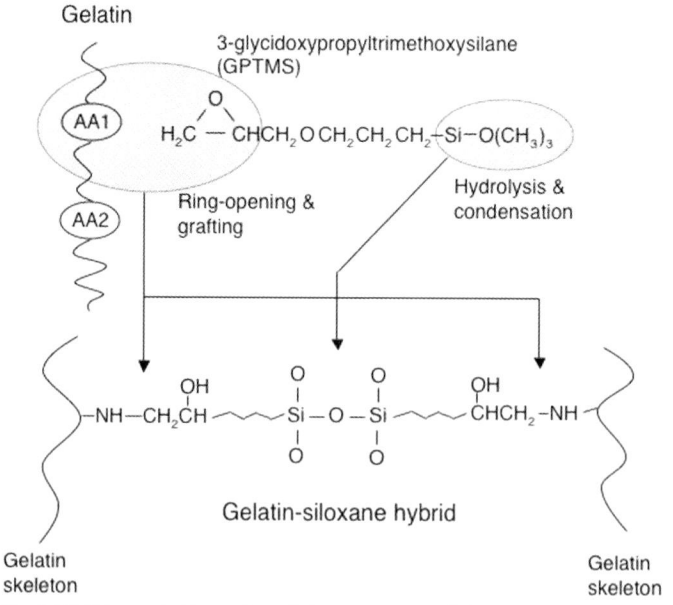

Fig. 8.12 Schematic illustration of the reactions and the resultant structures for gelatin-GPTMS hybrid system.

the range of 25~50 °C, while that of the GPTMS single hybrid is located at ~−25 °C.

Weight loss in a lysozyme solution may simulate the degradation of materials *in vivo*, while a TRIS (tris-(hydroxymethyl)aminomethane) buffer solution (pH = 7.4) might be a measure of biodegradation whose simple chemical durability is essential to evaluate. SBF is not a good medium because it yields apatite deposition. The hybrid of 33GPTMS–67Gelatin (in mass) is fully degraded in 42 days. In contrast, hybrids containing more than 50% GPTMS survive longer than 40 days, and 50% of the original volume remains undissolved. Considering that in clinical operation, an early stage of recovery, about 4 to 8 weeks, is important, the present gelatin–GPTMS hybrids are valuable. This indicates that the crosslink density in the gelatin–siloxane hybrids controls not only flexibility but also degradation.

The gelatin–GPTMS hybrids with calcium ions deposit apatite in SBF within 14 days while Ca-free hybrids are found to be not bioactive. Thus, incorporation of Ca ions is the essential factor for apatite deposition on the gelatin–siloxane hybrids in which those ions occupy sites similar to those in the organoalkoxysilane hybrids (Fig. 8.11), i.e. as being electrostatically bound to non-bridging oxygen atoms ($^-$O—Si).

8.2.6
Chitosan–Silicate Hybrids

Chitosan consists of polysaccharide chains, derived by eliminating acetyl groups (CH_3CO-) from chitin (= deacetylation). It is bioresorbable, biocompatible, nonantigenic, and nontoxic, as well as cytocompatible (= cell compatible). However, it is inferior in mechanical properties and too severely degraded in plasma for most medical applications. Hence, the polysaccharide chains are crosslinked using GPTMS molecules as in the gelatin hybrids via the sol–gel route so that its mechanical properties are improved and has controlled biodegradability. The GPTMS molecules are reacted with amino acid components (amino acid residues) of the chains, while the fraction of amino acid residues linked to a GPTMS molecule increases with the fraction of GPTMS. In an extreme case, almost all residues (~80%) participate in the linking, resulting in a highly crosslinked density and a more rigid structure. A silicate network similar to that in the gelatin hybrids is also established. The increase in crosslinking naturally stiffens the hybrids, resulting in an increase in elasticity, with tensile and compressive stress failure.

Chitosan might exhibit bioactivity since it adsorbs calcium ions well. Contrary to expectation, it deposits no apatite in SBF, probably because the affinity for calcium is too large and the calcium ions, once adsorbed, are not leached out into the surrounding medium. Therefore, as usual, calcium ions are incorporated into the chitosan–GPTMS hybrids via sol–gel route; they are added to the precursor solutions. Those hybrid membranes deposit apatite in SBF. Strangely, the hybrid membranes derived under open conditions, i.e. xerogel membranes, have exhibited little bioactivity. It is common, as indicated above, that closed gels involve better-developed networks than the xerogels. Therefore, a silicate network constructed by Si—O—Si bridging bonds is established, hence, Si(IV) or silicate ions would be leached out much less from closed-gel membranes into the aqueous medium than from xerogel membranes. The contrary has been observed. No interpretation of this phenomenon has been given up to now. It must be at least pointed out that such release of Si(IV) surely affects the promotion and stimulation of cell proliferation.

Several osteoblastic and fibroblastic cells are commercially available. They have been taken mostly from tumors generated in rats or mice, and they are also of human origin. *In vitro* (on the laboratory bench) examination of bone cell compatibility is often conducted by the use of MG63 (human osteoblastic) and MC3T3E-1 (mouse osteoblastic) cells. The MG63 culture has indicated that the chitosan–inorganic hybrids are also cytocompatible, regardless of the GPTMS content. In addition, the chitosan–GPTMS hybrids of xerogels exhibit better cell proliferation than those of closed gels. A possible reason is the release of Si(IV) from the hybrid into the culture medium. Although it is generally accepted that silicate components in plasma stimulate bone formation, too much Si(IV) is harmful. The Si(IV) leached from the closed-gel hybrid membrane might have exceeded some threshold value to result in poorer cell proliferation.

A similar effect has been observed on the calcium content in the chitosan–GPTMS membranes (xerogels). A lower number of MG63 cells are observed on hybrids with many calcium ions than on Ca-free hybrids. Optimization of the calcium ion content is necessary in order to achieve both bioactivity and cell proliferation.

Human bone marrow cells (primary cells; taken directly from human tissue in the course of surgery procedure) cultured on the chitosan–GPTMS hybrid (xerogel) membranes form a fibrillar extracellular matrix with numerous globular particles, regardless of the presence of dexamethasone in the culture medium. Here, dexamethasone is a substance that stimulates the growth of connective tissues (almost all cells and tissues in the living body), and is commonly added to culture media. This proves that the hybrids are excellent in tissue regeneration. That is, the chitosan–GPTMS hybrid is promising for bone–tissue engineering scaffolds or bone defect cover membrane applications.

8.3
Surface Modifications for Biocompatible Materials

8.3.1
Molecular Brush Structure Developed on Biocompatible Materials

Synthetic organic polymers are the mainstream materials in clinics by virtue of their properties, such as flexibility and ease of shaping, but they are not bioactive.

Metals are not active, either. Providing them with bioactivity and cytocompatibility (cell compatibility) is advantageous in fixation to hard and soft tissues, or in securing cell attachment, and several surface modification techniques have been proposed. Vinyltrimethoxysilane (VTMS; Table 8.2) is photochemically grafted on polyethylene (PE) substrates, and the methoxy groups is subsequently hydrolyzed by hydrochloric acid. Thus, a molecular brush structure is established where silanol groups are aligned at the top of the layer. Though silanol groups are accepted to favor apatite deposition, those of the VTMS brush layer on the PE substrates are not so active as to induce apatite deposition. More active silanol groups are precipitated using the mother glass or Cerabone A-W®; silicate components are leached from the glass granules in an aqueous medium to be deposited on the substrate to form a hydrated layer similar to that which forms on the Bioglass® surface in plasma (Fig. 8.5). Bioactivity is given by such procedure to γ-methacryloxypropyltrimethoxysilane (γ-MPS; Table 8.2) molecular brush layers on polyvinylchloride and a polyamide (Nylon 6®). The multilayer structure, a silicate layer on the MPS polymer brush, is hardly effective for high-density polyethylene (HDPE), probably because HDPE is so chemically inert and γ-MPS molecules would not be reacted to form anchoring bonds. Silicone is one of the mainstream polymers in current medical usage, and chemically inert, too. In contrast, the γ-MPS molecular brush has successfully been established on a silicone surface. This molecular brush layer improves the cytocompatibility of silicone; the mouse fibroblast cell L929 as well as the mouse osteoblastic cell MC3T3-E1 and human osteoblastic cell MG63 are so well proliferated and highly activated that they spread many pseudopodia (legs and arms), and reach a confluent state (the cells are united together to cover the whole surface) much earlier. Thus, the grafting of γ-MPS promotes cell attachment, proliferation, and activity.

8.3.2
Alginic Acid Molecular Brush Layers on Metal Implants

Metals and its alloys are applied to blood-contacting devices, such as the casing and inner impeller of an implant-type artificial heart and blood vessel stent. Their surface must be truly blood compatible and must not cause blood coagulation or form clots. Thus, their poor blood compatibility must be improved. At present, patients are often asked to take some anticlotting drugs such as heparin and ticlopidine HCl which may cause severe side effects in some cases.

Blood clotting is initiated by the activation of various factors, as presented in Fig. 8.6, most of which are proteins. Thus, one of the strategies is to firmly coat the metal surface with layers that prevent nonspecific adsorption of blood components. Polysaccharide is the name of a group of carbohydrates consisting of many units of sugar, and its coating does not adsorb such proteins. Alginic acid is a natural polysaccharide, found in seaweed, and is already applied to drug delivery or wound dressing materials. Moreover, according to one report [M. Morra (2001) in bibliography] one kind of fibroblast cell does not become attached to alginic acid film, referring to the importance of the conformation, hydration, and surface density of alginic acid molecules. Those issues strongly suggest that alginic acid is promising as a blood-compatible coating material. However, a polysaccharide is just a kind of hydrocarbon, and possesses no reactivity with metals. Therefore, an intermediate layer present between the polysaccharide layer and metal substrate is necessary for fixing the layer.

In this respect, a γ-aminopropyltriethoxysilane (γ-APS; Table 8.2) layer is first formed as the intermediate layer on SS316L (Table 8.5) and titanium substrates, then an alginic acid molecular brush layer is immobilized on the γ-APS layer. Immobilization is established by the condensation reaction between the free amino group on γ-APS and the carboxyl groups on alginic acid, similar to that for Eq. (8) above.

$$\text{(metal substrate surface)}-\text{O}-\text{Si}(-\text{O})_2(\text{CH}_2)_3-\text{NH}_2 + \text{HOOC}$$
$$-\text{(alginic acid skeleton)} \rightarrow \text{(metal)} \cdots \text{NH}-\text{CO}-\text{(alginic acid)} + \text{H}_2\text{O} \quad (9)$$

The multilayer coating is very effective to reduce the number of platelets adhering under *in vitro* conditions while it does not activate blood-clotting factors, indicated in Fig. 8.6. Moreover, BSA is adsorbed on titanium with one γ-APS molecular brush layer, whereas BSA is scarcely found on the titanium substrates with a multilayer structure. These *in vitro* experimental results directly indicate the significant role of alginic acid in suppressing BSA adsorption. The same results have been confirmed for the SS316L substrate. It is not inadequate to presume that *in vivo* blood protein adsorption is suppressed as well. In conclusion, immobilizing alginic acid leads to a highly blood-compatible layer as it does not activate blood clotting factors, nor adsorb such proteins (factors), nor adsorb platelets.

8.3.3
Organotitanium Molecular Layers with Blood Compatibility

A titanium oxide layer developed on titanium substrates based on chemical treatment with hydrogen peroxide is highly blood-compatible. However, since blood-contacting devices have a complex structure or deform extensively in use, the anti-blood-clotting layer needs to be thin and flexible as well as blood-compatible. Organotitanium compounds, for example, titanium methacrylate triisopropoxide (TMT; Table 8.2) have both organic chain skeletons and titanium alkoxy (Ti—OR) groups. The former provides flexibility, while the latter contributes to blood compatibility. Thus, a flexible layer involving Ti—O as a component is expected to be fabricated starting from such compounds. A TMT molecular layer developed on SS316L substrates consists of both a methacrylate chain and Ti—O—Ti bonds. The TMT-coated substrates adsorb only traces of fibrinogen, and scarcely affect blood clotting times, PTT and PT (Fig. 8.6). Thus, the TMT coating is presumably applicable to SS316L stents.

8.4
Porous Hybrids for Tissue Engineering Scaffolds and Bioreactors

8.4.1
PDMS–Silica Porous Hybrids for Bioreactors

The ability of materials to promote cell culture is essential not only for tissue engineering but also for applying cells to bioreactors. The latter case requires porous materials that are chemically stable but with better affinity for the pertinent cells. Porosity is introduced in a wide range of materials in many ways, by simple physico-chemical treatments, or using some pore-introducing agents (porogens). For biomedical use, if such an agent is used, a completely harmless one should be employed. In this respect, sugar (sucrose) and sodium chloride granules are two of the most adequate materials.

Most of the Ormosil-type (TEOS-PDMS) hybrids are chemically stable, while their biological activity is controllable with, e.g. impregnating calcium ions or other physiologically active agents in the matrices. Precursor sols for the Ormosil hybrids are mixed with sucrose granules as the porogen, and porous hybrids are yielded. The porosity is controllable with a volume fraction of the granules in the precursor mixture and their size; the porosity reaches about 90%, and the average pore size can be as large as 0.5 µm in diameter. In addition, the pore distribution can be considerably uniform when the sucrose granule size is controlled by sieving. These porous hybrids indicate a typical stress–strain behavior, characteristic of a cellular solid structure i.e. the S—S curves show a plateau (corresponding to lower elasticity) due to collapse of the pores.

The porous Ormosils derived from sols in which calcium ions are doped are bioactive enough to deposit apatite on their wall surfaces. Figure 8.13 illustrates the apatite-covered pore walls of a porous TEOS-PDMS hybrid. The Ca-free Ormosil hybrids do not deposit apatite yet they have excellent cell-attaching ability, which makes the Ormosil hybrids promising materials for biomedical scaffolds.

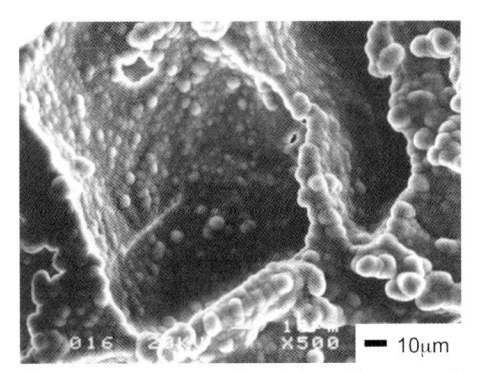

Fig. 8.13 Apatite covers the whole of the pore walls of calcium-containing PDMS-TEOS hybrids after soaking in SBF (Table 8.6).

A radial-flow bioreactor (RFB) is a highly functional 3-D cell culture system, applicable to an artificial liver. Such a 3-D RFB is applied either to harbor proliferated cells or to use the system as a bioreactor of specific functions that are determined by the cells. Currently, scaffold materials include porous silica and apatite beads or polyvinylalcohol and polyurethane foam. However, those ceramic beads are hard and it is difficult to monitor the conditions of the cells after culturing, while polymer foams involve pores with nonuniform size, which causes some difficulties in harboring the cells.

Taking advantage of cell compatibility, almost uniform pore size, and machinability, an attempt has been made to employ the Ormosil hybrids in a radial-flow bioreactor (RFB). It has been indicated that human hepatocellular carcinoma cells (denoted as HepG2) are proliferated actively and form cell clusters more efficiently in a porous Ormosil scaffold than in a polyvinylalcohol (PVA) one. Moreover, HepG2 cells cultivated on an Ormosil scaffold have secreted three times as much albumin as that secreted in a monolayer culture on a solid Ormosil hybrid with the same composition as that of a porous one. This confirms the efficiency of the 3-D cell culture.

For the potential application of RFB to future clinical use, it is essential to develop a method to propagate liver cells that maintain highly specific functions. Cell compatibility should be optimized by changing the composition or impregnating other functional agents: for example, an agent called dimercaprol (BAL), is effective in detoxicating several toxic metals such as As, Sb, Bi, and Hg, to name a few, due to chelation. Yet, the Ormosil-type hybrids are promising basic materials for developing such functional culture methods.

8.4.2
Gelatin–Silicate Porous Hybrids

Bioresorbable (biodegradable) scaffolds have been derived from hybridizing gelatin and GPTMS (section 8.2.5). For scaffold applications, porosity has been introduced by freeze-drying (lyophilizing) the wet hybrid gels. It is essential to soak the wet gels in an alkaline solution before the freeze-drying procedure. After a ^{29}Si CP-MAS NMR structure analysis, by doing so, a part of the silicate network (Fig. 8.12) is broken so that ice granules freely grow. This leads to a homogenous pore size distribution. Figure 8.14 shows scanning electron microphotographs presenting the microstructure of Ca^{2+}-free porous gelatin–GPTMS hybrids (GPTMS content = 33%), where the mean pore size is about 10, 40, and 400 µm, and the total porosity is 47, 62, and 80% for freezing temperatures of –176, –80 and –17 °C, respectively. Incorporating calcium ions for improving bioactivity little affects the pore characteristics.

Bimodal pore structure is attained due to soaking the hybrids containing 300- to 500-µm pores in the aqueous solutions (pH 3 to 11), and subsequently freeze-drying at –196 °C. According to SEM observation of the fracture surface of a Ca-containing hybrid, the second freeze-drying procedure introduces micropores 5–10 µm in diameter in the walls of the large pores, regardless of the Ca incorpo-

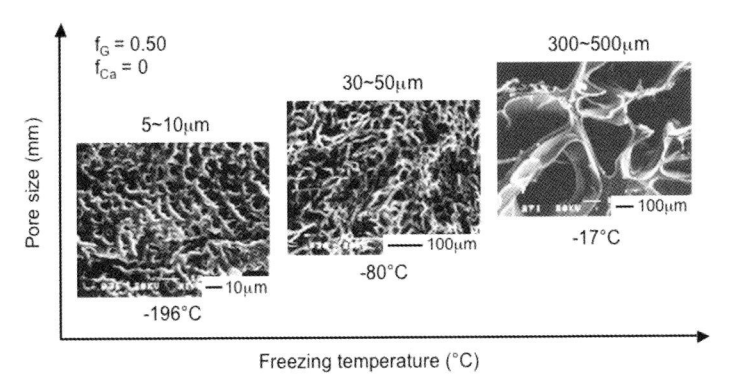

Fig. 8.14 Scanning electron microphotographs presenting the microstructure of Ca^{2+}-free porous gelatin–GPTMS hybrids.

ration. Such hybrids with bimodal pore distribution are more bioactive than bulk (solid) hybrids of the same composition. Apatite is deposited within 1 day in SBF not only on the outer surface and on the larger pore wall surface but also in the smaller pores.

Excellent cytotoxicity and cytocompatibility has been confirmed for the single-mode porous hybrids with respect to osteoblastic cell (MC3T3-E1) proliferation, ALP activity, or their responses to the hybrids and their extracts. In addition, appropriate incorporation of calcium ions stimulates cell proliferation and differentiation *in vitro*.

8.4.3
Chitosan–Silicate Porous Hybrids for Scaffold Applications

Other biodegrading porous hybrids have been synthesized from GPTMS and chitosan, instead of gelatin. The procedure is similar to that employed for the gelatin hybrids, but their precursor solutions should be freeze-dried once before gelation. The pore size is approximately 100 μm, and more than 90% porosity is attained, regardless of the GPTMS content, as in the gelatin hybrid analogues.

From stress–strain behavior, GPTMS molecules are responsible for stiffening the porous gelatin–GPTMS hybrids, suggesting chemical stability increase due to the hybridization. Indeed, biodegradation in a lysozyme solution is greater for porous chitosan than that for the chitosan–GPTMS hybrids, accordingly. The Ca-incorporated hybrids are also bioactive enough to deposit apatite in SBF. The apatite layer grows not only on the outer surface but also on the pore walls, as in the other bioactive porous hybrids. Osteoblastic cells (MG63) grow better on the porous hybrids than on porous chitosan. MG63 cells infiltrate into the pores and cover the pore walls. Those cell culture experiments confirm that these porous hybrids are also applicable to tissue engineering scaffolds.

8.5
Chitosan-based Hybrids for Drug Delivery Systems

Chitosan is one of the candidates for drug carriers because of its biodegradation and biocompatibility, as touched upon above. Various classes of drugs are attempted, including antihypertensive agents, anticancer agents, proteins, peptide drugs, and vaccines.

Tissue engineering scaffolds may incorporate some additives such as drugs or proteins, like growth factors, that have certain effects on cell growth, cell differentiation, and anti-inflammation either in their matrix or in the pores. As they degrade in plasma, they release those impregnated substances. Porous chitosan–GPTMS hybrids have been prepared through the same procedure as mentioned previously, but using two kinds of chitosan, different in the rate of degradation. Chemically, one is highly deacetylated and has a larger molecular weight, hence is less degradable, whereas the other is low in deacetylation with a smaller molecular weight and is much degraded. The hybrids from the high-molecular-weight chitosan would be hardly degraded either in a phosphate-buffered solution (PBS) or in a lysozyme solution, whereas those from the smaller-molecular-weight chitosan are likely to be degraded in those solutions. Needless to say, an increase in the GPTMS content chemically stabilizes the hybrid network, and reduces the rate of degradation, accordingly.

When dosing a drug to a patient, the drug is to be rigorously released to increase the concentration in blood immediately up to a threshold value. Then, it is continuously released to keep the concentration almost constant. The first sudden increase in the concentration is denoted as burst. With two hybrids, hybrids for controlled release of drugs in that way are designed: A drug is incorporated in the matrix of a porous carrier made of a sluggishly degrading hybrid, and then the pore walls are coated or lined with a rapidly degrading hybrid impregnated with the drug. The material concept is shown in Fig. 8.15. If the system worked as designed, it would be a kind of ideal drug-delivery system.

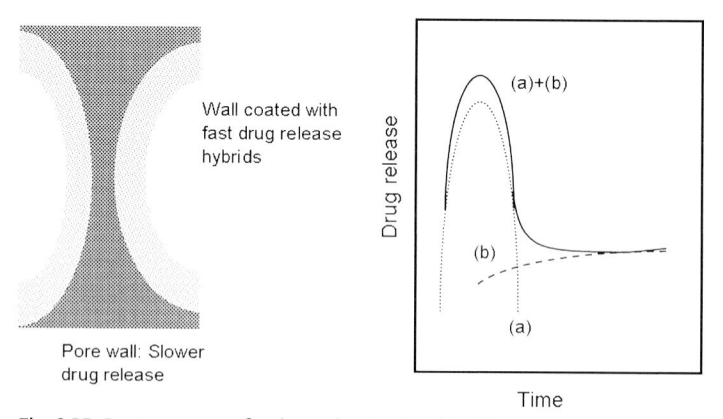

Fig. 8.15 Design concept for drug-releasing hybrids. The coating layer releases drugs fast to give a burst (a), while the pore wall slowly releases steadily (b).

8.6 Summary

Organic–inorganic hybrids consist of organic and inorganic entities that are mixed together at the molecular level: they are different from composites comprised of ingredients in larger size. A good example of a composite is a dental resin (resin cements) involving ceramic fillers (~μm in size) dispersed in a methacrylate-based polymer matrix. They are mosty derived from sol–gel processing, they can be shaped in any way: films and coatings, bulk solids, and even strings. Porosity is also introduced with the freeze-drying technique and by using porogen.

The hybrids exhibit a variety of properties. They can behave like ceramics (brittle) or polymers (ductile) like Ormosil-type hybrids which are derived from poly(dimethylsiloxane) and tetraethoxysilane. Other ingredients are incorporated via sol–gel processing. When calcium ions are incorporated, the hybrids can form strong chemical bonds to tissues as they induce spontaneous deposition of apatite in a body environment. The hybrids containing gelatin or chitosan as the organic components exhibit controlled biodegradation depending on the composition. Taking advantage of biodegradation and the flexible fabrication technique (sol–gel route), drugs or other substances are easily incorporated in them, and they are released according to a controlled release design.

Bibliography

On material-tissue bonding and ceramic materials in medicine

An introduction to bioceramics, Ed. L. L. Hench and J. Wilson, World Scientific, Singapore, **1993**.

L. L. Hench, *J. Am. Ceram. Soc.* **1991**, *74*, 1487–1510.

Biomaterials, artificial organs and tissue engineering, Ed. L. L. Hench and J. R. Jones Woodhead Publishing Limited, Cambridge, **2005**.

M. Neo, S. Kotani, Y. Fujita, T. Nakamura, T. Yamamuro, Y. Bando, C. Ohtsuki and T. Kokubo, *J. Biomed. Mater. Res.*, **1992**, *26*, 255–267.

C. Ohtsuki, T. Kokubo and T. Yamamuro, *J. Non-Cryst. Solids*, **1992**, *143*, 84–92.

S-B. Cho, K. Nakanishi, T. Kokubo, N. Soga, C. Ohtsuki, T. Nakamura, T. Kitsugi and T. Yamamuro, *J. Am. Ceram. Soc.* **1995**, *78*, 1769–1774.

On human blood clotting processes

S. R. Hanson, in *Biomaterials Science – An introduction to materials in medicine*, 2nd edn, Ed. B. D. Ratner *et al.* Elsevier, **2004**, p335.

On sol–gel processing

Handbook of Sol–Gel Science and Technology – processing, characterization, and applications, Volumes I, II, III, Ed. S. Sakka, Kluwer Academic, Boston, USA, **2005**.

G. Philip, H. Schmidt, *J. Non-Cryst. Solids*, **1984**, *63*, 283–292.

On novel hybrids for tissue substitutes and scaffolds

K. Tsuru, C. Ohtsuki, A. Osaka, T. Iwamoto and J. D. Mackenzie, *J. Mater. Sci.; Mat. Med.*, **1997**, *8*, 157–161.

Q. Chen, F. Miyaji, T. Kokubo and T. Nakamura, *Biomaterials* **1999**, *20*, 1127–1132.

L. Ren, K. Tsuru, S. Hayakawa and A. Osaka, *Biomaterials*, **2002**, *23*, 4765–4773

Y. Shirosaki, K. Tsuru, S. Hayakawa, A. Osaka, M. A. Lopes, J. D. Santos and M. H. Fernandes, *Biomaterials*, **2005**, *26*, 485–493.

On polysaccharide coatings

Water in Biomaterials Surface Science, Ed. M. Morra, John Wiley & Sons, Ltd, **2001**.

T. Yoshioka, K. Tsuru, S. Hayakawa and A. Osaka, *Biomaterials*, **2003**, *24*, 2889–2894.

9
Hybrid Materials for Optical Applications

Luís António Dias Carlos, R.A. Sá Ferreira and V. de Zea Bermudez

9.1
Introduction

The drive to miniaturization with the corresponding demand for smaller machines and components using less resources and energy that occurred during the last two decades, have been rapidly pushing industry into the atomic and nanometer scale. The development of new synthesis strategies for advanced materials with enhanced properties and affording an effective control at the nanometer level is therefore required. The sol–gel process is probably the utmost "soft" inorganic chemistry process that allows this chemical control and design. In fact, its unique characteristics, such as the low-temperature processing and shaping, high sample homogeneity and purity, availability of numerous metallo-organic precursors and the processing versatility of the colloidal state, permit the synthesis of multifunctional organic–inorganic hybrid structures through a molecular nanotechnology bottom-up approach based on a tailored assembly of organic and inorganic building blocks.

There is a widespread agreement within the sol–gel scientific community that the inherent flexibility of the "soft" chemistry approach allows the implementation of the design strategies that are the basis of photonic[1] hybrid materials. In particular, siloxane-based hybrids present several advantages for designing materials for optical and photonic applications:

- Flexible, relative facile chemistry and highly controlled purity, since they are synthesized from pure precursors.
- Versatile shaping and patterning depending on the foreseen application.
- Good mechanical integrity, excellent optical quality and easy simple control of the refractive index by changing the relative proportion of the different precursors.

---

1) Term coined for devices that work using photons or light. It is analogous to "electronic" for devices that work with electrons.

Hybrid Materials. Synthesis, Characterization, and Applications. Edited by Guido Kickelbick
Copyright © 2007 Wiley-VCH Verlag GmbH & Co. KGaA, Weinheim
ISBN: 978-3-527-31299-3

- Encapsulation of large amounts of emitting centers (organic dyes and inorganic chromophores) isolated from each others and protected by the organic–inorganic host, thus decreasing nonradiative decay pathways.
- As they are chemically produced at mild temperature, the dopants can be introduced in the original sol and thus be easily entrapped into the forming matrix and preserved without destruction (particularly relevant for organic dyes).
- As in most cases the hybrids are porous structures, they allow interaction of dopants with external liquids or gases; sintering is unnecessary in various systems, since the incorporated optically active molecules are not leached out of the matrix cages by solvents.
- Possibility of having energy transfer between the host and the emitting centers, which, in turn, undergo the corresponding radiative emitting process.

Several requirements must be fulfilled by sol–gel derived organic–inorganic hybrids for optics and photonics:
- The materials must exhibit chemical, mechanical, photo and thermal stabilities;
- They should be highly transparent in the whole visible range;
- Care must be taken to avoid close interaction between emitting centers;
- It is essential to ensure that they do not incorporate in their chemical composition a large number of organic groups prone to being involved in nonradiative pathways.

This last point is one of the main disadvantages of hybrid optical materials. The use of the sol–gel method at mild temperature (under 100 °C) allows the presence of many OH groups which contribute to propagation losses at 1310 nm and 1550 nm (the two regions, designated as second and third telecommunication windows, respectively, correspond to low-loss windows of commercial silica-based transmission fibers). In organic–inorganic hybrids incorporating lanthanide ions those groups are also responsible for the inhibition of the luminescence of these ions favoring the nonradiative decay by multiphonon emission.

A significant number of innovative and advanced siloxane-based hybrids have been thus synthesized in the past few years with mechanical properties tuneable between those of glasses and polymers and improved optical properties. We can refer, for instance, optical switching and data storage hybrid devices; photoelectrochemical cells and coatings for solar energy conversion; hybrid materials having excellent laser efficiencies and good photostability, fast photochromic responses, high and stable second-order nonlinear optical responses, or which can be used as original pH and fiber-optic sensors, photopattern waveguiding structures for integrated optics (IO), and electroluminescent diodes.

This chapter will focus on examples of functional siloxane-based hybrids for coatings, for light-emitting and electro-optic proposes, for integrated and nonlinear optics and for photochromic and photovoltaic devices. The chapter describes initially the major synthesis strategy for optical applications listing some illustrative examples. A brief review of several successful applications of hybrids in coatings for optics will be given. An explanation of the photoluminescence, absorption and electroluminescence processes and the quantification of the materials luminescence features (color emission, quantum yield and radiance) then follow. Examples of typical emitting centers and recombination mechanisms will be provided. The stimulated emission in solid-state dye-lasers is briefly described and two sections describing the requirements to develop photochromic, photovoltaic and IO devices (including representative examples) are presented before conclusions are drawn.

9.2
Synthesis Strategy for Optical Applications

The synthesis of organic–inorganic hybrid frameworks for optics and photonics should be oriented to slight different specific purposes as the materials incorporate or does not incorporate optically active external centers (metal ions or organic chromophores).

Optically passive hybrids The synthesis strategy has been essentially engaged on the optical transparency and the control of the index of refraction.

Optically active hybrids For materials lacking metal activator ions the strategy has been essentially directed towards the reduction or virtually suppression of groups responsible for quenching of the luminescence (e.g. silanol (Si—OH) groups, residual solvent, etc.). For hybrids doped with metal activators, the focus is the encapsulation of the emitting centers with their protection from nonradiative decays by the organic–inorganic host itself favoring potential energy transfer processes. The amount of ions incorporated does not depend only on the absorption ability of the matrix without phase separation but should be optimized attending that the interaction between closest ions also quenches the luminescence.

Obviously, more general requirements, such as the highly purity, the versatile shaping and patterning, and the good mechanical integrity, should be also taken into account. In more detailed terms the following five synthesis strategies have been proposed:
1. Fine control of the sol–gel processing conditions. This may be achieved by means of optimization of water content, use of reactive hydrophobic precursors and thermal curing.

2. Use of nonhydrolytic procedures, such as the solvolysis or carboxylic acid method. Compared to the hybrid materials synthesized from the conventional sol–gel process, the hybrids derived from this modified sol–gel process have the following advantages: (a) they exhibit higher emission quantum yield (approximately 30% higher); (b) compact and uniform gel is produced without any entrapment of liquid favoring the performance of functional devices.

Table 9.1 demonstrates the relevance of this synthetic procedure in the light of optics as the materials as prepared may display extremely high quantum yield even if they lack emitting centers.

The solvolysis process relies on the fact that the formation of hydroxyalkoxysilanes can occur in the absence of water, if the precursor molecule is reacted with a carboxylic acid. Scheme 9.1 illustrates the reaction mechanism proposed for the reaction of tetraethoxyorthosilicate (TEOS) with acetic acid (CH_3COOH). The condensation reactions that take place

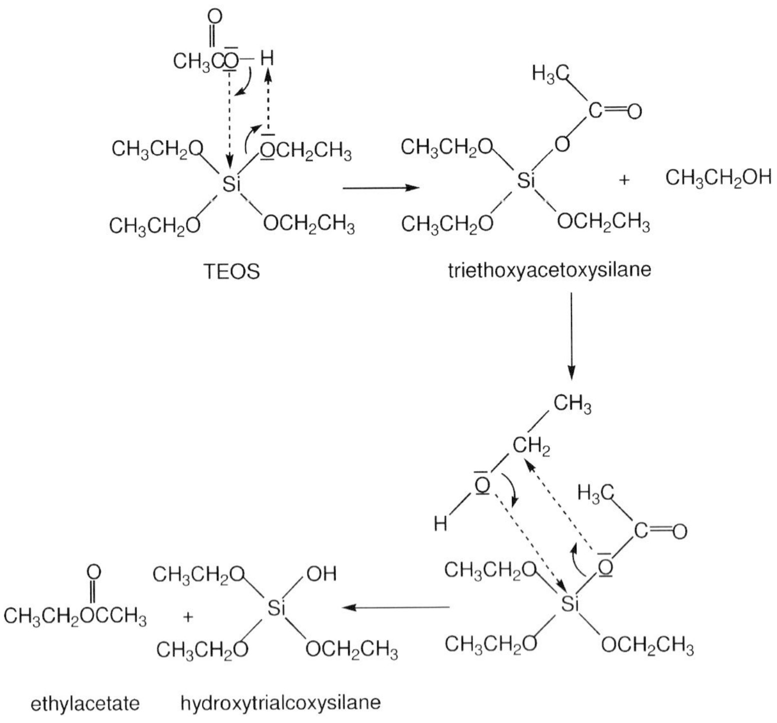

Scheme 9.1 Proposed mechanism for acetic acid solvolysis of TEOS. (Adapted from E. J. A. Pope and J. D. Mackenzie, *J. Non-Cryst. Solids* **1986**, *87*, 185.)

9.2 Synthesis Strategy for Optical Applications

Table 9.1 Hybrids lacking emitting centers derived from the solvolysis process.

Hybrid Matrix Formers	Carboxylic Acid	Quantum Yield (%)
3-aminopropyltriethoxysilane (APTES)	Formic (1) Acetic (2)	35 ± 1 12–21 ± 2
di-ureapropyltriethoxysilane (d-UPTES(600)) (n_{POE}=8.5, a+c=2.5)	Valeric (3)	10 ± 3
di-urethanepropyltriethoxysilane (d-UtPTES(300)) (n_{POE}=6)	Acetic (3)	20 ± 5
di-ureapropyltriethoxysilane (n_{POP}=3–68)	Formic (4) Acetic Valeric	—

(1) Green et al., Science 276, 1826 (1997); (2) Bekiari et al., Langmuir 14, 3459 (1998) and Chem. Mater. 10 3777 (1998); (3) Fu et al. Chem. Mater. 16 1507 (2004); (4) Stathatos et al., Adv. Mater. 14, 354 (2002) and Brankova et al., Chem. Mater. 15, 1855 (2003).

after this nonhydrolytic stage proceed in a way similar to that occurring in the conventional sol–gel method.
3. Introduction of inorganic emitting centers (e.g. lanthanide ions introduced as ionic salts) into the hybrid structure using strategies 1 or 2.
4. Combination of strategies 1 or 2 with the encapsulation of the emitting species through complexation or chelation. The complexes or chelates may be added directly to the matrix or be formed in situ. In the latter case, appropriate ligands are grafted or anchored to the hybrid network itself prior to the addition of the lanthanide ions. The encapsulation of rare earth ions within nanoparticles is also attempted involving the combination of microemulsion techniques with the sol–gel process (essentially strategy 1).
5. Introduction of organic chromophores into the hybrid structure using strategies 1 or 2. The list of candidates includes, for instance, dyes with photochromic, laser and nonlinear optical features, liquid crystals, and biological or enzymatic functional molecules. The synthesis of organic inorganic hybrids showing second order nonlinear effects should attend that high nonlinear optic performance is strictly related with the chromophore loading and alignment. The chromophore should be properly poled to avoid a random and centrosymetric orientation that will quench the performance. (Readers are referred to Innocenzi 2005).

It should be stressed that the classical sol–gel and the solvolysis chemical routes are the main pathways to produce Class I and Class II non-organized hybrids. These materials, not only exhibit an infinity of macrostructures, but they are generally polydisperse in size and locally heterogeneous in terms of chemical composition.

In the last few years, the creation of new types of hybrid photonic materials whose structure and function are organized hierarchically is emerging. The first reports of the optical properties of such materials suggested that these solids are promising candidates for optical applications, such as lasers, light filters, sensors, solar cells, pigments, optical data storage, photocatalysis and frequency doubling devices.

Ordered periodic micro-, meso-, and macroporous materials allow the construction of composites incorporating many guest species, such as organic molecules, inorganic ions, semiconductor clusters and polymers. These host/guest materials combine the high stability of the inorganic host, new structure-forming mechanisms due to confinement of guest species in well-defined pores and a modular composition.

The covalent bonding of dye molecules to ordered or semi-ordered porous MCM-41 or MCM-48 materials[2] is a theme that has been the subject of a recent interest. The organic molecules can be dispersed and separated from each other, thus reducing intermolecular quenching of fluorescent features. Covalently linked chromophores might be used as sensors in separating processes to detect the morphologies of molecules located within the channel-like pores of thin membranes of MCM-41 or MCM-48 materials. Moreover, as the optical characteristics of the chromophore are very sensitive to the local environment, they could be used to probe the internal structure of mesostructured silicas.

We have collected in Tables 9.2 and 9.3 some representative examples of lanthanide-doped organic dye-doped hybrids (amorphous and organized structures) that have been proposed in the literature in the context of the strategies 3, 4 and 5.

9.3
Hybrids for Coatings

Hybrid materials lend themselves to application as coatings (passive and active) in optics because they offer a series of particularly attractive features, in particular high transparency, good adhesion, corrosion protection and easy tuning of both the refractive index and the mechanical properties. Moreover, hybrid coatings find also important industrial applications in other domains in which the optical absorption and refractive index of the coating itself are relevant. Emphasis has been given on protection purposes against scratch, abrasion resistance and weather etching (polluted atmospheric conditions). There are examples (e.g. glass coloration for packaging purposes) in which the demand of protection or conservation purposes requires particular passive or active optical functionalities. Some of the most interesting examples of the use of hybrid coating exploiting their passive and active optical properties are listed bellow.

Passive optical properties: Decorative and functional coatings for glasses and glass coloration for packaging purposes
- The glass-like transparency of many sol–gel-based hybrid materials has rendered them attractive candidates to modify glasses by applying thin coatings. Due to the presence of Si—OH groups, which can react with their counterparts in the sol–gel-based hybrids, the latter materials display proper adhesion properties to glass surfaces. The large scale dip-coating process employed to prepare reflective and

---

2) MCM-41 and MCM-48 are hexagonal and cubic phases, respectively, of periodic mesoporous silica materials (so-called M41S) that have attracted considerable attention since their discovery in 1992, C. T. Kresge et al., Nature **1992**, 359, 710.

Table 9.2 Examples of hybrids incorporating emitting lanthanide centers.

Molecular Structure/Synthesis and Emission Details	Molecular Structure/Synthesis and Emission Details
Eu^{3+}, Gd^{3+} **(1)** Precursors: acid chlorides or 2,6-bis-(3-carboxy-1-pyrazolyl)pyridine and APTES; Eu(NO$_3$)$_3$·6H$_2$O and Gd(NO$_3$)$_3$·6H$_2$O; monolithic bulks. 250–420 nm absorption, 4f^6 lines and blue ligand broad band (Gd^{3+}-based hybrid).	Nd^{3+}, Dy^{3+}, Yb^{3+}, Er^{3+}, Pr^{3+}, Sm^{3+}, Ho^{3+} **(2)** Precursors: poly(ethylene glycol) (average molecular weight, MW, 200 gmol^{-1} PEG(200)) and tetramethylortosilicate (TMOS); fluorescein ligand (calcein, calc45). 4f^n and blue ligand broad band absorption. Dy^{3+}, Er^{3+}, Nd^{3+}, Yb^{3+}: near-infrared emission; Pr^{3+}, Sm^{3+}, Ho^{3+}: no emission.
di-urethanepropyltriethoxysilane d-UtPTES(300) (n$_{POE}$ = 6) (see Table 1) Eu^{3+} **(3)** Precursors: PEG(300) and 3-isocyanatepropyltriethoxysilane (ICPTES); Eu(CF$_3$SO$_3$)$_3$; transparent monolith. Ligand-to-metal charge transfer (LMCT) excitation (350 nm) and 4f^6 lines. Hybrid host emission band (350–650 nm) and 4f^6 transition. $\tau(^5D_0) \sim 0.20$–1.40 ms (300 K). $\phi \sim 0.7$–8.1% (345–395 nm).	DPS Eu^{3+}, Tb^{3+} **(4)** DPS precursors: 2,6-aminopyridine and ICPTES; transparent film. UV absorption (282 and 340 nm). 4f^8 and 4f^6 lines. $\tau(^5D_4) \sim 1.39$ ms, $\tau(^5D_0) \sim 0.35$ ms (300 K).
(5) Precursor: 5-amino-1,10-phenantroline (Phen-NH$_2$) and ICPTES; EuCl$_3$. UV/blue absorption (282 nm). 4f^6 lines. $\tau(^5D_0) \sim 0.524$ ms (300 K)	Eu^{3+}, Gd^{3+} **(6)** Precursors: dicarboxylic acid chloride and bis[3-(triethoxysilyl)propyl]amine. UV Absorption. 4f^6 lines and blue ligand broad band (Gd^{3+}-based hybrid, 14 K). $\tau(^5D_0) \sim 1.1$ ms (14 K).

9.3 Hybrids for Coatings | 345

Table 9.2 Continued

Molecular Structure/Synthesis and Emission Details	Molecular Structure/Synthesis and Emission Details
Eu^{3+}, Tb^{3+} **(7)**	Tb^{3+}, Gd^{3+} **(8)**
TAT precursor: trimellitic anhydride and APTES; transparent film. UV absorption (335 nm). $4f^8$ and $4f^6$ lines. $\tau(^5D_4)$ ~ 0.6–1.8 ms and $\tau(^5D_0)$ ~ 0.4–1.2 ms (300 K).	Precursors: p-aminobenzoic acid and ICPTES; Tb_4O_7 and $GdCl_3$; transparent film (Tb^{3+}), white precipitate (Gd^{3+}). UV absorption (292 nm). $4f^8$ and blue ligand band emission (Gd^{3+}). $\tau(^5D_4)$ ~ 1.2 ms (300 K).
(9)	TEOS + **(10)**
[Eu(TTA)$_3$(H$_2$O)$_2$] complex (TTA: 2-thenoyltrifluoroacetonate) and 5-(N,N-bis-3-(tri-ethoxysilyl)propyl)ureyl-1,10-phenantroline added to TMOS and diethoxydimethylsilane (DEDMS). Broad excitation band (389 nm). $4f^6$ lines.	Er(DBM)Phen complex (DBM: dibenzoylmethane) synthesized in-situ; $ErCl_3$; transparent, monolith. Broad excitation band (250–462 nm) and $4f^{11}$ lines. $4f^{11}$ emission (FWHM ~ 72 nm).

Table 9.2 Continued

Molecular Structure/Synthesis and Emission Details	Molecular Structure/Synthesis and Emission Details
di-ureapropyltriethoxysilane (see Table 1) + **Eu^{3+}, Gd^{3+}, Nd^{3+}, Ce^{3+}, Er^{3+} (12),(13)** Precursors: diamine(POE)s or diamine(POP)s (MW = 230–4000 g · mol^{-1}) and ICPTES, $Eu(ClO_4)_3$, $Eu(CF_3SO_3)_3$, $Ln(NO_3)_3$, (Ln = Eu, Tb, Nd, Ce), rubbery transparent gel. LMCT (340–365 nm) and 4f^6 lines. Hybrid host emission (350–650 nm) and 4f^n lines. $\tau(^5D_0)$ ~ 0.19–0.42 ms, hybrid host lifetime: 6.0–10.1 ns (300 K). ϕ ~ 1.4–13.0% (395 nm).	or + **TbBz$_3$ (Bz: benzoate) (11)** Broad excitation band (160–300 nm). 4f^8 lines. ϕ ~ 3–21% (254 nm).
(14) Precursors: TEOS, APTES, p-methylpyridine; $EuCl_3$ and $TbCl_3$. Ligands broad band excitation (336 nm). 4f^8 and 4f^6 lines (FWHM < 15 nm). $\tau(^5D_4)$ ~ 0.6–1.8 ms $\tau(^5D_0)$ ~ 0.12 ms (300 K).	**di-ureapropyltriethoxysilane d-U(600) (n$_{POE}$=8.5)** (see Table 1) + **Tb(NO$_3$)$_3$** + (15) Precursors: diamine(POE) (MW = 800 g · mol^{-1}) and ICPTES; $Tb(NO_3)_3$; terbium-enriched sample immersed in a solution of 2,2'-bipyridine (bipy). Excitation broad band (318 nm). 4f^8 lines and hybrid host broad band emission (350–650 nm). $\tau(^5D_4)$ ~ 1.19 ms, hybrid host lifetime: 8.7 ns (300 K). ϕ ~ 0.7–8.1% (345 and 395 nm).

Table 9.2 *Continued*

Molecular Structure/Synthesis and Emission Details	Molecular Structure/Synthesis and Emission Details
di-ureapropyltriethoxysilane d-U(2000) (n_{POE}=40.5) (see Table 1) + [structure of Ln complex with six chelating ligands bearing CF_3 and phenyl groups, hydrogen-bonded to NH groups] **Eu^{3+}, Tb^{3+}, Tm^{3+} (16)** [Ln(BTFA)$_3$(H$_2$O)$_2$] complex (BTFA: 4,4,4-trifluoro-1-phenyl-1,3-butanedione). Broad band excitation (318 nm). $4f^8$, $4f^6$ and near infrared (Eu^{3+}, Tm^{3+}) lines, hybrid host broad band (350–650 nm). $\tau(^5D_4) \sim 0.7$–0.1 ms, $\tau(^5D_0) \sim 0.6$ ms (14–300 K).	**TEOS** + [structure of MeO–Si(OMe)$_2$–(CH$_2$)$_3$–C(=O)–O–CH$_3$] **3-(trimethoxysilyl)propylmethacrylate (MAPTMS)** + [structure of [Eu(phen)$_2$]$^{3+}$ complex] [Eu(phen)$_2$]Cl$_3$ **(17)** Ligand broad band excitation (335 nm). $4f^6$ emission lines. $\tau(^5D_0) \sim 1.404$ ms (300 K). **or** + [structure of Tb complex with anthranilate-type ligand, 1/3 stoichiometry] **(18)** Ligand broad band excitation (280–325 nm). $4f^8$ and ligand emission (360–480 nm). $\tau(^5D_4) \sim 1.0$–1.2 ms (300 K).

Table 9.2 *Continued*

Molecular Structure/Synthesis and Emission Details	Molecular Structure/Synthesis and Emission Details
di-ureapropyltriethoxysilane (n_{POP}=3-68) (see Table 1) + [Eu^{3+} complex structure] ^{3}(19)	**MCM-48** + [(H$_2$O)$_2$Eu complex structure] ^{3}(20)
Precursors: diamine(POP)s and ICPTES; Eu(NO$_3$)$_3$. Ligands broad band excitation (350 nm). Tb^{3+} and Eu^{3+} emission lines. φ ~ 60% (370 nm).	MCM-48 precursors: cetyltrimethylammonium bromide (CTAB) and TEOS. Ligand broad band excitation (340–400 nm). 4f^6 emission. τ(^{5}D$_0$) ~ 0.370 ms (diluted samples).
[methyldiethoxysilane structure] **methyldiethoxysilane (MHTEOS)** + **colloidal silica** + [Erbium complex] Erbium-2,4-pentadionate **(21)**	**TEOS** + **MAPTMS** + [methacrylic acid structure] **methacrylic acid (MAA)** + **Zr(OPr^n)$_4$** zirconium n-propoxyde (ZPO) + **Er^{3+} (22)**
Buffer layer: methyltrimethoxisilane (MTMOS), active guiding layer: MHTEOS and colloidal silica, photoinitiator: Irgacure 819. Near infrared 4f^{11} emission lines.	Ormosils as buffer and guiding layers; ErCl$_3$.6H$_2$O; photoinitiator: Irgacure 1800. Near infrared 4f^{11} lines.

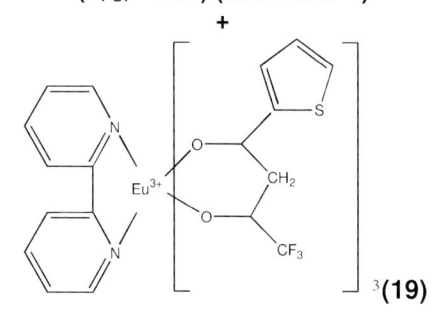

(1) Franville et al., *Solid State Sciences* 3, 221 (2001); (2) Driesen et al., *Chem. Mater.* 16, 1531 (2004); (3) Gonçalves et al., *Chem. Mater.* 16, 2530 (2004); (4) Liu et al., *New J. Chem.* 6, 674 (2002); (5) Li et al., *Chem. Mater.* 14, 3651 (2002); (6) Franville et al., *J. Alloys and Compd.* 275–277, 831 (1998) and *Chem. Mater.* 12, 428 (2000); (7) Dong et al., *Adv. Mater.* 12, 646 (2000); (8) Liu et al., *Thin Solid Films* 419, 178 (2002); (9) Binnemans et al., *J. Mater. Chem.* 14, 191 (2004); (10) Sun et al., *Adv. Funct. Mater.* 15, 1041 (2005); (11) Bredol et al., *Opt. Mater.* 18, 337 (2001); (12) Ribeiro et al., *J. Sol-Gel Sci. Tech.* 13, 427 (1998), Carlos et al., *Adv. Mater.* 12, 594 (2000), Sá Ferreira et al., *Chem. Mater.* 13, 2991 (2001) and *J. Sol-Gel Sci. and Tech.* 26, 315 (2001), Nunes et al., *Proceedings of the Materials Research Society*, vol. 847, 2005, p.EE.13.31.1; (13) Bekiari et al., *Chem. Mater.* 12, 3095 (2000); (14) Li et al., *New J. Chem.* 28, 1137 (2004); (15) Bekiari et al., *Chem. Phys. Lett.* 307, 310 (1999); (16) Carlos et al., *Adv. Funct. Mater.* 12, 819 (2002); (17) Li et al., *Thin Solid Films* 385, 205 (2002); (18) Li et al., *Mat. Lett.* 56, 597 (2002); (19) Moleski et al., *Thin Solid Films* 416, 279 (2002); (20) Meng et al., *Micropor. Mesopor. Mat.* 65, 127 (2003); (21) Etienne et al., *Optics Communications* 174, 413 (2000); (20) Xu et al., *Mat. Lett.* 57, 4276 (2003).

9.3 Hybrids for Coatings

Table 9.3 Examples of hybrids incorporating organic chromophores.

Molecular Structure/Synthesis and Emission Details	Molecular Structure/Synthesis and Emission Details
di-ureapropyltriethoxysilane (n_{POP}=3-68) (see Table 1) pyrene + **Coumarine 153 (1)** Precursors: diamine(POP)s (MW = 230–4000 g·mol^{-1}) and ICPTES. Excitation: 337.1 nm. Fluorescence and amplified spontaneous emission (ASE): 545 nm. Laser action tuned (525–580 nm).	**PEO$_{20}$POP$_{70}$POE$_{20}$** Pluronic P123 (non-ionic surfactant) + trifluoroacetic acid (TFA) + titanium(IV)ethoxide **Rhodamine 6G (2)** Mesostructured waveguide: evaporation induced self-assembled, TFA to prevent 3D TiO$_2$ cross-linking. Excitation: 532 nm, ASE: 545 nm. Laser action (modes with a spacing of 280 ± 30 cm^{-1}).
TEOS + (3) Dye-doped MCM-41: TEOS, 3-(2,4-dinitrophenylamino)propyl-(triethoxy)silane and hexadecyl-(trimethyl)ammonium bromide (C$_{16}$TMABr). Absorption: 317 and 417 nm.	**ZPO** + Glycidyloxypropyltrimethoxysilane (GLYMO) + **Polymethine dyes IR1051 and IR5 (4)** Stabilizers: acetic acid and 2-methoxyethanol. Absorption: 500, 980 and 1100 nm. Emission: 100–1300 nm.

(1) Sthatatos et al., *Chem. Phys. Lett.* **345**, 381 (2001); (2) Bartl et al., *J. Am. Chem. Soc.* **126**, 10826 (2004); (3) Fowler et al., *Chem. Commun.*, 1825 (1998); (4) Casalboni et al., *Appl. Phys. Lett.* **75**, 2172 (1999).

antireflective layers based on Pd containing TiO$_2$ or SiO$_2$/TiO$_2$ systems has become a successful technology. More recently the porous nature of silica-based sol–gel thin films was explored to impart antireflective properties to architectural glasses. In this case organic additives present in the coating solution lead to a gradient in porosity after thermal treatment which results in a gradient of the refractive index that takes the visible light transmissivity of the glass substrate to very high values (>99%). This system is highly suitable for solar applications as cover sheets for photovoltaic cells and collectors. Similar index gradients have been obtained through the generation of submicrometer structures on glass surfaces via patterning of appropriate coatings (Fig. 9.1).

- Dye-doped transparent hybrid sol–gel coatings containing organic dyes have been also used on cathode ray tubes to improve color TV image resolution. TEOS-based coatings doped with different types of organic pigments display the light-fastness and scratch resistance required for spin-coating of large TV screens.
- Coatings derived from 3-glyxidoxipropyltrimethoxysilane (GPTMS), aluminum-tris-(2-butylate), and

Fig. 9.1 Antireflective, nanostructural pattern generated in the surface of a hybrid polymeric matrix by a two-step embossing/UV-curing technique. (Taken from G. Shottner et al., *Thin Solid Films*, **1999**, *351*, 73).

phenyltrimethoxysilane exhibit excellent adhesion, abrasion resistance and chemical stability on glass surfaces. Commercially available organic dyes can be dissolved in the respective sols and colored coatings are obtained by spraying the sol onto the glass. The incorporation of hydrophobic organic epoxy resins copolymerizing with the functional groups present as substituents in the GPTMS molecule improves the chemical stability of the cured coating against alkaline media, thereby providing the coating with sufficient stability in dishwashing machines. In addition, by means of the aromatic epoxy pre-polymers, the refractive index, as well as abrasion resistance and optical appearance, can be adjusted to the substrate. Partial coating of objects or complete coloration is possible. The broad color range of organic dyes allows articles to be made in a wide variety of colors, which enhances consumer appeal. The procedure is environmentally friendly (recycling of the dye-colored glasses is easier, as no color classification is needed) and also cost-effective in comparison to the laborious traditional coloration techniques via molten glass batches containing toxic transition metal oxides.

Photoactive optical properties:
- An interesting cheap, new photo-electrochemical cell based on a chelating metal-complex dye and doped TiO_2 nanoparticles has been developed for solar energy conversion.
- Sol–gel-based hybrid coatings have been used to detect various species via fiber-optic sensors. Recently it was demonstrated that photochromic-doped sol–gel materials can be attached to optical fibers and that the properties of the light throughput may be modified and optically processed. Sol–gel waveguide (see the next two sections) photochromic material can be easily formed. Moreover the fabrication process of the hybrid fiber device can be adapted to complicated configurations. The material shape can be modified upon manufacturing; the adopted shape is kept after the curing process. Once cured, these devices behave as optically addressed variable delay generators.

The active optical features that support the use of hybrids as coatings (e.g. photoluminescence, photochromaticity and nonlinear effects) will be addressed in detail in the remaining of the chapter.

Apart from applications in glass sheets for furniture, sanitary appliances and building industry, hybrid coatings lend themselves to application for protection purposes against scratch, abrasion resistance and weather etching (polluted atmospheric conditions) are:

- *Automobile industry:* Hybrid coatings with superior scratch and environmental-etch resistance have been used as topcoats in automobiles since 1997. These coatings are composed of a complex mixture of two hybrid polymers crosslinked simultaneously during curing to yield a polymer network that is partially grafted and interpenetrated. A high-modulus and a scratch-resistant function are provided by a high density acrylate polymer core, organically modified with alkoxysilane groups and residual unsaturation, that is dispersed in a low crosslinked density polymer that provides the film-forming properties.
- *Ophthalmic lens:* Traditionally, most of the polymeric ophthalmic lenses (CR 39, polycarbonate, poly(methyl methacrylate) etc.) produced in the last few years have been coated with thermosetting films by wet chemical techniques involving spin- or dip-coating. One of the first hybrid sol–gel systems developed for CR 39 ($n_D = 1.498$) was obtained by sol–gel processing of GPTMS, tetramethoxyortosilane (TMOS), and titanium tetraethoxysilane (Ti(OEt)$_4$). The ophthalmic lens market is innovative and highly competitive. Recently polymer lenses showing high refractive indices ($n_D = 1.56, 1.60, 1.67$) have been commercialized. Therefore, high index coatings are necessary to avoid interference strings appearing with thin coatings and insufficient index matching between coating and substrate. The hybrid sol–gel coatings based on transition metals offer good prospects to develop adhesive, abrasion resistant, transparent, and index-matched materials.
- *Monument and art preservation:* Superfine organic pigments may be used to coat large sheets of glass. One of the most interesting examples is the long-term protection of the 14th century "Last Judgement" mosaic of the St. Vitus cathedral in Prague[3]. This coating, made from a combination of a hybrid nanocomposite and a fluoropolymer, is a transparent efficient barrier against corrosion.

---

[3] The completion of the ten-year project to conserve the Last Judgment mosaic, at St. Vitus Cathedral in Prague was published recently in *Conservation of the Last Judgement*, Getty Conservation Institute, Los Angeles, 2005.

9.4
Hybrids for Light-emitting and Electro-optic Purposes

9.4.1
Photoluminescence and Absorption

Luminescence is a general term which describes any nonthermal processes in which energy is emitted from a material at a different wavelength from that at which it is absorbed. The term broadly includes the commonly-used categories of fluorescence and phosphorescence. Fluorescence occurs where emission ceases almost immediately after withdrawal of the exciting source, whereas in phosphorescence the emission persists for some time after removal of that excitation. The distinction between the so-called types of luminescence is somewhat arbitrary and confusing. Confusion is avoided by using the term luminescence, and specifying the activating energy as a descriptive prefix. For instance, bioluminescence is related with light emission from lived animals and plants, cathodoluminescence results from excitation by electrons, chemicaluminescence is the emission occurring during a chemical reaction, roentgenoluminescence is produced by X-rays, triboluminescence is ascribed to rubbing, mechanical action, and fracture, electroluminescence is the conversion of electrical energy into light, and photoluminescence results from excitation by photons. In this chapter emphasis will be placed on photoluminescence and electroluminescence.

Photoluminescence requires the absorption of photons with energy $E = h\nu$ (where h is the Planck constant and ν is the frequency). The interaction mechanisms between the photon and the matter depend on the photon energy. When the photon energy of the incident radiation is lower than the energy difference between two electronic states, the photons are not really absorbed and the material is transparent to such radiation energy. For higher photons energy, absorption occurs and the valence electrons will make a transition between two electronic energy levels. The excess of energy will be dissipated through vibrational processes that occur throughout the near infrared (NIR) spectral region. Then, the excited atoms may return to the original level through radiative (with the spontaneous emission of a photon) and nonradiative transitions.

Examples of radiative and nonradiative processes will be given considering the typical distribution of electronic levels in a molecule with two electrons (Fig. 9.2). In the fundamental levels the electrons in the same orbital have opposite spins ($s_1 = +1/2$ and $s_2 = -1/2$); so that the total spin ($S = s_1 + s_2$) is equal to zero. Thus, the fundamental state multiplicity ($M = 2S + 1$) is one, and the ground state is designated as singlet (S_0). After optical absorption, the electrons will be excited. If this transition does not involve spin inversion, the excited state is also a singlet (S_1), i.e. it has the same state multiplicity as the ground level. However, if there is spin inversion, the two electrons have the same spin, $S = 1$ and $2S + 1 = 3$, and the excited state is called a triplet (T_1). It should be noted that such absorption involving a triplet state is forbidden by the spin selection rule: allowed transitions must

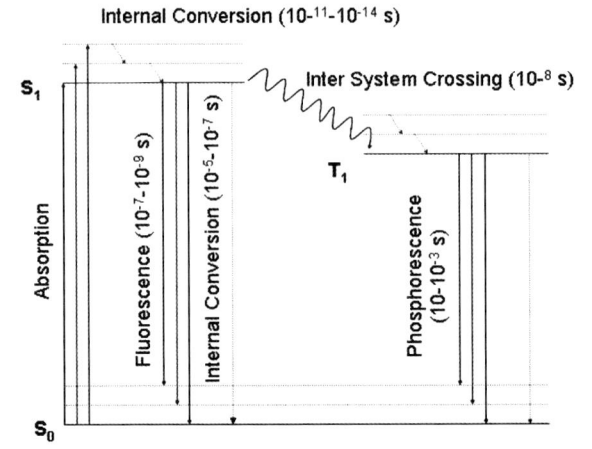

Fig. 9.2 Scheme of possible radiative and non-radiative transitions within a molecule. The S and T denote Singlet and Triplet states, respectively. The time-scale characteristic of each transition type is indicated in parenthesis.

involve the promotion of electrons without a change in their spin ($\Delta S = 0$). The relaxation of the spin selection rule can occur though strong spin-orbit coupling, which is for instance the case of rare earth ions. Fig. 9.2 summarizes the typical radiative and nonradiative transitions within a molecule.

Examples of nonradiative processes
- Internal conversion; an electron close to a ground state vibrational energy level, relaxes to the ground state via transitions between vibrational energy levels giving off the excess energy to other molecules as heat (vibrational energy).
- Intersystem crossing; the electron transition in an upper S_1 excited state to a lower energy level, such as T_1.

Examples of radiative processes
- Fluorescence; emission of a photon from S_1 to the vibrational states of S_0 occurring in a time scale of 10^{-10} to 10^{-7} s.
- Phosphorescence; emission of a photon from T_1 to the vibrational sates of S_0. This process is much slower than fluorescence (higher than 10^{-5} s) because it involves two states of different multiplicity. For very long luminescence time decays (seconds, minutes, even hours), the emission is called "glow in the dark". Phosphorescence is red-shifted relatively to fluorescence, because T_1 is excited via intersystem crossing.

9.4 Hybrids for Light-emitting and Electro-optic Purposes

Due to the nonradiative transitions the emission will occur at lower energy (longer wavelengths) than that of the absorbed photons. The energetic difference between the maximum of the emission and absorption spectra ascribed to the same electronic transition is known as Stokes–shift.

If radiative emission occurs, it is very useful to acquire an excitation spectrum. Such measurement can be done by monitoring a certain emission wavelength, under illumination of the material by light of different wavelengths. The resulting spectrum gives us the information of the excitation wavelengths that effectively contribute to the monitored emission. This measurement should not be confused with absorption spectra, although similar results may occur. As already mentioned, an absorption spectrum tells us about the spectral range absorbed by the sample, independently of the occurrence (or not) of radiative transitions. An excitation spectrum is a selective measurement that selects the part of the absorption spectrum which contributes to the observation of the monitored emission. Moreover, due to the nonradiative transitions mentioned above, it might be observed a redshift of the absorption spectrum with respect to the wavelengths extracted from the excitation spectra.

Measurement techniques We will detail further the experimental distinction between the above mentioned steady-state photoluminescence measurements, namely absorption, emission and excitation. Fig. 9.3 schematizes a possible photoluminescence experimental characterization setup. The key elements are the excitation source, the monochromators and the detection device.

- The excitation source can be monochromatic, such a laser or light emitting diodes (LEDs), or a broad spectrum lamp coupled to a monochromator. Xe arc lamps are ultraviolet to visible light sources, the Hg lamps have very strong peaks in the ultraviolet, whereas the tungsten-halogen lamps are a good choice for long wavelengths from the visible to the NIR. Fig. 9.4 shows the emission spectrum of different types of possible excitation sources.
- Monochromator enables spectral separation and it is formed by two mirrors and one diffraction grating aligned according to the Czerny–Turner layout (see Fig. 9.3).
- Concerning detector devices, there are single channel (photodiodes, photomultiplier tubes) and multichannel arrays (charged coupled device, CCD). The choice of a particular detector depends on the requirements of a specific application, in particular the wavelength of the light, the sensitivity needed and the speed of response required.

To record absorption spectrum the sample is illuminated within a certain wavelength range and the transmitted light through the sample is scanned in the same spectral region. The corrected absorption spectrum (S_{abs}) can be achieved by:

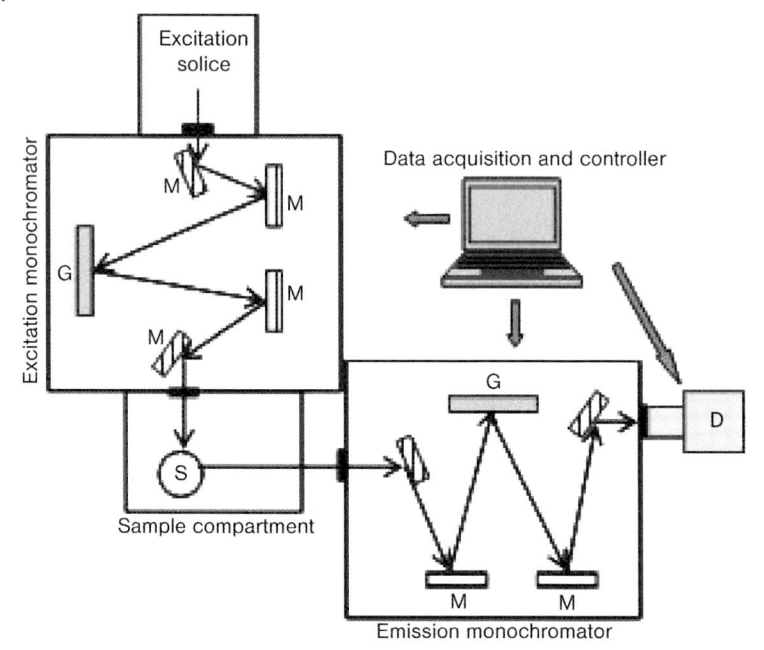

Fig. 9.3 Experimental layout for photoluminescence measurements. M, G, S and D stand for mirror, diffraction grating, sample and detector, respectively. The arrows indicate the light path. The layout presented for the mirrors and gratings inside each monochromator is called as Czerny–Turner configuration.

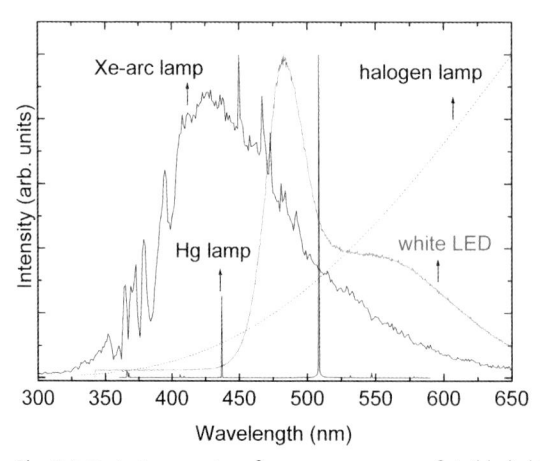

Fig. 9.4 Emission spectra of common sources of visible light.

$$S_{abs} = \frac{S_{exp}}{S_{source}} \quad (1)$$

where S_{exp} and S_{source} are the uncorrected absorption spectrum and the emission of the illuminating source, respectively.

An excitation spectrum is measured by setting the emission monochromator fixed in a given emission wavelength (for instant the one corresponding to the maximum of the emission spectrum). The excitation monochromator is then scanned in a given wavelength interval and the luminescence intensity corresponding to the monitored emission wavelength is measured.

An emission spectrum is acquired by exciting the sample with an absorbed wavelength, usually the maximum intensity absorption (or excitation) peak, and the emission monochromator scans the luminescence within a wavelength interval.

Typically organic-inorganic hybrids are almost transparent in most of the visible spectral region. The maximum absorption occurs essentially from the ultraviolet to the blue spectral region, see Fig. 9.5(a). The maximum of the excitation spectra (Fig. 9.5(b)) appears in the same spectral region of the absorption one, but deviated to longer wavelengths. This indicates that only the lower energy side of the absorption spectra is converted into efficient radiative emission (Fig. 9.5(c)). Such emission is characterized by a large broad band, usually Gaussian in shape, whose peak position occurs within the blue and green spectral regions, depending on the structural properties of the hybrid, as we will detail in Section 9.4.4. Another intrinsic property is related with the emission dependence on the excitation wavelength, in such a way that increasing the excitation wavelength results in an emission red-shift (deviation to lower energies), as it is well illustrated in the inset of Fig. 9.5(c). Such behavior is far from being completely understood and it is often ascribed either to the presence of (a) different size of the emitting monomers or to (b) defects that induced the presence of localized states. For further information on this advanced topic, readers are referred to Thorpe 1997, and Sing 2003.

So far we have discussed steady-state photoluminescence for which the excitation source is continuous and the photoluminescence detection takes place under continuous irradiation of the sample. Photoluminescence may also be acquired in time-resolved mode, a technique that involves the use of a pulsed excitation source and requires that the spectral detection takes place after a certain time interval subsequent to the excitation pulse. The photoluminescence time-resolved mode enables the evaluation of the time-scale in which the processes behind the emission occur (dynamical evaluation). Moreover, it is a powerful technique to permit the unequivocally establishment of different emission and excitation components with distinct time scales.

Figure 9.6 illustrates that dynamical evolution showing an example of a time-resolved spectrum of an organic–inorganic hybrid. The spectrum acquired at shorter delay time reveal the presence of three distinct main peaks centerd at ca. 425, 460 and 500 nm, whereas for starting delays higher than 5.00 ms, only the

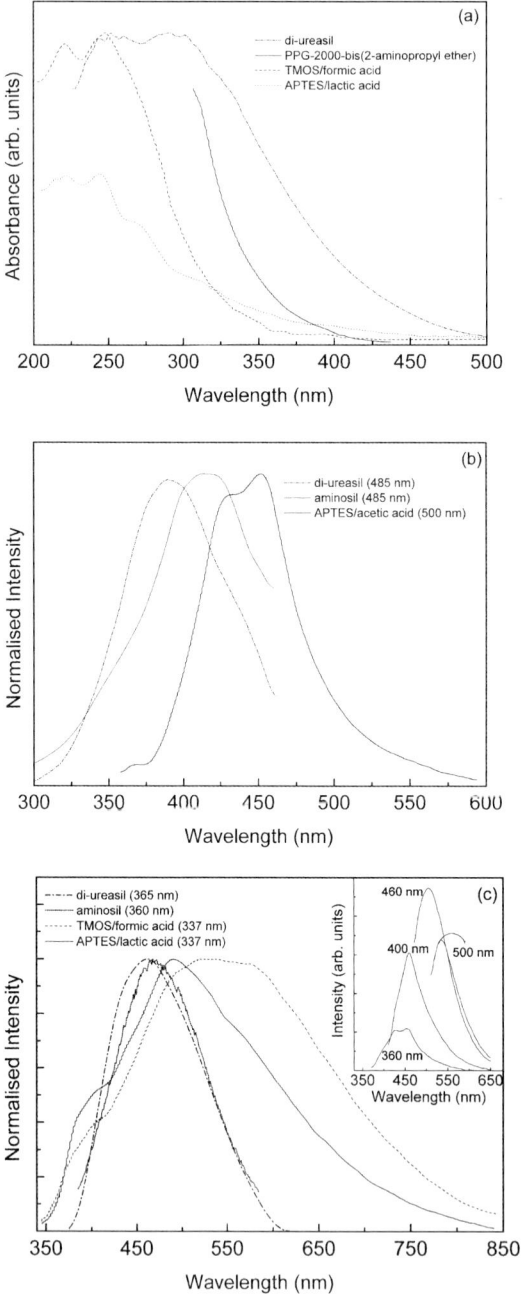

Fig. 9.5 (a) Absorption, (b) excitation, and (c) emission spectra of representative organic-inorganic hybrids.
Di-ureasils: (Carlos et al., *Adv. Func. Mater.* **2001**, *11*, 111)
Aminosil: (Carlos et al., *J. Phys. Chem. B* **2004**, *108*, 14924);
TMOS/formic acid and APTES/lactic acid: (Green et al., *Science* **1997**, *276*, 1826); PPG-2000-bis(2-aminopropyl ether) and APTES/acetic acid: (Brankova et al., *Chem. Mater.* **2003**, *15*, 1855).

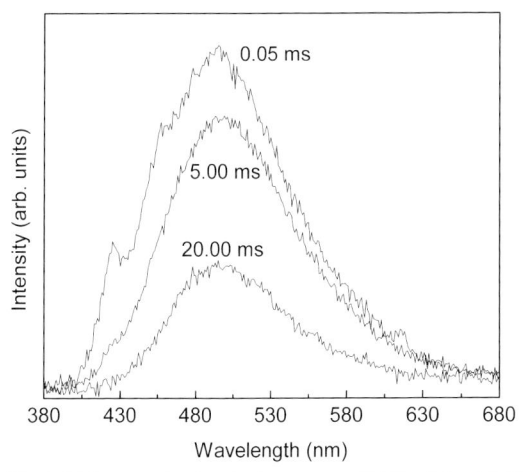

Fig. 9.6 Time-resolved emission spectra (14 K) of the d-A(8) di-amidosil at an excitation wavelength of 360 nm for different starting delays.

lower energetic peak can be discerned. This indicates that the components at 425 and 460 nm may have the same nature, whereas the band at c.a. 500 nm must have a different origin.

The accurate time scale behind each emission can be obtained by monitoring a decay curve around the two bands observed in the time resolved spectrum, and extract the lifetime of each emitting state. The lifetime of a certain excited level is the mean time that an atom or molecule stays in the excited state before returning to the ground level. The pulse excitation source illuminates the sample, and for a certain emission wavelength (for instance, the maximum intensity of the emission time-resolved spectrum) the detection is made after successive increasing delay times. For example, Fig. 9.7 shows the decay curves monitored around the maximum intensity of the bands in Fig. 9.6. The shape of the decay curve depends on the starting delay. Only for the higher starting delay (40.00 ms), the decay curve is well reproduced by a single exponential function revealing a lifetime value around 160 ms.

9.4.2
Electroluminescence

Electroluminescent materials are of great interest for several light-emitting applications, as, for instance, the new types of flat full-color displays. Traditionally, this field of application is governed by inorganic materials, Silicon and gallium arsenide being the most common ones. In the last decade, a growing research focused on organic materials, lead to the production of the so-called organic light emitting diodes (OLEDs). The real worldwide interest in this field starts after development of thin film double layer OLEDs in the Eastman Kodak's research

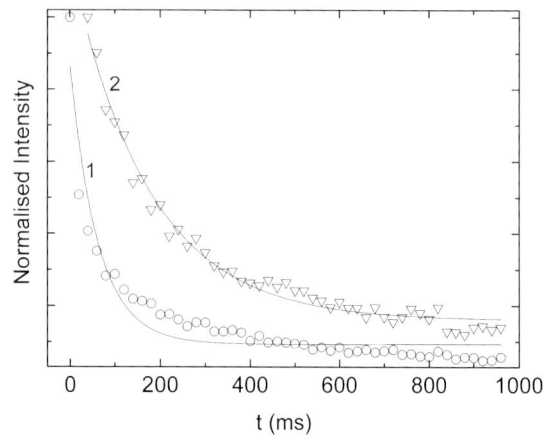

Fig. 9.7 Experimental decay curves for a di-ureasil hybrid measured at 14 K under 365 nm excitation wavelength and monitored around 500 nm for different starting delays (1) 0.05 ms, and (2) 40.00 ms. The straight lines represent a single exponential decay.

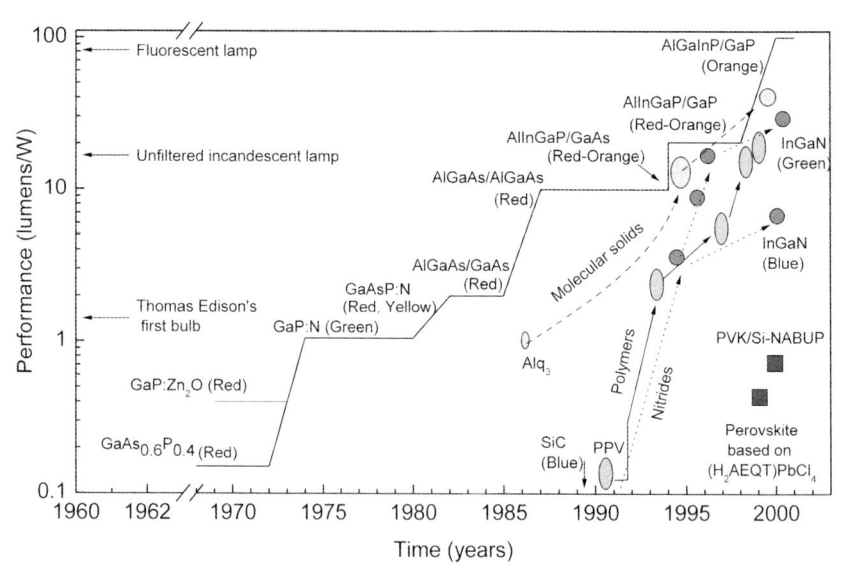

Fig. 9.8 Evolution of LED/OLED performance. (Adapted from J. R. Sheats et al., *Science* **1996**, *273*, 884.)

laboratories in 1987. The first polymer OLED device in 1990 has followed this device based on small–molecular weight materials. The evolution of OLEDs (both single molecule and polymer LEDS) is specifically interesting. The sharp increase in their performance has really made it a commercially viable alternate to conventional inorganic semi–conductor based LED technology (which suffered a great increase after development of inorganic high brightness LEDs using double heterojunction or multi quantum–wells active layers), Fig. 9.8.

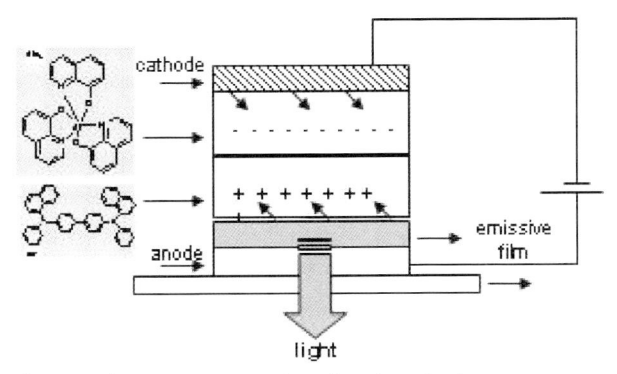

Fig. 9.9 Schematic representation of an electroluminescent device, using Alq3 (tris(8-hydroxyquinolinato)aluminum, as the electron transport, and NPB, N,N'-di(naphthalene-1-yl)-N,N'-diphenylbenzidine as the hole transport layer. (Adapted from J. M. Shaw and P. F. Seidler, *IBM J. Res. & Dev.* **2001**, *45*, 3)

The OLEDs present a great advance in this field, due to their high performances, such as small volume, high brightness, fast response time, and low driven voltage (working voltage). Luminous efficiencies in excess of 30 lm/W and operating voltages as low as 4V are known for OLEDs (Fig. 9.8). However, OLEDs also exhibit some drawbacks, namely the lack of long-term chemical and thermal stability and poor mechanical strength. To achieve electroluminescence, the emitting film should be sandwiched between a cathode and an anode (Fig. 9.9). When direct electric current is applied the cathode and the anode will release an electron and a hole, respectively, into the organic film, furnishing the energy necessary to excite the system. The emitted color will depend on the selected material of the film. The OLEDs are based on multilayer devices, where two layers of different materials work as electron and hole injectors. The dissolution between the two layers of different materials is a recurrent problem that renders difficult the processing of OLEDs. The combination in a single material of such organic materials with an inorganic counterpart my overcome such drawbacks.

The wide range of possible hybrids would include innovative combinations, such as inorganic clusters, fullerenes or metal nanoparticles dispersed in organic polymers, or macrocycles or polyethylene oxide chains intercalated into silicate materials. For organic–inorganic semiconductor hybrids few representative examples are listed as:

- Alternating organic–inorganic films, such as, for instance, amorphous multilayers of copper phthalocyanine and TiO_x.
- Nanoclusters and organic–inorganic quantum dots (QD) of a variety of semiconductor and metallic materials. Organic ligands, used as solvent and to control the QD growth rate during preparation, coordinate the surfaces of the QDs and may be exchanged for alternate organic molecules with

different lengths and electronic structures. (CdSe and CdS cores with trialkylphosphine chalcogenide capping layers, for example). The increased photoluminescence quantum yield of the QDs enables LEDs with external quantum efficiencies as high as 0.22% and operating voltages of ~4 V.

- Crystalline hybrids (perovskites). The basic layered perovskite structures are $(R-NH_3)_2MX_4$ and $(NH_3-R-NH_3)MX_4$ (where M is generally a divalent metal and X= Cl^-, Br^-, or I^-). While most of the perovskites studied to date contain relatively simple organic cations, such as alkylammonium or phenethylammonium, more complex organic molecules can be incorporated (e.g. oligothiophene derivatives). The organic–inorganic perovskites have demonstrated a Hall mobility of $50\,cm^2\,V^{-1}\,s^{-1}$ (larger than amorphous silicon) and a maximum efficiency of $0.16\,lm/W$ (at 8 V and 0.24 mA), for a perovskite structure based on $(H_2AEQT)PbCl_4$, AEQT is 5,5'''-bis(aminoethyl)-2, 2':5', 2'':5'', 2'''-quaterthiophene, providing a possible path to increased performance, Fig. 9.10.

- Siliceous-based networks. The functionalization of alkoxysilanes with carriers transport units is a new way of preparing new hybrids OLEDs. These materials can be obtained though sol–gel route offering the possibility of tailoring the final properties of the material and allowing the preparation of thin films though spin-coating techniques. The organic polymer becomes chemically stable due the grafting with the inorganic counterpart. An

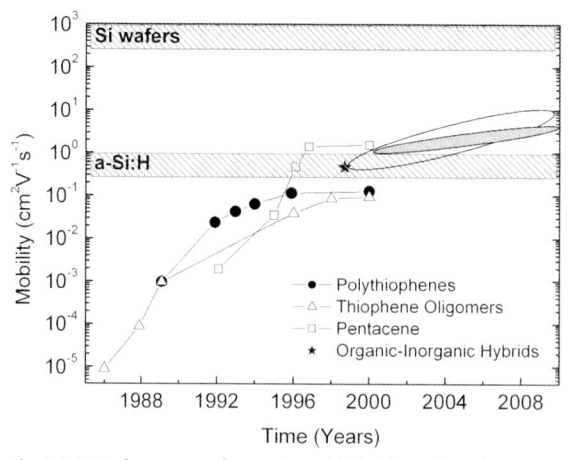

Fig. 9.10 Performance of organic and hybrid semiconductors. Adapted from J. M. Shaw, P. F. Seidler, *IBM J. Res. & Dev.* 45, 3 (2001).

additional argument supporting the effective interest of hybrid OLEDs is that the functionalization process does not affect substantially the polymer emitting species. Table 9.4 gives the molecular structure of silylated precursors that were prepared by the modification of active molecules of carbazole (Si—KH), oxidiazole (Si—BPD) and tetraphenylphenylenediamine (Si—TPD) derivatives with hole transporting units that have the higher reported mobility value. These hybrid OLEDs are composed of a hole transporting layer and of an emissive layer incorporating electron-transport guest molecules (for instance fluorescent molecules such as 4-dicyanomethylene-2-methyl-6-[p-(dimethylamino)styryl]-4H-pyran, DCM, or N-(4-butylphenyl)-4-[(N-2-hydroxyethyl)(methyl)amino]naphthalimide, NABUP) that are sandwiched between a transparent indium tin oxide electrode and metallic cathodes (aluminum or LiF/Al), Table 9.5. The single layer device involving Si—DCM doped with 50%

Table 9.4 Molecular structure and the mobility value at field strength $5 \times 10^5\,V\,cm^{-1}$ of silylated precursors prepared by the modification of active molecules of carbazole (Si–KH), and tetraphenylphenylenediamine (Si–TPD) derivatives with hole transporting units.

Molecular Structure of the Silylated Precursor	*Hole Mobility ($cm^2\,V^{-1}\,s^{-1}$)*
Si-TPD **(1)**	5.7×10^{-5}
Si-KH **(1)**	$\sim 2.9 \times 10^{-6}$

(1) Dantas et al., Adv. Mater. Opt. Electron. 10, 69 (2000).

Table 9.5 Two examples of hybrid OLEDs.

Chromophor	Molecular Structure of the Silylated Precursor	Colour	Luminance
Naphtalamine (1)		green	440 cd.cm^{-2} @ 24 V
4-dicyanomethyllene-2-methyl-6.(p-(dimethyl-amino)styryl)-4H-pyrane (2)		orange	4000 cdm^{-2} @ 27 V

(1) Dantas et al., *C. R. Acad. Sci. Paris* 4, 479 (2000) and (2) *Adv. Mater.* 11, 107 (1999).

PBD emits in the orange with an external quantum yield of 0.13%, a maximum luminance of 440 cd/m² (at 24 V) and a luminous efficiency of ca. 0.03 lm/W. An higher performance is obtained using a green emitting bylayer device of polyvinylcarbazole, PVK, and Si—NABUP with a LiF/Al cathode, external quantum yield of 1% (at 100 cd/m²), a maximum luminance of 4000 cd/m² (at 27 V) and a luminous efficiency of ca. 0.65 lm/W.

Figures 9.8 and 9.10 show the evolution of the hole mobility and the LED/OLED performance evidencing the improvement in the OLED performance due to the design of new materials. The efficiency of the preliminary hybrid LEDs (0.2–0.7 lm/W) is quite promising, and higher values are expected with further fine tuning of the device structure, especially through the incorporation of chromophores with fundamentally better fluorescence quantum yield. Therefore, the organic–inorganic hybrids are considered one of the most promising advanced materials for electroluminescence purposes in the coming years.

9.4.3
Quantifying Luminescence

9.4.3.1 Color Coordinates, Hue, Dominant Wavelength and Purity

The quantification of materials luminescence features, such as color emission, quantum yield and radiance, permits an accurate comparison between different materials, with respect to their light-emitting performance.

The color coordinates are usually calculated using the system proposed by the Commission International de L'Éclairage (CIE) in 1931. This procedure is based on the human eye response to the visible light, where there are three cone cells responsible for the color distinction. Each cone cell presents a different sensibility, designated as $\bar{x}_\lambda$, $\bar{y}_\lambda$ and $\bar{z}_\lambda$, with maximum values at ca. 419, 531 and 558 nm, respectively. The sum of the three cone sensitivity functions is called the photonic response and displays a maximum sensibility cantered at 555 nm, in the green spectral region. The $\bar{x}_\lambda$, $\bar{y}_\lambda$ and $\bar{z}_\lambda$ are called the CIE color-matching functions. Such color-matching functions were calculated considering a 2° or a 10° field of view (defining the 2° and the 10° standard observers, respectively). In addition, the CIE defined three new primaries colors X, Y and Z, needed to match any specific colors. These quantities, X, Y and Z, are known as tristimulus values. These primary colors established by the CIE have a great advantage, when compared with other color systems, since the color-matching functions do not display negative parts. The emission is weighted by the $\bar{x}_\lambda$, $\bar{y}_\lambda$ and $\bar{z}_\lambda$ functions in order to determinate the X, Y and Z primary color though the following set of equations:

$$\begin{cases} X = \int \bar{x}_i . E_i d\lambda \\ Y = \int \bar{y}_i . E_i d\lambda \\ Z = \int \bar{z}_i . E_i d\lambda \end{cases} \quad (2)$$

where E stands for the emission spectra intensity and i represents the emission wavelengths.[4] To simplify the calculus, the spectra can be divided in small wavelength intervals, $\Delta\lambda$ (usually $\Delta\lambda = 5$ or $10\,\text{nm}$), and Eq. (2) can be rewritten as:

$$\begin{cases} X = \sum_{\lambda=380}^{720} \bar{x}_\lambda . E_\lambda \Delta\lambda \\ Y = \sum_{\lambda=380}^{720} \bar{y}_\lambda . E_\lambda \Delta\lambda \\ Z = \sum_{\lambda=380}^{720} \bar{z}_\lambda . E_\lambda \Delta\lambda \end{cases} \quad (3)$$

To make the color perception easier, the tristimulus are converted into a two dimensional system, through a linear transformation. The tristimulus are transformed into chromaticity coordinates, (x,y), that are plotted in a two dimensional chromaticity diagram, Fig. 9.11.

The chromaticity coordinates are related to X, Y and Z by:

$$\begin{cases} x = \dfrac{X}{X+Y+Z} \\ y = \dfrac{Y}{X+Y+Z} \end{cases} \quad (4)$$

The (x,y) chromaticity diagram has particular characteristics, namely:
- The center of the diagram is taken as the white point, (0.33,0.33);
- The curve that is made of pure colors from the blue to the red, covering the entire spectral visible range (380–730 nm), is designated as spectral locus;
- The straight line that connects the two extremes of the spectral locus is denominated as purple boundary. The colors represented by the purple boundary are not pure colors, as they include a mixture of pure red and blue.
- The area circumscribed by the diagram, spectral locus and purple boundary, encloses the domain of all colors. The CIE tristimulus X, Y and Z do not belong to the spectral locus, and therefore they do not represent real colors. This is a direct consequence of the postulate that states that any visible color can be obtained by adding only positive quantities of X, Y and Z.

---

4) For color due to surface reflection Eqs. (2) and (3) should also include the spectral distribution of the illuminant as a multiplicative factor. For evaluating the color coordinates of an absorption or transmission the parameter E in Eqs. (2) and (3) should be replaced by the absorption or transmission intensity times the spectral distribution of the illuminating source. This spectral distribution of the excitation source is not relevant for emission color coordinates, since the emission spectra are acquired at a certain fixed excitation wavelength.

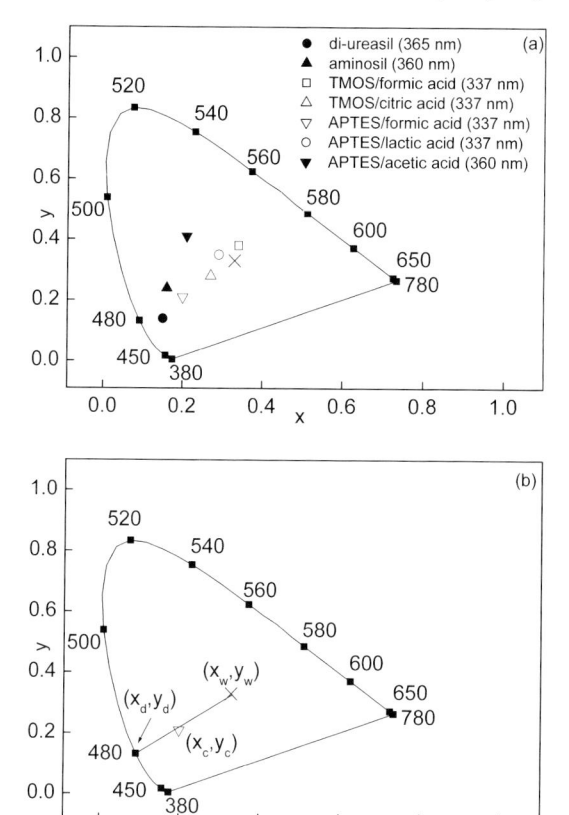

Fig. 9.11 CIE color coordinates represented on the (x,y) diagram proposed in 1931 for the emission of representative organic-inorganic hybrids. The excitation wavelength is indicated in parenthesis.

The (x,y) color diagram does not represent a linear space, e.g. the distribution of colors is not uniform. Two colors with the same perceptual difference have a higher spatial difference in the green spectral region of the (x,y) diagram, when compared with the red-spectral area. To minimize this effect, the (x,y) spectral locus is often transformed into other two dimensional color spaces most useful to discern colors, such as (u,v), (u′,v′) or (u*,v*).

With respect to color, three concepts are relevant:
- Hue is the property of colors by which they can be perceived as ranging from red through yellow, green, and blue, as determined by the dominant wavelength of the light.
- Dominant wavelength (λ_d) represents the pure color that is closer to that of the measured emission and it is strictly related with the definition of hue.

Table 9.6 Color coordinates and purity of the emission of several representative organic–inorganic hybrids, calculated according to the CIE using a 2° standard observer. The color coordinates of the dominant wavelength of the emission of each hybrid are also presented.

Hybrid	Excitation Wavelength (nm)	Emission Colour		Dominant Wavelength		Purity (%)
		x	y	x	y	
di-ureasil	365	0.15	0.14	0.10	0.09	78
aminosil	360	0.16	0.24	0.07	0.20	65
TMOS/formic acid	337	0.34	0.38	0.38	0.60	20
TMOS/citric acid	337	0.27	0.27	0.09	0.13	25
APTES/formic acid	337	0.20	0.21	0.10	0.12	57
APTES/lactic acid	337	0.29	0.35	0.03	0.48	13
APTES/acetic acid	360	0.21	0.41	0.01	0.54	38

- Color purity (p) quantifies the proximity between the emission color and the dominant wavelength.

The dominant wavelength can be determined by connecting, with a straight line, the color coordinates of the emission, (x_w, y_w), with the center of the diagram, (x_c, y_c). Then, the line is prolonged until it intercepts the spectral locus at a certain pure color. The intercepting point defines the dominant wavelength with coordinates (x_d, y_d). The color purity can be quantitatively inferred by:

$$p = \frac{y_c - y_w}{y_d - y_w} \quad \text{or} \quad p = \frac{x_c - x_w}{x_d - x_w} \tag{5}$$

as the line connecting the center of the diagram and the samples color coordinates is closer to the vertical (y axis) or horizontal (x axis), respectively.

The (x,y) chromaticity diagram in Fig. 9.11(a) illustrates the emission color coordinates for several representative organic–inorganic hybrids. In general, the emission color coordinates lies within the blue spectral region and close to the center of the diagram where the white color is defined. Table 9.6 lists the color purity and the color coordinates of the dominant wavelength for the emission of the hybrids mentioned in the diagram of Fig. 9.11(a). As an example, the determination of the color purity and the dominant wavelength is demonstrated in Fig. 9.11(b). Due to the large broad band shape characteristic of the emission spectra the color purity is low, when compared, for instance, with emission arising from atomic and atomic-like states. Thus, in general the emission of organic–inorganic hybrids may be used in applications where broad band emitters are requested.

9.4.3.2 Emission Quantum Yield and Radiance

The absolute emission quantum yield (ϕ) and the radiance (R) quantify in an absolute way the emission intensity (the two concepts, however, are different and should not be confused).

Let us analyze again Fig. 9.2 where the absorption, radiative and nonradiative paths that are possible to occur in a given molecule are schematized. If we are interested, for instance, in quantifying the phosphorescence arising from the T_1 state excited state, the quantum yield will be the ratio between the number of absorbed photons and the number of emitted ones. The radiance is the radiant intensity in a specified direction per unit of projected area, as represented in Fig. 9.12. Thus, radiance includes only the quantification of the emitted photons. The main difference between the radiance and quantum yield results from the fact that the quantum yield value is weighed by the number of absorbed photons. Thus, when we are interested in comparing radiance values similar experimental conditions with respect to the excitation power and concentration of emitting centers should be employed.

The radiance can be estimated by using an integrating sphere complemented with an excitation and emission spectrometers or a calibrated CCD camera, and it is often expressed in units of $W.m^{-2}.sr^{-1}$, where sr stands for steradian and is the S.I. unit of solid angle (Ω in Fig. 9.12). The solid angle corresponds to the observer's visual field. Radiance is a radiometric quantity, and should not be confused with luminance that is a photometric measure determined from the radiance spectral distribution with the spectral luminous efficiency of the human eye, being given in units of $cd.m^{-2}$. Table 9.7 lists different radiometric and respective photometric quantities.

An integrating sphere and calibrated detector setup is suitable for accurate absolute value light power measurement of emitting samples, by the spatial integration of radiant flux. The inside of the sphere is coated with a highly reflective material (e.g. $BaSO_4$). The theory of the integrating sphere originates in the theory of radiation exchange within an enclosure of diffuse surfaces. The integrating sphere radiance can be estimated through:

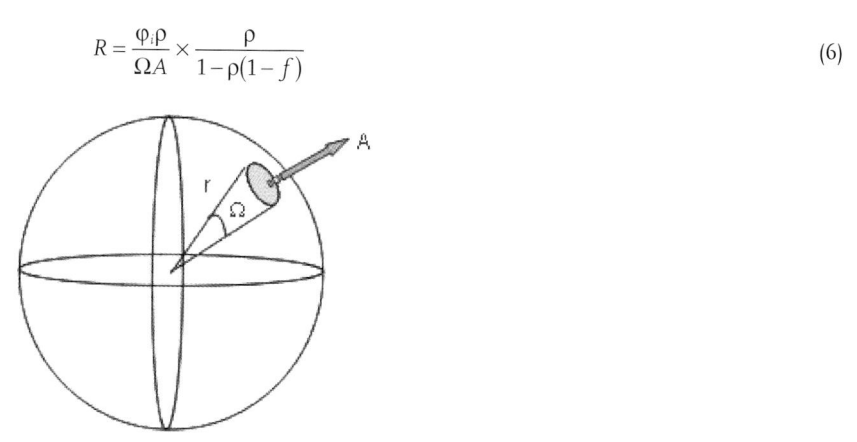

$$R = \frac{\varphi_i \rho}{\Omega A} \times \frac{\rho}{1-\rho(1-f)} \tag{6}$$

Fig. 9.12 The solid angle corresponds to the volume limited by the conic surface. The symbols r and A denote the radius of the sphere and the area on the sphere surface, respectively. The solid angle value is determined from $\Omega = A/r^2 sr$. A radiant power emitted by a surface element dA in the direction of the solid angle element $d\Omega$ is given by $L_e \cos(\alpha) dA d\Omega$ where α is the angle between the $d\Omega$ direction and the normal of the dA element.

Table 9.7 Radiometric and the respective photometric quantities. The units often used to express them are indicated in parenthesis.

Radiometric Quantity	Photometric Quantity
radiant flux (W)	luminous flux (lumen, lm)
radiant intensity (W.sr^{-1})	luminous intensity (candela, cd)
irradiance (W.m^{-2})	illuminance (lux, lx)
Radiance (W.m^{-2}.sr^{-1})	luminance (cd.m^{-2})

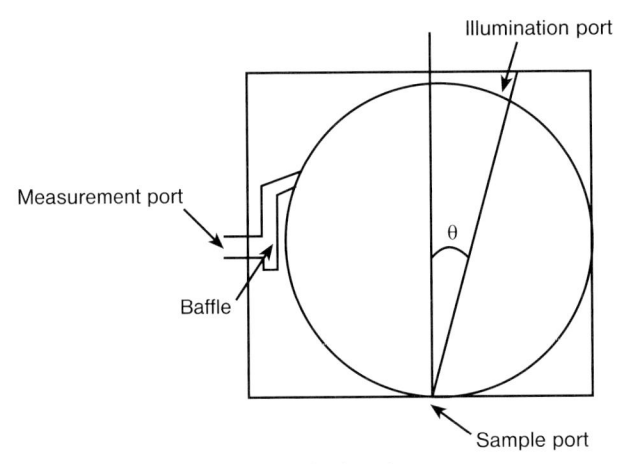

Fig. 9.13 Scheme of an integrated sphere for emission quantum yield measurements. The small ports allow the entrance of light and access to a detector. The sample is placed outside and near one of the ports.

where ϕ_i is the input flux, ρ represents the reflectance, A is the illuminated area, Ω is the total projected solid angle from the surface, and (1–f) corresponds to the fraction of flux received by the sphere that is not consumed at the port openings. Figure 9.13 depicts a scheme of an integration sphere evidencing the respective main experimental details. The first term in Eq. (6) represents the diffuse surface radiance and the second term, designated as sphere multiplier, accounts for R increase due to multiple reflections.

The evaluation of the quantum yield of thin films may be quite complex when compared with that of powdered samples because films usually have a high refractive index that may waveguide the luminescence. The use of an integration sphere is the most used method. For powdered samples there are two experimental methods for quantum yield quantification involving the use of i) a white reflecting standard and ii) a reflecting white standard combined with a phosphor standard. Their experimental errors are 25 and 10%, respectively.

a) The absolute emission quantum yield can be described through the following expression:

$$\phi = A_H / (R_s - R_H) \tag{7}$$

where A_H is the area under the hybrids emission spectra, and R_s and R_H are the diffuse reflectances (with respect to a fixed wavelength) of the hybrids and of the reflecting standard, respectively. Emission and diffuse reflectance spectra should be corrected for the detector and optical spectral responses. As reflecting standard, different white standards with high reflective coefficients (r > 0.9) can be selected, such as, for instance, potassium bromide, magnesium oxide, barium (or sodium) sulfates and sodium salicylate. Powder size and packing fraction are crucial factors because the intensity of the diffuse reflectance R_s and R_H depends on them. Thus, the diffuse reflectance is first measured at a wavelength not absorbed by the hybrids. The hybrids should be thoroughly ground until R_H is totally overlapped with R_s, indicating a similar powder size and packing fraction. To prevent insufficient absorption of the exciting radiation, a powder layer around 2 mm should be used and utmost care must be taken in order to ensure that only the sample is illuminated, in order to diminish the quantity of light scattered by the front sample holder. After this initial calibration procedure the diffuse reflectance and the emission spectra with respect to a certain excitation wavelength can be measured, in order to apply Eq. (7). Before estimating the parameters in this equation the wavelength units should be converted into the reciprocal energy space. Figure 9.14 illustrates the experimental steps described above.

b) Equation (7) should be replaced by:

$$\phi = \left(\frac{1 - R_{phos}}{1 - R_H} \right) \left(\frac{A_H}{A_{phos}} \right) \phi_{phos} \tag{8}$$

where R_{phos}, A_{phos}, ϕ_{phos} are the diffuse reflectance, the area under the emission spectra and the quantum yield of the standard phosphor at a certain excitation wavelength, respectively. The procedure is similar to that described above. The standard phosphor should be first calibrated with respect to grain size and packing fraction (Fig. 9.14(a)), and then the diffuse reflectance and emission spectra should be acquired (Fig. 9.14(a) and 14(b)). There are several compounds that can be used as standard phosphors, depending on the samples, namely excitation wavelength, color and full width at half maximum (FWHM) of the emission. The standard phosphor should be excited and should emit in the same spectral range of the hybrids. Sodium salicylate presents a large broad band peaking around 425 nm, with a constant φ value of about 60% for excitation wavelengths between 220 and 380 nm. These properties render sodium salicylate an adequate standard for ultraviolet absorbing samples, such as most of the emitting organic–inorganic

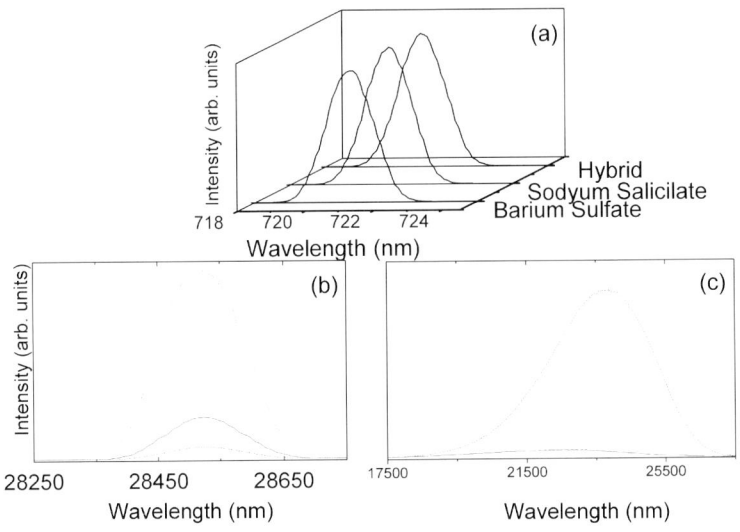

Fig. 9.14 Diffuse reflectance spectra of a representative organic–inorganic hybrid, sodium salicylate and barium sulfate at (A) a non absorbing wavelength (720 nm) and at (B) an excitation wavelength of 350 nm. (C) Emission spectra of the hybrid, and of the sodium salicylate at excitation wavelength of 350 nm.

hybrids. There are other standard phosphors for the red, green and blue spectral regions, such as, for example, Y_2O_3:Eu, Gd_2O_2S:Tb, and ZnS:Ag, respectively. All these phosphors have a quantum yield around 95% for an excitation wavelength around 262 nm.

Table 9.1 lists some examples of quantum yields in hybrid materials lacking metal activator ions. For siloxane-based organic–inorganic hybrids incorporating lanthanide ions (e.g. Ce^{3+} and Eu^{3+}) absolute emission quantum yields about 5–11% have been reported.

9.4.4
Recombination Mechanisms and Nature of the Emitting Centers

For semiconductors, a number of different phenomena may be associated with radiative emission. The most common phenomena are excitons and donor–acceptor (D–A) pairs. Excitons are the elementary excitation of a solid. When electrons absorb energy and are excited from the valence to the conduction band, a hole is created in the valence band. This Coulomb correlated electron/hole pair is so-called exciton. A donor-acceptor pair can be seen as the simplest defect combination. The presence of impurities with electrons excess (D_0) or electron deficiency (A_0) may recombine. These two transition types are schematized in Fig. 9.15.

The distinction between these two types of transitions can be made through experimental photoluminescence measurements, namely through the analysis of the

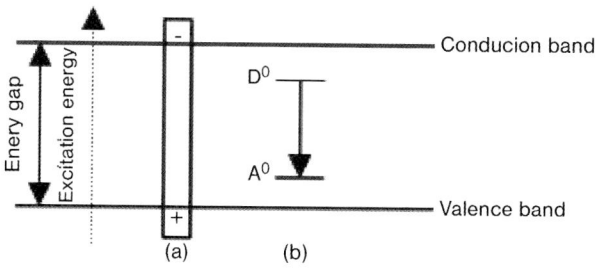

Fig. 9.15 Scheme with two types of transitions in a semiconductor after excitation into the conduction band: (a) excitonic transition, and (b) D-A pair transition.

behavior of the emission intensity as the excitation power is varied. The emission intensity, I, depends on the excitation power, L, according to the power law: $I \propto L^k$. When $1 < k < 2$ we are in the presence of exciton-like transitions and $k \leq 1$ is characteristic of D–A pairs. Deviations from this behavior are observed when L is varied more than two orders of magnitude or when the excitation wavelength energy is selected as resonant with the semiconductor band gap. For higher power excitation, we should expect a decrease in the k value, resulting from the fact that distant pairs that have a smaller transition probability are saturate and cannot accommodate more carriers. Another consequence of the lower transition probability of more distant pairs is the decrease in the steady-state emission intensity. Consequently, if it is possible to control the distance between the donor and the acceptor, the D–A pair behavior can be also evidenced.

Further experimental evidences of D–A pair transitions can be inferred from time-resolved spectroscopy. For longer delay times, it is expected a red shift in the spectrum because distant pairs have a smaller recombination probability. The respective energy levels have therefore larger lifetimes, a situation that is favored with the increasing delay time. As a consequence, the recombination takes place at lower energies.

Organic–inorganic hybrids can be treated as semiconductor materials as the energy gap is around 2.5–4.0 eV (Fig. 9.5(a)). Recalling Fig. 9.5(c) and Fig. 9.6, it is possible to discern an emission red-shift as the starting delay increases, which reveals typically D–A behavior. This time-resolved analysis can be complemented through the measurement of the decays curves along the emission band. By increasing the monitoring wavelength, the decays curves revealed a longer decay constant, resulting from the fact that more distant pairs, recombine slowly at longer wavelengths. In addition, for amine crosslinked hybrids it was found that the emission intensity versus low-intensity excitation power is approximately proportional to the excitation power, with a k value close to the unit, suggesting a recombination mechanism typical of D–A pairs.

Besides the identification of the emission recombination mechanisms, it is also important to tentatively identify the donor and acceptor chemical species. To unambiguously perform this assignment, results from distinct techniques should be joined due to the high complexity degree of the local structure of organic–inorganic hybrids. Electron paramagnetic resonance (EPR) spectroscopy is an experimental technique largely applied on the study of the electronic structure of defects in semiconductors enabling the direct investigation of the existence of paramagnetic centers. In a general way, the EPR technique may be applied whenever a system has unpaired electrons, such as the case of D–A pairs. With this technique, we can obtain information concerning the total angular momentum and the local symmetry of point defects. Moreover, it can also give precious insight into the chemical nature of centers, with a great deal of information provided by nuclear hyperfine interactions, which represents a clear chemical fingerprint of the atoms present in the center. For further information on this advanced topic, readers are referred to Pankove 1990, Thorpe 1997, and Sing 2003.

Table 9.8 lists some examples of organic–inorganic hybrids in which the chemical species that induces the photoluminescent features reported was tentatively identified.

9.4.5
Lanthanide-doped Hybrids

In recent years a significant amount of research has been focused on the possibility of using hybrid frameworks incorporating rare earth ions as matrices in optical active devices. The rare earth elements comprise the elements with atomic numbers 57, lanthanum (La), through 71, lutetium (Lu). Yttrium (Y) and scandium (Sc) are sometimes included in the group of rare earth elements and according to the IUPAC (International Union of Pure and Applied Chemistry) definition the lanthanide series encompass the elements from Ce through Lu. The lanthanides usually exist as trivalent cations with an electronic configuration corresponding to $[Xe]\,4f^n$, n varying from 1 (Ce^{3+}) to 14 (Lu^{3+}). The transitions between the energy levels of the $4f$ orbital are responsible for the interesting photophysical properties of the lanthanide ions, such as the long-lived luminescence (excited state lifetimes in the micro to millisecond range) and the atomic-like absorption and emission lines. The $4f$ electrons are indeed inner electrons and the $4f$ orbital is shielded from the interaction with the surroundings (called Ligand-field interaction) by the filled $5s^2$ and $5p^6$ orbitals. Although weak, the influence of the host on the optical transitions within the $4f^n$ configuration is essential to explain those interesting spectroscopic features.

The energy levels of the $4f$ electrons can be determined with good accuracy from a sum of the free ion interaction with the ligand-field Hamiltonian. The free ion part is composed by the central field with several other interactions, which are generally treated as perturbations. Among these interactions the interelectronic Coulomb repulsion and the spin-orbit interaction are the most relevant. The ligand field distorts the $4f$ shell removing, to a certain degree which depends on the local symmetry around the metal ion, the M_J degeneracy of the free ion $4f$ levels (the well-known Stark effect). A schematic representation of the intra-atomic and ligand-field interactions (including their typical order of magnitude) is presented

9.4 Hybrids for Light-emitting and Electro-optic Purposes

Table 9.8 Characteristics of the major optical active defects observed in organic–inorganic hybrids.

Hybrid	Possible Defect	Absorption or Excitation Peak (eV)	Emission Peak (eV)	Emission Decay Time (ms)/Temperature (K)	EPR Signal
di-ureasil (1)	•O—O—Si ≡ (CO_2) NH_2^+/NH^-	4.13–3.10 4.13–2.81	2.65–3.10 2.50–2.6	3–4/14 $1-5 \times 10^{-6}/300$ 150–160/14 $10-50 \times 10^{-6}/300$	$g_\parallel = 2.035 \pm 0.007$ $g_\perp = 2.018 \pm 0.007$ —
TEOS and methyltriethoxysilane with acetylacetone and ZPO (2)	≡Si—O—O—Si≡ or O_2^-	3.54–3.10	2.81–2.93	$3-3.7 \times 10^{-3}/300$	g = 2.033, under irradiation of 365 nm (2)
APTES,TEOS and acetic acid (heat treatment 20–250 °C) (3)	NH_3^+/ CH_3COO^-	4.54–2.91	2.81–2.54	—	—
TEOS and TMOS with carboxylic acids (heat treatment >400 °C) (4)	thermal decomposition of the carboxylic acid, carbon impurity O—C—O and/or —Si—C—	6.20–3.10	2.25–3.10	—	—

(1) Carlos et al. J. Chem. Phys. B 108, 14924 (2004); (2) Cordoncillo et al. Opt. Mater. 18, 309 (2001); (3) Han et al. Mater. Lett. 54, 389 (2002); (4) Green et al. Science 276, 1826 (1997).

in Scheme 9.2. For further information on this advanced topic, readers are referred to Hüfner, 1978 and Newman, 2000.

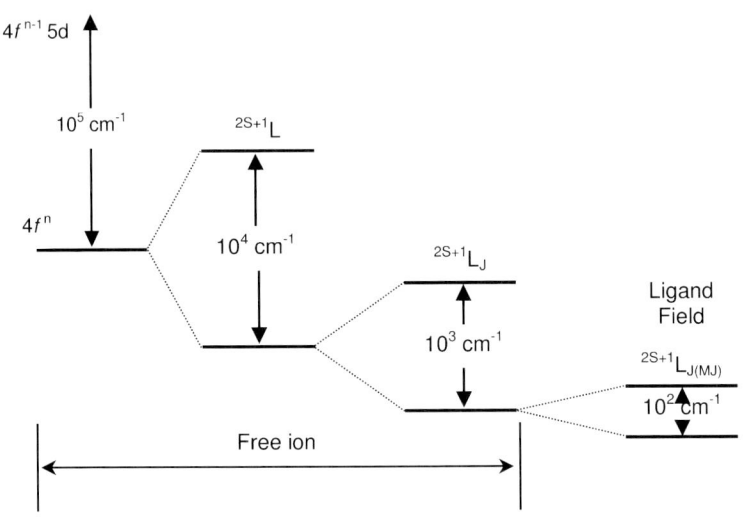

Scheme 9.2 Schematic representation of the free ion and ligand field interactions. The energy levels are labeled according to the well-known Russel–Saunders coupling scheme where L is the total orbital angular momentum, S the total spin angular momentum and J the total angular momentum, J ($M_i = -J, -J + 1, \ldots, +J$) must satisfy the condition $|L - S| \leq J \leq L + S$. $J = L + S$.

The direct photoexcitation of lanthanide ions is not very efficient due to the ions' poor ability to absorb light, a consequence of the forbidden nature of the f–f transitions (Laporte's selection rule). This can be overcome by using energy transfer from organic chromophores (for example) to lanthanide ions. The energy absorbed by such chromophores can be transferred to nearby lanthanide ions, which in turn undergo the corresponding radiative emitting process (the so-called sensitization of lanthanide luminescence or "antenna effect"). Lanthanide ions can form complexes with various organic molecules, such as β-diketonates, aromatic carboxylic acids, and heterocyclic ligands (see some examples in Table 9.2), and these materials are of great interest for a wide range of photonic applications such as tuneable lasers, amplifiers for optical communications, components of the emitter layers in multilayer OLEDs and efficient light conversion molecular devices (LCMDs). Recently, these LCMDs have found a series of useful applications, such as luminescent labels in advanced time-resolved fluoroimmunoassays, light concentrators for photovoltaic devices and antenna in photosensitive bioinorganic compounds.

Despite lanthanide complexes being characterized by a highly efficient light emission under UV excitation (some of them even exhibit laser action in solution), their low thermal and photochemical stability and the poor mechanical properties are important disadvantages concerning their technological applicability as tuneable solid-state lasers or phosphor devices. Moreover, most of those complexes are usually isolated as hydrates in which two or three water molecules are included in the first coordination sphere of the central ion, which quenches emission due to activation of nonradiative decay paths.

Among these generally undesirable properties, another serious drawback is the degradation of most of the lanthanide-based complexes under prolonged (in certain cases, only few hours) ultraviolet irradiation (which decreases luminescence intensity). A typical example is the family of lanthanide β-diketonates chelates. Whilst the origin of such degradation has not been yet completely understood, it is often attributed to photobleaching. The phenomenon of photobleaching (also commonly referred to as fading) occurs when the efficiency of the emission decreases due to photon-induced chemical damage and covalent modification of the complex. This mechanism can be used to assign the degree of photostability of the complex, and it has been used to develop lanthanide-thin-film based dosimeters with high sensitivity and selectivity to the three main UV regions related to skin damage effects: UV-A (365 nm), UV-B (315 nm), UV-C (290 nm).

One of the strategies adopted in the last few years to simultaneously improve the thermal stability, the mechanical features and the light emission properties of lanthanide complexes is their encapsulation into sol–gel derived organic–inorganic hybrids (particularly siloxane-based ones) through the simple embedding of the complexes within the sol–gel matrix, or through the use of complexing ligands covalently grafted in situ to the hybrid skeleton (see the examples of Table 9.2). Furthermore, this encapsulation into organic–inorganic hybrids may also contribute to decrease the rapid degradation of the complexes under UV exposure. The major interest of this approach is the possibility of preparing multifunctional nanoscale hybrid materials with tuneable design and suitable photonic features using the processing advantages of sol–gel matrices discussed in the introduction (e.g. flexible and versatile shaping and patterning, good mechanical integrity and excellent optical quality).

These organic–inorganic hybrids exhibit improved luminescence properties, with respect to the simple incorporation of lanthanide ions in silica-based matrices, essentially due to following factors:

- Better dispersion of the incorporated lanthanide ions within the matrix avoiding clustering and allowing larger concentrations of emitting centers;
- The protection of the central lanthanide ions by the ligands circumvent the quenching of the emission originating from residual water, silanol groups and dopant clustering, thus decreasing the nonradiative decay pathways.
- The encapsulation of the metal ions by light-absorbing ligand-cage type structures can give rise to efficient intramolecular and intermolecular ligand-to-metal energy transfer.

Figures 16a and 16b display the emission spectra of some organic–inorganic hybrids doped with Eu^{3+}-, Tb^{3+}-, Nd^{3+}- and Er^{3+}-based complexes that are representative of the structures depicted in Table 9.2. The spectra display a series of intra-$4f$ transitions between the energy levels schematically presented in the Figures. There is no emission either from the hybrid host or the ligand excited sates, indicating efficient energy conversion from the hybrid emitting levels to the ligands and/or to the lanthanide ions. However, organic–inorganic hybrids incorporating lanthanide salts present usually an emission spectra characterized by a broad white-light emission associated to the hybrid host overlapped to the typical intra-$4f$ transitions in the green (Tb^{3+}), red (Eu^{3+}) and infrared (Nd^{3+}, Er^{3+}, and Tm^{3+})

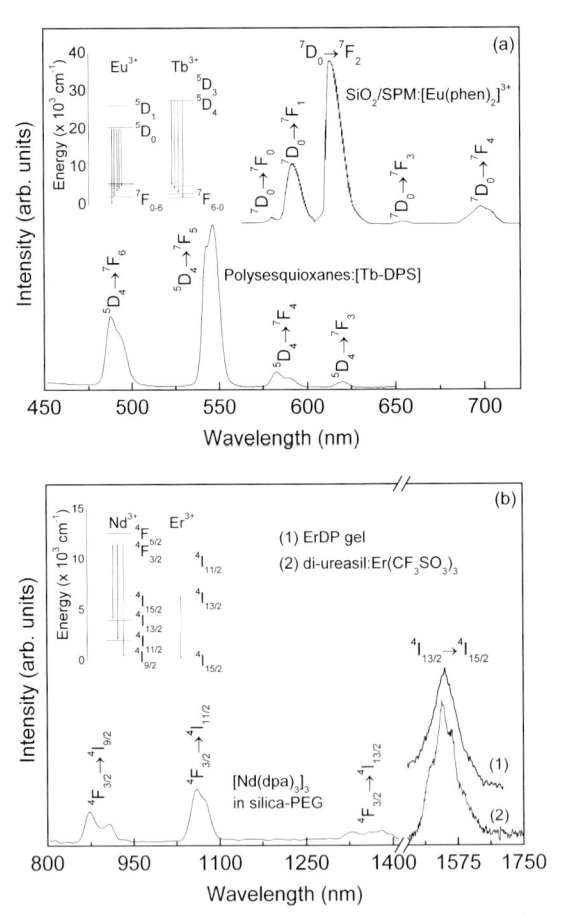

Fig. 9.16 Room-temperature emission spectra of (a) $SiO_2/SPM:[Eu(phen)_2]^{3+}$, Polysesquioxane:[Tb-DPS] (Li et al., *Thin Solid Films* **2002**, *385*, 205), (b) di-ureasil modified by $Er(CF_3SO_3)_3$ (Nunes et al., *MRS Symposium Proceedings Series*, Vol. 847, EE 13.31.1.) ErDPD gel (Sun et al., *Adv. Funct. Mat.* **2005**, *15*, 1041), and $[Nd(dpa)_3]^{3-}$ (Driesen et al., *Chem. Mater.* **2004**, *16*, 1531) in a silica-PEG excited under 300, 269, 488, 425 and 580 nm, respectively.

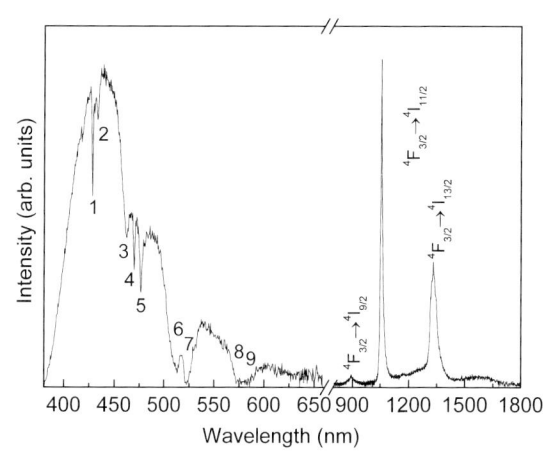

Fig. 9.17 Room-temperature emission spectra of di-urethanesil modified by $Nd(CF_3SO_3)_3$. (Gonçalves *et al.*, *J Phys. Chem. B* **2005**, *109*, 20093) The visible and NIR emission were acquired at an excitation wavelength around 365 and 514 nm, respectively. 1–9 denotes the intra-$4f^3$ self-absorptions: $^4I_{9/2} \rightarrow$ $^2P_{11/2}$, $^2D_{5/2}$, $^2G_{9/2}$, $^4G_{11/2}$, $^2K_{15/2}$, $^4G_{7/2}$, $^2K_{13/2}$, $^4G_{7/2}$, $^2G_{7/2}$, respectively.

regions. The relative intensity between these two emissions and therefore the emitted color may be readily tuned across the CIE chromaticity diagram by varying the amount of the lanthanide salt and the excitation energy. Figure 9.17 illustrates this aspect for a particular organic–inorganic hybrid for which the energy transfer processes between the hybrid host emitting centers and the lanthanide ions are much less efficient.

The atomic-like 4*f* transitions are also often used to get further information about the local environment around the metal ions. This is particularly evident for Eu^{3+} which is a powerful local ion probe. Thus, changes in the number of Stark components of each intra-4*f* manifold, variations in the relative intensity between them, differences observed in the energy of particular lines (induced by variations on the excitation wavelength and the temperature), and the analysis of the excited state decay curves, furnish important information about the local coordination of the lanthanide ions such as:

- Existence of more than one local site.
- Number of coordinated water molecules.
- Local symmetry group.
- Magnitude of the ligand field.
- Importance of the covalency effects.

9.4.6
Solid-state Dye-lasers

Solid-state lasers (acronym from Light Amplification by Stimulated Emission of Radiation) require stimulated emission of radiation, a concept slightly different

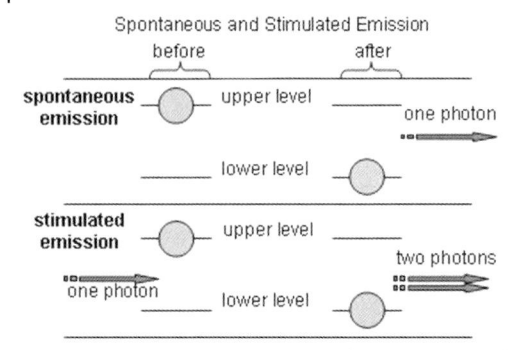

Fig. 9.18 Schematic representation of spontaneous and stimulated emission.

from spontaneous emission discussed before, Fig. 9.18. Spontaneous emission may occur when the electron in the excited state returns to the ground level. If while the atom is in the excited state, the material is illuminated with a photon with the same energy as that corresponding to the difference between the ground and the upper state the atom may be stimulated by the incoming photon to return to the ground state and simultaneously emit a photon at that same transition energy. A single photon interacting with an excited atom can therefore result in two photons being emitted. This process is known as stimulated emission. The main problem responsible for the non-occurrence of stimulated emission is that under normal conditions of thermodynamic equilibrium the number of atoms in the ground state is much higher than that in the upper excited state. Thus, when a photon illuminates the sample, the probability of occurrence of absorption is much higher and the stimulated emission will be insignificant when compared to spontaneous emission. To promote the stimulated emission occurrence it is therefore necessary to have more atoms in the upper excited states. This situation is termed as population inversion. If population inversion is asserted, the stimulated emission may dominate resulting in a cascade of photons with the corresponding amplification of the emitted light. The stimulated radiation is classified as spatial and temporal coherent due to the unique properties of the stimulated photons: they are in phase, have the same wavelength and travel in the same direction of the incident photon.

An important factor to produce stimulated emission is the lifetime of the excited level. The spontaneous emission probability is inversely proportional to the lifetime of the excited level, which means that materials with smaller lifetimes have greater probability that spontaneous emission will occur. For spontaneous lifetimes of the order of microseconds to a few milliseconds, the probability that a spontaneous transition will take place can be considered relatively low. Thus, the conditions which favor stimulated emission are enhanced.

The room temperature lifetime values of the emitting states of organic–inorganic hybrids is typically of the order of 10–100 ns, which makes the spontaneous emission probability much higher than that of stimulated one. Thus, to achieve

Table 9.9 Examples of active laser dyes molecular structure. The excitation and emission wavelengths were measured in methanol solution.

Chromophor	Excitation Peak (nm)	Fluorescence Peak (nm)
IR1051 (see Table 3)	900	800
IR5 (see Table 3)	1100	800
Coumarine 153 (see Table 3)	390	540
Rhodamine 6G (see Table 3)	510	590
Rhodamine B	567	610

room temperature laser emission using organic–inorganic hybrids it is necessary to modify them by the incorporation of laser active organic species (up to now essentially organic dyes). Examples of laser dyes molecules that can be successfully incorporated into organic–inorganic dyes are listed in Table 9.9. As examples of laser emission in dye-doped organic–inorganic hybrids it has been shown that Coumarine 153 dye incorporated into di-ureasil hybrid and titania films with Rhodamine dye (Table 9.3) have demonstrate efficient laser emission.

9.5
Hybrids for Photochromic and Photovoltaic Devices

Photochromism is defined as the reversible photocoloration of a single chemical species between two states having distinctly different absorption spectra, that results from the action of electromagnetic radiation in at least one direction. Similarly to many other photochemical reactions, photochromism corresponds to a photoequilibrium and, thus, the photoproduct does not appear as a separate phase. Photochromism displays, however, a unique feature that distinguishes it from

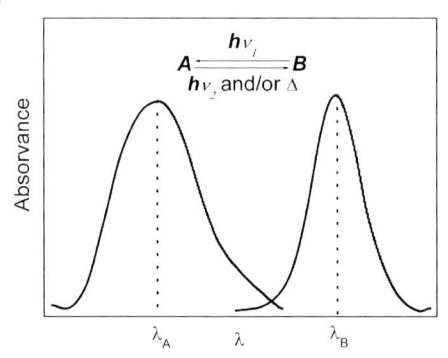

Fig. 9.19 The photochromic process undergone by compound A under the irradiation of light to yield product B (the reverse reaction can occur thermally or photochemically).

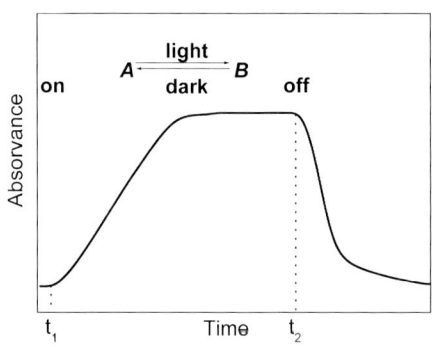

Fig. 9.20 Typical pattern for the absorbance prior, during and after irradiation of compound A at a fixed wavelength.

general photochemical reactions: reversibility. Most photochromic systems are based on unimolecular reactions, although bimolecular reversible processes are not excluded.

Figure 9.19 illustrates the photochromic process undergone by compound A under the irradiation of light to yield product B and shows that the reverse reaction can occur thermally or photochemically. Figure 9.20 reproduces the typical pattern for the absorbance prior, during and after irradiation of compound A at a fixed wavelength.

Photochromic systems can be classified into several groups:
- Photoreversible systems, in which the initial compound A absorbs at a shorter wavelength than product B (Fig. 9.19). These are further distinguished on the basis of the type of back reaction of the colored product B: (a) it undergoes a light-induced reaction to give rise to A, (b) it reverts thermally to form A and (c) it goes back to A both thermally and photochemically.

- Reversible photochromic systems, in which compound A absorbs at longer wavelengths than product B (B < A).
- Multireversible photochromic systems, in which more than one chromophore is present. In this case it is not always clear if one or more chromophores are active.

Classical photochromic glasses have been limited by the fact that only a few dopants can withstand the high temperatures of glass melting. In contrast, organic photochromism is considerably more interesting, since tens of thousands of molecules whose applicability may be tailored – through the shade of color change, the direction of the photochromism, the rate of change in color intensity, reversibility, or unidirectionality – are available.

Photochromic-doped sol–gel silica (SiO_2) gel-glasses can be synthesized by adding a photochromic molecule, such as spiropyrane dyes and its derivatives, to a polymerizing solution of alkoxysilane monomers. A common observation for all the photochromic compounds used has been that photochromism (pm) changes gradually to reversed-photochromism (rpm) along the gel–glass formation (the stable forms in the dark are the colored ones, which can be bleached then by UV irradiation). At the first stage of the sol–gel reaction, the photochromic molecule is dissolved in the starting liquid solution (pm behavior). As the reaction proceeds, this situation gradually changes during the formation of the wet gel surface environment to an rpm behavior. During the gel drying stage, as a result of UV irradiation, the molecular isomerizations and color change become restricted in the new solid environment (dry xerogel). This situation may be attributed to two factors: (1) the effective free volume diminishes, thus suppressing molecular rotation; (2) the photochromic molecule is stabilized by means of strong hydrogen bonding to the Si—OH groups of the SiO_2 cage. Therefore, the photochromic silica gel–glasses suffer from two problems: the photochromism is reversed (rpm) and, even more limiting, the photochromism stops at the final dry xerogel stage and obviously no practical applications can be envisaged.

The application of pore surface variations in the inner SiO_2 cage of the SiO_2 gel–glasses which contains the photochromic trapped molecules may solve these problems. The introduction of an organic phase on the SiO_2 surface through the use of additives (second phase) or by producing an organically modified ceramics may offer the key for a successful and adequate opportunity to fabricate novel photochromic stable materials for practical applications. For instance, in organically modified sol–gel glasses obtained from silane–ethyl monomers (SE gel–glasses, Table 9.3) and doped with spiropyrane-type photochromic dyes (derivatives of 6-nitro-1′,3′,3′-trimethylspirol[2H-1-benzopyran-2,2′-indoline], abbreviated 6-nitro-BIPS), the apolar cage surface composed of Si—CH_2CH_3 groups does not stabilize the colored form of the trapped molecules and, hence, normal photochromism results.

In addition, the low degree of crosslinking of this monomer in the final xerogel (only three of the Si bonds participate in the polymerization, compared to four in the pure SiO_2 SE gel–glasses) provides the grade of flexibility that is needed to ensure molecular rearrangements and rotations. Therefore, the direction of photochromism (pm or rpm) is controllable, allowing the photochromic properties to remain in the final material with good chemical stability.

An important feature regarding long-term technological or industrial applications of the photochromic materials is their thermal (chemical) and photochemical stability (photobleaching). As noted above, the photochromic molecules are matrix environmental (cage composition) sensitive. Consequently, considerable effort in the preparation of new photochromic materials based on hybrid matrices, using different photochromic compounds, is made on optimize the preparation mechanism to obtain better products of superior stability, i.e, with improved photochromic response, as well as fading rate, while keeping low levels of photofatigue and a high color stability.

These concepts have been used to fabricate a thermochromic memory cell for optical information storage media. Coloration can be made either by exposure to UV light or by heat and writing (selective decoloration) by irradiation with an adequate laser.

Photovoltaics is the direct conversion of light into electricity at the atomic level. Some materials exhibit a property known as the photoelectric effect that causes them to absorb photons of light and release electrons. When these free electrons are captured, an electric current results that can be used as electricity.

The diagram of Fig. 9.21 illustrates the operation of a basic photovoltaic cell, also called solar cell. Solar cells are made of the same semiconductor materials that are extensively used in microelectronics. In solar cells, a thin semiconductor wafer is specially treated to form an electric field, positive on one side and negative on the other. When light energy strikes the solar cell, electrons are knocked loose from the atoms in the semiconductor material. If electrical conductors are attached to the positive and negative sides, forming an electrical circuit, the electrons can be captured in the form of an electric current.

The most common photovoltaic devices employ a single junction, or interface, to create an electric field within a semiconductor such as a photovoltaic cell. In a single-junction photovoltaic cell, only photons whose energy is equal to or greater than the band gap of the cell material can free an electron for an electric circuit. Thus, the photovoltaic response of single-junction cells is limited to the portion

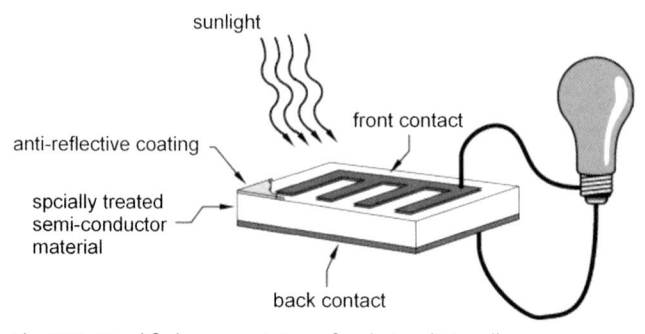

Fig. 9.21 Simplified representation of a photovoltaic cell.

of the sun's spectrum whose energy is above the band gap of the absorbing material.

One way to overcome this limitation is to use two (or more) different cells – with more than one band gap and more than one junction – to generate a voltage. These are named *multijunction* (or *cascade* or *tandem*) cells. Multijunction devices can achieve higher total conversion efficiency because they can convert more of the energy spectrum of light to electricity. Much of the research carried out at present on multijunction cells focuses on gallium arsenide as one (or all) of the component cells. Such cells have reached efficiencies of around 35% under concentrated sunlight. Other materials have been amorphous silicon and copper indium diselenide.

An important breakthrough in solar cell technology was the recent introduction of electrochemical dye-sensitized solar cells by Grätzel and co-workers at the beginning of the nineties. In this innovative solar cell, nanostructured titanium dioxide (TiO_2), a wide band gap oxide semiconductor, is made light-sensitive through the anchoration of a sensitizer, i.e. a monolayer of an organic dye, at its surface. The nanoporous morphology of the oxide (high surface to volume ratio) is a prerequisite to adsorb larger amounts of the dye ensuring therefore an efficient absorption of the solar light. The sensitizer is often bound to the oxide surface via a spacer bearing carboxylic acid or phosphonic functionalities. Thus, with respect to multicrystalline silicon solar cells, where a layer thickness of at least 50 μm is required for total light absorption, with the strongly enlarged semiconductor surface, in Grätzel cells a nanometer thin sensitizer film is enough for total light capture.

A Grätzel solar cell (Fig. 9.22(a)) simulates some features of the natural solar cell, which enables photosynthesis to take place. In a natural solar cell the chlorophyll molecules absorb light (most strongly in the red and blue parts of the spectrum, leaving the green light to be reflected). The absorbed energy is sufficient to free an electron from the excited chlorophyll. In the further transport of electron, other molecules are involved, which take the electron away from chlorophyll. A Grätzel cell consists of an array of nanometer-sized crystallites of TiO_2 welded together and coated with light-sensitive molecules that can transfer electrons to the semiconductor particles when they absorb photons. The light-sensitive molecules, which play a role equivalent to chlorophyll in photosynthesis, are complexes of ruthenium and organic bipyridine ligands, which absorb light strongly in the visible range. The TiO_2 nanocrystals carry the received photo-excited electrons away from electron donors. As the donor molecule must receive back an electron, in order to absorb another photon, this assembly is immersed in a liquid electrolyte containing molecular species (dissolved iodine molecules) that can pick up an electron from an electrode immersed in the solution and ferry it to the donor molecule. These cells can convert sunlight with efficiency of 10% under direct sunlight and they are even more efficient under diffuse daylight (Fig. 22(b)).

Although these systems compete with the conventional solid-state devices in terms of light conversion efficiency, they still suffer from some limitations that have hindered their industrial development and commercialization:

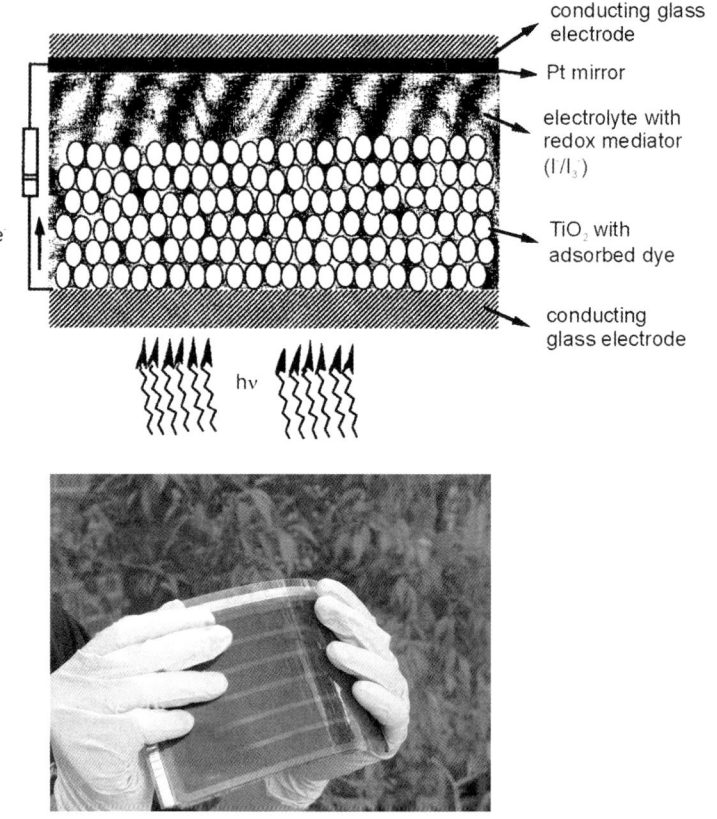

Fig. 9.22 (a) Schematic representation of a Grätzel cell. (Adapted from K. Kalyanasundaram and M. Grätzel, *Coord. Chem. Rev.* **1998**, *77*, 347). (b) Flexible dye-sensitized solar cell. (Taken from http://www.sta.com.au/)

- Non-optimized morphology of the nanoporous oxide layers;
- Low stability under UV light relating to the semiconducting features TiO_2;
- Weak stability of the oxide-sensitizer bridge (leading to the desorption of the dye with the concomitant drastic decrease of the cell efficiency);
- The presence of an organic liquid electrolyte in the most efficient devices (the main drawback). This liquid electrolyte is based on highly volatile solvents with low viscosity and is responsible for severe problems, such as degradation with time, cell sealing and handling.

New developments in this domain are directed towards the synthesis of structured new materials with desirable morphology, such as mesoporous channels or nanorods aligned perpendicular to the transparent conducting oxide glass. The use of an inorganic mesophase structure in a photovoltaic cell is extremely attractive, considering the presence of numerous accessible porous in the organized structure that can be optimized for enhancement of energy conversion efficiency and solar cell performance. The development of the covalent grafting of organic dyes leading to stable dye–oxide crosslinkages and the design of quasi-solid or solid electrolytes to replace the organic liquid ones are the other two expected improvements.

9.6
Hybrids for Integrated and Nonlinear Optics

The optical integrated circuit presents several advantages, when compared to those typical of electric integrated circuits, in particular, larger bandwidth, low-loss coupling, smaller size, weight and power consumption and it is immune to vibrations. The major disadvantage is related with the high cost involved in the developing and implementation of a new technology. One potential cost effective solution is the use of sol–gel derived organic–inorganic hybrids for the future optical integrated circuits. The combination in a single material of the organic and inorganic counterparts permits to overcome some of the disadvantages associated with the use of transparent inorganic glasses or organic polymers in optical applications. For instance, the low mechanical flexibility and high brittleness of inorganic glasses, together with their high-temperature processing requirements are not compatible with optoelectronic integration. On the other hand, polymers can exhibit relatively low heat and water resistance.

9.6.1
Planar Waveguides and Direct Writing

Passive IO devices split signals into more than one part and route them in previously set directions, enabling the interconnection between various devices in an optical integrated circuit. A simple representation of a waveguide can be done considering a three layer scheme, in which each layer in the middle has a smaller refractive index (n_1) than that of the outside layers (n_2), as presented in Fig. 9.23. The refractive (or refraction) index is the ratio between the speed of light in vacuum and the speed of light in a particular material medium. If we consider that the propagation occurs along the $\hat{z}$ direction and that the layers are confined only in the $\hat{x}$ direction, when the light enters in the layer with a certain angle higher than a critical angle value (θ_c), total internal reflection will occur, enabling the light guiding. The critical angle depends on the refractive indexes difference, according to:

$$\sin \theta_c = n_2/n_1 \tag{9}$$

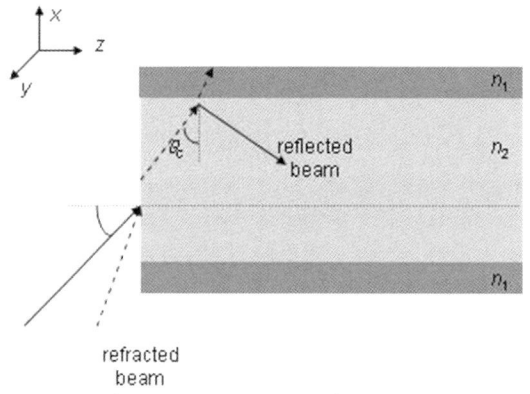

Fig. 9.23 Schematic representation of a planar waveguide structure.

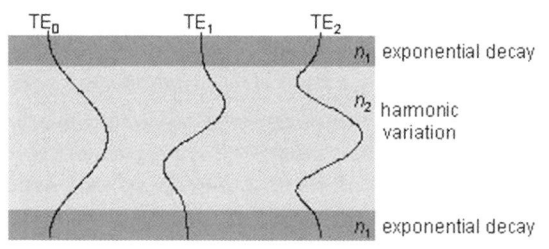

Fig. 9.24 Scheme of possible modes in a planar waveguide.

Upon increasing the confinement in the $\hat{x}$ direction the propagation of light cannot be described though the total internal reflection model, so that the light travels parallel to the middle layer boundaries.

The optical waveguides can be divided into planar, linear and cylindrical waveguides, in accordance with the geometry of the propagating region in the material. Another important property of optical waveguides is related with the number of allowed propagating modes. Considering that a propagating mode is a spatial distribution of electromagnetic energy in one or more dimensions, we should expect that a planar waveguide is multimode and the linear and cylindrical waveguides are monomodes. In a multimode waveguide the output pulse will be broader than the incoming one due to the different group velocity of each propagating mode. The different modes can be TE (electric transverse, with electric field in the x direction $E_y = 0$ and $E_z = 0$), TM (magnetic transverse, $H_y = 0$ and $H_z = 0$) and HE or HM that are hybrid modes ($E_y \neq 0$, $H_y \neq 0$). For a perfect cylindrical waveguide the TE and TM modes are degenerated and, thus cannot be distinguished. Figure 9.24 illustrates some TE modes possible to occur in a planar waveguide. The integer number used in the TE label in Fig. 9.24 indicates order of the allowed modes that depends on the refractive indexes of the interfaces, on the thickness of the layer and on the frequency of the propagating light. The order of

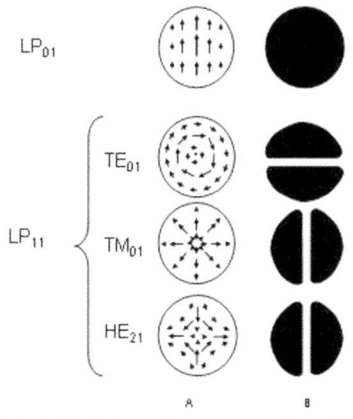

Fig. 9.25 Schematic representation of the LP_{01} and LP_{11} (TE_{01}, TM_{01}, and HE_{21}) propagating modes. (A) electric and magnetic fields vectors and (B) respective, light spot outside the waveguide.

each mode corresponds to the number of null intensity of the electric field within the channel. Another often used notation to identify the exact propagating modes is represented by the LP_{jm} notation, where LP refers to linear polarized mode and the indexes j and m are related with the mode number. Figure 9.25 illustrates different propagating modes and the respective notation.

In most applications, in particular those involving the propagation of high-debit signals, monomode waveguides are preferable to the multimode ones. In a monomode waveguide the signal degradation due to the broadening of the pulse is avoided, enhancing the quality of the output signal. Other factors, generally identified as the waveguide attenuation (or loss) may also contribute to degradation of the propagating information. The attenuation is usually related with scattering, absorption and radiation mechanisms. Those mechanisms depend on the material and are also function of the wavelength. The attenuation due to scattering is ascribed to the presence of chemical impurities or mechanical imperfections within the channel.

The absorption losses for waveguides based in organic–inorganic hybrid are an important issue and present the higher disadvantage for the commercial application of such materials. There is a larger attenuation in the NIR spectral region ascribed to both organic and inorganic counterparts, particularly due to the vibrations of organic groups such as CH, NH, CO and OH oscillators of silanol moieties. The attenuation in the ultraviolet/visible spectral region is also not negligible, as the absorption spectra of Fig. 9.5 illustrate.

Table 9.10 lists several organic–inorganic hybrids that have demonstrated their capability of being processed as passive optical waveguides, together with their refractive index and attenuation coefficient at 1310 and 1550 nm. The control over the refractive index of the films can be achieved, for instance, through zirconia and titania addition.

Table 9.10 Molecular structure of several organic–inorganic hybrids processed as waveguides and the respective refractive index and attenuation coefficient.

Hybrid	Non-hydrolysed Precursor	Refractive Index	Attenuation Coefficient (dB/cm)
ORMOCER®(1)	F-C₆H₄-Si(OCH₃)₃	1.418	<0.02@1310 nm 0.18@1550 nm
	(F-C₆H₄)₂Si(OMe)₂	1.450	<0.02@1310 nm 0.10@1550 nm
	CH₂=CH-C₆H₄(F)-Si(OCH₃)₃	1.504	0.24@1310 nm 0.32@1550 nm
	CH₂=CH-CH₂- / F-C₆H₄-Si(OCH₃)₃	1.436	0.18@1310 nm 0.24@1550 nm
	C₆H₅-Si(OCH₃)₃	1.473	0.14@1310 nm 0.36@1550 nm
	(C₆H₅)₂Si(OMe)₂	1.545	0.18@1310 nm 0.36@1550 nm
	CH₂=CH-C₆H₄-Si(OCH₃)₃	1.505	0.22@1310 nm 0.46@1550 nm
	CH₂=CH-C₆H₄-Si(OCH₃)₃	1.492	0.44@1310 nm 0.44@1550 nm
	di-ureasil + ZPO + MAA (see Table 2) **(2)**	1.5212–1.5246 1.5050–1.5090	0.4–0.6@632.8 nm 0.6–1.1@1550 nm
	MAPTMS modified by ZPO and MAA (see Table 2) **(3)**	1.42	0.13@1310 nm

(1) Kahlenberg *et al. Mater. Res. Soc. Symp. Proc.* 847, EE14.4.1 (2005); (2) Molina *et al. J. Mater. Chem.* 15, 3937 (2005); (3) Najafi *et al. J. Lightwave Technol.* 16, 1640 (1998).

Active waveguides may be used as basic components, such as lasers and amplifiers, in the field of all-optical telecommunications technology. The recent explosive increase in the traffic of optical communications has spurred research efforts to develop highly efficient broad-band fiber optical amplifiers working in the low-loss windows of commercial silica-based transmission fibers located in the NIR region around 800, 1310 and 1550 nm, I, II and III telecommunication transmission windows, respectively. The optical features of organic–inorganic hybrids with technological potential for IO must include low attenuation coefficients and optically active centers with efficient emission lines around such wavelength regions. For instance, the NIR emission lines of Nd^{3+} and Er^{3+} (see Fig. 9.16) lie within the wavelength interval of the three mentioned telecommunication transmission windows.

The performance of lanthanide emission for amplification has been demonstrated to enhance with higher doping levels. However, emission quenching due to concentration effects (energy transfer between close ions followed by nonradiative deactivation in some recombination centers, lattice defects, for instance) is often observed. Thus the amount of lanthanide ions incorporated into the hybrid matrix should be optimized avoiding therefore these effects. Table 9.3 contains examples of active waveguides based on lanthanide-doped organic–inorganic hybrids, in particular organically modified ceramics and MHTEOS/SiO_2 hybrids doped with Er^{3+} ions.

Usually, functional IO devices require radiation confinement in the two orthogonal directions transverse to the waveguide axis of propagation, and consequently some form of channel waveguide structure is necessary (instead of a simpler planar waveguide configuration). Thus, it is necessary, to write closely-spaced channel waveguides in the material. The first step involves the selection of substrate on which the film will be deposited. Such choice is determined by the desired final application for the final waveguide and also by the refractive index value that should be higher than that of the film. Table 9.11 lists the most conventional substrates and the respective refractive index.

Other constrains are also related with the adherence between the substrate surface and the film. If the material contains a photopolymerizable chemical species

Table 9.11 Refractive index and respective wavelength range of materials for waveguide substrates.

Material	Wavelength (nm)		
	543.5	632.8	1550
Silicon	3.5	3.5	3.5
Soda-lime glass	1.5127	1.5090	1.497
Borosilicate glass	1.47	–	–
Silica glass	1.4603	1.4575	1.445

the channel can be written through direct ultraviolet patterning. The film is covered by an amplitude or phase mask and when the film is irradiated with ultraviolet light the channels are directly written. If the film does not contain a photopolymerizable organic group, an amplitude mask should be used and laser radiation will dig channels. Another method often used involves the chemical etching of the films.

The film is covered with a photoresist material and an amplitude printing mask. In the presence of ultraviolet light (or X-rays) the mask pattern is written on the photoresist. The exposed photoresist materials is removed and the etch is performed on that regions not protected by the photoresist. After the achievement of a waveguide structure, a replication tool for the subsequent molding process can be achieved by replication of the photoresist structures into a thermoplastic ultraviolet transparent polymer (for instance, polydimethylsiloxane) on a glass substrate. Figure 9.26 illustrates the process of pattern waveguides through ultraviolet direct-patterning method.

Due to the flexible processing methods, the final waveguides can be used in various optical devices. Figure 9.27 shows several examples for ORMOCER® (trademark of the Fraunhofer-Gesellschaft zur Förderung der angewandten Forschung e.V. München) organic–inorganic hybrids.

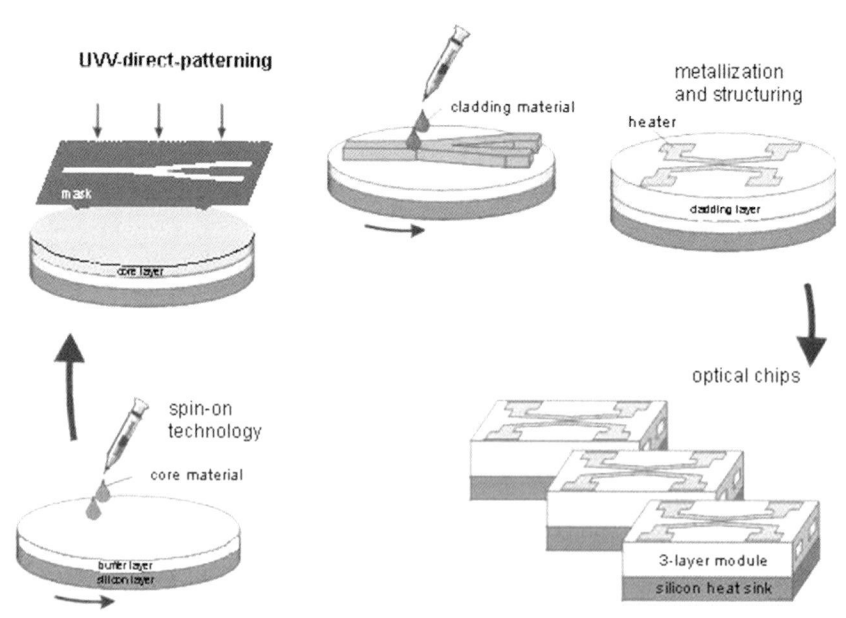

Fig. 9.26 Ultraviolet direct-patterning for integrated optics. (Adapted from R. Buestrich et al. Sol-Gel Sci. Technol. **2001**, 20, 181).

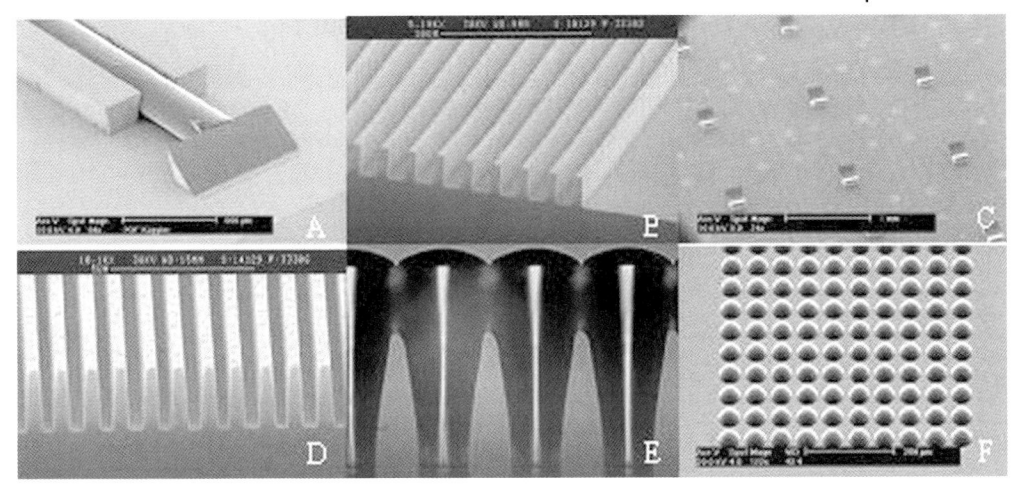

Fig. 9.27 Micro-optical elements (lenses, lens arrays, gratings, prisms) fabricated by replication, lithography, or a combination of both. (A) Deflection prism (B) UV-molded high aspect ratio grating in ORMOCER®, (C) Replicated prisms on detector chip wafers, (D) UV-moulded grating, (E) Side view of 750 μm high ORMOCER lenses® on a GaAs wafer, (F) Replicated 10 × 10 micro lens array. (Taken from www.microresist.de)

9.6.2
Nonlinear Optics

Many applications in optics and photonics involving light detection and modulation are based on nonlinear processes. The nonlinear properties are only exhibited for high intensity external electric fields ($\vec{E}$) that induce a charge displacement within the material. For low intensity electric fields, the polarization vector ($\vec{P}$) induced is proportional to the electric field:

$$\vec{P} = \varepsilon_0 \chi \vec{E} \tag{10}$$

where ε_0 is the permittivity in vacuum and χ is constant without dimensions termed as electrical susceptibility of the medium. For high intensity external fields, the polarization vector is no longer proportional to the electric field. Assuming an isotropic medium, where $\vec{E}$ and $\vec{P}$ have the same direction, the polarization vector ($\vec{P}$) can be described by a power series.

$$\vec{P} = \varepsilon_0 \cdot \left(\chi^{(1)} \vec{E} + \chi^{(2)} \vec{E}\vec{E} + \chi^{(3)} \vec{E}\vec{E}\vec{E} + \dots \right) \tag{11}$$

where $\chi^{(i)}$ ($i = 1, 2, 3, \dots$) is the ith order of the electrical susceptibility. To account for the polarization effects, $\chi^{(i)}$ is a tensor of rank ($i + 1$). For low intensity electric fields, the high order terms of the susceptibility ($\chi^{(i)}$, $I = 2, 3, \dots$) are negligible and the material exhibit linear optical properties. For high intensity electric fields, those terms are no more negligible and will be responsible for the nonlinear optical properties. The second order susceptibility is responsible for the second order nonlinear effects (e.g. Pockels, second harmonic generation). In materials whose structure is characterized by an inversion center, this term is zero. The third order susceptibility is responsible for the third order nonlinear effects: Kerr, Raman, four wave mixing, two-photons absorption and optical limiting power. High order susceptibilities are usually ignored. For further information on this advanced topic, readers are referred to Guenther 1997.

- The Pockels and the Kerr effects are responsible for the refractive index changes (Δn) with the applied electrical field,

$$n(E) - n = -\frac{1}{2} r_{ij} n^3 E + \frac{1}{2} s_{i'j'} n^3 E^2 \qquad (12)$$

where n is the refractive index for low intensity electric fields, r_{ij} and s_{ij} are tensors of rank 3 representing the electro optical Kerr and Pockels coefficients, respectively. These tensors contain information about the direction of the polarization of the output wave with respect to the polarization and propagation direction of the light incident. These nonlinear effects are used to produce devices such as electro optical modulators. For crystals, r_{ij} and s_{ij} have typical values of $10^{-12} - 10^{-10}$ m/V and $10^{-18} - 10^{-14}$ m²/V², respectively. The nonlinear refractive index due to Kerr effect is usually designated by n_2.

- The Raman effect results from the molecular vibrations originated by changes in the polarization of the medium due to the high intensity of the electrical field. This phenomenon may be used to optical amplification.
- At high laser intensities it is possible that the absorption of two photons produces only one hole-electron pair. This is the most probable non linear effect in light absorption. The change in intensity (a) of a beam as it propagates through a medium with linear, and two-photon absorption, is given by $-(\alpha I + \beta I^2)$, where α and β are the linear and two photons absorption coefficients, respectively. After two-photon absorption occurs the medium may emit light with twice the frequency (half the wavelength) of the incident light. This phenomenon is called as second harmonic generation and it is characterized by the d_{lm} parameter which is a tensor of rank 3, which contain information about the direction of the polarization of the output wave with respect to the polarization and propagation direction of the light incident.
- Optical limiters are smart devices in which the efficiency of signal limiting increases with the signal level and should be able to absorb, scatter or refract incident light via nonlinear effects.

Nowadays, the most common material for electro-optical devices with second order nonlinear optical properties is the inorganic crystal LiNbO$_3$ with an electro-optical coefficient (r_{33} = 30.8 pm V^{-1} and d_{33} = 60 pm V^{-1}) As alternative cost effective solutions, several organic chromophores have been proposed (Table 9.12).

The poling of the chromophores (Fig. 9.28) can be done by an electric field, but the ordered state of the molecules obtained by this process normally decays to an equilibrium isotropic state when the electric field is removed. Other drawbacks usually ascribed to organic chomophores are poor thermal and chemical stability, low glass transition temperature and high optical loss which enable them to be used as optical devices. To overcome such problems, the incorporation of the non linear molecules into different matrix is a good solution. The matrix may be an inorganic network (e.g. silica or silica-titania), an organic–inorganic hybrid or chromophores specifically functionalized with alkoxy groups that form the hybrid network itself though hydrolyzes and condensation reactions. The first materials investigated are related with the physical entrapped of the organic molecule in the host (class I hybrid). Examples of such materials and the respective electro-optic coefficient are listed in Table 9.12. Although good optical poling of the nonlinear molecules has been achieved, fast relaxation to a random state is achieved after the electric field is removed. It has been demonstrated that thermal treatments delay the structural relaxation. However, the poor thermal stability of the chromophore limits its use.

The incorporation via covalent bonds of the chromophores into perfectly ordered noncentrosymmetric lattices (class II hybrids), may induce permanent molecular alignment, contributing to enhancement of the electro-optic coefficient. Moreover, the matrix may also enhance their thermal and chemical stability. It has been demonstrate that the number of bonding sites and geometries affects the value of the electro-optics coefficient. The best results were achieved for two anchor sites in the vertical position with respect to the main chain. The data available in the literature indicate that the poling is quite efficient in organic–inorganic hybrids, with electro coefficients as high as 150 pm V^{-1} and temporal stability up to 10 000 hours with azo dyes in Table 9.12 (DR1, DR17, DR19, DO).

Third order nonlinear properties have been observed in organic dyes containing conjugated π electrons, due to their high polarizability. To accomplish with the limiting structural properties of the dyes (e.g. low chemical and thermal stability, clustering) for commercial devices, the incorporation of such dyes into inorganic or organic–inorganic hybrids is one of the adopted strategies. The host also contributes to enhance the magnitude of the third order nonlinear features. Table 9.13 lists several organic–inorganic hybrids with the respective third order nonlinear susceptibility value. Examples of optical limiters include the C$_{60}$ molecule and its derivatives.

Another important class of organic–inorganic hybrids with third order nonlinear properties involves the combination of nano-sized organic and inorganic segments, taking advantage of the quantum confinement and dielectric confinement effects, and the possibility of incorporate high doping levels of nanoparticles. This

Table 9.12 Examples of second order non-linear hybrids.

Non-linear Chromophor	Matrix	Second Order Non-linear features (pm/V)
N-(4-nitrophenyl)-(s)-prolinol	$CH_3SiO_{1.5}$-ZrO_2 **(1)**	$d_{33} = 0.16$
	SiO_2-TiO_2 **(2)**	$\chi^{(2)} = 10.9$
4-(dimethylamino)-4′-nitrostilbene (DANS)	TiO_2 **(3)**	$\chi^{(2)} = 11$ $d_{33} = 11$
	SiO_2 **(4)**	$d_{33} = 75$
	poly(methyl methacrylate) (PMMA)-SiO_2 **(5)**	$\chi^{(2)} = 4$
Dispersed Red1 (DR1)	SiO_2 **(6)**	$d_{33} = 55$
Dispersed Red 17 (DR17)	Red 17 functionalized with ICPTES **(7)**	$d_{33} = 78$–150 $r_{33} = 20$
	(8)	$\chi^{(2)} = 62.83$–96.36
Dispersed Red 19 (DR19)	AEAPTMS or 3-glycidoxypropyltri-methoxysilane-N-(3-(trimethoxysilyl)propyl]-ethylenediamine (GPTMS) **(9)**	$d_{33} = 0.66$ $d_{11} = 0.335$

(1) Toussaere et al., *Nonlin. Opt.* 1, 349 (1991); (2) Zhang et al. *Chem. Mater.* 4, 851 (1992); (3) Nosaka et al., *Chem. Mater.* 5, 930 (1993); (4) Izawa et al., *Jpn. J. Appl. Phys.* 32, 807 (1993); (5) Xu et al., *Phys. Lett.* 24, 1364 (1999); (6) Riehl et al., *Phys. Lett.* 245, 36 (1995); (7) Lebeau et al., *Curr. Opin. Solid State Mater. Sci.* 4, 11 (1999) and *New J. Chem.* 20, 13 (1996); (8) Jiang et al., *Adv. Mater.* 10, 1093 (1998); (9) Innocenzi et al., *Chem. Mater.*, 14, 3758 (2002) and *J. Sol–Gel Sci. Technol.* 26, 967 (2003).

9.6 Hybrids for Integrated and Nonlinear Optics | 397

Random orientation: virtual cancellation of the second order Non-linear optical properties $3^{(Z)} = 0$

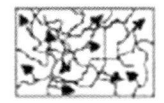

NLO chromophore

Oriented molecules: distribution of the centrosymmetric order. The single contributions are added as vectors, $3^{(Z)} = 0$

After polarization

Axis of polarization

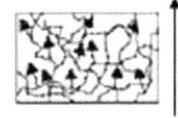

Fig. 9.28 Schematic drawing of the alignment of the chromophores during the poling process. (Taken from P. Innocenzi and B. Lebeau, *J. Mater. Chem.* **2005**, *15*, 3812).

Table 9.13 Examples of third order non-linear hybrids.

Third Order Non-linear Chromophor	*Matrix*	*Non-linear features*
Dispersed Red 1 (DR1) — trans and cis forms	TEOS **(1)**	$n_2 = -2.37 \times 10^{-5}\,\text{cm}^2\,\text{W}^{-1}$
Rhodamine-6G (see Table 3)	SiO_2-TiO_2 **(2)**	$\chi^{(3)} = 10^{-9}\,\text{esu}$
4′-dimethylamino-N-methyl-4-stilbazolium iodide (DMSI)	SiO_2 **(3)**	$\chi^{(3)} = 4\text{–}5 \times 10^{-14}\,\text{esu}$
pseudisocyanine dyes (PIC)	SiO_2 **(4)**	$\chi^{(3)} = 10^{-7}\,\text{esu}$
1,1′-diethyl-2,2′-cyanine bromide	SiO_2 **(5)**	$\chi^{(3)} = 5 \times 10^{-7}\,\text{esu}$
metal dithiolene and phthalocyanine dyes	SiO_2 + PMMA **(6)**	$\chi^{(3)} = 10^{-11}\,\text{esu}$
0.8 (Poly(1,4-phenylene vinylene) (PPV)) + 0.2 (Poly(2,5 dimethoxy-1,4-phenylene vinylene)) (DMPPV)	SiO_2 **(7)**	$\chi^{(3)} = 7.9\text{–}59 \times 10^{-11}\,\text{esu}$

(1) Rosso *et al.*, *J. Non-Cryst. Solids* 342, 140 (2004); (2) Zhang *et al.*, *Proc. SPIE* 3175, 302 (1998) and; (3) M. Nakamura *et al.*, *J. Non-Cryst. Solids* 135, 1 (1991); (4) Wanabe *et al.*, *J. Sol–Gel Sci. Technol.* 19, 257 (2000); (5) Zhou *et al.*, *J. Sol–Gel Sci. Technol.* 19, 803 (2000); (6) Gall *et al.*, *Proc. SPIE* 2288, 372 (1994); (7) Han *et al.*, *J. Non-Cryst. Solids* 259, 107 (1999).

Table 9.14 Examples of third order non-linear nanosized hybrids.

Third Order Non-linear Nano-sized Hybrid	Non-linear features
nanocrystalline TiO$_2$ + PMMA **(1)**	$\chi^{(3)} = 0.14\text{--}1.93 \times 10^{-9}$ esu $n_2 = 0.17\text{--}1.0 \times 10^{-11}$ cm^2 W^{-1} $\beta = 160\text{--}550 \times 10^{-9}$ cm W^{-1}
titania-polymethyl methacrylate (TiO$_2$–PMMA) **(2)**	$\chi^{(3)} = 1.05\text{--}5.27 \times 10^{-9}$ esu $n_2 = 1.3\text{--}6.2 \times 10^{-11}$ cm^2 W^{-1} $\beta = 885\text{--}2260 \times 10^{-9}$ cm W^{-1}
poly(styrene maleic anhydride)/TiO$_2$ **(3)**	optical limiting lower continuous laser light Intensity (10^7 W/m^2). $\beta = 200\text{--}960 \times 10^{-4}$ cm W^{-1} $\alpha = 0.45\text{--}1.2$ mm^{-1}
nanometer-sized CdO CTAB **(4)**	$\chi^{(3)} = 1.67 \times 10^{-10}$ esu $n_2 = -155 \times 10^{-16}$ cm^2 W^{-1} $\beta = 0.91 \times 10^{-9}$ cm W^{-1}
dodecylbenzene sulfonate (DBS) **(4)**	$\chi^{(3)} = 1.22 \times 10^{-10}$ esu $n_2 = -697 \times 10^{-16}$ cm^2 W^{-1} $\beta = 8.64 \times 10^{-9}$ cm W^{-1}
CdS + PDMS-TEOS or APTES **(5)**	$\chi^{(3)} = 10^{-6}$ esu
CdS + NaOAc + B(OC$_2$H$_5$)$_3$ + Si(OCH$_3$)$_4$ + Cd(OAc)$_2 \cdot$ 2H$_2$O **(6)**	$\chi^{(3)} = 1.2\text{--}6.3 \times 10^{-7}$ esu
PbS ǀ TMSPM ǀ MAA ǀ ZPO **(7)**	$\chi^{(3)} = 1.8 \times 10^{-10} - 1.3 \times 10^{-7}$ esu $n_2 = -(3.1 \times 10^{-16} - 6.7 \times 10^{-15})$ m^2 W^{-1}

(1) Yuwono et al., *J. Mater. Chem.* **13**, 1475 (2003); (2) Yuwono et al., *J. Mater. Chem.* **14**, 2978 (2004); (3) Wang et al., *Phys. Lett. A* **281**, 59 (2001); (4) Wu et al., *Appl. Phys. Lett.* **71**, 2097 (1997); (5) Li et al., *Proc. SPIE* **2288**, 151 (1994); (6) Takada et al., *J. Non-Cryst. Solids* **147**, 631 (1992); (7) Martucci, et al., *J. Non-Cryst. Solids* **244**, 55 (1999).

kind of materials can be easily prepared by sol–gel route and good quality transparent films have been produced (Table 9.14).

9.7
Summary

Organic–inorganic hybrids are a technologically key class of advanced multifunctional materials in one of the main challenge of the beginning of this century: the drive towards a miniaturization at the nanometer scale (requiring small components and devices and less resources and energy). In this chapter, we focused on functional siloxane-based hybrids with applications in optics and photonics, namely for coatings, for light-emitting and electro-optic proposes, for photochromic and photovoltaic devices and for integrated and nonlinear optics. We have shown how the synthesis strategy can decides and controls the materials features for optical and photonic applications. The concepts of photoluminescence, absorption and electroluminescence and the quantification of the materials luminescence features (color emission, quantum yield and radiance) were described and illustrated with representative examples. The emitting centers typical of hybrid materials and the associated recombination mechanisms were discussed for the cases where information is available. The concept of stimulated emission in solid-state dye-lasers and the requirements to develop photochromic, photovoltaic and IO devices were presented and exemplified with some symbolic examples.

The expected advances in the synthesis processes (using even bio-inspired strategies) and in the level of knowledge and understanding in nanosciences will allow the design of multiscale structured hybrids, hierarchically organized in terms of structure and optical functions, permitting, therefore, to anticipate significant improvements in the properties of photonic organic–inorganic hybrids in terms of i) the mechanical integrity, transparency and optical loss of the host matrix; ii) the emitting center characteristics such as chemical, thermal, and photostability and optical efficiency; and iii) the interaction between the emitting centers and the host matrix due to a cage-type encapsulation with better chemical, thermal and photostability and with efficient host-to-emitting center energy transfer processes.

Bibliography

Synthesis strategy for optical applications

D. Avnir, S. Braun, O. Lev, D. Levy, M. Ottolenghi, in *Sol-Gel Optics, Processing and Applications* (Ed. L. Klein), Kluwer Academic Publishers, Dordrecht, **1994**, pp. 539.

C. J. Brinker, G. W. Scherer, *Sol-gel Science, The Physics and Chemistry of Sol-Gel Processing*, Academic Press, San Diego, **1990**.

Functional Hybrid Materials (Eds.: P. G. Romero, C. Sanchez), Wiley-VCH, Weinheim, **2003**.

C. Sanchez, G. J. de A. A. Soler-Illia, F. Ribot, T. Lalot, C. R. Mayer, V. Cabuil, *Chem. Mater.* **2001**, *13*, 3061.

Hybrids for coatings

B. Arkes, in *Hybrid Organic-Inorganic Materials* (Ed. D. A. Loy), Materials Research Society Bull. **2001**, *26*, 402.

G. Schottner, *Chem. Mater.* **2001**, *13*, 3422.

Hybrids for light-emitting and electro-optic purposes

C. Sanchez and B. Lebeau, in *Hybrid Organic-Inorganic Materials* (Ed. D. A. Loy), Materials Research Society Bull. **2001**, *26*, 377.

L. D. Carlos, V. de Zea Bermudez, R. A. Sá Ferreira, in *Handbook of Organic-Inorganic Hybrid Materials and Nanocomposites* Vol. 1 (Ed. H. S. Nalwa), American Scientific Publishers, Los Angeles, **2003**, pp. 353.

R. A. Sá Ferreira, L. D. Carlos, V. de Zea Bermudez, in *Encyclopedia of Nanoscience and Nanotechnology*, Vol. 4 (Ed. H. S. Nalwa), American Scientific Publishers, North Lewis Way, California, **2004**, pp. 719.

C. Sanchez, B. Lebeau, F. Chaput, J.-P. Boilot, in *Functional Hybrid Materials* (Eds. P. G. Romero, C. Sanchez), Wiley-VCH, Weinheim, **2003**, pp. 122.

R. Houbertz, G. Domann, C. Cronauer, A. Schmitt, H. Martin, J.-U. Park, L. Frohlich,

R. Buestrich, M. Popall, U. Streppel, P. Dannberg, C. Wachter, A. Brauer, *Thin Solid Films* **2003**, *442*, 194.

Organic Electronics, (Eds. J. M. Shaw, P. F. Seidler) *IBM J. Res. & Dev.* **2001**, *45*.

Colour coordinates

R. W. Hunt, *Measuring Colour*, Ellis Horwood Series in Applied Science and Industrial Technology, John Wiley & Sons, New York, **1987**.

K. Narisada and S. Kanaya in *Phosphor Handbook* (Eds. S. Shionoya, W. M. Yen), CRC press, Boca Raton, **1999**, p. 799.

Emission quantum yield and radiance

Handbook of Applied Spectroscopy (Ed. C. DeCusatis), Optical Society of America and Springer-Verlag, **1998**.

W. B. Fussel, in *Optical Radiation Measurements: Approximate Theory of the Photometric Integrating Sphere*, National Bureau of Standards Publishers, **1974**.

C. de Mello Donegá, S. J. L. Ribeiro, R. R. Gonçalves, G. Blasse, *J. Phys. Chem. Solids* **1996**, *57*, 1727.

M. S. Wrigton, D. L. Ginley, J. Morse, *Chem. Phys.* **1974**, *78*, 2229.

L.-O. Parson, A. P. Monkman, *Adv. Mat.* **2003**, *9*, 230.

Recombination mechanisms

J. I. Pankove, *Optical Processes in Semiconductors*, Dover Publications, Inc., New York, **1990**.

C. F. Klingshirn, in *Semiconductor Optics*, Springer-Verlag, Berlin, **1995**.

J. Sigh, K. Shimakawa, in *Advanced in Amorphous Semiconductors*, CRC PRESS, **2003**.

G. E. Pake and T. L. Estle, in *The Physical Principles of Electron Paramagnetic Resonance*; W. A. Benjamin Inc., London, **1973**.

M. F. Thorpe, M. I. Mitkova, in *Amorphous Insulators and Semiconductors*, Kluwe Academic Publishers, **1997**.

Lanthanide-doped hybrids

G. Blasse, B. C. Grabmaier, *Luminescent Materials*, Springer-Verlag, Berlin, **1994**.

Crystal Field Handbook (Eds. D. J. Newman and B. Ng), Cambridge University Press, **2000**.

S. Hüfner, *Optical Spectra of Transparent Rare Earth Compounds*, Academic Press, New York, **1978**.

R. Reisfeld and C.K. Jørgenson, in *Handbook on the Physics and Chemistry of Rare Earths*; Gschneider (Eds. K. A. Gschneidner Jr., L. Eyring); Elsevier Science Publishers: Amsterdam, **1987**, Vol. 9, pp. 61.

Solid-state dye-lasers

R. Reisfeld and C.K. Jorgensen, *Optical Properties of Colorants or Luminescent Species in Sol-Gel Glasses*, (Eds. R. Reisfeld and C.K. Jorgensen) Structure and Bonding, Springer-Verlag, **1992**, Vol. 77, pp. 207.

Hybrids for photochromic and photovoltaic devices

D. Levy, *Chem. Mat.* **1997**, *9*, 2666.

C. Sanchez, B. Julián, P. Belleville, M. Popall, *J. Mat Chem.* **2005**, *15*, 3559.

F. Chaumel, H. Jiang, A. Kakkar, *Chem. Mater.* **2001**, *13*, 3389.

Hybrids for integrated and nonlinear optics

M. P. Andrews, in *Integrated Optics Devices: Potential for Commercialization*, Vol. 2997 (Eds. S. I. Najafi, M. N. Armenise), SPIE The International Society of Optical Engineering Series, Bellingham, WA, **1997**, p. 48.

R. G. Hunsperger, *Integrated Optics Theory and Technology*, Springer-Verlag, Berlin, **1995**.

Robert Guenther, in *Modern Optics*, John Wiley and Sons, **1990**.

A. B. Seddon, in *Sol-Gel and Polymer Photonic Devices*; Critical Review of Optical Science and Technology Series, The International Society of Optical Engineering Series, Bellingham, WA, **1997**, Vol. CR68, pp. 143.

F. Chaumel, J. Hongwei, K. Ashok, *Chem. of Mater.* **2001**, *13*, 3389.

P. Innocenzi, B. Lebeau, *J. Mat Chem.* **2005**, *15*, 3821.

10
Electronic and Electrochemical Applications of Hybrid Materials

Jason E. Ritchie

10.1
Introduction

The goal of this chapter is to illustrate for the reader several interesting electronic and electrochemical applications for hybrid organic/inorganic and composite materials. Examples have been chosen which are particularly illustrative of the hybrid materials application. This chapter is not intended to be a thorough review of the literature in this area, rather as a description of interesting applications of hybrid materials. A somewhat broad variety of electrochemically active organic–inorganic hybrids have been covered including the traditional incorporation of organic species into sol–gel materials, and also the incorporation of conductive particles into sol–gel materials, some siloxane based materials, and the so-called "polyether hybrid redox melts" produced through the organic modification of inorganic redox molecules.

New methods of materials synthesis must be developed in order to realize the next generation of electrochemically-active materials. This is a very broad field in which many challenges remain unsolved. There are many materials challenges in electrochemistry that may potentially find solutions in hybrid materials, including finding new materials that display substantial anhydrous proton conductivity at elevated temperatures for proton-exchange membrane (PEM) fuel cell applications.

Nanocomposite and hybrid materials are of special interest in the field of electrochemistry because of the potential synergistic interaction of the component materials. That is, these hybrids may exhibit properties that are significantly different from the average of the individual components. For example, while most sol–gel materials are dimensionally stable electrical insulators, the addition of a liquid-like conductive material can create a dimensionally stable, ionically conductive hybrid material. The rational design of hybrid materials allows for specific physical and electrochemical properties to be incorporated into the resulting material. As mentioned above, sol–gel oxide materials are frequently incorporated into materials in order to increase the dimensional stability of the resulting hybrid. In many cases, the organic components of the hybrid materials can be chemically tailored by covalent attachment of functional groups in order to add desired functionality to the material.

Sol–gel chemistry is a convenient, rapid, and extremely flexible method for creating advanced ceramics, glasses, and thin films. There are several methods for producing electrochemically-active hybrid materials using sol–gel chemistry. These hybrid materials can be described according to the Sanchez and Ribot scheme as either Class I, which contain weak bonds between organic and inorganic components, and Class II materials, which contain strong covalent bonds between organic and inorganic components.

Electrochemically active hybrid materials generally provide several advantages over their constituent components. The properties of a hybrid material are frequently synergistic combinations of the properties of its constituent components. For example, a hybrid material may offer advantages in processibility and stability (like a solid), while maintaining the ion and small molecule diffusion properties of a liquid. The ability to design a material that contains desirable properties while minimizing undesirable properties is very attractive. Hybrid materials can also combine the electrical conductivity properties of solid materials with the ionic conductivity properties of liquid materials. This combination can lead to hybrid materials with applications in electrochemical power sources such as fuel cell and battery electrolytes and catalytic electrode materials.

This chapter will discuss several electrical applications of hybrid materials including their use in H^+- and Li^+-conducting electrolytes, for fuel cell and battery applications, the use of hybrid materials in electrochemical sensors, and the use of hybrid materials in the creation of electrochemically generated luminescence.

10.2
Historical Background

One of the first applications of electrochemically active hybrid materials was the formation of chemically modified electrodes consisting of either siloxane polymers labeled with redox active groups, or trichlorosilane labeled redox molecules in the late 1970s. These chemically modified electrodes were prepared in order to examine how redox molecules behaved when confined to an electrode's surface. These systems showed surface confined redox waves, and electron hopping through mixed valent layers of the redox films. In the mid-1980s, both class I and class II redox-active hybrids consisting of redox molecules encapsulated in a sol–gel glass and a redox-active ormocers were prepared in order to study the behavior of redox molecules in solid materials. In the last 10–15 years, electrochemically active composite materials have become a popular way to create ion conducting electrolytes, chemical sensors, and electrochromic devices with tailored physical and chemical properties. This chapter will discuss many of the electrochemical applications of this exciting class of hybrid materials.

10.3
Fundamental Mechanisms of Conductivity in Hybrid Materials

There is great interest in the basic mechanisms of ionic and electron transport in polymers and hybrid materials. These mechanisms are especially relevant to the design and synthesis of the next generation of electrolytes and electrode materials for fuel cell and battery applications. This section will discuss the basic issues and mechanism(s) of electrical conductivity in composite materials, and of the ionic conductivity of Li^+ and H^+ ions.

The electrical properties of solid, inorganic materials can be classified into four categories: insulator, semiconductor, metal, and superconductor. Generally, solid-state inorganic materials do not show appreciable ionic conductivity due to the extremely slow mass transport of atoms within the solid-state structure. There are countless examples of liquid organic-based systems that display large ionic conductivities. Frequently, these organic systems are electrical insulators and generally respond to applied voltages with ionic motion (in organic species, the HOMO-LUMO energy gap is frequently large, giving large activation energies to delocalized electron motion). The combination of inorganic and organic materials allows for the formation of hybrid materials that combine electrical and ionic conductivities in interesting ways. For example, the introduction of ionic conductivity to solid materials allows for the synthesis of semi-solid electrolytes for lithium-polymer batteries, and PEM fuel cells. In addition, the introduction of electrical conductivity to insulating polymers allows for the creation of stable electrochromic films and electrochemical sensors.

10.3.1
Electrical Conductivity

While many inorganic oxides are insulators, there are several examples of inorganic oxides which display substantial electrical conductivity (e.g. tungsten oxide, vanadium pentoxide, and manganese oxide). In these conducting oxides, the electronic conductivity results from a mixture of oxidation states on the metal atoms. In vanadium oxide, this mixture of oxidation states allows an electron to hop from a reduced vanadium site to an adjacent vanadium site in a higher oxidation state. This hopping typically occurs with a small activation barrier of less than 0.5 eV. This small activation barrier to conductivity indicates that you would expect to see semiconducting (not-metallic) behavior in this material (similarly to other small band gap semiconductors). Furthermore, the bulk redox state of these conducting oxides can be adjusted electrochemically, and typically produce a color change in the material. For example, vanadium oxide (V_2O_5) changes from a yellow color to deep blue upon reduction to V_2O_4, or upon intercalation of Li to form $Li_xV_2O_5$. In addition, tungsten oxide (WO_3) changes from a yellow color to a greenish-bronze color upon intercalation of Na forming the well-known sodium tungsten bronze (Na_xWO_3).

In addition to the mixed-valent oxides, there are several examples of semiconducting oxides including ZnO, $BaTiO_3$, and TiO_2 that are wide band gap intrinsic semiconductors. An intrinsic semiconductor has a valence band, whose electronic states are completely occupied with electrons, separated from a empty conduction band (i.e. electronic states that are unoccupied by electrons) by an energy gap (no electronic states). In order for electrons to participate in electrical conductivity, an electron must be promoted from the valence band across the band gap to the empty conduction band. Once the electron is transferred to the conducting band, it is free to move from electronic state to state, and therefore be mobile throughout the material (the "hole" left behind in the valence band is also mobile and contributes to electrical conductivity).

Intrinsic semiconductors can frequently be easily converted to much more conductive narrow band gap, extrinsic n-type semiconductors by the addition of n-type dopants. These n-type dopants have more electrons than the parent atoms (atoms to the right of the parent atom on the periodic table) and add electron density into bands near the vacant conduction band. For example, n-type dopants for intrinsic silicon include P, As, and Sb as they each contain one more valence electron than Si. The addition of a small n-type dopant band effectively narrows the band gap of the material and leads to larger electrical conductivities. These materials are extensively used in photoelectrochemical applications such as sensitizers in photovoltaic cells.

In addition to materials that display bulk electrical conductivity, composites containing electrically conducting components have been synthesized by the entrapment of small particles of conductive material within an insulating matrix. In one example of this type of material, carbon-ceramic composite electrodes have been constructed through the incorporation of conductive graphite particles into a silicate matrix. In these types of hybrid materials, the electrical conductivity of the composite is dependant on the formation of an electrical percolation network of conducting particles. That is the concentration of the conductive particles must be high enough that they form a closely connected and continuous network that spans the hybrid material. A composite formed with a low concentration of conducting particles would not show electrical conductivity because each particle would be isolated and not part of a continuous electrical network. While these materials appear to be electrical conductors, it would be more accurate to say that they contain electrically conductive pathways within the composite material. In addition, the large number of particle-particle junctions can create a significant amount of electrical resistance, which may limit the overall conductivity. Furthermore, the electrical conductivity is also sensitive to the number of parallel conduction pathways, that is, the more concentrated the conductive particles, the more likely it is that the random distribution of the particles in the material will form a large number of conductive pathways, which increases the measured conductivity.

Similar conducting hybrid composites that have the components reversed (i.e. electrically insulating inorganic particles with electrically conducting organic polymers) can also be easily prepared. In one example of this type of material, a composite of oxidized polypyrrole (PPy) and ZrO_2 nanoparticles lead to a composite material with significantly improved electronic conductivity over unmodified PPy (conductivity went from 1 to 17 S cm^{-1}, see Table 10.1).

Another type of electron conductivity has been described in polyether-based redox materials. Here, the redox chemistry has been extensively studied in semi-solid poly(ethylene glycol) based materials in order to understand the microscopic effects of rigid environments on mass transport and electron transfer dynamics. So far, the attachment of low molecular weight methyl poly(ethylene glycol) oligomers (MePEGs) with redox molecules has proven to be a convenient route to the synthesis of semi-solid (i.e. very viscous liquids), high concentration solutions of these so-called "hybrid redox polyethers". For example, the attachment of 6MePEG$_7$ units (2 per bipy) to the redox active Ru(bipy)$_3^{2+}$ cation results in a room temperature, high viscosity molten salt. These molten salts are essentially solu-

Table 10.1 Comparison of band gaps, electrical conductivities, and characteristics for a variety of materials and hybrid materials.

Material	Band Gap (eV)	Electrical Conductivity 25 °C (S cm^{-1})	Character / Notes
C (diamond)	5.47	<10^{-12}	insulator
SiO$_2$	9	10^{-12}–10^{-18}	insulator
Si	1.12	2 × 10^{-5}	intrinsic semiconductor
Ge	0.66	2 × 10^{-2}	intrinsic semiconductor
TiO$_2$ (rutile)	3.0	10^{-12}–10^{-18}	intrinsic semiconductor
BaTiO$_3$	3.2	10^{-8}–10^{-15}	intrinsic semiconductor
ZnO	3.2	–	intrinsic semiconductor
V$_2$O$_5$	0.2 (E$_A$)[a]	8 × 10^{-4}	mixed-valent conductor
WO$_3$	2.6 (E$_g$)	–	mixed-valent conductor
Au MPCs	0.04–0.15 (E$_A$)[b]	10^{-3}–10^{-5}	mixed valent film of monolayer protected gold nanoparticles
[Ru(bpy)$_3$(PEG)$_2$](ClO$_4$)$_2$	–	3 × 10^{-8}	mixed valent redox polymer film based on PEG-modified Ru(bipy)$_3$ cations
Pt	–	9 × 10^4	metallic
Au	–	5 × 10^5	metallic
C (graphite)	–	1000	metallic
Polypyrrole (PPy)	–	1–100[c]	metallic
PPy / ZrO$_2$ hybrid	–	17	metallic
C / SiO$_2$	–	1	CCE electrode composite

a indicates activation barrier to electronic conductivity.
b activation barrier depends on the charge state of the mixed valent film.
c dependent on the preparation of the polymer and oxidation.

tions that have very high concentrations of redox molecules (sometimes in excess of 1 M). A more general method for the synthesis of these redox melts has been developed where a MePEG chain is attached to an anion or cation, and then ion-exchanged with the counterion of a charged redox molecule. This method allows for the easy creation of melts of any charged redox molecule.

When this "redox melt" is oxidized electrochemically (e.g. during cyclic voltammetry of the neat melt) a mixed-valent region is created near the electrode. In this region, the $Ru(bipy)_3^{2+}$ cations closest are oxidized to the $Ru(bipy)_3^{3+}$ form. Since this material is very viscous, the physical diffusion of the oxidized $Ru(bipy)_3^{3+}$ away from the electrode is slow. The $Ru(bipy)_3^{3+/2+}$ redox couple has a relatively large electron self-exchange coefficient (homogeneous electron transfer). That is, an adjacent unoxidized $Ru(bipy)_3^{2+}$ cation can transfer an electron to the $Ru(bipy)_3^{3+}$ cation closer to the electrode. The now reduced $Ru(bipy)_3^{2+}$ cation next to the electrode can then be reoxidized by the electrode. This "electron hopping" process (shown schematically in Fig. 10.1) leads to an enhancement of the electrochemically determined diffusion coefficients, such that the apparent diffusion coefficient (D_{APP}) is greater than the physical diffusion coefficient (D_{PHYS}). This enhancement is called the electron diffusion coefficient (D_E).

Electrical conductivity has also been produced in a hybrid system of monolayer protected gold nanoclusters (MPCs). In this system, a film of mixed-valent gold nanoparticles are cast onto an interdigitated array electrode. The mixed valent charge state of the gold nanoparticle cores was established by quantized electrochemical double layer charging, or by a chemical oxidation with Ce^{4+} ions. The as

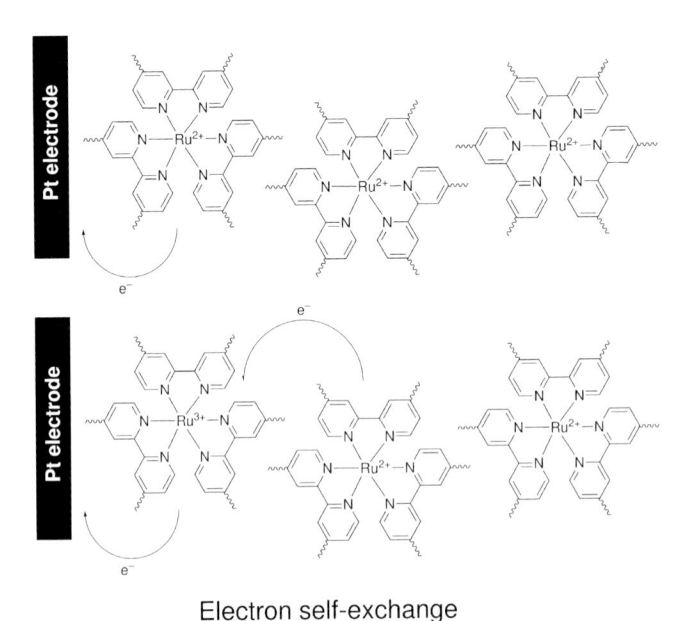

Fig. 10.1 Electron self-exchange in $Ru^{2+}(bipy)_3$ based hybrid redox polyethers.

synthesized Au nanoparticles have negative charges on the cores (likely from the BH_4^- reductant), resulting in an open circuit potential of about $-0.61\,V$ (vs. Ag/Ag^+). The open circuit potential is measured as the potential difference between the working electrode and a reference electrode, here the large negative value indicates that the Au clusters have the potential to transfer their excess electrons to other species and is a measure of their effectiveness as a reducing agent. The open circuit potential can be adjusted to more positive values through exposure to the Ce^{4+} oxidant. This oxidation of the Au cores allows the average charge state of the cores to be adjusted. Electrical conductivity results from an electron self-exchange reaction between adjacent Au nanoparticles in different charge states

$$MPC^0 + MPC^+ \xrightarrow{k_{EX}} MPC^+ + MPC^0 \quad \text{Electron self-exchange reaction} \quad (1)$$

Furthermore, the conductivity is a function of mixed-valent charge state, such that a rest potential which corresponds to a single valent average charge state (e.g. -2, -1, 0, $+1$) has a lower conductivity than a sample with a rest potential corresponding to a mixed valent average charge state (e.g. -1.5, -0.5, $+0.5$). In this system an average charge state of -0.5 would correspond to a mixture of 50% negatively charged MPC^- and 50% neutral MPC^0. The loss of electrical conductivity in single valent MPC films is because electron self-exchange is not possible when all the nanoclusters are in the same charge state (electron hopping in a single valent film is not a self-exchange reaction, $\Delta G^0 > 0$).

Molecular conductors capable of switching on and off their conductivity based on an external stimulus (such as light, temperature, and pressure) have been prepared from hybrid materials. These materials generally incorporate an organic radical cation such as tetrathiafulvalene (TTF) or bis(ethylenedithio)tetrathiafulvalene (ET) complexed with an anion inorganic coordination complex. In one example, the charge-transfer salt $ET_4K[Fe(CN)_5NO]_2$, which consists of a layered structure, displayed metallic electrical properties down to very low temperatures. The irradiation of this material leads to a structural reorganization which allows for electron localization in the conduction band leading to metallic-like conductivity under irradiation.

10.3.2
Li$^+$ Conductivity

Polymer electrolytes are frequently based on an organic macromolecule that is capable of dissolving inorganic salts of this ion of interest (i.e. PEO dissolves $LiPF_6$). In the poly(ethylene oxide) system, each ethylene oxide segment contains a Lewis basic oxygen atom which can coordinate to the Li^+ cation. This formation of coordinative bonds between the polymer and the Li^+ ion is critical to overcoming the lattice energy of the salt, because otherwise the salt will crystallize in the polymer matrix which will lead to a dramatic loss of Li^+ mobility. Since there is a limit to the energy released by the formation of these coordinative covalent bonds, inorganic salts with very strong lattice energies are unsuitable for polymer electrolyte applications (e.g. LiF has too large a lattice energy to be dissolved in most polymer systems). In addition, the use of very weakly coordinating anions such as PF_6^- and $TFSI^-$, $(CF_3SO_2)_2N^-$, tends to lower the lattice energy of the lithium salt, and serves to reduce ion-pairing in the electrolyte.

Poly(ethylene oxide) based materials, and the similar poly(ethylene glycol) materials, are known to conduct small cations through segmental motions of the ethylene oxide units. Ionic conductivity in these types of polymer electrolytes occurs primarily in the amorphous phase of the polymer through these segmental motions of the polymer's units. In order for the Li^+ cation to move, these polymer segments have to reorganize to "hand off" the Li^+ cation from one coordination site to another site. The rate of reorganization of the polymer segments is dependant on the glass transition temperature (T_g) of the material. The T_g is the temperature where an amorphous polymer transitions from a glassy state (polymer segments are "frozen" – immobile but not crystalline) to a rubbery state (polymer segments are free to move and rearrange).

The glass transition temperature is a measure of the freedom of the polymer segments to reorganize, and can be compared between different polymers. The T_g of a polymer is strongly affected by the flexibility of the polymer backbone and the nature of the pendant groups. Siloxane and phosphazene polymers have very flexible backbones and generally have very low T_g values. Large bulky pendant groups are more likely to become entangled and can raise the T_g of the polymer. However, large pendant groups can also lower the T_g by increasing the separation between polymer chains creating "free volume" in which the polymer can more easily reorganize.

Electrolytes based on high molecular weight PEO frequently contain crystalline regions at room temperature that generally do not participate in ionic conductivity which leads to a drop in the room temperature ionic conductivity (the conducting pathway has to go around these crystalline regions). One strategy to maximize the ionic conductivity is to design polymers with large volume fractions of amorphous regions within the polymer. The addition of inorganic components into polymer electrolytes can substantially increase the volume fraction of amorphous regions in polymer electrolytes, leading to electrolytes with potentially larger ionic conductivities. In this regard, the influence of inert oxide fillers on the conductivity of PEO-based lithium electrolytes has been explored. It was found that small amounts of added nanoscale fillers such as SiO_2 and Al_2O_3 stabilized the amorphous phase of the PEO giving a completely amorphous polymer electrolyte. Addition of these nanoscale fillers to already amorphous polymer phases did not improve the conductivity. Thus, the addition of oxide fillers would only be expected to help a crystalline, or partially crystalline, electrolyte to become more amorphous. Since the lithium ion conduction occurs exclusively in the amorphous regions, making the polymer more amorphous should increase the ionic conductivity.

While Li^+ conducting polymer electrolyte systems based on poly(ethylene oxide), poly(acrylonitrile), poly(methyl methacrylate), and poly(vinylidene fluoride) have been developed, very little work has been done on hybrid polymeric systems based on polymers other than poly(ethylene oxide). This is likely due to the ease with which PEO and PEG based polymers can be synthetically modified. However, some recent work has explored hybrid systems composed of polyfluorosilicone polymers reinforced with nanosized silica. The polyfluorosilicone polymer was chosen for its high chemical and thermal stability. These hybrid electrolytes could be swollen in the presence of conventional liquid lithium electrolytes to obtain a gel polymer electrolyte. In addition, hybrids of poly(methyl methacrylate) and TiO_2 particles swollen with ethylene carbonate and lithium perchlorate were studied as gel polymer electrolytes. Here, the addition of TiO_2 particles to the gel electrolytes resulted in enhanced diffusion of cations, while the diffusion of carbonate solvent molecules remained constant.

10.3.3
H⁺ Conductivity

There are two general mechanisms responsible for proton mobility in proton-conducting electrolytes: the vehicle mechanism (which relies on the physical transport of a vehicle to move protons) and the Grotthus mechanism (which involves the proton being handed-off from one hydrogen bonding site to another). The Grotthus mechanism is superficially similar to Li⁺ conductivity in that it depends on a site-to-site hopping mechanism, while the vehicle mechanism depends on the rate of physical diffusion of the vehicle. Vehicles are simply molecules that are capable of forming a bond to the free proton and of freely diffusing. Under certain conditions, species such as H_3O^+ and H_3PO_4 can serve as proton carrying vehicles.

Some of the most successful proton-conducting electrolytes are water swollen, acidic polyelectrolytes. For example, water-swollen Nafion (Fig. 10.2) typically

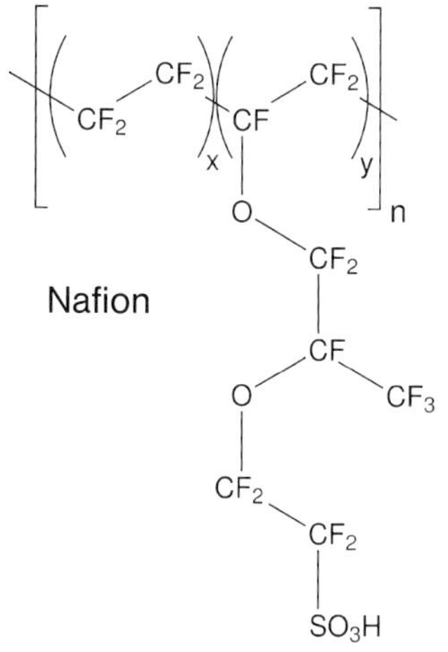

Fig. 10.2 Chemical structure of Nafion.

reaches a room temperature ionic conductivity of about 0.01 S/cm. Nafion is a fluoropolymer with pendant —SO_3H groups. This polymer tends to aggregate to form nanoscale hydrophobic regions composed of the polymer backbone and nanoscale hydrophilic regions occupied by the pendant acid groups. Absorbed water migrates to the hydrophilic regions forming a continuous conducting pathway for the H^+ ions. Thus, water is critical to the conductivity as proton mobility only occurs in the water-swollen hydrophilic regions of the polymer. In these acidic polyelectrolytes, the water molecules are likely protonated by the pendant acid groups forming H_3O^+ ions. These materials are especially interesting for fuel cell applications because the anions are covalently attached to the polymer backbone and therefore immobile. The ionic conductivity of these materials essentially represents pure H^+ diffusion. Proton transport in these materials likely occurs through either a physically diffusion of the H_3O^+ molecule (vehicle mechanism) or through a homogeneous H^+ transfer to an adjacent hydrogen bonded water molecule (Grotthus mechanism).

In addition to Nafion, there are several other systems that have been explored as proton conducting membranes for PEM fuel cell applications. Ballard Advanced Materials Corporation has synthesized a proton-conducting membrane by sulfonating a copolymer primarily based on α,β,β,-trifluorostyrene. Here, the acidic —SO_3H groups are introduced to the fluorinated polystyrene polymer after the polymerization step. Similarly, proton-conducting membranes have been prepared by sulfonation of poly(arylene ether)s such as poly(arylene ether ether ketone) (PEEK). These sulfonated poly(arylene ether)s have the potential advantages of wide availability, ease of processing, and chemical and thermal stability. Polymers based sulfonated poly(imide)s have also been used as proton-conducting electrolytes, unfortunately, these materials have a propensity to be quickly chemically degraded. Substituted polyphosphazenes, which are based on the $[-P(R)_2=N-]_x$ backbone, have very flexible backbones, and have been employed as Li^+ conducting electrolytes. Proton-conducting polyphosphazene polymers have been synthesized by attaching sulfonated aryl groups to the polymer backbone.

The one-way transport of H^+ cations (from anode to cathode) in a fuel cell application can lead to the unfortunate problem of electro-osmotic drag of water molecules. This motion of water molecules from the anode to the cathode also causes methanol to move at the same rate in a direct methanol fuel cell. This "cross-over" of fuel molecules can lead to a direct reaction between the methanol fuel and oxidizer at the cathode, which severely decreases the efficiency of the fuel cell.

In systems that display anhydrous proton conductivity, there are no water molecules around to serve as vehicles for physical diffusion. These systems typically show a large degree of Grotthus conductivity due to the hopping of the H^+ ions from one coordination site to another.

10.4
Explanation of the Different Materials

10.4.1
Sol–Gel Based Systems

Electrochemically-active hybrid materials have been prepared by covalently attaching redox-active groups to siloxane (or other inorganic) polymers. These are typically prepared through a condensation of a redox-active molecule containing a —Si(OR)$_3$ group to form a sol–gel network, or by hydrosilation of a vinyl-containing redox-active molecule to an inorganic polymer containing a Si—H bonds. Strictly speaking, however, materials prepared by hydrosilation of Si—H groups are generally not considered sol–gel based materials. Since the redox-active groups are covalently immobilized on the inorganic polymer, and are immobile, charge transfer must occur through an electron hopping mechanism. In addition, the oxidation or reduction of these redox-active groups must be accompanied by a diffusion of counterions into or out of the redox-active polymer in order to counteract the change in oxidation state. For example, the oxidation of a vinylferrocene film (Fc^0 to Fc$^+$) must be accompanied by an influx of negatively charged counterions to maintain charge neutrality. However, in electron conducting systems, where electrons move through this hopping mechanism, the oxidation state of individual redox centers change, but the overall oxidation state of the film is unchanged. In this system no counterion flux is needed as there is no change in to the overall oxidation state of the material. It turns out that the electrochemical properties of these polymer systems tend to be very dependant on the nature of the counterions and their ability to diffuse through the inorganic matrix in order to compensate for the charges created during redox cycling.

A similar class of redox active hybrids which are based on sol–gel synthesized silicates instead of siloxane polymers have also be prepared. These hybrid materials can be referred to as hybrid redox silicates and have been formed by mixing a solution of a redox active molecule, such as quinone, with carbon particles and a silica sol, and allowing the mixture to gel into a carbon ceramic composite electrode (CCE) forming a class I hybrid. Class II redox active hybrids have also been synthesized through sol–gel condensation of a trimethoxysilane-modified ferrocene (i.e. Fc—Si(OMe)$_3$) with TMOS (shown in Fig. 10.3).

In addition to redox active systems, the sol–gel synthesis of ion-conducting electrolytes shows considerable promise due to the easy processability of sol–gel materials, especially for thin-film and chemically modified electrode applications. In this area, there has been considerable effort in the sol–gel synthesis of hybrid H$^+$- and Li$^+$-conducting electrolytes for fuel cells and batteries. For example, class I Li$^+$-conducting electrolytes have been synthesized by the addition of silica particles to conventional Li$^+$-conducting liquid electrolytes, such as ethylene and propylene carbonate based solvents. These composite electrolytes show a considerable increase in the dimensional stability of the resulting electrolyte allowing for the construction of lithium battery cells with varied geometric designs. While, these

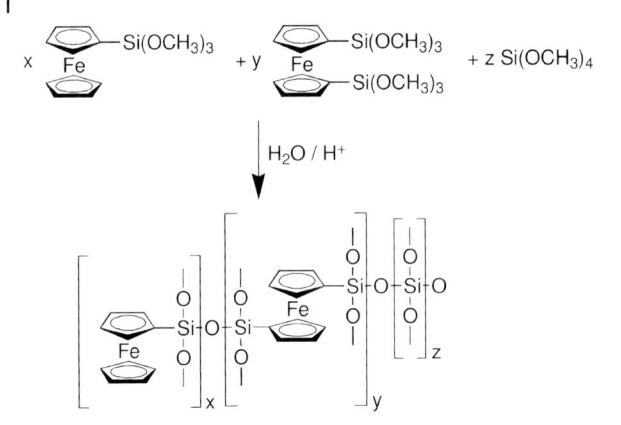

Fig. 10.3 A Class II redox-active silicate formed from trimethoxysilylferrocene and 1,1′-bis(trimethoxysilyl)ferrocene copolymerized with tetramethoxysilane.

composite electrolytes offer advantages over all liquid electrolytes, they are still liquid mixtures which are susceptible to leakage and separation. Class II Li⁺ conducting electrolytes can be prepared by attaching oligomeric poly(ethylene) glycol segments to alkoxysilanes, and then hydrolyzing the resulting PEG-modified silanes and allowing the resulting material to condense into an organically modified silicate (see Section 10.6.4). These class II materials are rubbery polymers, not viscous liquid gels, and are not susceptible to leakage problems. The design and synthesis of new ion conducting polymer electrolyte materials is fast becoming a lively area of research, with a few groups pursuing sol–gel based, hybrid inorganic–organic materials.

10.4.2
Nanocomposites

One of the simplest hybrid materials with electrochemical applications is a dispersion of conducting particles into a solid matrix, such as a gel, polymer, or inorganic solid. These systems typically show good electrical conductivity when the concentration of the conductive particles exceeds the percolation threshold. This threshold is reached when the particles are at a high enough concentration such that they can come into close enough contact with each other to create a network of connected particles throughout solid matrix (section 3.1). These carbon–ceramic composite electrodes (CCEs see Fig. 10.4) materials can be synthesized from the incorporation of conductive particles of graphite into a solid silicate matrix forming a brittle but porous structure. Electrodes were synthesized by mixing carbon powder with a silica sol and drawing the mixture into a glass capillary (the capillary serves to support the brittle material when dried). The mixture was then allowed to dry, and electrically contacted with silver paint. These composite

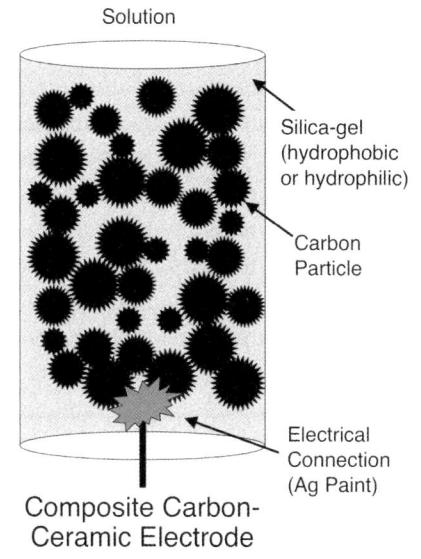

Fig. 10.4 Diagram of a composite carbon-ceramic electrode (CCE).

electrodes typically showed excellent conductivities of about 1 S/cm. The choice of alkoxysilanes can be used to control whether the resulting electrode would be hydrophilic, $Si(OCH_3)_4$, or hydrophobic, $CH_3—Si(OCH_3)_3$. The hybrid CCEs combine the structural properties of the silica matrix, and the conductivity properties of the carbon powder forming an effective hybrid material.

A similar type of material can be synthesized from carbon-black powder and insulating organic polymers. These hybrid carbon/polymer composites swell in the presence of organic vapors as the organic molecules partition into the polymer (shown in Fig. 10.5). This incorporation of organic molecules increases the volume of the composite (swelling) which disrupts the percolation network of conductive carbon particles by both decreasing the number of electrical connections and by increasing the tortuosity of the surviving connections. This disruption of the network leads to a decrease in the electrical conductivity of the composite. By constructing an array of these composites from different polymers, which each have their own swelling properties with different vapors (i.e. their partition coefficients for different organic vapors are different for each of the different polymers), a sensor can be developed which can discriminate between different organic vapors (this is the so-called "electronic nose").

Nanocomposites of conducting polymers and inorganic particles can also be used to prepare functional electroactive materials. For example, a composite of conducting polyaniline (PAn) and silica forms a solid electrochromic material (see Section 10.6.2). This electrochromic composite experiences a reversible color change from pale yellow to green when oxidized. Conducting polymers have also been used in nanocomposites with catalytically active metal particles. A com-

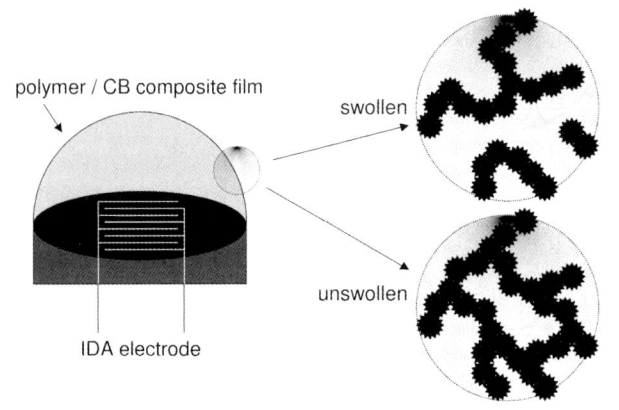

Fig. 10.5 Schematic of the "Electronic Nose". As the insulating polymer / carbon black composite becomes swollen with the analyte organic vapor, the concentration of CB particles decreases, yielding a smaller number of electrical connections, and a higher tortuosity within the composite which is measured as a decreased conductance.

posite of polythiophene (PT) and platinum particles have been used as the oxygen reducing electrode in a fuel cell application.

10.4.3
Preparation of Electrochemically Active Films (and Chemically Modified Electrodes)

The chemical modification of an electrode's surface is a powerful technique for altering the electrode's surface properties for specific electrochemical applications. Electrodes that are chemically modified with inorganic materials are especially attractive to electrochemists because of their stability (thermal, mechanical and chemical), their durability, their rigid structures, and their potential catalytic properties. The combination of inorganic and organic materials allows electrochemists to combine the properties of each material to create a new functional hybrid material. One of the simplest methods for producing an electrode chemically modified by a hybrid material is to spread a sol solution onto an electrode's surface and allow that sol to form a gel. This sol solution can contain organic dopants (to form a class I material) or contain organic modifiers labeled with trialkoxysilane groups (to form a class II material). These types of modified electrodes allow the organic dopants to interact with both the surface of the electrode and the solution. The combination of the structural and encapsulating properties of the silica, with the electrochemical properties of the organic dopants, results in a functional hybrid material combining the properties of the two components.

One of the most common types of challenges with the class I design of chemically modified electrodes is preventing the dopant molecule from leaching out of the hybrid film on the electrode into the solution. The dopants in a class I material are retained in the silica film through simple physical entrapment. Here, the leaching of electroactive species from the silica film can be prevented by covalently attaching it to the silica, however this will severely limit the electroactive species mass transport within the film.

Electrocatalytic applications are a powerful demonstration of the advantages of chemically modified electrodes by preparing electrodes with tailored electrochemical properties. In one example of a hybrid electrocatalytic application, zeolite films were attached to electrodes in order to selectively bind O_2 gas near the electrode for reduction. Frequently, electrocatalysis is slow at bare electrodes, meaning that a large overpotential is needed to achieve rapid redox chemistry. In this case, a mediator can speed up the electrode reaction by oxidizing the analyte, and becoming reduced in the process. The mediator can then be rapidly re-oxidized at the electrode surface, effectively mediating the oxidation of the analyte. With this mediation, the oxidation of the analyte occurs at the formal potential of the mediator. One example of this type of chemistry is the o-quinone mediated oxidation of NADH at a carbon electrode. Here, o-quinone can be attached to a carbon electrode through either spontaneous adsorption of a monolayer film, or encapsulation in a polymer film. The immobilized o-quinone then oxidizes solution-phase NADH (becoming reduced). The reduced o-quinone is then immediately re-oxidized by the electrode.

10.5
Special Analytical Techniques

The electrical and electrochemical properties of hybrid materials are typically characterized through electrochemical techniques designed to understand how ions and electrons are transported in this material. In this section, a brief introduction to these special techniques will be given.

10.5.1
Electrochemical Techniques

One of the most interesting parameters of a solid electrolyte is the total ionic conductivity and the fraction of that conductivity that is carried by each ion. In most electrolytes, ionic conductivity occurs through the motion of charged species (i.e. ionic mobility). The electrochemical measurement of ionic conductivity is difficult due to the resistance to ion motion at the electrode's surface (as compared to the relatively easy measurement of DC electrical conductivity in solids). In order to overcome this capacitance problem, a technique called ac-impedance spectroscopy is frequently used. In this measurement, the liquid (or semi-solid) electrolyte is placed between two parallel electrodes (either in a special ionic conductivity cell or on an interdigitated array electrode), and a sinusoidal potential is applied between the electrodes. The current flow is monitored, and the phase difference between the applied potential and the resulting current is determined (shown in

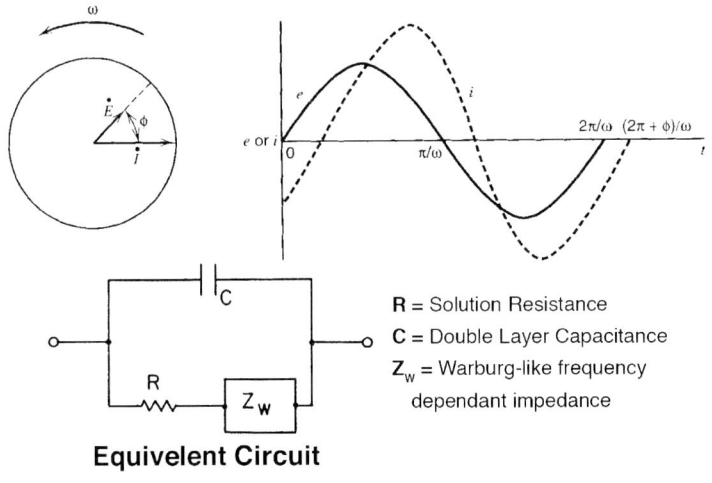

Fig. 10.6 Phasor diagram showing the phase relationship between the applied ac-potential (e) and observed current (i), and equivalent circuit diagram of an ion conducting electrolyte. (Adapted from *Electrochemical Methods: Fundamentals and Applications* 2nd edn by A. J. Bard and L. R. Faulkner, **2001**, Wiley, New York City).

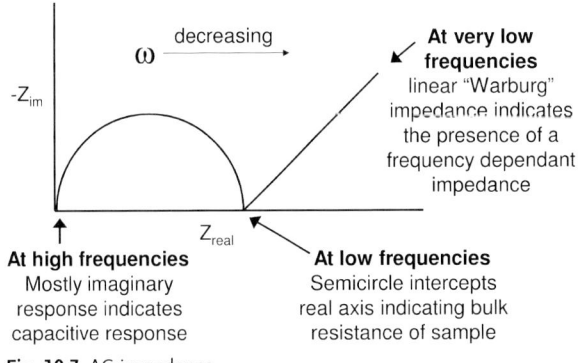

R = Solution Resistance
C = Double Layer Capacitance
Z_w = Warburg-like frequency dependant impedance

Equivelent Circuit

Fig. 10.7 AC impedance.

At high frequencies: Mostly imaginary response indicates capacitive response

At low frequencies: Semicircle intercepts real axis indicating bulk resistance of sample

At very low frequencies: linear "Warburg" impedance indicates the presence of a frequency dependant impedance

Fig. 10.6). This phase difference between the potential and current is determined for a series of different frequencies and the impedance is separated into real (resistive – Z_{real}) and imaginary (capacitive – Z_{im}) components.

Impedance measurements are typically presented on a Cole-Cole plot, which plots the imaginary component of the impedance ($-Z_{im}$) on the y-axis, and the real component (Z_{real}) on the x-axis for a series of different frequencies (shown in Fig. 10.7). In this experiment, the phase difference is generally measured at 20–50 different frequencies and presented on a Cole-Cole plot, which produces a semicircle. If the material being analyzed is a pure ion conductor (i.e. where none of the measured current is Faradaic current due to redox reactions) the low frequency intercept with the x-axis equals the bulk resistance of the electrolyte. This resistance is then converted to the material specific (and geometry-independent) conductivity values using the calculated or measured cell constant of the impedance electrode.

While, ac-impedance measurements are very helpful at determining the total ionic conductivity, it is difficult to separate out the individual contributions of each ion. According to the Nernst-Einstein equation (2), the total conductivity (σ_{ION}) is equal to the sum of the contributions from the individual ions (in this case an anion and a cation).

$$\sigma_{ION} = \frac{F^2}{RT}(z_+^2 D_+ C_+ + z_-^2 D_- C_-) \quad \text{Nernst-Einstein equation} \tag{2}$$

In the Nernst-Einstein equation, the contribution to the overall ionic conductivity from each ion is the product of the ion's charge squared (z^2), the ion's diffusion coefficient (D), and the concentration (C) of each ion. Since, the total ionic conductivity is a sum of contributions from all the ions present, the contributions from each of the ions must sum to give the total conductivity. Another way to describe this is to say that the sum of the fractions of charge carried by the individual ions (the transference numbers: t_+ and t_-) must be equal to unity (equation 3).

$$t_+ + t_- = 1 \tag{3}$$

The fraction of charge carried by any one ion (t_i) can be described as the ratio of that ion's contribution to conductivity to the sum of the contributions from all the ions (equation 4).

$$t_i = \frac{z_i^2 D_i C_i}{\sum_j z_j^2 D_j C_j} \quad \text{Transference Number} \tag{4}$$

In order to calculate the fraction of charge carried by a specific ion (i.e. the transference number t_i), you first need to be able to measure the diffusion coefficient for that specific ion. With the diffusion coefficient of one of the ions, and the ac-impedance measured total ionic conductivity (σ), you can then solve for the diffusion coefficient of the other ion through the Nernst-Einstein equation. For example, transference numbers have been calculated through electrochemical measurements of polyether-tailed $Co(bipy)_3^{2+}$ hybrid molten salts. In this system, both the ionic conductivity of the melt, and the electrochemical diffusion of the $Co(bipy)_3^{2+}$ cation can be measured electrochemically. These values can then be used to calculate the diffusion coefficient of the ClO_4^- anions, and then determine the transference number of each of the ionic species.

The measurement of electrical conductivity in hybrid materials is somewhat simpler than the measurement of ionic conductivity because the electrical conductivity does not suffer the same large resistances at the electrode interface as ionic conductivities. Electrical conductivities are typically measured using the direct current technique of four-point probe resistivity. This technique has the strong advantage of automatically canceling out the contact resistances that are due to the frequently large impedances at the interfaces between the probes and the material. Four-point probe measurements are capable of quickly and effectively measuring a geometry independent, material-property specific resistivity value (which is easily converted to conductivity). The key to the four point probe technique is that current and voltage are delivered and sensed on separate electrodes (two probes deliver current, while the other two probes sense voltage). This technique is especially useful for solid samples, but is difficult to apply to liquids where the presence of double layer capacitances greatly complicate the interpretation of the resistivity data.

10.5.2
Pulsed Field Gradient NMR

Nuclear magnetic resonance techniques have recently been developed that are capable of measuring the self-diffusion of NMR-active nuclei. The basic experiment for this measurement is called the Pulsed Gradient Spin-Echo (PGSE) experiment. In this experiment, an initial pulsed-field gradient pulse phase-encodes the nuclear spins according to their physical position along the z-axis (the intensity of the pulse varies along the z-axis gradient). After a period of delay time, a second pulsed-field gradient pulse is applied which is exactly opposite to the first pulse. If the nuclei have not translated (e.g. diffused) during the delay, then the second pulse exactly undoes the encoding from the first pulse, and the obtained NMR signal is just as if there had been no initial pulses. However, if the nuclei have moved, the second gradient pulse does not exactly reverse the first pulse, and the resulting NMR signal is attenuated. This loss of intensity can be described by an equation in terms of the diffusion coefficient of the nuclei. Typically, a series of different delays are employed, and the resulting data is fit to extract the self-diffusion coefficient.

Pulsed field gradient NMR techniques have been used to directly measure the diffusion coefficients of Li$^+$ cations and fluorine containing anions to determine mobilities and transference numbers in lithium-conducting polymer electrolytes. This technique works well for systems that contain NMR active nuclei, which have NMR signals that are clear of interferences. For example, ^{7}Li is a NMR active nuclei, and in most lithium conducting polymer electrolytes, the mobile Li$^+$ ions' signal is uncomplicated by the presence of any other chemical forms of lithium in the electrolyte. These pfg-NMR measurements can be made using both solid-state or conventional liquid NMR instruments and produce very reliable diffusion data.

In one example of this type of measurement, pulsed field gradient NMR (pfg-NMR) measurements (section 5.2) were used to directly measure the ion-transport properties in a of composite electrolyte composed of poly(ethylene glycol) dimethyl ether, and hydrophobic fumed silica doped with Li(CF$_3$SO$_2$)$_2$N (LiTFSI). In this experiment, ion diffusion coefficients could be determined by measuring the Li$^+$ cation and the fluorine in the TFSI$^-$ anion. By directly measuring the diffusion coefficients of both anions and cations, the transference numbers for the ions could be calculated.

10.6 Applications

10.6.1 Electrochemical Sensors

Carbon-ceramic composite electrodes (CCEs) have been designed and synthesized as reference electrodes and for electrocatalytic applications. As described above, these electrodes were synthesized by mixing carbon powder with a silica sol. Electrodes containing only carbon particles in the silica gel can be used as "normal" indicating electrodes. When a hydrophilic silica precursor such as tetramethylorthosilicate (TMOS, Si(OCH$_3$)$_4$), is used, the hybrid composite of carbon particles in silica is hydrophilic, allowing an aqueous solution to permeate into the CCE. This leads to a very large effective surface area of the electrode as a large number of carbon particles are in contact with the solution. However, if a hydrophobic silica precursor such as methyltrimethoxysilane, CH$_3$—Si(OCH$_3$)$_3$, is used, the surface of the composite is hydrophobic, leaving only the carbon particles on the outer interface in contact with the solution. This leads to a small effective surface area as only the carbon particles on the surface are in contact with the solution. Fortunately though, this hydrophobic material acts as an array of microelectrodes as the carbon particles are dispersed, and on average are far from each other. This type of electrode geometry can be very favorable for general electrochemical applications.

This CCE chemistry can also be extended to prepare biosensing electrode assemblies. The bioactive hybrid electrode containing glucose oxidase was prepared by the entrapment of this enzyme in the above describe CCE geometry. This film was cast upon a glassy carbon electrode and dried (shown schematically in Fig. 10.8). In the case of a liquid sensor, the electrode design must allow for fast equilibration with the analyte solution. Here, the large surface area film on the glassy carbon electrode allows for fast mass transport between the solution and the composite film on the electrode's surface. The modified electrode can then be placed into an analyte solution. Upon addition of glucose, the hydrogen peroxide products of the electrochemical oxidation of glucose to gluconolactone can be detected electrochemically in the hybrid electrode.

While graphite typically exhibits poor electrocatalytic activity, the addition of electrocatalytically active dopants, to form a modified CCE electrode, shows electrocatalytic activity. The addition of cobalt tetramethyoxymesoporphyrin (TMMP) to a CCE allows this composite electrode to anodically detect SO$_2$, and cathodically detect CO$_2$ and O$_2$. In this configuration (shown schematically in Fig. 10.8), the Co-TMMP serves as the electrocatalyst and the carbon particles are the conductive

Gas Sensor

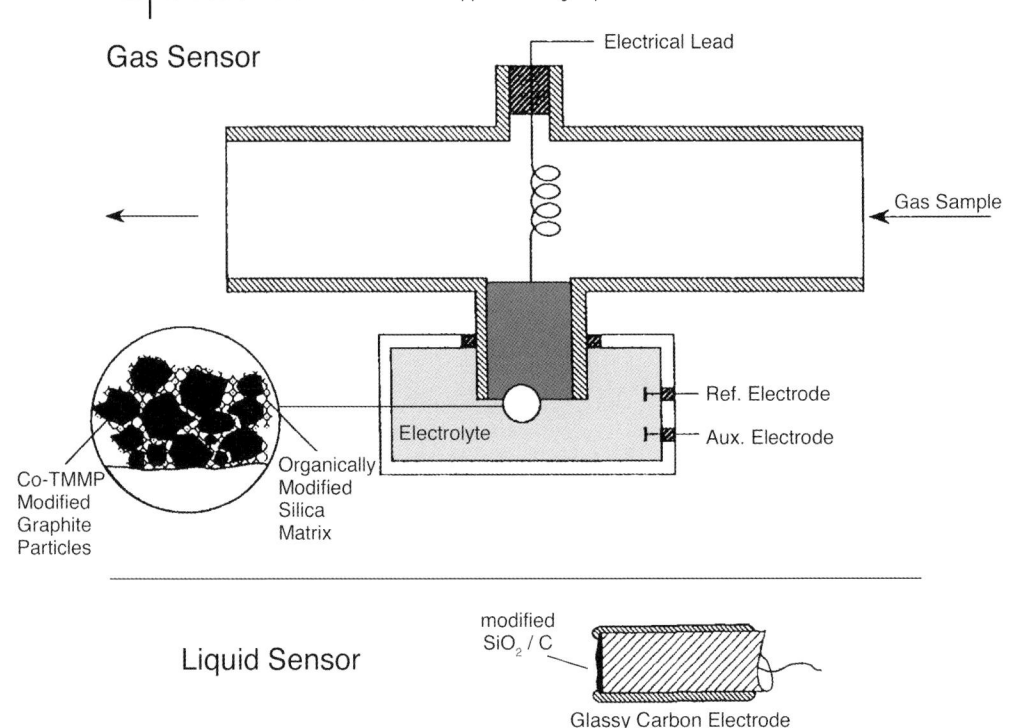

Fig. 10.8 Scheme of CCE-based gas and liquid sensors (adapted from Tsionsky and Lev, *Anal. Chem.* **1995**, *67*, 2409 and Tsionsky et al. *Anal. Chem.* **1994**, *66*, 1747).

medium to electrically connect the porphyrin molecules to the modified electrode. The design of the gas sensor must allow for rapid absorption of the gaseous analyte onto the catalytic surface, which is also in contact with the electrolyte solution. The Co-TMMP is incorporated to the hybrid electrode by mixing a solution of the porphyrin with the carbon particles, evaporating the solvent, and then pyrolyzing the carbon/porphyrin composite. The pyrolyzed carbon/porphyrin is then added to the silica sol.

Electrochemically active sensors have been constructed from hybrid materials through the incorporation of electrocatalysts into inorganic matrices. In general, these types of sensors typically incorporate an electrocatalyst that is capable of selectively reacting electrochemically with the analyte of interest. The electrical signal is typically conducted through a conductive inorganic matrix like TiO_2 or through a composite of a conducting material such as graphite particles in SiO_2. Conducting polymers have also been used in nanocomposites with catalytically active metal particles. Hybrid nanocomposites of polythiophene (PT), poly(styrene sulfonate), and Pt have been used as the oxygen reducing electrode in a fuel cell application. Here, the PT conducting polymer is used to bind and electrically connect the catalytically active Pt nanoparticles. These composites have excellent electrical and H^+ conductivities.

The electrochemical reduction of $Fe^{3+/2+}$ in a room-temperature melt of iron tetraphenylporphyrin with four covalently attached short poly(ethylene glycol) chains can be used to sense the presence of volatile Lewis bases such as pyridine. This iron porphyrin has an open axial coordination site and coordination of a Lewis base at this axial site causes a shift in the electrochemical reduction potential of the Fe^{3+} porphyrin. The presence of a Lewis base can be detected by applying a potential of 0 V which is not sufficient to reduce the uncoordinated Fe^{3+}, but is sufficient to reduce the base-coordinated Fe^{3+}. In this experiment, the flow of current would indicate the coordination of a Lewis basic gas molecule. This material has been shown to be an effective and selective detector in a gas chromatographic application for Lewis basic molecules, and is minimally responsive to analytes that do not coordinate to the Fe porphyrin.

Hybrid materials can also be adapted to ion-selective applications through the formation of hybrid materials doped with ionophores. Typically, these materials consist of an ionophore immobilized within a porous ceramic matrix. In one example, a composite silicate was synthesized containing a covalently attached small 12-crown-4 ether and a tetraphenyl borate anion in order to sense K^+ and Na^+ ions. The incorporation of these ions into the film was used to generate a field-effect transistor effect. Effective ionophores are typically bulky species which are immobilized in porous supports with sufficient pore size to allow the influx of ionic species.

10.6.2
Optoelectronic Applications

Electrogenerated chemiluminescence (ECL = the generation of light from electrochemically produced chemical species) has been created in hybrid composites of swollen silica gels containing the $Ru(bipy)_3^{3+}$ ion and the reductant tripropylamine (TPA). The ECL emission arises from the injection of an electron into a ligand-based excited state on the $Ru(bipy)_3^{3+}$. This electron can then fall down in energy into a metal-based orbital releasing a photon of light (the ECL). In this system, the ECL is generated by either a direct electrochemical reduction of $Ru(bipy)_3^{3+}$ or by a reaction of $Ru(bipy)_3^{3+}$ with the TPA radical. The silica gel serves to establish a microscopically rigid structure that cuts down (but does not eliminate) diffusion of the $Ru(bipy)_3^{3+}$ and TPA species. This hybrid system has shown a significant increase in the stability of the ECL which is likely due to the rigidity of the system provided by the silica gel.

Electrogenerated chemiluminescence has also been generated in a system composed of a redox polyether melt of $Ru(bipy)_3^{2+}$. In this system, two short poly(ethylene glycol) chains are attached to a $Ru(bipy)_3^{2+}$ cation making the perchlorate salt into a room temperature viscous molten salt (only two PEG chains were attached in order to increase the T_g of the resulting hybrid). This material was then placed on an interdigitated array, and a voltage of 2.4 V was applied. This voltage is sufficient to oxidize $Ru(bipy)_3^{2+}$ at the anode and reduce it at the cathode forming concentration gradients of the 3+ ion at the anode, and the 1+ ion at the cathode.

These concentration gradients expand into the bulk solution until they meet somewhere near the midpoint between the two electrodes. In this system, an interdigitated array electrode is used to minimize the distance between electrodes, and give an open geometry that is conducive to detecting luminescence between the electrodes. When the concentration gradients meet, the reduced Ru(bipy)$_3^+$ transfers an electron to the oxidized Ru(bipy)$_3^{3+}$ species which up accepting the electron forms the excited Ru(bipy)$_3^{2+*}$ state, which decays to ground state Ru(bipy)$_3^{2+}$ by emitting a photon. Since this material was designed to have a low T_g, the authors were able to "freeze" the established concentration gradients by simply lowering the temperature below the material's T_g of –5 °C. The frozen film retained its ability to generate ECL, showed diode-like rectification, and even responded rapidly to applied voltages.

Dye-sensitized solar cells show great promise for solar energy conversion. In this system, an electronically conducting, mesoporous semiconducting anode (typically TiO_2) is modified by an adsorbed layer of photosensitizing dye molecules. Photoexcitation of the dye molecule results in the injection of an electron into the conduction band of the TiO_2. This electron then moves through an external circuit to the cathode. The photooxidized dye molecule is reduced by a species in the electrolyte (typically the I^-/I_3^- redox system). In this case, $3I^-$ are oxidized to I_3^- which diffuses to the cathode. At the cathode, I_3^- is reduced by the photoexcited electrons that have traveled through an external circuit. The voltage generated under illumination corresponds to the difference between the Fermi level of the electron in the solid and the redox potential of the solution redox couple. Overall, electricity is generated without permanent chemical transformation. In addition, this system has been constructed using a hybrid gel electrolyte composed of fumed silica and the organic ionic liquid, 1-methyl-3-propylimidazolium iodide. In this electrolyte, the silica particles serve to solidify the ionic liquid electrolyte. This semi-solid electrolyte may enable the fabrication of flexible, solid-state devices free of leakage and available in varied geometries.

Electrochromic Materials can also be prepared from hybrid materials. For example, a composite PAn, PMMA, and silica can be synthesized by the hydrolysis of TEOS in a solution containing the preformed PAn and PMMA polymers. This electrochromic composite is then brushed onto a transparent conductor such as indium-doped tin oxide (ITO). When the PAn conducting polymer containing composite is oxidized, a reversible color change from pale yellow to green is observed. The incorporation of the conducting polymer into the composite shows similar electrochemistry to that of an electrochemically synthesized film of PAn on an electrode. In addition, the incorporation of PAn into the composite greatly increases the adhesion of the film to the electrode by preventing the oxidized form of PAn from dissolving from the surface of the electrode.

The incorporation of WO_3 particles into a PAn film offers the opportunity to access the clear to dark blue color change of WO_3 when reduced. In this device, alternating layers of WO_3 and PAn were electrodeposited onto a transparent conductor. This multilayer composite film can then be reduced forming the dark blue form of the WO_3, and then oxidized changing the WO_3 layer back to colorless and changing the PAn layer to green. Thus, this device represents and electrochromic window that can change color between dark blue and green.

10.6.3
H$^+$-conducting Electrolytes for Fuel Cell Applications

There is a strong desire for a power source that can be used in automotive and miniaturized applications. The proton exchange membrane (PEM – shown schematically in Fig. 10.9) fuel cell design provides high power densities, low physical volumes and weights, operates at relatively low temperatures (60–120 °C), and has quick start-up times. The proton conducting electrolyte is a critically component in PEM fuel cells, and needs to have a mobile H$^+$ ions, and must be dimensionally, chemically, and thermally stable. Hybrid materials are very likely to find applications in PEM fuel cells, in which the proton exchange membrane is typically a polymer based material separating the two electrodes of the fuel cell.

Currently, Nafion (a sulfonated fluoropolymer) is the most widely used proton exchange membrane in PEM fuel cells, because of its mechanical properties, chemical stability, and high proton conductivity when hydrated. However, Nafion membranes require hydration which limits its maximum operating temperature to around 100 °C. At these relatively low temperatures, the tolerance of the electrocatalyst against carbon monoxide poisoning is low. Thus, new electrolyte materials are needed that are capable of operating at high temperatures and low relative humidities. These electrolytes would be able to tolerate a H$_2$ fuel stream which was synthesized from coal, and would likely contain a small concentration of carbon monoxide.

Because Nafion-based H$_2$/O$_2$ PEM fuel cells tend to suffer from low proton conductivities at temperatures near the boiling point of water, new proton conducting hybrid electrolytes have been developed to address this and several other

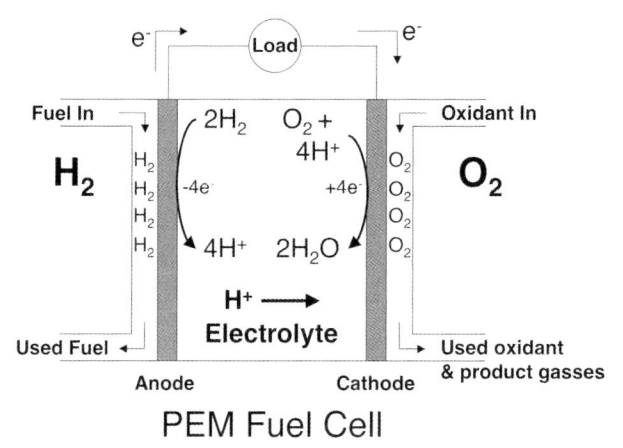

Fig. 10.9 Schematic of a PEM fuel cell.

shortcomings of current polymer electrolytes. In one general example of this chemistry, inorganic components have been employed to improve the self-humidification of the membrane (especially at the anode) by incorporating a hydrophilic inorganic oxide into the membrane that serves to attract and retain water. In addition, these inorganic components can also contain a catalyst that promotes the reaction between dissolved oxygen and the fuel in order to stop the fuel from crossing over in direct methanol fuel cell applications. Furthermore, this self-humidification application has been extended to use solid oxide proton conductors as the inorganic oxide component in order to both retain water and to increase proton conductivity by offering an addition proton conducting path. Inorganic materials can also be incorporated into proton conducting membranes to improve the thermal and mechanical stability of composite membranes without losing ionic conductivity.

This strategy has been employed to create advanced H^+-conducting hybrids based on Nafion and silica or titania that are capable of self-humidification. Specifically, a silica or titania sol was added to a solution of Nafion, then cast and dried to form the hybrid polymer electrolyte membrane. Platinum nanoparticles, in the form of $Pt(NH_3)_4Cl_2$, were also added to the hybrid membrane. The membrane was then capable of self-humidifying when any H_2 fuel crossed over and reacted with O_2 at the Pt nanoparticles. The product of this reaction is water, which can then be absorbed by the hygroscopic silica or titania particles. The addition of the inorganic components into this membrane allows for a thinner electrolyte, which increases the performance of the fuel cell.

In addition to self-humidification, the addition of oxide materials to proton conducting membrane has been explored for high-temperature operations in order to allow the composite electrolyte to remain hydrated at temperatures near or above the boiling point of water. Here, a sol–gel technique has been used to add nanoscale silica particles into a PFSA membrane. The silica particles must be very small such that they can enter the conducting channels of the Nafion membranes and increase the water uptake of the electrolyte composite. This strategy was tested in a H_2/O_2 proton exchange membrane (PEM) fuel cell where it was determined that the presence of ~10% by weight of the nanoscale silica improved the water retention of the electrolytes, which served to increase the proton conductivity at high temperatures. Furthermore, the PEM fuel cell equipped with the composite electrolyte was able to provide four times the current density as a fuel cell with an unmodified Nafion at 130 °C and 3 atm of pressure.

While in the previous example, the nanoscale silica particles were only there to improve the water retention, other examples of the use of bi-functional additives, where the added component serves both water management and proton conduction duties, has been described. Typically, these composite materials take advantage of the ability of certain solid oxide proton conductors to transport protons at high temperatures, while using their oxide particle nature to improve water management. In one example, zirconium phosphate, a solid inorganic proton conductor at high temperatures, was incorporated into a Nafion film. Here, zirconium phosphate particles were formed inside the Nafion membrane by impregnated the membrane with zirconyl chloride and phosphoric acid. The resulting hybrid showed an approximately four fold increase in current density (at 130 °C and 3 bars of pressure) over unmodified Nafion membranes. In another example, silicotungstic acid (SiWA – also an inorganic solid proton conductor) was incorporated into Nafion membranes. The water uptake (~2x) and ionic conductivity (~8x) of the silicotungstic acid/nafion-117 composite membrane were both significantly better than plain nafion-117 membranes. In addition, the mechanical and chemical stability of the composite membranes were as good as plain Nafion.

The addition of particles of solid proton conductors to a Nafion membrane assist proton conduction because they provide the water another pathway for conduction within or on the surface of the added particle. In addition, the added particles are hydrophilic and provide addition hydrogen bonding sites. The formation of hydrogen bonds between the water and the added inorganic particle serves to slow down the evaporation of water and maintain the membrane's hydration. This ability of the added particles to slow evaporation allows the membrane to operate at higher temperatures where water would normally evaporate from the membrane.

In addition to the above described class I hybrid composites, class II hybrids containing strong covalent bonds between the organic and inorganic components, have also found applications in fuel cells. A class II, H^+-conducting hybrid electrolyte has been synthesized by the sol–gel condensation of a species containing sulfonic and sulfonamide functionalized trialkoxysilanes. In this hybrid material, the sulfonamide groups are basic (i.e. H^+ acceptors), and provide a site for the proton to "hop" to (the H^+ hops from the sulfonate donor site to the sulfonamide acceptor site). The incorporation of both donor and acceptor sites facilitates the hopping mechanism of proton transport in this material leading to faster transport and higher ionic conductivities.

One of the first class II, H^+-conducting hybrid materials was a silicate material modified with a trialkoxysilane-modified alkylamine, NH_2—R—$Si(OR)_3$. The amine group acts as a base and becomes protonated in the presence of added acid. The acid loading is kept at less than 50% of the available amine groups in order to leave a significant fraction of amine groups unprotonated. This is the key, because in order to allow for a Grotthus-type proton conductivity, there must be acceptor sites available for the H^+ ion to hop into. This hybrid proton conducting electrolyte, called an "aminosil", is hard and nonporous when dried and shows a liquid-like conductivity of 10^{-5} S cm^{-1} at room temperature. In another example, anhydrous proton conducting electrolytes consisting of a hybrid PEG/silicate polymer have been prepared through sol–gel chemistry. When this material is rigorously dried, a hopping mechanism that depends on the volume fraction of PEG in the polymer is responsible for transporting the H^+ cations from one PEG coordination site to another.

While a large concentration of sulfonate groups in the proton conducting membrane is desirable for high conductivity (more charge carriers gives greater conductivity), the addition of an excessive amount of sulfonate groups to a polymer electrolyte may be accompanied by too much swelling, potentially leading to a water soluble polymer. Too much swelling leads to a loss of mechanical strength of the electrolyte membrane. However, the introduction of an inorganic component into the polymer electrolyte can compensate for the loss of mechanical stability. This strategy has been employed to create hybrid proton conducting electrolytes by attaching the acidic sulfonate group to the organic component after forming a hybrid gel. In an example of this type of chemistry, a mixture of benzotrialkoxysilane and an organotrialkoxysilane were hydrolyzed and allowed to gel forming an organic–inorganic hybrid. The phenyl component can then be sulfonated, adding pendant and acidic —SO_3H groups to the phenyl groups. The resulting hybrid material showed an excellent room temperature conductivity of $10^{-2}\,S\,cm^{-1}$ and a thermal stability up to 250 °C.

10.6.4
Li$^+$-conducting Electrolytes for Battery Applications

Lithium conducting electrolytes have seen a great deal of attention, primarily due to the need for very high energy density batteries for portable electronics. Lithium-ion batteries (shown in Fig. 10.10) are currently the battery of choice for mobile electronics including: cellular phones, laptop computers, and digital cameras. However, most consumers would like longer battery life in these types of applications. Lithium-ion batteries contain a potentially flammable liquid electrolyte contained inside a porous separator. One alternative to the lithium-ion battery is the lithium-polymer battery, which employs a solid polymer electrolyte instead of a liquid electrolyte. The lithium-polymer battery offers a simplified design that is easier to fabricate, more rugged, and has a very thin profile (~1 mm) which gives this material a very flexible form factor. While some "lithium-polymer" batteries are reaching the market (mostly in very thin cellular telephones), these batteries are really hybrids that contain a gelled electrolyte that contains some amount of plasticizing solvent (these are sometimes referred to as lithium-ion-polymer batteries). There remains a strong demand for new lithium-conducting solid electrolytes to maximize the energy density of this system.

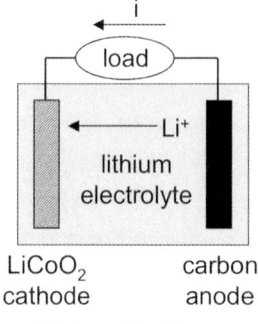

Fig. 10.10 Li$^+$ ion battery schematic.

One general route to making class I hybrids for electrolyte applications is to immobilizing a liquid electrolyte in a solid-state matrix (this would be structurally analogous to what is currently being called a lithium polymer battery). This has been accomplished by adding a silica sol to a solution of $LiBF_4$ in ethylene carbonate and propylene carbonate. The mixture of the silica sol and liquid electrolyte was allowed to gel, forming a dimensionally stable, semi-solid hybrid material which maintains a substantial Li^+ conductivity. The addition of the inorganic silica to this otherwise conventional liquid electrolyte adds mechanical stability, while maintaining a large room temperature ionic conductivity ($\sim 4\times 10^{-3}\,S\,cm^{-1}$). In addition, thermal gravimetric analysis shows that this material is thermally stable up to 90 °C, and that the thermal stability is limited by the liquid EC/PC electrolyte. Thus, this material may have applications as a lithium-ion-polymer battery electrolyte as it could easily be cast into very flexible geometries.

In another example of this type of material, a composite electrolyte was prepared from a mixture of low molecular weight poly(ethylene glycol), $LiClO_4$, and Al_2O_3 and SiO_2 particles. The authors note an increase in the ionic conductivity in this material over mixtures of $LiClO_4$ in poly(ethylene glycol). Interestingly, the formation of ion pairs was examined by FT-IR spectroscopy and determined to be decreased in the Al_2O_3 and SiO_2 composites over the noncomposite materials.

Similar class II, Li^+ conducting systems have been synthesized by attaching organic poly(ethylene glycol) side chains to siloxane based polymers through hydrosilation reactions. For instance, a double comb-polymer with two oligomeric poly(ethylene glycol) side chains, that is capable of dissolving Li^+ salts has been prepared through a sol–gel condensation as a lithium electrolyte (shown in Fig. 10.11). This di-substituted polysiloxane polymers, $[—Si(MePEG_n)_2O—]_x$

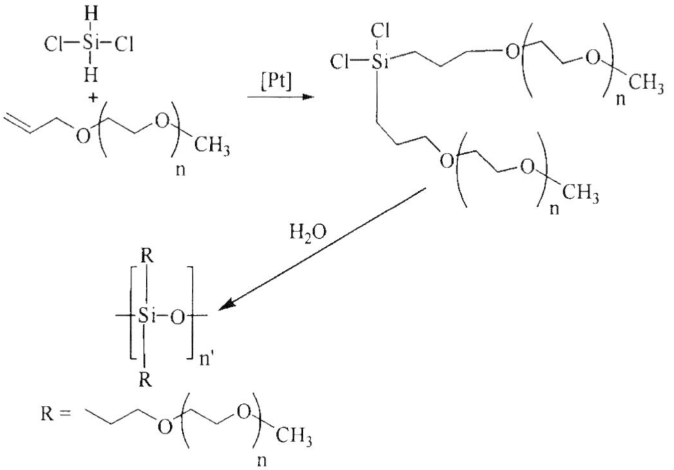

Fig. 10.11 Li^+ conducting siloxane-based polymer. This di-substituted polysiloxane polymer forms completely amorphous, homogenous solutions with $LiN(SO_2CF_3)_2$ (LiTFSI). (Adapted from Hooper et al. *Macromolecules* **2001**, 34, 931).

(where $MePEG_n$ = —$(CH_2)_3O(CH_2CH_2O)_nCH_3$), formed completely amorphous, homogenous solutions with $LiN(SO_2CF_3)_2$ (LiTFSI). The ionic conductivity reached a room temperature peak of ~ $5 \times 10^{-4}\,S\,cm^{-1}$ with $MePEG_6$ side chains (i.e. the side chains were 6 ethylene oxide repeat units long). Unfortunately the dimensional stability of these materials is fairly poor (they are viscous liquids that flow at room temperature). However, a method was developed for cross-linking a similar material to form a free-standing polymer with greater dimensional stability.

Siloxane-based polymers have also been prepared with a mixture of poly(ethylene glycol) side chains, and a second side chain containing a covalently attached anion. These materials have been prepared with trifluoromethylsulfonamide (—NSO_2CF_3) anions, and sulfonate terminated perfluoroethers. These polyelectrolyte materials are rather unique, because the weakly interacting anions are covalently linked to the polymer's backbone. In polysiloxane-based hybrid materials, the siloxane polymer backbone adds a significant amount of flexibility to the hybrid which serves to decrease the glass transition temperature (T_g) and increase the frequency of segmental motions of the polymer, which leads to increased ionic conductivity. Furthermore, the low energy barrier to rotation about the Si—O bond in siloxane polymers (~$0.8\,kJ\,mol^{-1}$) contributes a substantial amount of free volume to the resulting polymer, which serves to further increase the rate of polymer reorganization, and consequently, ionic conductivity.

The applications for polyethylene oxide (PEO) based lithium conducting polymer electrolytes are limited by the inability to achieve both large ionic conductivities and mechanical stability at temperatures below 70 °C. Addressing one of these shortcomings tends to degrade the performance in the second area. For instance, the addition of a plasticizer can increase the ionic conductivity of a PEO-based electrolyte, but generally tends to dramatically decrease the mechanical stability. This problem has been addressed by creating hybrid composites of low molecular weight, end-capped PEO, a lithium salt, and nanoscale surface-modified fumed silica. This hybrid composite is a semi-solid material which displays room temperature ionic conductivities in excess of $10^{-3}\,S\,cm^{-1}$. Here, the surface modification of the nanoscale fumed silica particles is critical to the formation of an open network structure that can increase the mechanical stability of the composite electrolyte. The surface modifying groups need to make the surface of the silica particles somewhat incompatible with the polar PEO, so that they will associate with other silica particles to form the network. This very polar network then increases the mechanical stability of the electrolyte without decreasing the high conductivity and Li^+ transport gained from using the low molecular weight PEO.

Hybrid nanocomposites of conducting polymers and inorganic solids have been prepared and utilized as high energy density electrode materials in Li^+ ion battery applications. For example, composites of polyaniline (PAn) and V_2O_5 have been combined to form a redox-active composite with Li^+ acting as the primary ionic participant in the charge/discharge reactions. Here, the organic conducting polymer serves as both a electrical conductor to cycle the oxidation state of the V_2O_5 to $Li_xV_2O_5$. As the polymer is oxidized, the Li^+ ions diffuse out of the composite as the V_2O_5's negative charges compensate the positive charges on the PAn.

10.6.5
Other Ion Conducting Systems

Mixed ionic/electronic conductors have been synthesized from a hybrid nanocomposite xerogel of V_2O_5 and poly(ethylene oxide). This material is synthesized by addition of an aqueous solution of PEO and lithium triflate to a vanadia sol. The solution was then evaporated to form a thin, flexible film. The addition of the PEO component increased the spacing between the V_2O_5 layers. In order to determine the relative contributions of electronic and ionic conductivity in this material, the authors measured both the electrical conductivity by a four-point probe method, and ionic conductivity by AC-impedance methods. This composite material shows high electronic conductivity through the V_2O_5 regions parallel to the film, as expected. In addition, a lower ionic conductivity was measured perpendicular to the composite film. While the total ionic conductivity was substantially lower than the electrical conductivity, the ionic conductivity of the composite film was 10x the ionic conductivity of the plain V_2O_5 aerogel. This indicates that the conducting of Li^+ ions in the PEO regions contributes to the ionic conductivity in the polymer electrolyte regions of the hybrid composite.

The mechanism of ionic conductivity has also been studied in a composite material of sodium montmorillonite clay and poly[bis(methoxyethoxy)ethoxyphosphazene] (MEEP). Montmorillonite clay is an interesting material because it is composed of sheets of magnesium aluminosilicate separated by intercalated mobile Na^+ cations. This hybrid material is prepared by combining a solution of MEEP and a suspension of the clay, and allowing the MEEP polymer to become intercalated between the sheets of the montmorillonite clay. The conductivity of the hybrid composite is substantially increased from the pristine sodium montmorillonite clay, with a substantial anisotropy. This material has a substantial preference for ionic conductivity parallel to the montmorillonite sheets and a coupling of the polymer segmental motions and long-range ionic conductivity.

Polyoxometalates have been incorporated with conducting polymers to form hybrid materials with electrochemical applications in batteries, supercapacitors, and organic solar cells. For example, molecular inorganic clusters of phospomolybdate ($PMo_{12}O_{40}$) have been combined with PAn to form an electrochemical supercapacitors. The combination of the two components yields a hybrid material with a remarkable improvement in the cyclability (a prerequisite for functional supercapacitors) of the device. In this hybrid material, the formation and redox cycling of the materials are synergic because the activity of the molecular inorganic species can only be formed into a practical electrode because of its integration and anchoring within the conducting polymer network.

The conduction of magnesium ions has been described in a system that has potential to offer a higher power density alternative to lithium based systems. Here, a magnesium electrolyte has been developed that is based on a magnesium organohaloaluminate salt, such as $Mg(AlCl_3R)_2$ where R=alkyl, and which is dissolved in a short oligomeric poly(ethylene glycol) solvent. In addition, the development of a $Mg_xMo_3S_4$ cathode was reported which reversibly intercalates Mg^{2+} ions, and that shows an efficiency close to 100% for chemical deposition.

10.7
Summary

Electrochemically active hybrid materials have a wide-range of applications and are a technologically important class of materials. The ability to chemically combine the properties of different amorphous, crystalline, liquid, and solid inorganic and organic components into one hybrid material allows the physical and electrochemical properties of the resulting hybrid to be tailored in order to be suitable for specific electrochemical applications. These combination of these physical and electrochemical properties frequently combine synergistically, yielding hybrid materials that are especially useful.

The fundamental mechanisms of electrical and ionic conductivity have been described in these types of materials as these topics are fundamental to the understanding how these materials are applied to real-world problems. Many different types of electrochemically active materials have been described in terms of how these materials are formed. In addition, the principles of analyzing these types of materials have also been described. The specific compositions and properties of selected examples of these materials have been described and explained in terms of mechanisms and chemistry of the specific application.

While many hybrid materials have been developed to address specific electrochemical applications, there is still quite a bit of room to develop a deeper understanding of how the electrical and ion-transport properties of hybrid materials are influenced by the properties of the inorganic and organic constituents. In summary, with many potential electrochemical applications awaiting materials with tailored physical and electrochemical properties to be developed, the future of electrochemically-active hybrid materials seems exceedingly bright.

Bibliography

Introduction
A. Walcarius, *Chem. Mater.* **2001**, *13*, 3351–3372.
O. Lev, Z. Wu, S. Bharathi, V. Glezer, A. Modestov, J. Gun, L. Rabinovich, S. Sampath, *Chem. Mater.* **1997**, *9*, 2354–2375.
M. A. Hickner, H. Ghassemi, Y. S. Kim, B. R. Einsla, J. E. McGrath, *Chem. Rev.* **2004**, *104*, 4587–4612.
C. Sanchez, F. Ribot, *New J. Chem.* **1994**, *18*, 1007.

Historical background
P. Gómez-Romero, C. Sanchez, Eds. *Functional Hybrid Materials*; Wiley: Weinheim, **2004**.
J. E. Mark, C. C.-Y. Lee, P. A. Bianconi, Eds. *Hybrid Organic-Inorganic Composites*; Oxford University Press, **1995**.

Fundamental mechanisms of conductivity in hybrid materials
J. H. Perlstein, *J. Sol. State Chem.* **1971**, *3*, 217–226.

M. Tsionsky, G. Gun, V. Glezer, O. Lev, *Anal. Chem.* **1994**, *66*, 1747–1753.

R. Gangopadhyay, A. De, *Chem. Mater.* **2000**, *12*, 608–622.

T. T. Wooster, M. L. Longmire, H. Zhang, M. Watanabe, R. W. Murray, *Anal. Chem.* **1992**, *64*, 1132–1140.

J. E. Ritchie, R. W. Murray, *J. Am. Chem. Soc.* **2000**, *122*, 2964–2965.

J. E. Ritchie, R. W. Murray, *J. Phys. Chem. B.* **2001**, *105*, 11523–11528.

H. Masui, R. W. Murray, *Inorg. Chem.* **1997**, *36*, 5118–5126.

E. Dickinson, M. E. Williams, S. M. Hendrickson, H. Masui, R. W. Murray, *J. Am. Chem. Soc.* **1999**, *121*, 613–616.

W. P. Wuelfing, S. J. Green, J. J. Pietron, D. E. Cliffel, R. W. Murray, *J. Am. Chem. Soc.* **2000**, *122*, 11465–11472.

E. Coronado, J. R. Gala-Mascaró, *J. Mater. Chem.* **2005**, *15*, 66–74.

M. A. Ratner, D. F. Shriver, *Chem. Rev.* **1988**, *88*, 109–124.

P. Johansson, M. A. Ratner, D. F. Shriver, *J. Phys. Chem. B.* **2001**, *105*, 9016–9021.

G. B. Appetecchi, F. Alessandrini, S. Passerini, G. Caporiccio, B. Boutevin, F. Guida-Pietrasanta, *Electrochim. Acta.* **2004**, *50*, 149–158.

J. Adebahr, N. Byrne, M. Forsyth, D. R. MacFarlane, P. Jacobsson, *Electrochim. Acta.* **2003**, *48*, 2099–2103.

K. D. Kreuer, *Chem. Mater.* **1996**, *8*, 610–641.

K.-D. Kreuer, S. J. Paddison, E. Spohr, M. Schuster, *Chem. Rev.* **2004**, *104*, 4637–4678.

Explanation of the different materials

P. Audebert, G. Cerveau, R. J. P. Corriu, N. Costa, *J. Electroanal. Chem.* **1996**, *413*, 89–96.

M. C. Lonergan, E. J. Severin, B. J. Doleman, S. A. Beaber, R. H. Grubbs, N. S. Lewis, *Chem. Mater.* **1996**, *8*, 2298–2312.

R. W. Murray, A. G. Ewing, R. A. Durst, *Anal. Chem.* **1987**, *59*, 379A–390A.

Special analytical techniques

A. J. Bard, L. R. Faulkner, *Electrochemical Methods: Fundamentals and Applications 2nd edition*, John Wiley, **2001**.

M. E. Williams, L. J. Lyons, J. W. Long, R. W. Murray, *J. Phys. Chem. B.* **1997**, *101*, 7584–7591.

P. T. Callaghan, *Principles of nuclear magnetic resonance microscopy*, Oxford University Press, **1991**.

H. J. Walls, P. S. Fedkiw, T. A. Zawodzinski, S. A. Khan, *J. Electrochem. Soc.* **2003**, *150*, E165–E174.

Electrochemical sensors

M. Tsionsky, O. Lev, *Anal. Chem.* **1995**, *67*, 2409–2414.

Z. Qi, P. G. Pickup, *Chem. Comm.* **1998**, 2299–2300.

J. W. Long, R. W. Murray, *Anal. Chem.* **1998**, *70*, 3355–3361.

K. Kimura, T. Sunagawa, M. Yokoyama, *Chem. Comm.* **1996**, 745.

Optoelectronic applications

M. M. Collinson, J. Taussig, S. A. Martin, *Chem. Mater.* **1999**, *11*, 2594–2599.

K. M. Maness, H. Masui, R. M. Wightman, R. W. Murray, *J. Am. Chem. Soc.* **1997**, *119*, 3987–3993.

A. Hagfeldt, M. Gratzel, *Acc. Chem. Res.* **2000**, *33*, 269–277.

P. Wang, S. M. Zakeeruddin, P. Comte, I. Exnar, M. Gratzel, *J. Am Chem. Soc.* **2003**, *125*, 1166–1167.

G.-W. Jang, C. Chen, R. W. Gumbs, Y. Wei, J.-M. Yeh, *J. Electrochem. Soc.* **1996**, *143*, 2591–2596.

P. K. Shen, H. T. Huang, A. C. C. Tseung, *J. Electrochem. Soc.* **1992**, *139*, 1840–1845.

H⁺-conducting electrolytes for fuel cell applications

Q. Li, R. He, J. O. Jensen, N. J. Bjerrum, *Chem. Mater.* **2003**, *15*, 4896–4915.

M. Watanabe, H. Uchida, Y. Seki, M. Emori, *J. Electrochem. Soc.* **1996**, *143*, 3847–3852.

Q. Deng, R. B. Moore, K. A. Mauritz, *Chem. Mater.* **1995**, *7*, 2259–2268.

K. T. Adjemian, S. J. Lee, S. Srinivasan, J. Benziger, A. B. Bocarsly, *J. Electrochem. Soc.* **2002**, *149*, A256–A261.

P. Costamagna, C. Yang, A. B. Bocarsly, S. Srinivasan, *Electrochim. Acta.* **2002**, *47*, 1023–1033.

B. Tazi, O. Savadogo, *Electrochim. Acta.* **2000**, *45*, 4329–4339.

L. Depre, M. Ingram, C. Poinsignon, M. Popall, *Electrochim. Acta.* **2000**, *45*, 1377–1383.

Y. Charbouillot, D. Ravaine, M. Armand, C. Poinsignon, *J. Non-Cryst. Solids* **1988**, *103*, 325–330.

B. D. Ghosh, K. F. Lott, J. E. Ritchie, *Chem. Mater.* **2005**, *17*, 661–669.

J. E. Ritchie, J. A. Crisp, *Anal. Chim. Acta.* **2003**, *496*, 65–71.

I. Gautier-Luneau, A. Denoyelle, J. Y. Sanchez, C. Poinsignon, *Electrochem. Acta.* **1992**, *37*, 1615–1618.

Li$^+$-conducting electrolytes for battery applications

P.-W. Wu, S. Holm, A. Duong, B. Dunn, R. Kaner, *Chem. Mater.* **1997**, *9*, 1004–1011.

M. Marcinek, A. Zalewska, G. Zukowska, W. Wieczorek, *Solid State Ionics* **2000**, *136–137*, 1175–1179.

D. Swierczynski, A. Zalewska, W. Wieczorek, *Chem. Mater.* **2001**, *13*, 1560–1564.

R. Hooper, L. J. Lyons, M. K. Mapes, D. Schumacher, D. A. Moline, R. West, *Macromolecules* **2001**, *34*, 931–936.

Z. Zhang, D. Sherlock, R. West, R. West, K. Amine, L. J. Lyons, *Macromolecules* **2003**, *36*, 9176–9180.

D. P. Siska, D. F. Shriver, *Chem. Mater.* **2001**, *13*, 4698–4700.

J. F. Snyder, J. C. Hutchison, M. A. Ratner, D. F. Shriver, *Chem. Mater.* **2003**, *15*, 4223–4230.

S. Kuwabata, T. Idzu, C. R. Martin, H. Yoneyamaa, *J. Electrochem. Soc.* **1998**, *145*, 2707–2710.

Other ion conducting systems

G. M. Kloster, J. A. Thomas, P. W. Brazis, C. R. Kannewurf, D. F. Shriver, *Chem. Mater.* **1996**, *8*, 2418–2420.

J. C. Hutchinson, R. Bissessur, D. F. Shriver, *Chem. Mater.* **1996**, *8*, 1597–1599.

P. Gómez-Romero, K. Cuentas-Gallegos, M. Lira-Cantú, N. Casañ-Pastor, *J. Mater. Sci.* **2005**, *40*, 1423–1428.

D. Aurbach, Z. Lu, A. Schechter, Y. Gofer, H. Gizbar, R. Turgeman, Y. Cohen, M. Moshkovich, E. Levi, *Nature* **2000**, *407*, 724–727.

11
Inorganic/Organic Hybrid Coatings
Mark D. Soucek

11.1
General Introduction to Commodity Organic Coatings

Organic coatings are thin film composites, consisting of an organic polymeric binder, with a variety of other constituents including pigments, fillers, solvents, water, organic crosslinkers, and myriad of additives, depending on usage. The components of a coating are mixed together into a coating formulation, which is afterwards applied to a substrate. There are liquid coatings based on organic solvent, water, or both; and powder coatings in which powders are applied directly to substrates. Each constituent has a function connected to end usage or final film properties. The organic binder is the continuous phase that binds the pigment and filler together. Pigments are used for color, corrosion protection, or to change barrier properties. Fillers such as calcium carbonate are used to replace some of the more expensive coating components (binder and pigment) to reduce cost. For liquid coatings, solvents are used to: 1) dissolve or make miscible all the organic components including binder, crosslinker, and additives; 2) control application viscosity; and 3) control the film formation process. Additives are the most versatile, and arguably the most interesting of the coating formulation components. There are additives that control flow or rheology, provide interface stabilization (dispersants, surfactants), promote adhesion, kill microbes and higher life forms (antibacterial, antifungal, antifouling), prevent foaming, etc.

Organic coatings are classified by the state of matter, continuous phase, polymeric binder, means of curing, and method of application. State of matter is either liquid or powder. Continuous phase for liquid coatings is either organic solvent or water. Solvent-borne are coatings in which the organic binders and the other organic components are dissolved in an organic solvent (usually mixture of solvents). A special sub-category is of reactive diluents where small organic molecules act as solvents, yet also participate in the chemical reactions of the curing process and are not emitted into the atmosphere as are most organic solvents. Dispersions of either liquid (water-reducible) or a soft solid (latex) in water are by definition water-borne systems. Both solvent-borne and water-borne coatings can

Hybrid Materials. Synthesis, Characterization, and Applications. Edited by Guido Kickelbick
Copyright © 2007 Wiley-VCH Verlag GmbH & Co. KGaA, Weinheim
ISBN: 978-3-527-31299-3

be thermosetting or thermoplastic. The vast majority of solvent-borne coatings are thermosetting, however. The classification of coatings based on means of curing is radiation cured coatings known either as UV-curable or electron beam curable coatings (E-beam). The method of application is an electrophoric process called E-coat which has usage in the automotive primer industry.

For a chemist, classification by polymeric binder is the most logical. There are 13 major coatings classifications for binders: drying oils, alkyds, polyesters, epoxides, polyurethanes, polyureas, acrylics, vinyl acetates, silicones, silicates, polyvinyl chlorides, phenolics, and vinylidene fluorides. It is interesting to note that the most ubiquitous organic group is the ester group. Drying oils, alkyds, polyesters, acrylics, and vinyl acetate all have ester groups in the repeat unit either in the backbone or pendent to the backbone. Alkyds are aromatic dicarboxylic acids reacted with monoglycerides and were named as a combination of alcohols with acids spelled (alcids) with a k and y instead of c and i for marketing purposes. The largest usage organic binders are acrylics, alkyds, and polyurethanes; while the vinylidene fluoride, silicates, and silicones are specialty coatings. This is contingent on the broadness of the definition of coatings. Polymeric binders can be formulated to be used in multiple media such as organic solvents (solvent-borne) or water (water-borne), and by different means of processing or application such as liquid UV-curable coatings or powder coatings as shown in Table 11.1.

Coatings can also be described by their processing. There are several processes in coatings most importantly including the application and film formation processes. There are multiple application processes for coatings. For liquid, there is spray, dip, and transfer of liquid from roller to roller. Spray can be a single nozzle to a curtain. Both dip and spray utilize gravity, surface tension, or columbic forces to aid in the coating process. For powder, there is spray and fluidized beds. The powder spray process is dependent either on static or applied columbic forces to attach the powder to the substrate.

In the realm of organic coatings, both thermoplastic and thermosetting systems are used. Thermoplastics coatings fall into the broad categories of latexes and plastisols. Latexes are principly acrylic or vinyl acetate based with the exception of paper coatings which are styrene–butadiene based. Plastisols are primarily polyvinyl

Table 11.1 Categories of coatings with polymeric binders systems.

Solvent-borne	Water-borne	UV-curable	Powder
Alkyd	Alkyd	Thiol-ene	Polyamide
Polyurethane	Polyurethane	Polyurethane	Polyurethane
Polyester	Styrene–butadiene	Polyester	Polyester
Epoxide	Epoxide	Epoxide	Epoxide
Acrylic	Acrylic	Acrylic	Acrylic
Phenolic	Vinyl acetate	Vinyl ether	Phenolic
Vinylidene fluorides	Vinylidene fluorides	Bismaleimide	Epoxide/polyester hybrid

chloride dispersions in either organic solvent or water. The only other exception is powder nylon coatings used for dishwasher racks. For the coatings scientist, "K" denotes the number of cans or containers that are needed to be mixed just prior to application of the coating. For example, epoxides are typically 2 K systems with the epoxide in one container and a hardener or curative in another. The two parts are mixed in a predescribed ratio, and applied to a substrate afterwards. Thermosetting coatings can be 1 K or 2 K, all the components in one can (1 K), or a 2-can system where two separate cans or feed streams are mixed shortly before or during the application process. For 2 K processes, potlife stability (time in which an application viscosity is maintained) and processing window are a constant concern.

Coatings are used as protective, decorative, or both. Inorganic/organic hybrid coatings have the possibilities in both the protective and decorative sides of coatings science. Thus far, most of the emphasis has been in the protective aspect, however, inorganic compounds also have the possibility of color due to size (quantum dots), d–d transitions (metals), ligand transfer bands (organic ligand to central atom, or attached dyes), and doping of semiconductors for light interference pigments. There are a myriad of inorganic pigments which as used in organic coatings for a variety of functions including color, corrosion and abrasion resistance. It is anticipated that intercalation of metals into sol–gel precursors would result in colored materials similar to presently used inorganic pigments. For protective purposes, abrasion and corrosion resistance are common themes for inorganic components in organic coatings.

11.2
General Formation of Inorganic/Organic Hybrid Coatings

Traditionally, the formation of ceramic materials from solid materials requires excessive heating at temperatures greater than several hundred degree centigrade (1,2). It was not until the 1930s that metal alkoxides were recognized as potential molecular precursors in the preparation of ceramic materials (3). The formation of metal oxide materials via the sol–gel process has provided a relatively new approach in the formation of ceramics and glasses and in particular coatings (4). The sol–gel process allows low temperature synthesis while yielding high purity homogenous ceramic-type materials (5–7). The preparation of multicomponent ceramic materials by the sol–gel process was first developed by careful control of the hydrolysis and condensation steps of metal alkoxides (5,8). This approach has been applied to thin coatings for the decoration of crystal glassware as a replacement for coloration by melting techniques (9). Sol–gel type hybrid coatings provide here several advantages they can be sprayed onto the substrate and can contain organic dyes, which is not possible with the high temperature procedures. Compared with conventional pure organic systems the sol–gel hybrids show a high abrasion resistance, almost perfect adhesion, refractive index matching and sufficient stability in dishwashing procedures.

These advantages lead to the development of a plethora of inorganic/organic coatings called "ormocers" (Organically Modified Ceramerics) (10). The coatings have a high inorganic content (95–40 wt%) and a low organic content (5–60 wt%). The inorganic component in these coatings is usually silica typically prepared from alkoxy silanes via the sol–gel process (Chapter 1). The alkoxysilane has an organic functional group that crosslinks via homopolymerization or copolymerization with a complementary group. Examples of reactive groups are glycidyl epoxides such as GPTMS (glycidyl propyl trimethoxysilane) and VTES (vinyl triethoxysilane) (11). The epoxide has also been reacted with aminoplast groups which can be attached to another alkoxysilane group or an organic resin. A whole host of usages including abrasion resistant and corrosion resistance has been widely reported for these types of inorganic/organic coatings. This chapter will focus on coatings where organic polymers or silicones are the continuous phase and are >80 wt% and not on ormocers.

Inorganic/organic coatings have received much attention during the past 20 years (12,13,14,15). In the beginning, inorganic/organic coatings based on sol–gel precursors were called ceramers. The term "ceramer" is defined as a material with both ceramic and polymeric characteristics (16). These materials are based on the concept of incorporating colloidal inorganic particles into an organic phase via the sol–gel process forming a continuous polymeric network with substantially better physical properties than conventional organic polymers. Ceramer coatings have also been developed utilizing organically modified metal alkoxides (17). In general, the organic matrix is the continuous phase in ceramer materials and the inorganic matrix is the continuous phase in ormocer materials. The resulting materials provide a combination of physical properties found in both ceramic and polymeric materials by producing a homogenous material with both inorganic and organic characteristics at low cure temperatures. These materials are often considered high performance on account of the ability to withstand high temperature applications. A pictorial representation of the ceramer concept is shown in Fig. 11.1 (18).

A few examples of sol–gel precursors commonly used in the sol–gel process are shown in Fig. 11.2. The reactivity is also influenced by the molecular complexity of the metal alkoxides. Titanium ethoxide exhibits an oligomeric structure, while titanium isopropoxide is usually in the monomeric form. However, the reaction kinetics and structure can be controlled by the choice of the organic matrix (19). For *in situ* formation of sol–gel colloids within an organic matrix the choice of the organic phase and curing conditions is critical. Most of the organic matrices have inherent coupling groups, nascent water concentration, and an acid or base catalyst. All three factors have limited adjustment within any of the organic matrices, including the introduction of auxiliary coupling groups.

11.2.1
Acid and Base Catalysis within an Organic Matrix

Acid and base catalysts have been widely used in the sol–gel process (19,20,21,22). The reaction rate is highly dependent on the pH of the media. It is important to

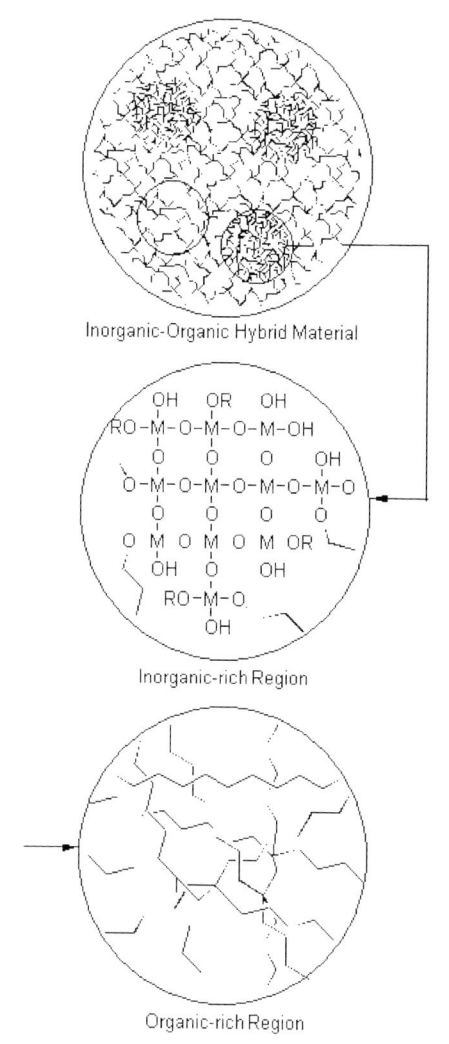

Fig. 11.1 Pictorial representation of the ceramer concept.

note that within an organic matrix, typically used for coatings, pH is in a nonaqueous and nonalcoholic environment, which has a large effect on catalyst mobility and strength. A further limitation of catalyst mobility exists when the catalyst is part of the polymer either in the backbone, tethered as a side group, or at the chain end. Besides being part of the organic matrix acid or base catalyst can be added externally. Depending on the sol–gel precursor, acids and bases can influence how fast the hydrolysis, and condensation steps proceed. Acid catalysts protonate the alkoxy group by enhancing the reaction kinetics, producing alcohol as a leaving group, whereas bases produce strong nucleophiles of the hydroxyl ligands. Depending on the catalytic conditions, the structure of the inorganic matrix

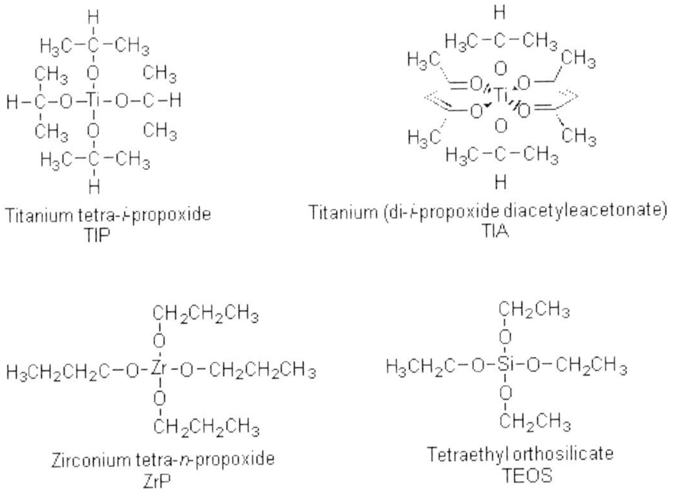

Fig. 11.2 Examples of sol–gel precursors commonly used in coatings.

can be tailored accordingly. Under acidic conditions, the rate of condensation is slow relative to the rate of hydrolysis (4). Generally, a linear more open structure is attained under acidic conditions. Under basic conditions, the rate of condensation is much faster than the rate of hydrolysis and as a result, highly condensed three-dimensional metal oxide clusters form resulting in more closed structures (4). Although simplistic, the final structures for an acid and base catalyzed sol–gel reaction can be depicted in Fig. 11.3.

The sol–gel process has been widely used in the synthesis of novel inorganic/organic materials (12,13,14,15). However, depending on the starting materials a variety of synthetic techniques involving the sol–gel process can be used to produce

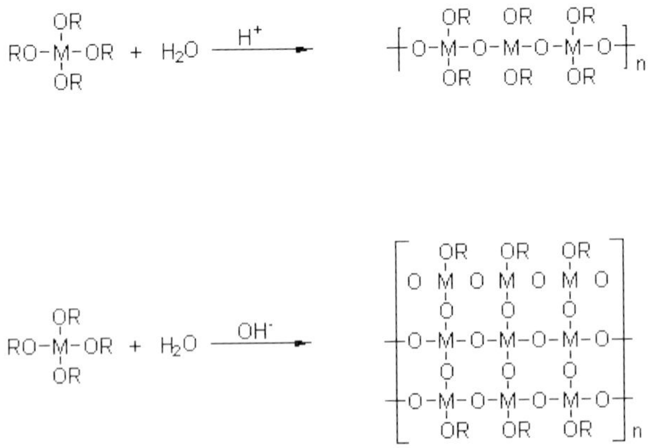

Fig. 11.3 Structural variations of acid and base catalysis *in situ* or *ex situ* formation in coatings.

various interactions between the inorganic and organic phases. In most cases, inorganic/organic materials can be classified into four major classes (14):

1. Inorganic domains incorporated into an organic matrix.
2. Organic domains incorporated into an inorganic matrix.
3. An interpenetrating network between the inorganic and organic phases with no covalent interactions.
4. An interpenetrating network between the inorganic and organic phase with covalent interactions.

Ceramers are best defined as materials where the organic matrix is the continuous phase and ormocers are materials where the inorganic matrix is the continuous phase. However, depending on the size of the inorganic or organic domains, the materials can also be classified as a nanocomposite (14). As a result, nanocomposites are best classified as a subdivision in any class described above. Furthermore, true hybrid inorganic/organic materials can only be obtained when covalent bonding between the two phases exist.

A wide variety of ceramer systems contain either tetramethyl orthosilicate (TMOS) and tetraethyl orthosilicate (TEOS) as the inorganic precursors (23,24). The hybrid ceramer materials are formed by co-condensation of functionalized oligomers with sol–gel precursors in which a covalent bond is established between the two phases as shown in Fig. 11.4. Some systems, for example, contain low molecular weight poly(dimethylsiloxane) (PDMS) oligomers terminally functionalized with silanol groups as the organic moiety and TEOS as the inorganic component (23,24). The inorganic/organic solution is usually catalyzed under acidic conditions in which monolithic and optically clear hybrid materials were produced. Various parameters such as pH, PDMS:TEOS ratio, and molecular weight of PDMS were varied in order to optimize the final structure and properties. It was found that the pH was the most important parameter in controlling the inorganic domain size in the final materials.

Optical transparency was initial evidence that the inorganic domain size was smaller than the wavelength of visible light. It was later verified by small angle X-ray scattering (SAXS) data that the morphology of these materials was relatively homogeneous with no major phase separation (25,26,27). Unfortunately, the PDMS–TEOS hybrid materials did not provide adequate tensile properties. Ceramer materials based on hydroxyl terminated poly(tetramethylene oxide) (PTMO) oligomer as the subsequent organic moiety with TEOS as the inorganic phase (28,29,30,31). The PTMO–TEOS materials provided better tensile properties while still maintaining optical transparency. The hydroxyl terminated PTMO provided functional groups with reactivity comparable to TEOS in which a more homogeneous material was obtained. Again, the reaction conditions such as pH, temperature, and PTMO:TEOS ratio ultimately governed the final properties.

Although most ceramer materials were based on either TEOS or TMOS, titanium isopropoxide and zirconium isopropoxide were also used as the inorganic phase with PDMS or PTMO as the organic component forming mixed metal hybrid materials (18,28,32,33,34). The subsequent materials were found to have

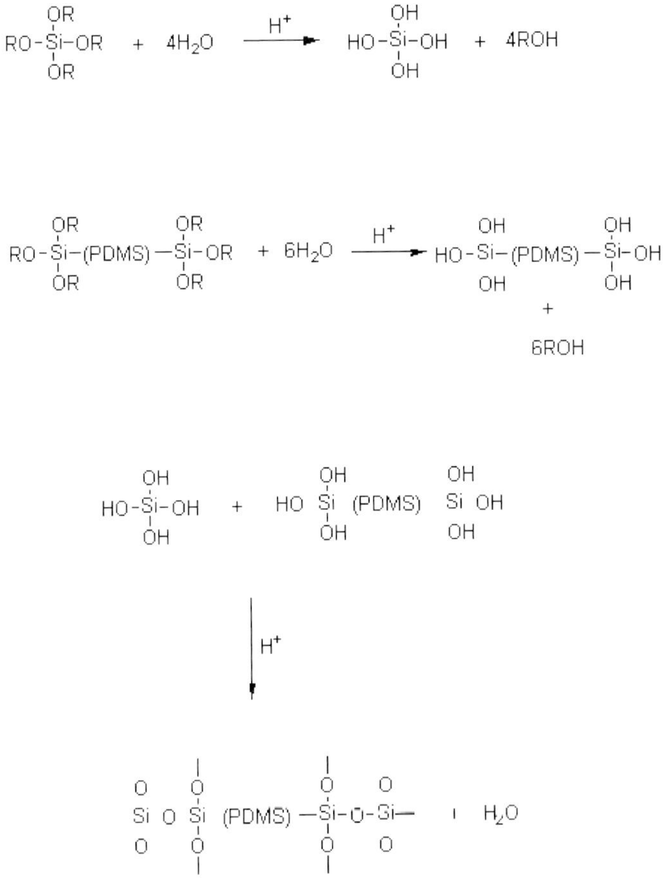

11 Inorganic/Organic Hybrid Coatings

$$\text{RO-Si(OR)}_3\text{-OR} + 4H_2O \xrightarrow{H^+} \text{HO-Si(OH)}_3\text{-OH} + 4ROH$$

$$\text{RO-Si(OR)}_2\text{-(PDMS)-Si(OR)}_2\text{-OR} + 6H_2O \xrightarrow{H^+} \text{HO-Si(OH)}_2\text{-(PDMS)-Si(OH)}_2\text{-OH} + 6ROH$$

$$\text{HO-Si(OH)}_2\text{-OH} + \text{HO-Si(OH)}_2\text{-(PDMS)-Si(OH)}_2\text{-OH}$$

$$\downarrow H^+$$

$$\text{Si(O)}_3\text{-O-Si(O)}_2\text{-(PDMS)-Si(O)}_2\text{-O-Si(O)}_2\text{-} + H_2O$$

Fig. 11.4 Ceramer materials formed by co-condensation of functionalized PDMS oligomers and TEOS.

some microphase separation between the inorganic and organic phases. Ceramer materials based on triethoxysilane terminated with poly(ether ketone), poly(arylene ether phosphene oxide), as well as other functionalized oligomers as the organic components. A detailed description of each hybrid material based on these organic moieties can be found in literature (34,35,36,37).

Inorganic/organic materials were derived using low molecular weight organoalkoxysilanes as the sol–gel precursor (17,38,39,40,41,42). In these hybrid materials, organic moieties were incorporated into the inorganic network via Si—C bonds. The properties of a number of other inorganic/organic-modified materials in which they concluded that the materials tend to behave more like inorganic glasses than organic polymers (43). Two groups of ceramers were of interest; materials with excellent mechanical surface properties and materials that exhibit thermoplastic properties that can be used as functional or protective coatings (44). These coatings were based on aluminum, silicon, titanium, and zirconium oxide networks with epoxy or methacrylate functionalities. It was found that the sol–gel synthesis of organically modified metal alkoxides leads to materials with inorganic and organic properties with minimal crack formation. One can assume that the organic phase in these materials improved the densification behavior and provided relaxation throughout the polymer matrix. In addition, the diffusion of oxygen and water as well as chemical and electrical properties are dependent on the organic functionality present. An application for these hybrid inorganic/organic materials for hard contact lenses (38).

A number of bioceramic materials have been reported including biosensors, and bioactive sol–gel glasses with active enzymes or antibodies (45). Biocompatible and biodegradable ceramer materials have also been prepared by the sol–gel process utilizing α,ω-hydroxyl poly(ε-caprolactone) (PCL) or triethoxysilane end capped PCL as the organic phase and TEOS as the inorganic phase (46,47,48,49) The hybrid aliphatic polyester–silica ceramers were found to have excellent scratch resistance while maintaining optical transparency. These materials were developed in expectation as a coating material for medical applications (47). Another bioceramic application is in the development of bioactive hydroxyapatite (50). It was shown that hydroxyapatite crystals were grown *in vivo* or *in vitro* in various sol–gel derived environments.

Polyimide–silica materials using TEOS and TMOS as the inorganic phase was extensively studied (51,52,53,54). These materials were derived by utilizing poly(amic acid) (PAA) as the organic phase. The approach consisted of a concomitant imidization and the *in situ* sol–gel process yielding the hybrid material. Unfortunately, this approach resulted in polyimide–silica ceramers containing 70 wt% silica with major phase separation and poor adhesion characteristics between the silica particles and the polyimide phase (55). Attachment of diamine functional groups to the inorganic matrix and were found to dramatically improve the mechanical properties of these films (56). The incorporation of diamine groups into the inorganic matrix implemented covalent bonding between the two phases. Again, solvent type and content were found to have a significant effect on the morphology.

Another method used to developed novel ceramer materials is in the formation of inorganic particles within an organic matrix via the sol–gel process forming an elastomeric reinforced material (57,58,59,60,61). Applying this approach the reinforcement of poly(dimethylsiloxane) (PDMS) networks by silica gel particles (62). The incorporation of silica particles within the PDMS matrix increased the tensile properties, and toughness. An acid or base catalyst primarily controlled the size and structure of the inorganic domains entrapped within the organic matrix. The acid-catalyzed system was shown to yield densely packed linear metal oxide structures while base catalyzed systems tended to yield three-dimensional colloidal domains with porous film structures. They concluded that well-distributed inorganic particles within the organic matrix could be obtained in the presence of a base catalyst.

There are many articles on novel inorganic/organic hybrid ceramer materials which emphasize a protective coating on metal substrates (63). Polymetallosiloxane (PMS) polymers have been studied for use as corrosion–protection coatings on aluminum substrates (64,65,66). These coatings were developed utilizing monomeric organofunctional silane derivatives with other metal alkoxides based on Ti, Zr, Si, and Al (67). In each case, an acid catalyzed reaction using HCl was necessary in order to produce a clear sol forming a smooth, uniform coating. It was observed that the type of monomeric organofunctional silane and ratio of organosilane to metal alkoxide controlled film formation such as wettability and uniformity on aluminum substrates. In addition, low concentration of Si—O—M linkages within the PMS matrix was necessary in order to minimize stress cracks within the films. Under basic conditions, phase separation occurred resulting in a clouding effect. Under acidic conditions, the solution remained clear signifying no phase separation between the condensed Si—O—M particles and PMS matrix. However, a pH in the range of 7 to 8 was found to provide the best overall protective properties.

Alkoxysilanes have also been used as adhesion promoters on aluminum substrates (68). The alkoxysilane acts as a coupling agent by providing a covalent bond between the organic coating and the aluminum substrate as depicted in Fig. 11.5. Ideally, a Si—O—Al linkage is achieved through a co-condensation mechanism between Si—OH and Al—OH (68). However, the mechanism is not fully understood and remains a current topic of interest. It is also possible that the increase in adhesion may be due to polar–polar interactions as well as hydrogen bonding. In addition, functionalized alkoxysilane monomers have also been used for corrosion protection (69). TEOS was used in conjunction with an aminosilane adhesion promoter. The solution was used as a pretreatment on aluminum in which excellent adhesion and corrosion protection was observed.

Hybrid ceramer coatings based on polyester resins as the organic phase and TEOS as the inorganic phase were also investigated (70). Hydroxyl and acid-terminated polyesters were used to create a covalent bond to TEOS upon curing. However, during the curing steps, it was found that TEOS evaporated before hydrolysis and condensation could take place. A prehydrolysis and condensation step was found necessary in order to minimize loss of TEOS prior to curing. This was initiated by water under acidic or basic conditions in which the latter condition was found to provide the best overall coating properties. The preformation of hydrolyzed silicon particles within the organic matrix promoted covalent interactions between the two phases. The ceramer coatings were found to have excellent

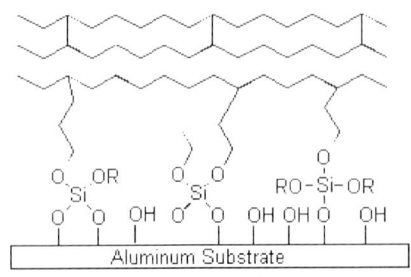

Fig. 11.5 Alkoxysilanes as adhesion promoters on aluminum substrates.

hardness and scratch resistance. Unfortunately, the Si—O—C crosslinking sites could be easily hydrolyzed leading to premature degradation and ultimately failure as a protective coating.

11.2.2
Thermally Cured Inorganic/Organic Seed Oils Coatings

Drying oils have been used for centuries as binders in paints but their usage has diminished due to the development of synthetic polymers during the past 40 years. Drying oils are mostly vegetable oils, such as linseed, soybean, safflower oil or wood oils such as tall oil, or even a tung oil. The term drying oil is derived from the ability of the oil to transform from a liquid to a solid when applied as a thin film. The term drying is misleading on account that the drying process usually is an evaporative one, not an autoxidative chemical crosslinking reaction as in this case. The synthetic polymers widely used today in coatings include acrylics, epoxies, and urethanes. However, these paints are generally applied using solvents that are slowly being eliminated under governmental regulations outlined in the 1990 Clean Air Act (71). Drying oils on the other hand are natural binders and are considered environmentally benign. The utilization of drying oils has an environmental impact by minimizing air pollution and also has an impact on the agricultural economy. The idea of a biomass feedstock that does not wide price fluctuations has appeal. Unfortunately, drying oils do not possess the necessary properties capable of protecting metal substrates compared with conventional coatings used today. However, when used as inorganic/organic hybrid materials the seed oil phase and sol–gel inorganic phase create a superior material (72).

Drying oils are among the oldest binders used in paints and are still used today as raw materials in alkyds, epoxy esters, and uralkyds (72,73). The physical properties of drying oils are governed by their nature and fatty acid composition. Their coating properties depend on the degree of unsaturation since it is through the double bonds that the polymerization takes place. Drying oils are fatty acid triglycerides as shown in Fig. 11.6. Oils such as linseed, tung, and tall oils are among

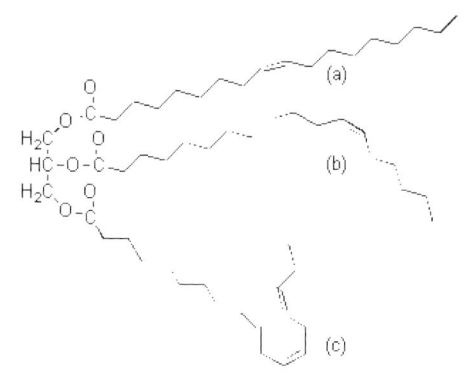

Fig. 11.6 Idealized triglyceride structure containing (a) oleic, (b) linoleic, and (c) linolenic fatty acids.

Table 11.2 Typical fatty acid composition of various oils used in the coatings industry.

	Fatty acid			
Oil	Saturated	Oleic	Linoleic	Linolenic
Linseed	10	22	16	51
Safflower	11	13	75	1
Soybean	15	25	51	9
Sunflower	13	26	61	Trace
Tung	85	8	4	3

the most popular oils once used as the primary binder in protective coatings. Oils with pronounced drying properties usually contain highly unsaturated acid groups such as linolenic acid. Table 11.2 shows the typical fatty acid composition and content of various oils (73).

11.2.3
Drying Oil Auto-oxidation Mechanism

A drying oil applied as a thin film will gradually change from a liquid to solid state forming a hard continuous coating. The transformation was termed "drying" for its resemblance to the change that occurs when water evaporates from an aqueous solution (74). However, this definition is rather misleading since minimal solvent evaporation takes place. The drying process is a molecular growth via polymerization initiated by absorption of oxygen (75). The essential process for which drying of an oil occurs is through an auto-oxidation process. Many reviews of the autoxidative polymerization process are available (76,77,78). The mechanism for which drying takes place has been extensively investigated but the entire process is still not fully understood. There are many parameters that can be varied in this process, including light, heat, moisture, catalysts, and inhibitors.

Drying oils crosslink through unsaturated fatty acid residues via an autoxidative process (74). Typically, the autoxidative process is separated into three steps; initiation, propagation, and termination as illustrated in Fig. 11.7. In the initiation step, naturally present hydroperoxides decompose forming free radicals. These free radicals have high reactivity toward antioxidants, which form peroxy free radicals. As the concentration of antioxidants decrease, propagation proceeds by abstraction of hydrogen atoms on methylene groups between double bonds forming free radical **1**. These free radicals react with oxygen forming a conjugated peroxy free radical such as **2**. Regeneration of free radical **1** can proceed by abstracting additional hydrogen atoms from other doubly allylic methylene groups. Crosslinking then occurs by radical–radical combination forming carbon–carbon **3**, ether **4**, and peroxide bonds **5**. In addition to the autoxidative process, the polymerization process of drying oils can also proceed under elevated temperatures in which C—C link-

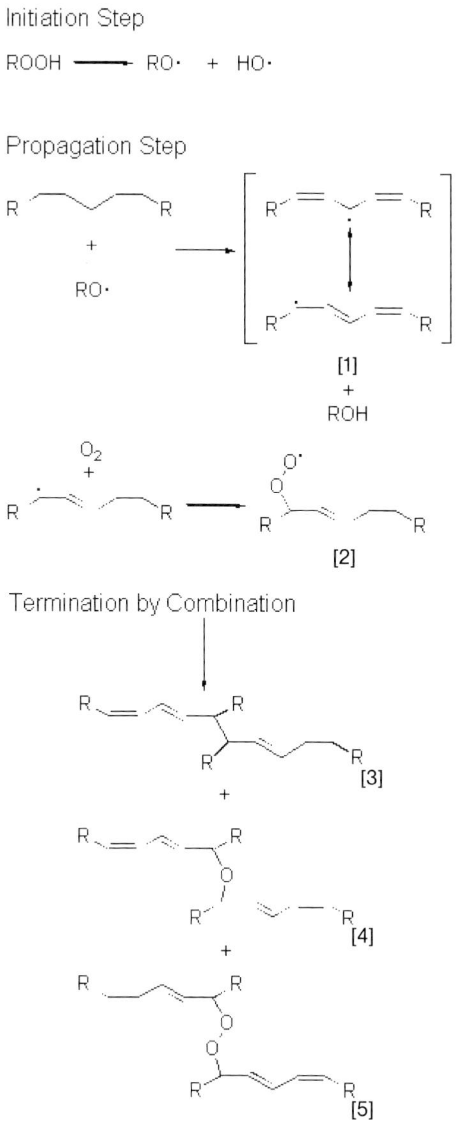

Initiation Step

ROOH ⟶ RO· + HO·

Propagation Step

[reaction scheme [1] + ROH]

[reaction scheme with O₂ giving [2]]

Termination by Combination

[structures [3], [4], [5]]

Fig. 11.7 The auto-oxidation process of a drying oil.

ages have been reported through a Diels–Alder reaction. The mechanism for this reaction can be found in literature (79,80,81).

11.2.4
Metal Catalysts

Metal catalysts also known as "driers" can be added to accelerate the drying process (73,82,83). A drier is usually a metal soap with acid moieties that provide solubility in the oil medium. Driers are typically used in quantities of less than 1 wt% of the oil.

Metal driers act in two ways such that the driers catalyze the uptake of oxygen and promote the formation of peroxide radicals, which initiate the autoxidative process as shown in Figs 11.8 and 11.9 (82). Metallic compounds such as naphthenates of cobalt, calcium, lead, manganese, and zirconium are usually the metals of choice. A typical drier is cobalt octanoate, or naphthenate as shown in Figs 11.8 and 11.9. Cobalt, manganese, and iron are known as top driers while zirconium and calcium soaps are known as auxiliary driers. A package of cobalt, zirconium, and calcium drier are usually used together for their synergistic effects together in coating systems.

It should also be noted that in the late 1970s, metal alkoxides were discovered as a potential candidate in the evolution of drier technology (83). Aluminum alkoxide derivatives were discovered to react with hydroxyl and carboxyl groups of alkyds forming a crosslinkable site bonding multiple alkyd molecules to the aluminum atom. The interest in this is on the basis that the hybrid inorganic/organic coatings utilize various metal alkoxides with titanium and zirconium that have potential to act as driers.

Ceramer coatings based on seed oils as the organic phase and metal alkoxides as the inorganic phase have been developed in our laboratory (84,85,86). The primary interest in the development of these coatings was to reintroduce seed oils as a renewable resource for coating applications. A variety of seed oils such as linseed, safflower, soybean, and sunflower oil have been used in ceramer formulations. The metal alkoxides used in these formulations include $Ti(Oi-Pr)_4$,

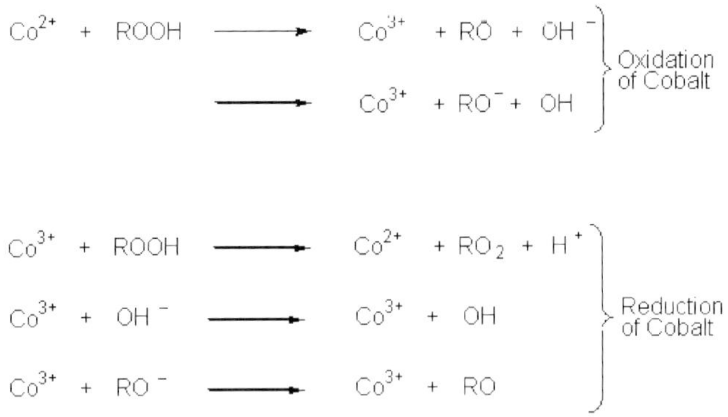

$$Co^{2+} + ROOH \longrightarrow Co^{3+} + RO + OH^-$$
$$\longrightarrow Co^{3+} + RO^- + OH$$
$$\left.\begin{matrix}\end{matrix}\right\} \text{Oxidation of Cobalt}$$

$$Co^{3+} + ROOH \longrightarrow Co^{2+} + RO_2 + H^+$$
$$Co^{3+} + OH^- \longrightarrow Co^{3+} + OH$$
$$Co^{3+} + RO^- \longrightarrow Co^{3+} + RO$$
$$\left.\begin{matrix}\end{matrix}\right\} \text{Reduction of Cobalt}$$

Fig. 11.8 The decomposition of hydroperoxides by oxidation and reduction of a cobalt drier.

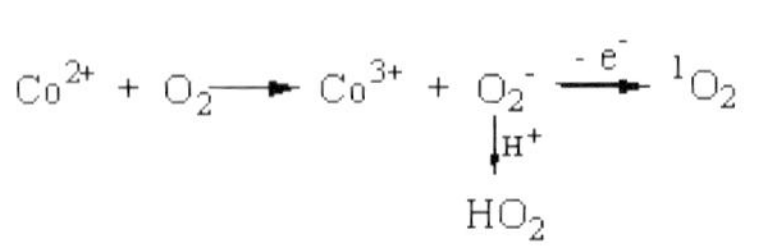

$$Co^{2+} + O_2 \longrightarrow Co^{3+} + O_2^- \xrightarrow{-e^-} {}^1O_2$$
$$\downarrow H^+$$
$$HO_2$$

Fig. 11.9 Generation of singlet oxygen via cobalt surface drier.

Ti(O*i*—Pr)$_2$(acac)$_2$, Zr(O*n*—Pr)$_4$, and Zn(Ac)$_2$ · 2H$_2$O. The incorporation of metal alkoxides in seed oil coatings significantly increased the mechanical and physical properties relative to the parent seed oil coatings. However, the synergistic effects present between the seed oils and metal alkoxides were not fully understood.

It was shown that the metal–oxo clusters once formed catalyzed the autoxidative reactions of the drying oil. Thus, the metal–oxo clusters performed as a drier. Since there had been studies indicating that the metal–oxo clusters derived from the sol–gel precursors had catalytic activity within the autoxidatively curing process of the drying oil matrix. It was thought that the addition of traditional cobalt driers to these systems might result in faster drying or a higher crosslink density. A cobalt/ zirconium naphthenate system was used with three sol–gel precursors, TIA, TIP and ZrP. (see Fig. 11.2) The titanium sol–gel precursors had a synergistic effect on curing with driers, while the more reactive zirconium sol–gel precursor intercalated the cobalt ions deactivating the catalytic activity of the driers.

Mixed metal sol–gel systems were investigated with an initial focus of corrosion protection, either by intercalation of a metal ion into a growing metal–oxo-cluster or by two or more sol–gel precursors reacting with each other. Titanium, zirconium, silicon were used as sol–gel precursors, and zinc was added as either a hydrate of an acetate or a phosphate. It was thought the addition of zinc ions into a sol–gel/seed oil system would provide superior corrosion protection. The corrosion protection of the inorganic/organic seed oil hybrid systems are outstanding almost universally, thus secondary ions such as zinc are not needed. However, two interesting observations were made. There was a catalytic effect of either adding the zinc ions to the titanium sol–gel precursors or the titanium and zirconium sol–gel precursors added together on the seed oil drying and final film properties. Secondly, there were visible spectral differences in the colors of the seed oil films when zinc was added either as a acetate or a phosphate. The acetate had a blood red color, and the phosphate had a bright orange color in comparison with the usual dark brown of the seed oils with or without the single sol–gel precursors. The color was attributed to ligand transfer bands of the intercalated zinc with the acetate or the phosphate. It was not ascertained whether the ligand transfer was between the titanium ligand (acetate or phosphate) interaction or the intercalated zinc ligand interaction.

Both soybean and linseed oil based inorganic/organic coatings were evaluated for corrosion protection on both steel and aluminum substrates using a salt spray prohesion protocol. The addition of sol–gel precursor had a considerable influence on the corrosion inhibition and adhesion of coatings. The barrier properties of several formulations were comparable to epoxide-based primer coatings. Corrosion performance was a function of sol–gel content and type. High levels (10–15 wt%) of sol–gel precursor afforded a better corrosion performance than lower levels of sol–gel precursor (5 wt%). The TIP and ZrP based ceramer coatings afforded more effective barrier properties than TIA. It was proposed that the preceramic metal–oxo layer passivated the aluminum surface via a self-assembly mechanism. Furthermore, the preceramic metal–oxo clusters impeded the transport through the coating to the metal surface.

The other most studied oil for inorganic/organic hybrid seed oils is soybean oil. Blown, bodied, and epoxidized soybean has been used with a variety of sol–gel precursors (87,88,89). It did not seem to matter which kind of type of modified oil was used with respect to the end properties. Most of the work was performed with blown oil out of convenience. As described previously, the inorganic/organic formulations were dependent on the amount of water, oligomeric acid content, and viscosity of the seed oil, thus soybean oil could not be used without modification. This paradigm held for the blown and bodied seed oils. For epoxidized soybean oils, there was not a dependence on the aforementioned variables and mixing was straightforward (87). It was observed that the epoxidized soybean oil-based ceramer dried faster than the blown soybean oil-based ceramer coatings at the same sol–gel precursor loading. It was speculated that the sol–gel precursors were presumably more reactive toward the epoxidized oil. The homopolymerization of epoxidized soybean oil was catalyzed by the titanium–metal–oxo clusters of inorganic/organic network. When TEOS oligomers [condensation products derived from TEOS] were used, an additional catalyst had to be used for curing.

There were more studies performed to elucidate the possible interactions between the inorganic and organic phases especially with respect to fatty acids within the drying oil (linseed oil) systems. There are many possibilities for which a chemical interaction can occur between the inorganic and organic phases. First, metal carboxylate formation may occur during the cure cycle. Secondly, the sol–gel precursor $Ti(Oi—Pr)_4$ is commonly used as an esterification catalyst in the synthesis of polyesters (14). Conversely, if $Ti(Oi—Pr)_4$ can be used as an esterification catalyst, it may also be capable in acting as a transesterification catalyst in this system resulting in free fatty acid formation. Another possibility is that the sol–gel precursors simply form inorganic domains within the organic phase and enhance the mechanical properties by acting as rigid particles encapsulated by the crosslinked organic phase (15). However, the morphology of these ceramer films were characterized in a previous study using SAXS (11). The data indicated that inorganic domains were relatively small compared with pigments commonly used in coatings. As a result, additional interaction between the inorganic and organic phases must be present in order to account for the physical properties of these ceramer coatings.

To elucidate possible interactions between fatty acids and sol–gel precursors and metal–oxo clusters with a seed oil-based matrix another more controlled attempt to react fatty acids directly with sol–gel precursors was undertaken. This time the fatty acids and titanium sol–gel precursors were reacted in toluene to form small controlled metal–oxo–carboxylate clusters. These metal carboxylates were added into the drying oils (instead of sol–gel precursors) and cured under the same curing conditions as the ceramer films for comparison. Both autoxidatively active and inactive fatty acids (carboxylates) were used for the comparison. Also, under similar reaction conditions as the mono-, di-, and triglyciderides were reacted with the titanium sol–gel precursors to evaluate whether the titanium–oxo clusters interacted with the hydroxyl groups on the glycerol, or for the triglyceride case acted as a transesterification catalyst.

A theoretical model depicting the interaction between the inorganic and organic phases is shown in Fig. 11.1. The primary interaction proposed is the metal carboxylate formation during the curing stages of these inorganic/organic coatings. This study has shown that Ti(Oi—Pr)$_4$ can act as a transesterification catalyst and subsequently form a mixture of free fatty acids and esters which can further react to form metal carboxylates. The metal carboxylates have the capability of forming covalent bonds between the inorganic and organic phases to form a homogeneous network. The proposed interaction between the inorganic and organic phases occur *in situ* crosslinking between the triglycerides and the unsaturated metal carboxylates. The incorporation of a titanium carboxylate containing linolenic acid in linseed oil coatings had similar mechanical and physical properties relative to linseed oil-based ceramer coatings containing Ti(Oi—Pr)$_4$. As a consequence, a crosslinking mechanism between the inorganic and organic phases was proposed as shown in Fig. 11.1.

11.3
Alkyds and Other Polyester Coatings

During the 1940s and 1950s, alkyds were the major binder system for high-performance coatings (90). As new resins were introduced, alkyds were gradually replaced by these newer binder systems. However, recent legislation has severely restricted VOCs (volatile organic compounds) resulting in a resurgence in alkyd use (91). There are several benefits inherent in the use of alkyds. First of all, alkyds comprised much of the early research performed with "high solids" coatings. A second benefit is the use of naturally occurring oils as raw materials. The resultant coatings are more environmentally compatible than other coating binder systems (91).

Alkyds are composed of dibasic acids, polyols, and fatty acids. The fatty acids are derived from natural oils such as linseed, soybean, and sunflower seed oil. There are two main methods of alkyd production: the monoglyceride and the fatty acid processes (90). The fatty acid process involves the use of free fatty acids in the synthesis of an alkyd. This process allows more control of the fatty acid content in the final alkyd. The fatty acid process is also simpler than the monoglyceride process in that it consists of only one "step". The monoglyceride process consists of two steps, which incorporates the seed oils as fatty acids. In the first step of the process, the seed oils are transesterified with additional triol, usually glycerol, to form a monoglyceride. In the second stage, the diacid is added to the monoglyceride and the alkyd is cooked. The lower cost of seed oils in relation to free fatty acids results in lower raw materials costs and therefore lower production costs.

Typical polyesters for coatings are prepared from aromatic diacids such as phthalic anhydride (PA) and isophthalic acid (IPA), and an aliphatic diacid such as adipic acid. However, the phenyl ring of aromatic diacids can cause yellowing the cured enamel and the aliphatic diacids have poor hydrolytic stability. To overcome this detrimental effect, cycloaliphatic diacids can be used instead of the combination of aromatic diacids and aliphatic diacids (92,93,94,95,96,97,98,99,100). Approaches to obtain low VOC for polyesters include (1) controlling molecular weight and molecular weight distribution, (2) using hydrogen-bond acceptor solvents, and (3) reducing the ratio of aromatic/aliphatic diacids (101). Another approach which was emphasized by Jones and co-workers (102) was lowering VOC using mixtures of diacids. The mixture of diacids can disorder uniform structure and thus reduce the intermolecular interaction. A mixture of three linear diacids was used instead of one diacid to suppress the melting point of oligoester below room temperature. The crystallinity of single diacid was avoided and very low viscosity was achieved. A mixed cycloaliphatic diacid was also used in our previous work to reduce the viscosity (103).

11.3.1
Inorganic/Organic Alkyd Coatings

The drying oil based inorganic/organic hybrid systems had long drying times and demanded high temperature curing which deterred commercial utilization. The concept of using alkyds instead of drying oils was based on the precept the alkyds would be more suited as a commercially viable system. Consequently, with the knowledge obtained from drying oil hybrids, alkyds were pursued. Four studies were reported using alkyds as the organic phase in inorganic/organic hybrid coatings (104). A commercially available alkyd was used and alkyds derived from pentaerythirtol, phthalic anhydride, soybean oil and tall oil was later reported. The third study was focused on the viscoelastic properties of the cured alkyd inorganic/organic hybrids, and a fourth study was focused on fracture toughness of alkyd ceramers.

In the first study, a commercial alkyd was formulated with three sol–gel precursors [titanium isopropoxide, titanium (diisopropoxide) bis(acetylacetonate), zirconium n-propoxide]. The physical properties were evaluated with respect to alkyd formulation, and sol–gel precursor type and content. A standard drier package of cobalt, zirconium, and calcium carboxylates was used. For the more reactive sol–gel precursors, a retardation of organic phase cure was noted until the sol–gel content exceeded 15 wt% after which the curing was no longer retarded. It was hypothesized that, like the drying oil based coatings, the cobalt drier was intercalated into the metal–oxo cluster phase, and when the concentration was high enough for the metal–oxo clusters to catalyze the autoxidative drying of the alkyds the retardation of cobalt drier was overcome by the catalytic activity of sol–gel derived the metal–oxo clusters. Perhaps most importantly, it was demonstrated that the alkyd as the organic phase did provide a commercially viable cure schedule as a hybrid coating platform.

The second study was more controlled comparing linseed to soybean oil based alkyds. The same sol–gel precursors were used and the effect of reactivity of the organic phase was investigated. Again, higher loading of sol–gel precursors in the alkyds increased hardness while maintaining sufficient flexibility. In addition, an increase in linolenic acid content in the alkyd formulation also enhanced the overall hardness. As the sol–gel precursor content increased beyond 15 wt%, there was a loss of adhesion to metal substrate. This was attributed to two reasons: 1) embrittlement, 2) interfacial weakness. As the sol–gel precursor content was increased, the ceramic character of the hybrid system increased, creating an embrittled coating. Especially for crosshatch adhesion residues, stress and inflexibility of the coating contributes to an observed loss of adhesion. Secondly, with more sol–gel precursor loading it was hypothesized that the interfacial preceramic layer between the continuous organic phase of the coating and the binder was increased. If the preceramic layer increases from a nano- to a meso-scale, this will result in a brittle failure at the film–substrate interface.

Using the DSC and DMTA data from the two studies, crosslink density calculations and a model of interaction between the phases was investigated. Although there is no model for calculating the crosslink density for two phase systems, the concept that the metal–oxo clusters were crosslinking sites was proposed, and consequently the hybrid systems could be approximated by an isotropic one-phase model as a first approximation. To support this approximation, direct bonding between the metal–oxo phases and the carboxylic acids in the alkyd were observed in the cured films using Raman spectroscopy. This proved to be only the beginning with respect to understanding how the two phases interact in the cured films.

11.4
Polyurethane and Polyurea Coatings

Polyurea resins are derived from the reaction of amino-functionalized monomers or polyamine with isocyanate. For coatings, the direct reaction with an amine is too rapid for most usages. It is an almost instantaneous cure requiring mixing of a two component system inside a spray gun. Consequently, either the amine or the isocyanate is blocked. One of the means to block an amine is to rely on water to react with an isocyanate. This reaction is part of the moisture curing process for isocyanates, which react with water to afford "moisture-cured polyurea coatings." The reactions for the moisture-curing process are depicted in Eqs. (1–3) (105).

$$R-NCO + H_2O \longrightarrow \left[RHN-\overset{O}{\underset{\|}{C}}-OH \right] \quad (1)$$

$$\mathbf{1} \longrightarrow R-NH_2 + CO_2 \quad (2)$$

$$R-NCO + R-NH_2 \longrightarrow RHN-\overset{O}{\underset{\|}{C}}-NHR \quad (3)$$

In Eq. (1), the isocyanate group reacts with water to form a carbamic acid intermediate. The carbamic acid is unstable and spontaneously decarboxylates into an amine and carbon dioxide, as shown in Eq. (2). This amine reacts with another isocyanate group rapidly producing a urea crosslink (Eq. 3). Water is similar to a secondary alcohol in reactivity. The reaction rate is dependent on temperature, relative humidity, water concentration and catalyst. The reaction of isocyanate with water can be catalyzed using the same the catalysts used in the reaction of isocyanate with alcohol. Commonly used catalysts are tertiary amines, such as diazabicyclo[2.2.2]octane (DABCO), and organotin compounds, such as dibutyltin dilaurate (DBTDL) (106).

Generally, solvent-borne polyurethane coating formulations, comprise approximately 70% by weight polyol, the other component being the crosslinker, a polyisocyanate (107). Polyisocyanates usually have low viscosity, therefore, the polyol controls the viscosity. Low viscosity is essential to achieve a low VOC. Hydroxy-terminated polyester and hydroxy-functional acrylic resins are the most common polyols, followed by polyethers which are seldom used in the coating industries due to the limited properties. Generally, polyesters can achieve higher solids, greater solvent resistance and better adhesion to metals compared with acrylic resins. In contrast, acrylic resins have much higher average molecule weight, and thus are more difficult to achieve high solids.

Typical low solids formulations of polyesters are prepared from aromatic diacids The largest volume of urethane coatings are two-package (2 K) coatings. These coatings are typically used for wood (108), plastics (109), automotive topcoat (110,111,112) and aircraft topcoat (113,114). Polyurethane resins are derived from reaction of a hydroxyl-functionalized oligomer or polyol with an isocyanate as shown in Eq. (4) (115):

$$R-NCO + R'-OH \longrightarrow R-\underset{H}{\overset{}{N}}-\underset{O}{\overset{\|}{C}}-OR' \qquad (4)$$

Hydroxy-terminated polyester and hydroxy-functional acrylic resins are the most common polyols which are crosslinked through the isocyanate group. Generally, polyesters can achieve higher solids, greater solvent resistance and better adhesion to metals compared with acrylic resins (116).

11.4.1
Polyurea Inorganic/Organic Hybrid Coatings

One of the goals of utilizing an inorganic/organic hybrid coating is to protect the metal surface from corrosion. In the sol–gel process, metal alkoxides as sol–gel precursor produced a condensed film as barrier layer on the metal substrate through the hydrolysis and condensation reaction. Silicon alkoxides are the most used sol–gel precursors, which were applied both as corrosion inhibitors and as adhesion promoters. The tetraethyl orthosilicate (TEOS) was reported to have better corrosion protection property compared with organofunctional alkoxysilane (69). TEOS has been applied as a binder in coating formulation for preparing zinc-rich primer (117). After application, TEOS absorbs water from atmosphere and undergoes hydrolysis and condensation reactions. It is a widely used method for corrosion protection. TEOS in conjunction with an aminosilane was used to prevent corrosion on aluminum substrates (118). Anisotropic coatings were developed using several alkoxysilanes to form multilayered films. The results of adhesion and corrosion resistance showed that a combination of a TEOS layer on the substrate and then attaching a layer of aminosilane gave optimal adhesion and corrosion protection.

Two inorganic/organic hybrid coatings based on alkoxysilane functionalized isocyanurates have been developed (119,120,121,122,123). In one study, an alkoxysilane functionalized isocyanurate and 1,6-hexamethylene diisocyanate (HDI) trimer were used formulated as coatings. The coatings were moisture cured with both the isocyanate and the alkoxyl silane participating in the crosslinking process by forming urea and siloxane linkages, respectively. The resulting coatings showed that the alkoxysilane functionalized isocyanurate content significantly increased the adhesion and crosslink density.

The inorganic/organic hybrid coatings were developed using the alkoxysilane-functionalized isocyanurate as the organic phase and prehydrolyzed TEOS oligomers as the inorganic phase as shown in Scheme 11.1. TEOS oligomers are hydrolyzed TEOS with sufficient molecular weight to render the oligomers not volatile under ambient conditions. The polyurea inorganic/organic coatings were cured via a moisture-curing process, in which isocyanate, alkoxysilane, and TEOS groups react with water affording an organic/inorganic crosslinking network. Scheme 11.1 depicts the polyurea/polysiloxane inorganic/organic network. In this system, the polyurea organic phase provides general mechanical properties and polysiloxane functions as adhesion promoter and corrosion inhibitor. The corrosion protection of the self-priming coatings was also compared with the state of the art 2 K epoxy primer with passivating pigments and chromate conversion coatings.

The ratio of silane functionalized isocyanurate to TEOS oligomer content was found to have a significant effect on the adhesion to aluminum substrates. The silane functionalized isocyanurate functioned as a coupling agent for the organic and inorganic phases promoting miscibility by covalently bonding the two phases. However, minor phase separation was still notable in these coatings. The addition of an acid catalyst (*p*-toluene sulfonic acid monohydrate) to the system reduced the phase separation in which an optically transparent coating was attained.

One of the goals of inorganic/organic polyurea coatings was to replace chromates as corrosion inhibitors for aluminum or steel substrates. Chromates are used both as a pretreatment for aluminum aircraft, and also are added as water-soluble pigments in aircraft primers. Although outstanding at providing corrosion inhibition, chromates are a known carcinogen and consequently are in the process of being phased out of use. A polyurea, polyurea/alkoxylsilane, and polyurea/alkoxysilane with TEOS oligomers (inorganic/organic hybrid) coatings were cured on aluminum panels and subjected to ASTM accelerated corrosion testing in a prohesion (salt-spray) protocol. The coated panels were scribed and subjected to a saline mist. The difference of creepage between the polyurea, and polyurea

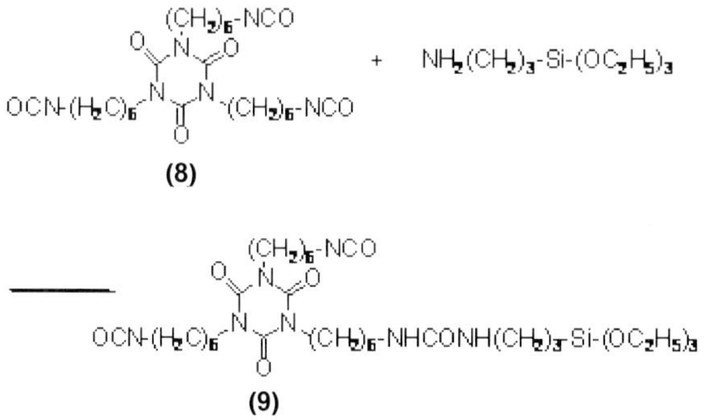

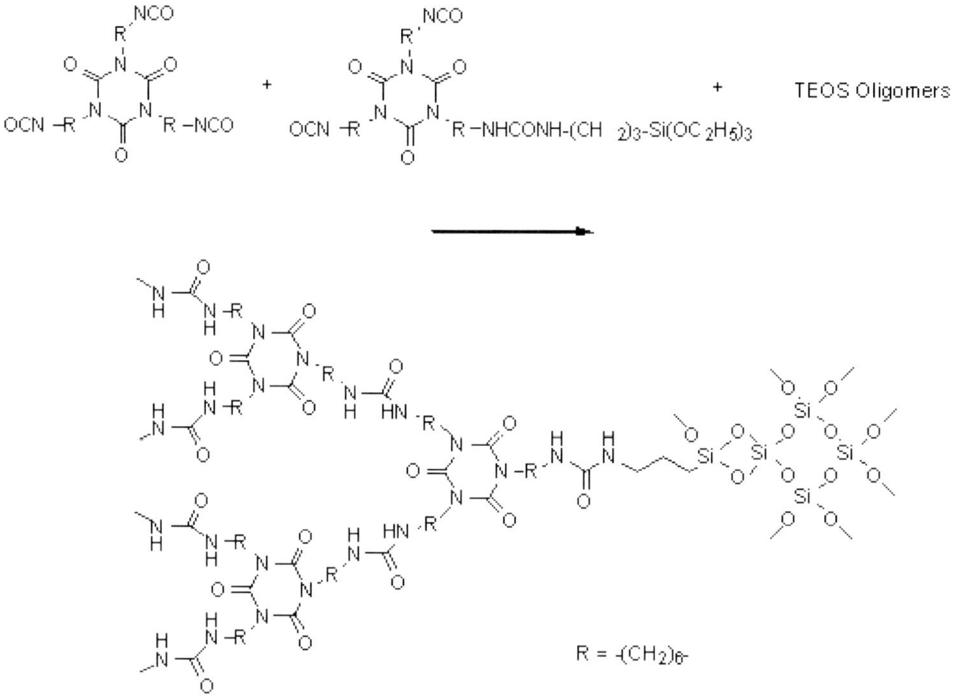

Scheme 11.1 The coupling mechanism of organic and inorganic phases coupled by alkoxysilane-functionalized isocyanurate for polyurea/polysiloxane inorganic/organic hybrid coatings network.

inorganic/organic hybrid coatings was observed (124). A small area of blistering was observed for the polyurea inorganic/organic hybrid coatings with high TEOS oligomer content (formulations T5 and T10). When the exposure time further increased to 2400 h (~14 weeks), the polyurea and polyurea/alkoxysilane coatings were heavily corroded and the ratings as shown in Fig. 11.10. The polyurea inorganic/organic hybrid coatings only slightly corroded. The blistering area increased

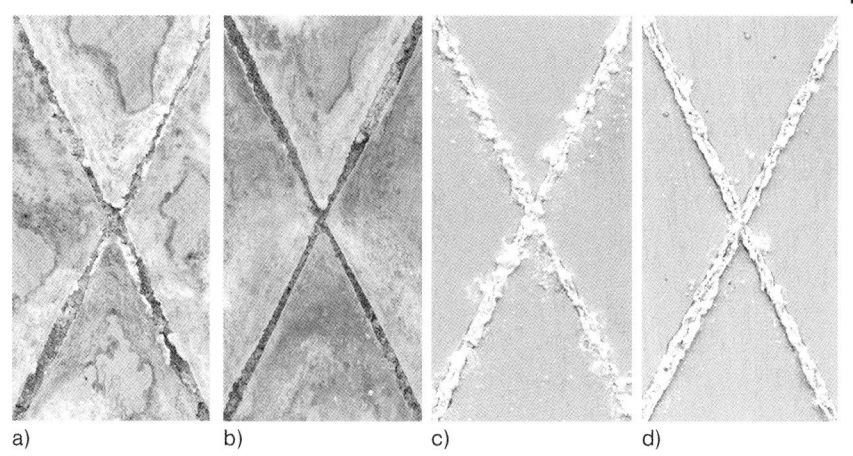

Fig. 11.10 The photographs of aluminum panels (2024 T3 bare) coated with the formulations (a) 0/100A; (b) T0; (c) T2.5; and (d) T10 at the exposure time 2400 h (dry film thickness: 40–45 μm).

from 0.1% to 0.3% for formulation T5 with the concentration of TEOS oligomers 5 wt% and increased from 0.3% to 3% for the formulation T10 with concentration of TEOS oligomers 10 wt%.

11.4.2
Polyurethane/Polysiloxane Inorganic/Organic Coating System

Isocyanate-terminated poly(ethylene oxide) and isocyanate-terminated polybutadiene were reacted with γ-aminopropyltriethoxysilane forming novel inorganic/organic hybrid alkoxysilane-terminated urethane materials (125,126,127). The hydrolysis and condensation mechanism was investigated using ^{29}Si NMR and size exclusion chromatography (SEC). The results showed an increase in the condensation reaction rate with respect to acid catalyst content. The morphology of these materials was dependent on the acid-to-alkoxide ratio. The acid concentration not only altered the condensation step in the sol–gel process but also the final structure of the material. By utilizing dynamic mechanical spectroscopy, two glass transition temperatures (T_g) were observed. One is attributed to the organic rich region and the second transition is due to the interfacial region between the organic-rich region and the inorganic domains. Adjusting the acid-to-alkoxide ratio can alter the second transition temperature. It was found that the second transition becomes more profound at a low acid/alkoxide ratio. However, sufficient acid catalyst must be present in order to minimize phase separation. The gelation process was observed to occur too slowly at low acid-to-alkoxide ratios and as a consequence, microphase separation was observed by an intensity increase in the second T_g.

A polyurethane/polysiloxane TEOS based self-priming coating or "Unicoat" system was reported (123,128). In this system, polyurethane provides the general mechanical properties and polysiloxane functions as an adhesion promoter and corrosion inhibitor (123,128). The seminal work was the synthesis and characterization of alkoxysilane-functionalized isocyanurate as shown in Scheme 11.1 (120). A hybrid coating system, polyurea/alkoxysilane, was then prepared using the alkoxysilane-functionalized isocyanurate and HDI isocyanurate (119,121,122). On the basis of the hybrid coating system, the organic/inorganic polyurea/polysiloxane coatings were formulated with the addition of TEOS oligomers to form an inorganic/organic hybrid coatings. In the inorganic/organic hybrid coatings system, the effect of acid catalyst on the inorganic/organic hybrid coatings properties was also evaluated. These previous studies showed that the alkoxysilanes have a dramatic effect on the adhesion enhancement. The polysiloxane formed by the sol–gel precursor, tetraethyl orthosilicate (TEOS) oligomers, has a strong effect on the corrosion inhibition.

The inorganic/organic hybrid coatings concept was applied to aircraft coatings. Cycloaliphatic polyesters were added into the polyurea/polysiloxane coating system to formulate polyurethane/polysiloxane coating systems. To improve the UV-resistance, a series of cycloaliphatic polyesters were synthesized in two reported studies (103,129). The cycloaliphatic polyesters (130,131) have been reported to provide better flexibility and yellowing resistance than aromatic diacids usually used in polyester formulations. Compared with mixtures of linear aliphatic and aromatic diacids, cycloaliphatic polyesters offered improved hardness/flexibility balance, hydrolytic stability, and corrosion resistance (132,133,134). The chemical structure of the cycloaliphatic polyesters are shown in Scheme 11.2, and the polyurethane crosslinking network in Scheme 11.3. The mixtures of diacids were shown to reduce the polyester application viscosity, and therefore a higher solids formulation was achievable (130).

Photographs of two TEOS based inorganic/organic coatings, BEPD(4)T2.5 and BEPD(4)T10, chromate pretreated and epoxy primer coatings at the time of 2400 h are shown in Fig. 11.11. The chromate pretreated sample showed the same rating loss as the TEOS based inorganic/organic coatings. The epoxy primer

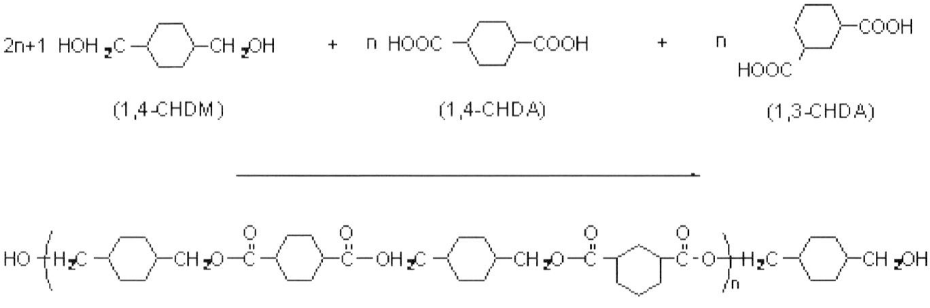

Scheme 11.2 Example of typical oligoester used in polyurethane inorganic/organic hybrid coatings.

Scheme 11.3 Formation of polyurethane crosslinking network.

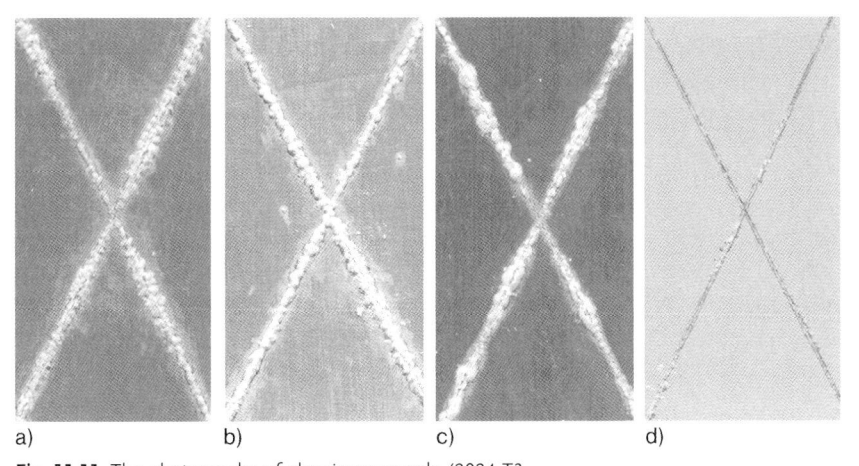

a)   b)   c)   d)

Fig. 11.11 The photographs of aluminum panels (2024 T3 bare) coated with the formulations (a) BEPD(4)T2.5; (b) BEPD(4)T10; (c) chromate pre-treatment; and (d) epoxy–polyamide primer, at the exposure time 2400 h (dry film thickness: 40–45 μm).

exhibited only slightly better corrosion inhibition than chromate pretreated formulation coatings. The polyurethane inorganic/organic network with the polyurethane as the organic phase has a similar structure to the polyurea/polysiloxane hybrid coatings. It was proposed that functionalized isocyanurate couples organic phase (polyurea or polyurethane) (122,123,124,135) with the inorganic phase via a urea and a siloxane linkage as shown in Scheme 11.4. The effect of TEOS oligomers on the corrosion is significant. Above 2.5 wt% TEOS loading, corrosion protection was competitive with chromate, and pigmented epoxy-primer. The TEOS oligomers provided the aluminum with better corrosion protection than just the triethoxysilane group (R—$Si(OC_2H_5)_3$). There was no undercutting of the scribe for the TEOS based inorganic/organic systems.

(P) = Polyurea/Alkoxysilane, Polyurethane/Alkoxysilane, Polyurea Ceramer or Polyurethane Ceramer Crosslinking Network

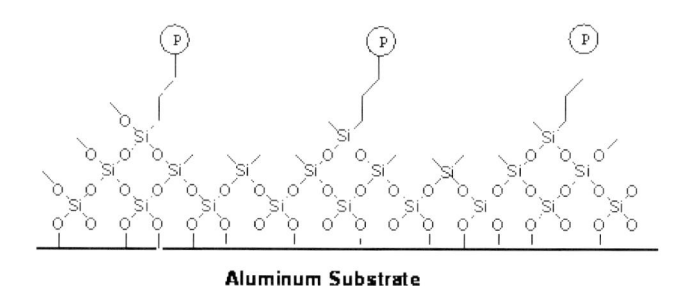

Scheme 11.4 Proposed mechanism of the interaction between inorganic/organic hybrid coatings and aluminum substrate and the condensed SiO_2 barrier layer formed from TEOS oligomers.

Similar to the polyurea/polysiloxane inorganic/organic coating systems, the polyurethane TEOS oligomers have significant corrosion inhibition on the aluminum alloy 2024-T3 as a function of TEOS content. The difference is that no blistering was observed in the polyurethane inorganic/organic coating systems even at high concentration of TEOS oligomers. This could be attributed to the polyurethane films having fewer defects than the corresponding polyurea systems. The curing mechanism for the polyurea inorganic/organic hybrid coatings involves the evolution of CO_2 (123). Film defects can occur if the curing is too rapid (128). On the other hand, the polyurethane coating is cured without the evolution of a gas and has a relatively soft polyester segment which can plasticize the film during the curing process resulting in a more defect free film.

In polyurea work, a self-assembled protective oxide layer has been proposed (123,128), but not experimentally supported. In polyurethane work, the proposed self-assembling formation was substantiated by the XPS data. The addition of TEOS oligomers with the alkoxysilane clearly enhanced the corrosion performance of both the polyurethane inorganic/organic hybrid coatings. The addition of TEOS oligomers into the coating formation clearly inhibited corrosion competitively with chromate conversion coating. As a consequence, the inorganic/organic hybrid approach has considerable promise as a chromate replacement coating. Moreover, the concept of combining metal pretreatment with a primer represents a significant leap forward in coating technology. If fully successful, the combination of metal pretreatment, primer, and topcoat into the one easy-to-apply coating would be even more substantial system (136). There is also presently an ongoing effort into using these coatings for other metal and plastic substrates, in particular, polycarbonate.

11.5
Radiation Curable Coatings

Ultraviolet-curable (UV-curable) coatings offer the advantages of fast cure response, high energy efficiency, and low volatile organic contents (VOCs) (137). As the pressure to reduce VOCs continues to mount, the advantages of UV coatings are becoming more attractive. There are two classes of UV-curable coatings, free radical and cationic. In comparison with free radical UV-curing, the cationic initiated UV-curing technology offers the advantages of insensitivity to oxygen and lower film shrinkage (138). The cationic initiated UV-curable coatings are also particularly useful in a variety of applications, including: paper coatings, wood coatings, plastic substrate coatings, lithographic and screen printing inks, and decorative metal varnishes (139).

Epoxides are one of the most important and widely used classes of resins in the field of cationic UV-curable coatings. The three major types of epoxides used are glycidyl ether, epoxidized seed oil (soybean or linseed oil), and cycloaliphatic epoxide. Of the three, cycloaliphatic epoxides are the most widely used due to their fast cure response. In addition to the fast cure response, cycloaliphatic epoxides provide a number of other important advantages in coating applications, including excellent adhesion to a wide variety of substrates, flexibility, good color stability, excellent gloss, low potential for skin irritation, low shrinkage, good weathering, and good electrical properties (140,141).

The homopolymer of cycloaliphatic epoxide is usually too brittle to be used as a coating. Typically, flexible crosslinkers such as di- and tri- functional polyols (especially ε-caprolactone derived polyols) are added into the formulations to improve the toughness and impact resistance (142). The super-acid catalyzed crosslinking reaction is shown in Scheme 11.5. The super-acid catalyzed UV-curing reaction of cycloaliphatic epoxide and polyol consists of four steps initiation, propagation, chain transfer, and termination. The initiation step involves two reactions: the formation of super-acid, and the addition of the super-acid to the monomer molecules to produce the chain initiating species and subsequent propagation of successive epoxide molecules to the chain initiated species. If there are nucleophiles present, such as water or alcohol, both chain transfer or termination reaction can occur (143).

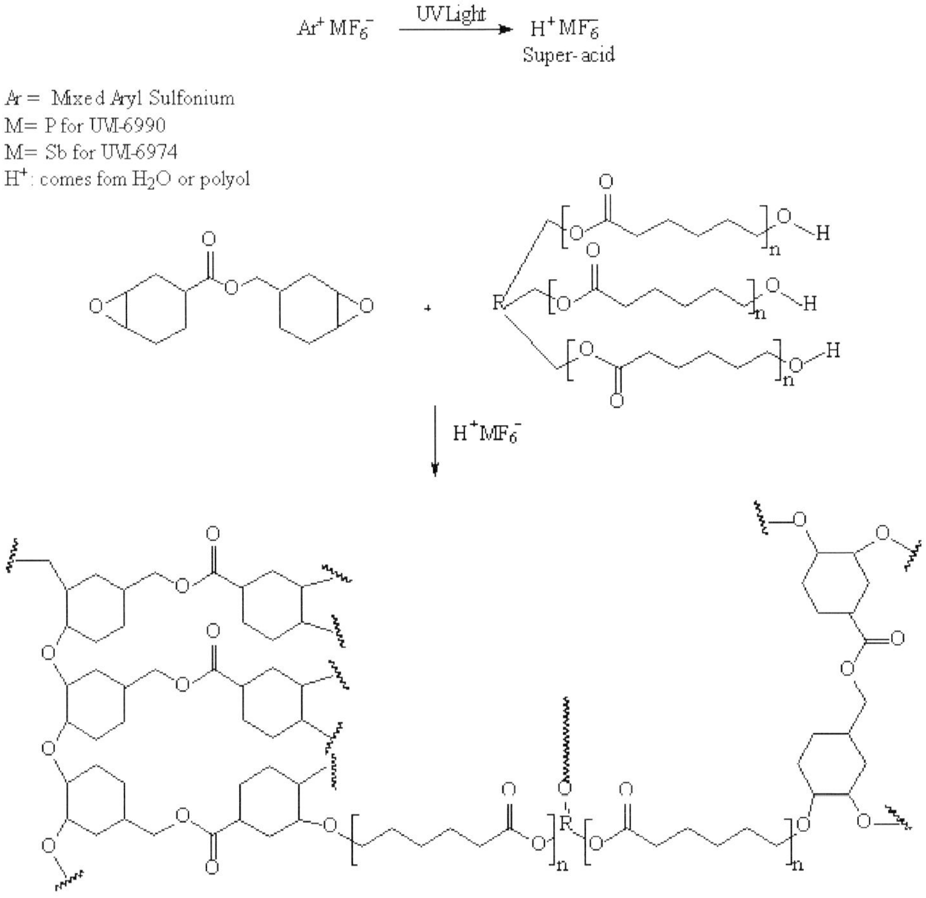

Scheme 11.5 Step-growth crosslinking reaction of cycloaliphatic epoxide with polyol.

Epoxy/Silanol/Curing/Acrylic hybrid coatings known as ESCA were previously reported (144). Most organic/inorganic hybrid network materials, including ESCA coatings were thermally cured (14,145,146,147). Alternatively, hybrid network materials can be prepared by radiation curing (148). In the early 1990s, UV-radiation has been used as a promising technology to anneal sol–gel films (149). Antireflective coatings on cathode ray tubes were based on the UV-annealing of TEOS and TEOTi-based (tetraethyl orthotitanate) sol–gel films (150,151). The photo-annealing of sol–gel film is described as an enhanced condensation reaction between two hydroxyl groups in which the UV-radiation induced the hydroxyl radical formation when water or hydroxyl groups are present (149).

11.5.1
UV-curable Inorganic/Organic Hybrid Coatings

The low cure temperature enables inorganic/organic hybrid coatings to be used as a protective barrier on organic polymer substrates. However, in order to use inorganic/organic hybrid coatings on plastic substrates the processing temperature must be lower than the substrates melting temperature. UV-curable inorganic/organic coatings have also been developed and are finding new applications on thermally sensitive substrates (172). In UV-curable inorganic/organic hybrid coatings, a highly crosslinkable inorganic/organic material can be achieved through free radical crosslinking reactions between the organic functional groups. UV-curable acrylic vinyl-based siloxane coatings have also been reported (172). Since free radical based UV-curing systems do not have a curing mechanism for sol–gel reactions inorganic/organic hybrid systems for these systems need secondary curing. Thus, a strictly UV-curable *in situ* inorganic/organic curing system for free radical systems is not practical. An *ex situ* approach will not be covered here.

An inorganic/organic hybrid coating was prepared by a sol–gel method using acrylate end-capped polyester or polyurethane oligomer, and TEOS (152). The coating was prepared with TEOS that is prehydrolyzed under acidic conditions, in order to facilitate a rapid formation of the inorganic network. The hybrid materials were cast onto a polycarbonate substrate and cured by UV-radiation to give a transparent organic matrix in which the polymeric sol was embedded. It was found that the incorporation of TEOS greatly improved the abrasive resistance of an acrylate end-capped resin.

A reactive diluent for UV-curable inorganic/organic hybrid coatings was synthesized using caprolactone polyols and TEOS as shown in Scheme 11.6

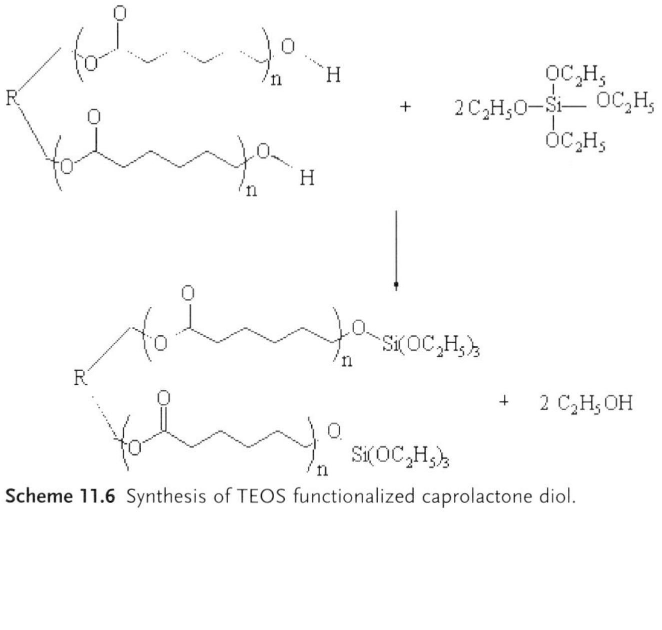

Scheme 11.6 Synthesis of TEOS functionalized caprolactone diol.

(153,154,155). The resulting siloxane functionalized polyols were used to formulate cationic UV-curable coatings with 3,4-epoxycyclohexylmethyl-3,4-epoxycyclohexane carboxylate. The structure–property relationship between the siloxane functionalized polyol and the overall film properties was established (156,157,158). The siloxane functionalized polyols effectively reduced the viscosity of the coating formulation. The coating films showed improved solvent resistance, hardness and adhesion in comparison with unmodified coatings. The siloxane modified polyols also showed to impart higher tensile properties, and higher T_g than the unmodified polyol/epoxide coatings.

The silicate modification of the polyols in inorganic/organic hybrids resulted in a number of beneficial properties and characteristics before, during, and after curing. First of all, the siloxane functionality removed the pervasive hydrogen-bonding of the parent polyols, lowering the viscosity for formulations. The silicate modification of the polyols provided the TEOS oligomers with a coupling agent which made the two phases compatible. There was also some evidence that the silanol groups were nucleophilic enough to directly attack the epoxide groups as shown in Scheme 11.7. Secondly, the silicates in conjunction with TEOS oligomers proved to be an excellent moisture scavenger to reduce the dependence of the cationic photoinitiators on relative humidity. This process in ameliorated by the siloxane and TEOS oligomers groups intercepting the water for hydrolysis reactions to start the sol–gel process, resulting in a well-cured organic phase almost regardless of relative humidity. Thirdly, surface properties of the coating were changed. Without the siloxane groups, surface active silicone based surfactants had to be added into the formulations to provide wetting to coat relatively high energy surfaces.

Unfortunately, the previous work did not include attempts to define the inorganic phase. Since the size of the inorganic phase is dictated by the speed of the organic phase formation, it was thought that UV-curing of the organic phase could result in nanosized inorganic phase formation within the continuous organic phase. It was proposed that UV-curing of the organic phase would result in smaller inorganic domains. This was based on the precept that an almost instantaneous cure speed of the organic phase would preclude sol–gel oligomers mobility. Several systems were attempted, and TEOS, and TIP based systems reported. For a similar organic system with a titanium sol–gel precursor, it was shown that very small nano-domains of titanium oxides were formed (159).

The TIP system was interesting for the overall approach. A clear coating containing TIP sol–gel precursor and an epoxidized was UV-cured concomitantly into a coating which blocked all UV-light. After curing, the UV-cured sol–gel based titanium inorganic/organic hybrid film was also optically transparent as shown in Fig. 11.12. The UV-absorbance spectrum of the cured inorganic/organic hybrid films as a function of sol–gel precursor concentration is shown in Fig. 11.13. The films appear to be dependent on concentration of TIP, and are comparable in absorbance to the films cured with titanium dioxide nanoparticles. When the percentage of TIP is increased from 5 to 10wt% the peak widens, absorbing strongly between 280 and 360 nm. The broadening peak can be attributed to sum of

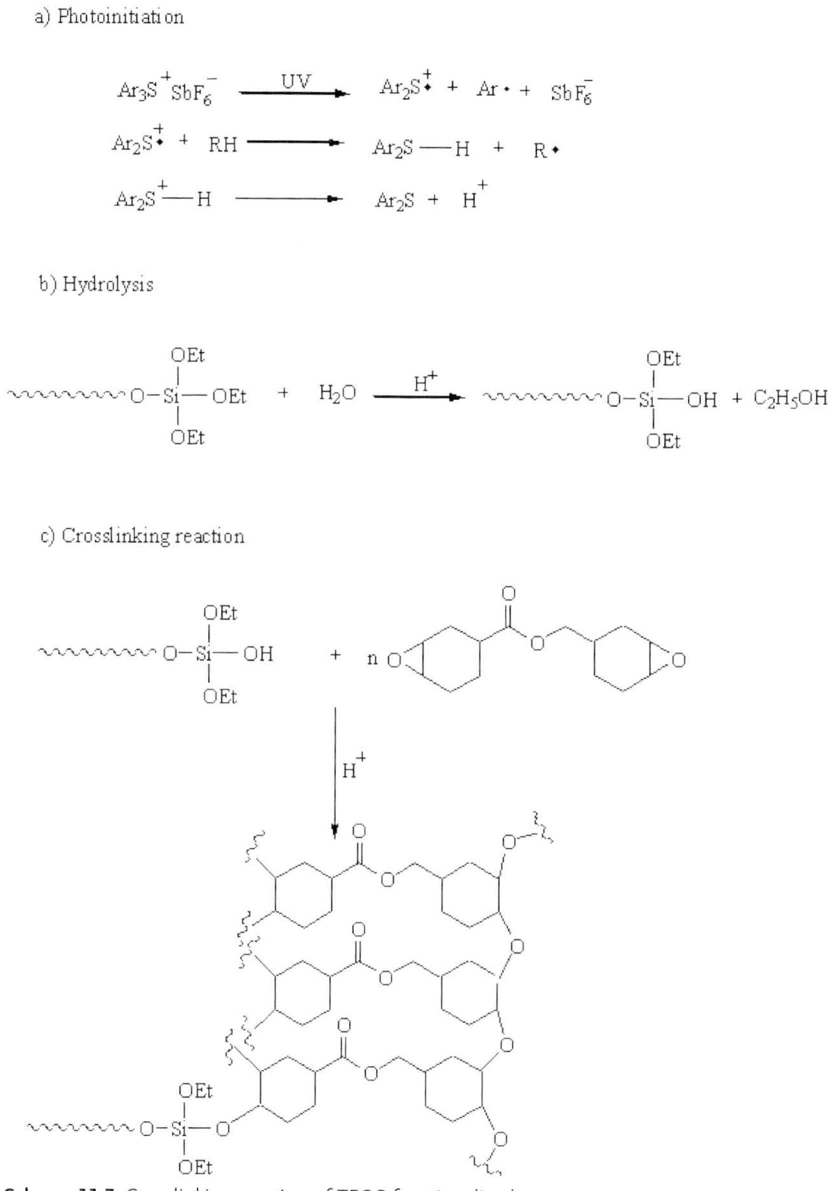

a) Photoinitiation

$$Ar_3S^+ SbF_6^- \xrightarrow{UV} Ar_2S^{+\bullet} + Ar\bullet + SbF_6^-$$

$$Ar_2S^{+\bullet} + RH \longrightarrow Ar_2S-H^+ + R\bullet$$

$$Ar_2S^+-H \longrightarrow Ar_2S + H^+$$

b) Hydrolysis

~~~O-Si(OEt)(OEt)-OEt + H$_2$O $\xrightarrow{H^+}$ ~~~O-Si(OEt)(OEt)-OH + C$_2$H$_5$OH

c) Crosslinking reaction

Scheme 11.7 Crosslinking reaction of TEOS functionalized polyol with cycloaliphatic epoxide under cationic UV conditions.

Fig. 11.12 Photograph of 10 wt% TIP ELO photo-cured film.

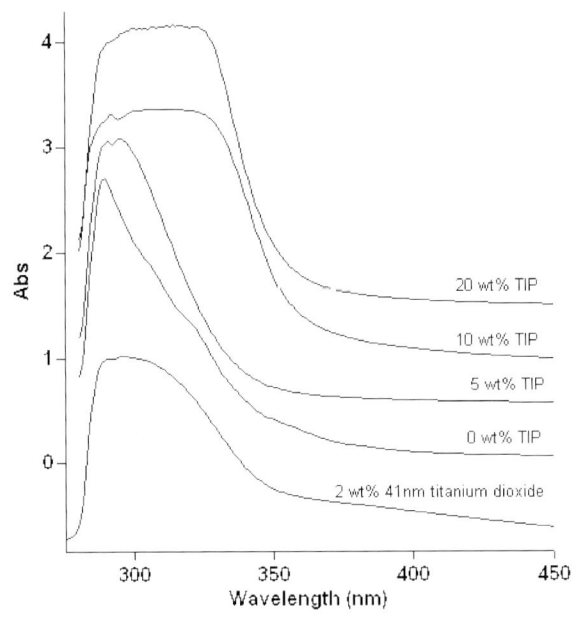

Fig. 11.13 UV Spectra of UV-cured epoxidized linseed oil containing 0 to 20 wt% titanium (IV) *iso*propoxide (TIP) and 2 wt% 41-nm TiO_2 particles.

individual auxochromic shifts associated with variety structures formed during the organic titanium oxide oligomerization. At greater than 5 wt%, Ti—O—Ti bonds can form resulting in auxochromic shifts.

This approach provided a method to form nanosized titanium–oxo cluster with cluster sizes <2 nm without measurable agglomeration. This is the first time to our knowledge that a titanium sol–gel precursor has been photo-cured to form a UV-blocking/absorbing inorganic/organic hybrid coatings film. The co-continuous inorganic/organic network maintains physical permanence, improved toughness, and optical clarity (visible light). Amorphous inorganic/organic hybrid coatings composites that are transparent to visible light, and block UV-light can exhibit an array of applications such as a UV-filter for optics. Also, control of titanium–oxo cluster size can lead to exploration of the edge of TiO_2 photonic properties.

Another important usage of nano-size titanium dioxide is for blocking of UV-light in coatings (160). When dispersed in a coating, the nano-sized titanium dioxide provides a coating with transparency to visible light, and blocking of all UV-light for diverse light sensitive substrates such as skin and plastics. Plastics which need UV-protection are usually also sensitive to heat with respect to deformation of the plastic sheeting or part (137,138,161). As a consequence, there has been ongoing research in the area of UV-curable coatings with UV-protection for plastic substrate in particular, polycarbonate and polyvinylchloride molded resin parts.

Further research in UV-curable inorganic/organic hybrids using TEOS has also been reported. Similar to the previously reported TIP based system, a UV-curable cationically cured norbornyl epoxide was used as the organic phase. The structure of the inorganic phase was unusual when compared with the previously reported thermally cured inorganic/organic hybrid systems (159,86,123,124,135,162, 163,88,164,165,166,87,167,168,169,170). For this system, a column or iceberg structure for the inorganic phase was observed. This could be attributed to the photoinitiator absorbing light on the surface of the film and the top layers of the film and curing first. As the photoinitiator was depleted near the surface the lower layers were cured in a stratified fashion resulting in columns of inorganic phase perpendicular to the film substrate interface.

11.5.2
Models for Inorganic/Organic Hybrid Coatings

A number of different models and structures for inorganic/organic hybrid materials have been reported. For a thermally cured system with a relatively long ramped curing schedule (1 h at 130, 180, and 210 °C), and at a high cure temperature, the structure on the inorganic phase was large diffuse regions almost blister like. It was not particulate by nature and had a large amount of organic phase interspersed within the organic phase. Consequently, an open interpenetrating inorganic/organic hybrid model (162) was proposed for long reaction times and highly crosslinked organic matrix with potentially a high molar ratio of coupling groups (163). For polyurethane based inorganic/organic hybrid thermally cured systems where the inorganic phase is coupling agent starved, the inorganic phase forms particulate silicon–oxo clusters with particles size being dependent on T_g of the polyol segment of the polyurethane. Finally, the structure of the UV-curing based on titanium sol–gel precursors in a seed oil based epoxy organic phase was observed to be ~2–5 nm. It was postulated that the size of the inorganic phase was controlled by the speed of the UV-curing, hence the rapid vitrification of the organic phase.

The inorganic/organic system reported herein, does not possess an aggressive coupling agent as a functional group (secondary hydroxyl) such as carboxyl or alkoxylsilane groups, even though there certainly is a large excess in stoichiometry favoring the coupling group. Also, the TEOS oligomers are considerably less reactive than the titanium based sol–gel precursors, thus the relative rate of reactivity of the inorganic phase is not the controlling factor in this case. Therefore, the inorganic domain size and structure is not fully dependent on the relative reaction kinetics of the organic versus inorganic phases. The accumulated data seems to point to the concentration and aggressiveness (reactivity) of the coupling agent as the primary factor for determining the structure and size of the inorganic domains.

The model describes small particles dispersed in a continuous organic phase with chemical bond linking the inorganic and organic phases together (28). In effect, this model depicts a co-continuous two system where the organic phase is linked together via inorganic particles, and the inorganic particles are linked together via flexible organic oligomers. A more structurally detailed model of the polyurethane/polysiloxane inorganic/organic hybrid coatings film is proposed in Scheme 11.8. Since the TEOS was prehydrolyzed and condensed via an acid catalyst, a more linear open silicon–oxo alkoxide structure would be anticipated. This model structure is collaborated by ESI-MS of the TEOS oligomers, and EDAX of the SiO_2 particles. For the 2.5 wt% TEOS films, the inorganic structure can be modeled by a co-continuous two phase system depicted by Schemes 11a and 11b where the silicon–oxo cluster are depicted as a open structure on a nanoscale, 2–10 nm. As the TEOS oligomers concentration increases to 10 wt%, the same model can be used up to 300–1000 nm scale. However, as the size of the silicon domain grows it is assumed that the structures are no longer so open, and as a consequence phase separation occurs. Presumably, if more coupling agent is added a higher TEOS content can be tolerated without phase separation.

The approach used in the polyurea/polysiloxane coating system was successfully extended into a polyurethane system. The polyurea/polysiloxane coatings lacked the flexibility (9–12) which was not surprisingly evident in the polyurethane/polysiloxane coatings reported in this study. However, the coatings proved to be both chemically and corrosion resistant. The inorganic/organic hybrid coatings showed considerable improvement in adhesion, corrosion resistance, and hardness over polyurethane coatings. With respect to flexibility and impact resistance, the polyurethane ceramer coatings were superior to the polyurea ceramer coatings.

A schematic model for the UV – curable hybrid films is proposed and shown in Fig. 11.14. The scattered small dark regions represent the silicate-rich domain and the connecting lines represent the organic crosslinked polymers formed by the photopolymerization of Epoxidized Norbornyl Linseed Oil (ENLO). A similar model of such hybrid materials has been proposed (18). The model was a highly condensed silica "cluster" connecting with organic-rich matrix (tether structure), which may not truly illustrate the hybrid materials structures. Since the UV-

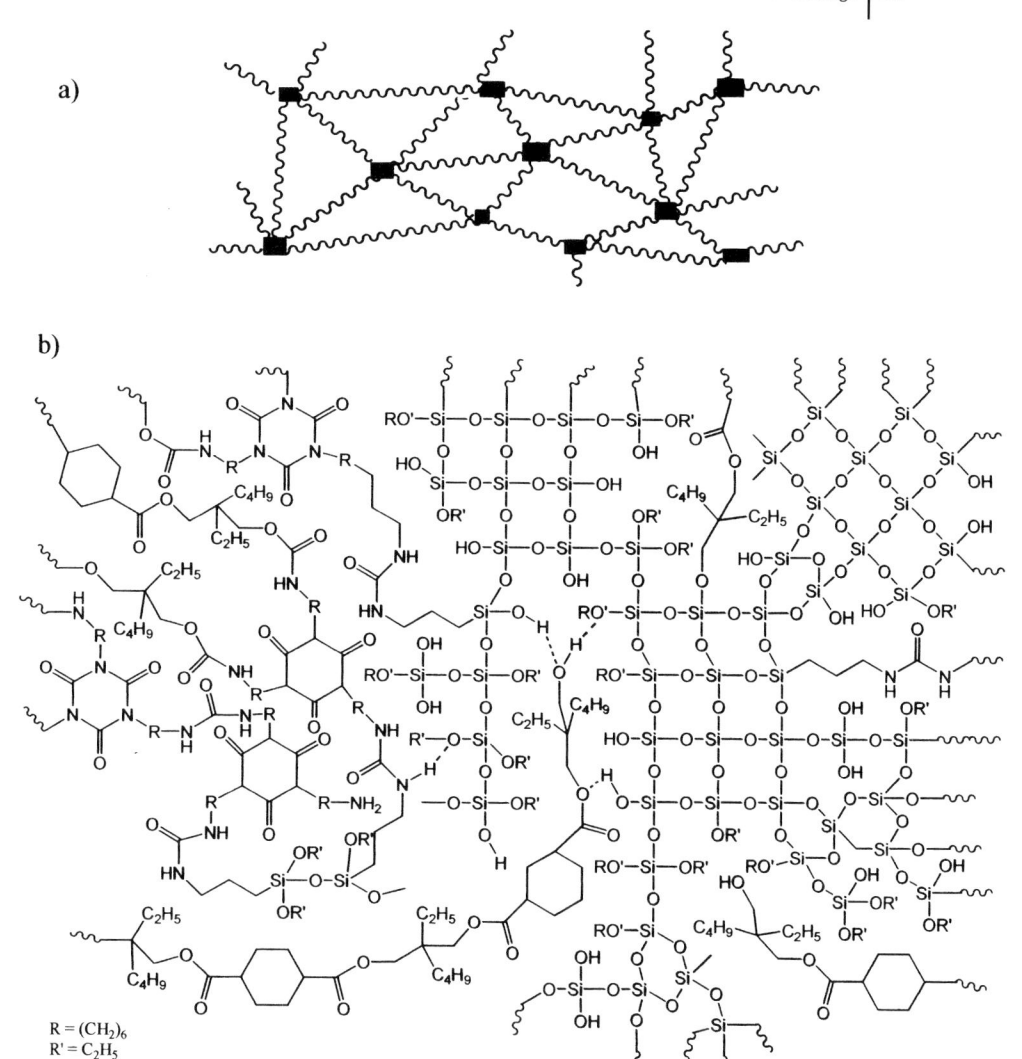

Scheme 11.8 Proposed model of polyurethane/polysiloxane inorganic/organic hybrid coatings film including organic phase, inorganic phase and various interactions between the two phases.

curing process is a very fast process in comparison with thermal processes, not all of the reactive groups such as ethoxyl and silanols would react with hydroxyl groups. The results from AFM, TEM, and SALS data support a moderately dense inorganic structure. However, other previously reported models include a more open structure with more intimate contact between the two phases.

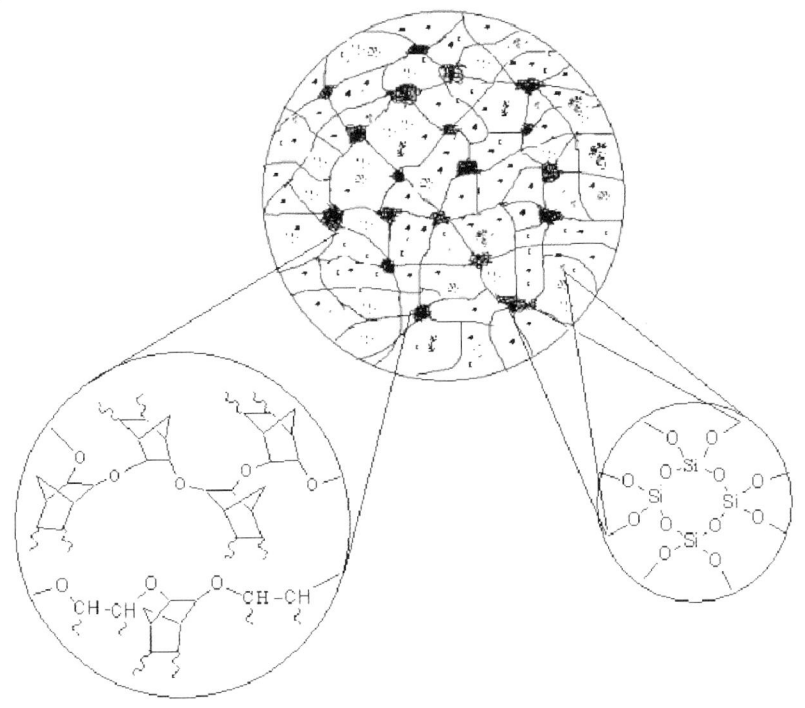

Fig. 11.14 A model structure of the UV-curable organic–inorganic hybrid films.

11.5.3
Film Morphology

The morphology of the films was investigated using SEM combined with an X-ray analysis (EDAX). Fig. 11.15 shows the morphology of cross section of polyester oligomers which are crosslinked with HDI isocyanurate (**1**) and alkoxysilane-functionalized isocyanurate (**2**) and as a function of TEOS oligomers content (0 wt%, 2.5 wt%, 10 wt%). With 2.5 wt% TEOS oligomers into the system, no phase separation was observed within the SEM. When the TEOS oligomers increased to 10 wt% spherical aggregates was observed in the cross-section for the inorganic/organic hybrid coatings as shown in Fig. 11.15.

An elemental analysis of silicon content was performed using EDAX for both the aggregate and continuous area. The silicon content by weight is 3.3 ± 1.2 Si wt% for continuous phase. The silicon content for the aggregates was dependent on the selected position on the aggregate ranging from 8.5 to 31.4 Si wt%. Although much higher than the continuous area, the silicon content in the aggregates is still much less than the silicon content of SiO_2 (46.7 wt%). It was surmised that the aggregate also contains polyurethane. A random area including about 20–30 aggregates was chosen to calculate the particle size for the inorganic/organic hybrid coatings film T10. Assuming all the aggregate shapes are

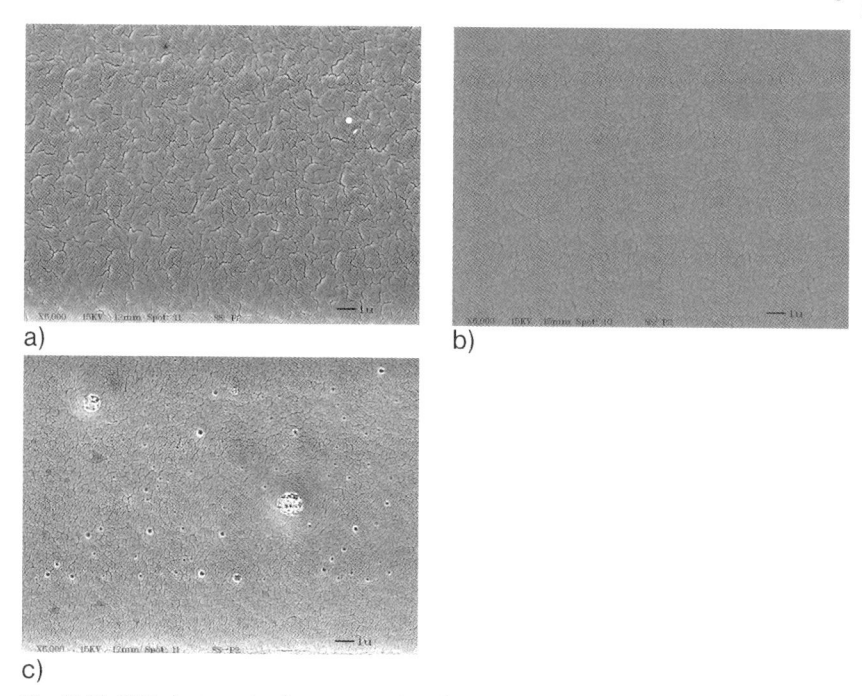

Fig. 11.15 SEM photograph of a cross section of inorganic/organic polyurethane hybrid coatings with a Tg of 80 °C at 6000×. (a) 0 wt% TEOS, (b) 2.5 wt% TEOS, (c) 10 wt% TEOS.

spherical, the visually detectable diameter ranges from the 147 nm to 1.35 µm. The calculation shows that the number average diameter was 317 ± 229 nm.

With the addition of 10 wt% TEOS oligomers, large aggregates are observed in the cross section of film as shown in Fig. 11.16. The EDAX analysis indicated that the silicon content (6.0–37.2 Si wt%) in the aggregates was enriched compared with the continuous area (2.8 ± 0.5 Si wt%). The aggregate size of T10 ranged from 240 nm to 5.48 µm in diameter. The number average of diameter was 1.03 µm with a standard deviation 1.54 µm. The average diameter of lower T_g polyurethane at T10 was three times larger than the higher T_g polyurethane at T10. The difference in particle size was attributed to T_g. The lower T_g afforded the sol–gel precursor more easily aggregated forming larger particles during the curing process. In contrast, at higher T_g, continuous phase impeded the silicon–oxo cluster from aggregating during the curing process resulting in smaller inorganic domains.

As demonstrated by the micrographs, not only the inorganic domain size but the morphology of inorganic phase is dependent on the glass transition of the organic phase. The difference in T_g between of the polyurethanes is 20–25 °C (~55 °C for BPED, ~80 °C for the CHDM). The curing temperature was 95 °C ~1 h. If the curing conditions were >30 °C below curing conditions the silica colloids had sufficient mobility to form discrete particles and also larger inorganic phases.

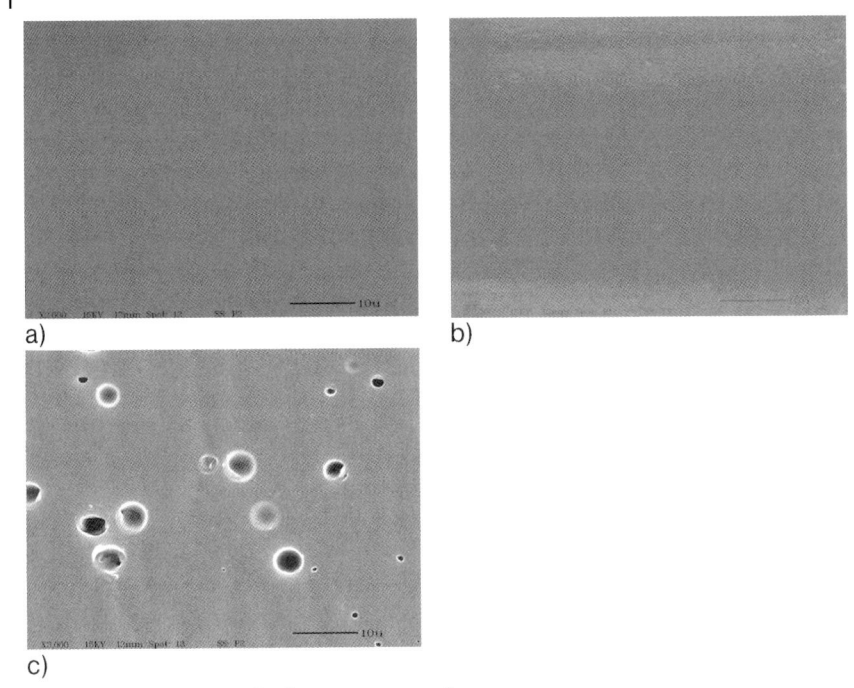

Fig. 11.16 SEM photograph of a cross section of inorganic/organic Polyurethane hybrid coatings with a Tg of 55 °C at 2000×. (a) 0 wt% TEOS, (b) 2.5 wt% TEOS, 10 wt%.

Within 20 °C of the cure temperature, mobility of the silica colloids is impaired, resulting in smaller inorganic domains and a more open inorganic-rich area, more comparable to fractal growth. It can be surmised that the open inorganic domains are a precursor or precondition of the particle nucleation process. The SEM study of the morphology of hybrid coating was reported by Van der Linde and coworkers (171) for a polyester–melamine/silica inorganic/organic hybrid coatings system. Similar results were observed in their study. Unfortunately, the aggregates and continuous area were not identified, and the particle size was not measured by a statistical method.

11.6
Applications

Hybrid inorganic/organic materials have been used to generate coatings on a variety of substrates such as metals, and polymers. They are currently being evaluated as antireflective, antistatic, diffusion barriers for food packaging coatings (172,173). Depending on their purpose, inorganic/organic hybrid coatings can range from soft thermoplastic-type materials to hard ceramic-type materials

(41). There have been many other usages thus far for inorganic/organic hybrid coatings, and hybrid materials in general. Although this chapter has focused on coatings, the same approaches can be readily used for adhesives, membranes, and composite matrices. As coatings, hybrids are used for abrasion resistance on plastics, eyeglasses, automotive topcoats, and polycarbonate for automotive windows. Hybrid coatings are extensively used in corrosion resistance. Indeed, this area has dominated the research of sol–gel hybrid coatings with the phasing out of chromates as corrosion inhibitors. Other usages involve higher temperature encapsulation to protect stained glasses windows, or colored glazing for leaded crystal. A newer usage is blocking or filtering light. This does not take into account all of the ablative or sacrificial inorganic/organic coatings which purposely decompose the organic phase for various usages.

The more specialized usage of modulating light is a prospective usage for inorganic/organic hybrid materials especially in the area of photonics where differential of refractive indices is important. There are other perspective applications in coatings for OLED screens and membrane technology for fuel cells. Hybrid coatings have the potential to provide oxygen and water vapor barriers for flexible displays. The barrier performance would be based on the ability to make a fractal two phase morphology. For fuel cells, controlling permeation of small molecules, thermal stability, and proton exchange properties are very important. Hybrids have all the characteristics necessary to balance and optimize for usage as fuel cells.

11.7
Summary

Inorganic/organic hybrid coatings are thermosetting coatings when connected to a continuous organic network. For any thermosetting system, pot-life stability is always a concern. Moisture curing systems such as the sol–gel precursors of the inorganic phase have inherent pot-life stability problems. Most coatings processes have to function well in almost any environment including a wide range of relative humidity. Although dip coating in a laboratory environment is acceptable, dip coating in an industrial environment is quite challenging. Spray application is a better technology for inorganic/organic coatings in an industrial environment. Both 1 K or for the more reactive sol–gel precursors a 2 K spray system would be more appropriate.

Inorganic/organic hybrid coatings have had a variety of applications mostly focusing on abrasion, chemical, and corrosion resistance. The inorganic phase of the coatings brings a number of features to organic coating: foremost in focus is protection, whether it is abrasion protection at the film–air interface or corrosion protection at the film substrate interface. Utilizing the mobility of the sol–gel precursors to aggregate at the interface is unique to sol–gel precursors or oligomeric silicon or metal–oxo clusters, thus far. With respect to corrosion protection, there are multiple coatings, cleaning, and pretreatment processes used to obtain the same level of corrosion protection afforded by a single inorganic/organic hybrid coating on relatively dirty substrates. It has been shown that these coatings inhibit further corrosion on substrates that already show corrosion. The industrial importance of this technology is the ability to replace multiple coating processes with a single coating process resulting in a savings of time, energy, space, and material, all of which means a substantial saving in cost.

Bibliography

Z. W. Wicks, F. N. Jones, S. P. Pappas, *Organic Coatings Science and Technology, 2nd Ed.;* John Wiley and Sons: New York, 1999.

S. Paul, *Surface Coatings Science and Technology 2nd Ed.;* John Wiley and Sons: New York, 1996.

Paint and Surface Coatings Theory and Practice, Lambourne, R. Ed.; John Wiley and Sons: New York, 1987.

References

1 D. L. Segal, *Chemical Synthesis of Advanced Ceramic Materials*, Cambridge University Press, New York, **1991**.
2 P. W. McMillan, *Glass Ceramics*, 2rd ed., Academic press, New York, **1979**.
3 W. Geffcken, E. Berger, German Patent 736 411, **1939**.
4 C. J. Brinker, G. W. Scherer, *Sol-Gel Science*, Academic press, New York, **1990**.
5 H. Dislich, *Angew. Chem., Int. Ed.*, **1971**, *10*, 363.
6 S. Sakka, K. Kamiya, *J. Non-cryst. Solids*, **1982**, *48*, 31.
7 B. J. Zelinski, D. R. Uhlmann, *J. Phys. Chem. Solids*, **1984**, *45*, 1069.
8 L. Levene, I. M. Thomas, U. S. Patent 3,640,093, **1972**.
9 G. Schottner, J. Kron, A. Deichmann, *J. Sol-Gel Sci. Technol.*, **1998**, *13*, 183.
10 J. D. Mackenzie, E. P. Bescher, *J. Sol-Gel Sci. Technol.*, **2000**, *19*, 23.
11 G. Schottner, *Chem. Mater.*, **2001**, *13*, 3422.
12 J. Wen, G. L. Wilkes, *Chem. Mater.*, **1996**, *8*, 1667.
13 U. Schubert, N. Husing, A. Lorenz, *Chem. Mater.*, **1995**, *7*, 2010.
14 B. M. Novak, *Adv. Mater.*, **1993**, *5*, 422.
15 S. Komarneni, *J. Mater. Chem.*, **1992**, *2*, 1219.
16 G. L. Wilkes, *Polym. Prepr.*, **1985**, *26*, 300.
17 H. Schmidt, *J. Non-cryst. Solids*, **1985**, *73*, 681.
18 H. Huang, G. L. Wilkes, *Polym. Bull.*, **1987**, *18*, 455.
19 J. Livage, M. Henry, C. Sanchez, *Prog. Solid State Chem.*, **1988**, *18*, 259.
20 R. Aelion, A. Loebel, F. Eirich, *J. Am. Chem. Soc.*, **1950**, *72*, 1605.
21 G. C. Frye, A. J. Rao, S. J. Martin, C. J. Brinker, *Mat. Res. Soc. Symp. Proc.*, **1988**, *121*, 349.
22 C. J. Brinker, A. J. Hurd, K. J. Ward, *Ultrastructure Processing of Advanced Ceramics*, J. D. Mackenzie, D. R. Ulrich, (Eds.), Wiley, New York, **1988**.
23 G. L. Wilkes, H. Huang, R. H. Glaser, *Silicon-Based Polymer Science: A Comprehensive Resource*, Advances in Chemistry Ser. 224, ACS, Washington, D.C., **1990**, 207.
24 G. L. Wilkes, A. B. Brennan, H. Huang, D. Rodrigues, B. Wang, *Mater. Res. Soc. Symp. Proc.*, **1990**, *171*, 15.
25 H. Huang, B. Orler, G. L. Wilkes, *Macromolecules*, **1987**, *20*, 1322.
26 R. H. Glaser, G. L. Wilkes, *Polym. Bull.*, **1988**, *19*, 51.
27 A. B. Brennan, B. Wang, D. E. Rodrigues, G. L. Wilkes, *J. Inorg. Organomet. Polym.*, **1991**, *1*, 167.
28 H. Huang, G. L. Wilkes, *Polymer*, **1989**, *30*, 2001.

29 D. E. Rodrigues, A. B. Brennan, C. Betrabet, B. Wang, G. L. Wilkes, *Chem. Mater.*, **1992**, *4*, 1437.
30 D. E. Rodrigues, G. L. Wilkes, *J. Inorg. Organomet. Polym.*, **1993**, *3*, 197.
31 C. S. Betrabet, G. L. Wilkes, *J. Inorg. Organomet. Polym.*, **1994**, *4*, 343.
32 B. Wang, G. L. Wilkes, *J. Polym. Sci., Part A*, **1991**, *29*, 905.
33 B. Wang, H. Huang, A. B. Brennan, G. L. Wilkes, *Polym. Prepr.*, **1989**, *30*, 227.
34 J. L. Noell, G. L. Wilkes, D. K. Mohanty, J. E. McGrath, *J. Appl. Polym. Sci.*, **1990**, *40*, 1177.
35 J. Wen, V. J. Vasudevan, G. L. Wilkes, *J. Sol-Gel Sci. Technol.*, **1995**, *5*, 115.
36 B. Wang, G. L. Wilkes, C. D. Smith, J. E. McGrath, *Polym. Commun.*, **1991**, *32*, 400.
37 B. Wang, G. L. Wilkes, J. C. Hedrick, S. C. Liptak, J. E. McGrath, *Macromolecules*, **1991**, *24*, 3449.
38 H. Schmidt, G. Philipp, *J. Non-cryst. Solids*, **1984**, *63*, 283.
39 H. Schmidt, H. Scholze, H. Kaiser, *J. Non-cryst. Solids*, **1984**, *63*, 1.
40 H. Schmidt, *Mater. Res. Soc. Symp. Proc.*, **1984**, *32*, 327.
41 H. Schmidt, *J. Non-cryst. Solids*, **1990**, *121*, 428.
42 H. Schmidt, *J. Sol-Gel Sci. Technol.*, **1994**, *1*, 217.
43 H. Schmidt, B. Seiferling, G. Philipp, K. Deichmann, *Ultrastructure Processing of Advanced Ceramics*, J. D. Mackenzie, D. R. Ulrich, (Eds.), Wiley, New York, **1988**.
44 H. Schmidt, H. Scholze, G. Tunker, *J. Non-Cryst. Solids*, **1989**, *80*, 557.
45 L. L. Hench, West J. K. *Life Chemistry Reports* **1996**, *13E*, 187.
46 D. Tian, P. H. Dubois, R. Jerome, *J. Polym. Sci., Polym. Chem.*, **1997**, *35*, 2295.
47 D. Tian, P. H. Dubois, R. Jerome, *Polymer*, **1996**, *37*, 3983.
48 D. Tian, P. H. Dubois, C. H. Grandfils, R. Jerome, P. Viville, R. Lazzaroni, J. L. Bredas, P. Leprince, *Chem. Mater.*, **1997**, *9*, 871.
49 D. Tian, S. Blacher, P. H. Dubois, R. Jerome, *Polymer*, **1997**, *39*, 855.
50 a) D. B. Haddow, S. Kothari, P. F. James, R. D. Short, P. V.Hatton, R. Van Noort, *Biomaterials*, **1996**, *17*, 501; b) D. B. Haddow, P. F. James, R. Van Noort, *J. Mater. Sci. Mater. Medicine*, **1996**, *7*, 250; c) P. Li, K. De Groot, T. Kokubo, *J. Sol-Gel Sci. Technol.*, **1996**, *7*, 127; d) M. U. Filliaggi, R. M. Pilliar, R. Yakubovich, G. Shapiro, *J. Biomedical Mater. Res. (Applied Biomaterials)*, **1996**, *33*, 225.
51 Y. Iyoku, M. Kakimoto, Y. Imai, *High Perform. Polym.*, **1994**, *6*, 43.
52 Y. Yoshitake, M. Kakimoto, Y. Imai, *High Perform. Polym.*, **1994**, *6*, 53.
53 Y. Iyoku, M. Kakimoto, Y. Imai, *High Perform. Polym.*, **1994**, *6*, 95.
54 A. Morikawa, H. Yamaguchi, M. Kakimoto, Y. Imai, *Chem. Mater.*, **1994**, *6*, 913.
55 A. Morikawa, M. Iyoka, M. Kakimoto, Y. Imai, *Polym. J.*, **1992**, *24*, 107.
56 A. Morikawa, M. Iyoku, M. Kakimoto, Y. Imai, *J. Mater. Chem.*, **1992**, *2*, 679.
57 J. E. Mark, C. Jiang, M. Tang, *Macromolecules*, **1984**, *17*, 2613.
58 Y. Ning, M. Tang, C. Jiang, J. E. Mark, W. Roth, *J. Appl. Polym. Sci.*, **1984**, *29*, 3209.
59 J. E. Mark, *Chemtech*, **1989**, *19*, 230.
60 S. B. Wang, J. E. Mark, *Polym. Bull.*, **1987**, *17*, 231.
61 J. E. Mark, S. B. Wang, *Polym. Bull.*, **1988**, *20*, 443.
62 J. Wen, and J. E. Mark, *Polym. J.*, **1995**, *27*, 492.
63 M. L. Zhelodkevich, I. M. Salvado, M. G. S. Ferreira, *J. Mater. Chem.*, **2005**, *15*, 5099.
64 T. Sugama, L. E. Kukacka, N. Carciello, *Prog. Org. Coat.*, **1990**, *18*, 173.
65 T. Sugama, *Mater. Lett.*, **1995**, *25*, 291.
66 T. Sugama, J. E. DuVall, *Thin Solid Films*, **1996**, *289*, 39.
67 T. Sugama, J. R. Fair, A. P. Reed, *J. Coat. Technol.*, **1993**, *65*, 27.
68 G. L. Witucki, *J. Coat. Technol.*, **1993**, *65*, 57.
69 S. D. Holmes-Farley, L. C. Yanyo, *J. Adhes. Sci. Technol.*, **1991**, *5(2)*, 131.
70 S. Frings, Ph.D. Thesis, University of Eindhoven, Holland, **1999**.
71 1990 Clean Air Act, refer to www.epa.gov.
72 M. D. Soucek, S. J. Tuman, *J. Coat. Technol.* **1996**, *68(854)*, 73.
73 Z. W. Wicks, F. N. Jones, P. S. Pappas, *Organic Coatings Science and Technology, Vol. 1*, John Wiley & Sons, New York, **1992**.

74 M. W. Formo, *Bailey's Industrial Oil and Fat Products*, Vols. 1 & 2, D. Swern, (Ed.), John Wiley, New York, **1979**.

75 A. E. Rheineck, R. O. Austin, *Treatise on Coatings*, Vol. 1(2), R. R. Myers, and J. S. Long, (Eds.), Marcel Dekker, New York, **1968**.

76 C. D. Miller, *J. Am. Oil Chem. Soc.*, **1959**, 36, 596.

77 O. S. Privett, *J. Am. Oil Chem. Soc.*, **1959**, 36, 507.

78 D. Swern, J. T. Scanlan, H. B. Knight, *J. Am. Oil. Chem. Soc.*, **1948**, 25, 193.

79 C. Henderson, W. W. Nawar, W. Witchwoot, *J. Am. Chem. Soc.*, **1980**, 57, 409.

80 D. H. Wheeler, J. White, *J. Am. Chem. Soc.*, **1967**, 44, 298.

81 C. P. A. Kappelmeier, *Farben-Ztg.*, **1933**, 36, 1018.

82 S. J. Bellettiere, D. M. Mahoney, *J. Coat. Technol.*, **1987**, 59, 101.

83 G. P. A. Turner, *Introduction to Paint Chemistry and Principles of Paint Technology*, 3rd Edition, Chapmann and Hall: New York, **1988**.

84 S. J. Tuman, M. D. Soucek, *J. Coat. Technol.*, **1996**, 68, 73.

85 S. J. Tuman, D. Chamberlain, K. M. Scholsky, M. D. Soucek, *Prog. Org. Coat.*, **1996**, 28, 251.

86 R. L. Ballard, S. J. Tuman, D. J. Fouquette, W. Stegmiller, M. D. Soucek, *Chem. Mater.*, **1999**, 11, 726.

87 G. Teng, M. D. Soucek, *J. Am. Oil Chem. Soc.*, **2000**, 77(4), 381.

88 D. Deffar, M. D. Soucek, *J. Coat. Technol.* **2001**, 73(919), 95.

89 D. Deffar, G. Teng, M. D. Soucek, *Comparison of TiO$_2$ Filler and Sol-Gel Precursor Effects on Coating Properties of Drying Oils*, Unpublished.

90 K. Holmberg, *High Solids Alkyd Resins*, Marcel Dekker, New York, **1987**.

91 A. Hofland, *Surf. Coat. Int.* **1994**, 77(7), 270.

92 L. K. Johnson, W. T. Sade, *J. Coaing. Technol.* **1993**, 65(826), 19.

93 P. C. Heidt, M. L. Elliott, "Aliphatic Dibasic Acid-Modified Polyesters Thermoset Industrial Coatings;" Water-borne, High-Solids, and Power Coatings Symposium; New Orleans, LA; **February 1995**.

94 Eastman Publication N-341A, (**October, 1994**), "1,4-CHDA Cycloaliphatic Intermediate for High-Performance Polyester Resins."

95 Eastman Publication N-327A (**February, 1995**), "Improved Coil Coating Performance with Eastman 1,4-CHDA."

96 Eastman Publication N-342 (**September, 1994**), "Improved Hydrolytic Stability of Water-borne Polyester Resins with 1,4-CHDA and BEPD."

97 Eastman Publication N-335A (**April, 1996**), "Eastman 1,4-CHDA vs. IPA, AD, and HHPA In a High-Solids Resin System Based on TMPD Glycol."

98 H. Ni, A. D. Skaja, P. R. Thiltgen, M. D. Soucek, W. J. Jr. Simonsick, W. Zhong, *Polym. Prepr.*, Am. Chem. Soc., Div. Polym. Chem., **1999**, 40(1), 615.

99 P. C. Heidt, "Cycloaliphatic-Based Thermoset Industrial Coatings;" Water-borne, Higher-Solids, and Powder Coatings Symposium; New Orleans, LA, **February 1994**.

100 T. E. Jones, J. M. McCarthy, "A Statistical Study of Hydrolytic Stability in Amine Neutralized Water-borne Polyester Resins as a Function of Monomer Composition;" Water-borne, High-Solids, and Power Coatings Symposium; New Orleans, LA, **February 1995**.

101 Z. W. Wicks, F. N. Jones, S. P. Pappas, *Organic Coatings Science and Technology*, Chapter 8, *Volume I: Film Formation, Components, and Appearance;* John Wiley and Sons: New York, **1992**.

102 F. N. Jones, *J. Coat. Technol.* **1996**, 68(852), 25.

103 H. Ni, J. L. Daum, A. D. Skaja, M. D. Soucek, *Prog. Org. Coat.* **2002**, 45(1), 49.

104 a) R. A. Sailer, M. D. Soucek, *Prog. Org. Coat.* **1998**, 33, 36. b) R. A. Sailer, M. D. Soucek, *Prog. Org. Coat.* **1998**, 33 (2), 117. c) R. A. Sailer, M. D. Soucek, *J. Appl. Poly. Sci.*, **1999**, 73, 2017. d) R. L. Ballard, R. A. Sailer, B. Larson, M. D. Soucek, **2001** "Fracture Toughness of Inorganic-Organic Hybrid Coatings" *J. Coat. Technol.* **2001**, 73(913), 107.

105 J. M. Borsus, R. Jérôme, P. H. Teyssié, *J. Appl. Polym. Sci.* **1981**, 26, 3027.

106 K. C. Frisch, in *Polyurethane Technology*; Bruins, P.F., Ed., pg. 1, Interscience: New York, **1969**.

107 E. Charrière-Perroud, in Water-borne & Solvent Based Surface Coating Resins and their Applications, Chapter 5, Volume III Polyurethanes, Thomas, P., Ed.; John Wiley and Sons: New York, **1998**.

108 W. Kubitza, *Surface Coatings International*, **1992**, 75(9), 340.

109 S. H. Shoemaker, *J. Coat. Technol.* **1990**, 62(787), 49.

110 B. V. Gregorovich, I. Hazan, *Prog. Org. Coat.* **1994**, 24(1–4), 131.

111 V. J. Andrieu, P. Laurent, *J. Coat. Technol.* **1998**, 70(882), 67.

112 R. R. Roesler, S. A. Grace, *Polym. Mater. Sci. Eng., Am. Chem. Soc., Div. Polym. Mater. Sci. Eng.*, **2000**, 83, 327.

113 C. R. Hegedus, D. F. Pulley, S. A. T. Eng, D. J. Hirst, *J. Coat. Technol.* **1989**, 61(778), 31.

114 A. K. Chattopadhyay, M. R. Zentner, *Aerospace and Aircraft Coatings*; Federation of Societies for Coatings Technology, Philadelphia, Pennsylvania, USA, **1990**.

115 G. Mennicken, W. Wieczorrek, in *Polyurethane Handbook*; pg. 510–529, G. Oertel, Ed.; Hanser: New York, **1985**.

116 J. D. Hood, W. W. Blount, W. T. Sade, *J. Coat. Technol.* **1986**, 58(739), 49.

117 D. M. Berger, *Met. Finish* **1979**, 77(4), 27.

118 S. R. Holmes-Farley, L. C. Yanyo, *Mat. Res. Soc. Symp. Proc.* **1990**, 180, 439.

119 H. Ni, A. D. Skaja, R. S. Sailer, M. D. Soucek, *Polym. Prepr.*, (Am. Chem. Soc., Div. Polym. Chem.), **1998**, 39(1), 367.

120 H. Ni, D. J. Aaserud, W. J. Simonsick Jr., M. D. Soucek, *Polymer*, **2000**, 41, 57.

121 H. Ni, A. D. Skaja, M. D. Soucek, *Polym. Mater. Sci. Eng., Am. Chem. Soc., Div. Polym. Mater. Sci. Eng.*, **1998**, 79, 21.

122 H. Ni, A. D. Skaja, R. A. Sailer, M. D. Soucek, *Macromol. Chem. Phys.* **2000**, 201(6), 722.

123 H. Ni, W. J. Jr. Simonsick, A. D. Skaja, J. P. Williams, M. D. Soucek, *Prog. Org. Coat.* **2000**, 38(2), 97.

124 H. Ni, A. H. Johnson, M. D. Soucek, J. T. Grant, A. J. Vreugdenhil, *Macromol. Mater. Eng.* **2002**, 287, 470.

125 H. Kaddami, F. Surivet, J. F. Gerald, T. M. Lam, J. P. Pascault, *J. Inorg. Organomet. Polym.* **1994**, 4, 183.

126 F. Surivet, T. M. Lam, J. P. Pascault, C. Mai, *Macromolecules*, **1992**, 25, 5742.

127 F. Surivet, T. M. Lam, J. P. Pascault, Q. T. Pham, *Macromolecules*, **1992**, 25, 4309.

128 H. Ni, A. D. Skaja, M. D. Soucek, *Prog. Org. Coat.* **2000**, 40, 175.

129 H. Ni, J. L. Daum, M. D. Soucek, *J. Coat. Technol.*, **2002**, 74(928), 49.

130 L. K. Johnson, W. T. Sade, *J. Coat. Technol.* **1993**, 65(826), 19.

131 P. C. Heidt, "Cycloaliphatic-Based Thermoset Industrial Coatings," Water-borne, Higher-Solids, and Power Coatings Symposium Sponsored by The University of Southern Mississippi, February 9–11, **1994**.

132 Eastman Publication N-341A.

133 Eastman Publication N-335A.

134 Eastman Publication N-327A.

135 H. Ni, M. D. Soucek, *J. Coat. Technol.* **2002**, 74(933), 125.

136 C. R. Hegedus, D. F. Pulley, A. T. E. Spadafora, D. J. Hirst, *J.Coat. Technol.* **1989**, 61(778), 31.

137 E. C. Hoyle, F. J. Kinstle, in: *Radiation Curing of Polymeric Materials*; American Chemical Society, Washington, D.C. **1989**, Chapter 1.

138 N. L. Price, *J. Coat. Technol.* **1995**, 67(849), 27.

139 T. R. Clever, N. R. Dando, P. L. Kolek, "New Monomers and Applications for UV-Curable Coatings" in: *Water-borne, High-Solids, and Powder Coatings Symposium*, University of Southern Mississippi, Hattiesburg **1995**, pp. 440–449.

140 M. Tokizawa, H. Okada, N. Wakabayashi, T. Kimura, *J. Appl. Polym. Sci.* **1993**, 50, 627.

141 M. Tokizawa, H. Okada, N. Wakabayashi, *J. Appl. Polym. Sci.* **1993**, 50, 875.

142 J. V. Crivello, R. Narayan, *Macromolecules*, **1996**, 29(1), 339.

143 P. C. Hupfield, S. R. Hurford, J. S. Tonge, *Proc. Rad. Tech.* **1998**, 98, 468.

144 T. Takahashi, S. Sano, M. Ishihara, *Water-borne, High-Solids, and Powder Coatings Symposium*, Univ. of Southern Mississippi, Hattiesburg, **1994**, pp. 606–624.

145 P. Judeinstein, C. Sanchez, *J. Mater. Chem.* **1996**, 6, 511.

146 Y. Wei, R. Bakthavatchalam, C. K. Whitecar, *Chem. Mater.* **1990**, *2*, 337.
147 M. In, C. Gerardin, J. Lambard, C. Sanchez, *J. Sol-Gel Sci. Technol.* **1995**, *5*, 101.
148 G. F. Medford, G. A. Patel, EP Patent 576 247 A2 **1993**.
149 R. E. Van de Leest, *Appl. Surf. Sci.* **1995**, *86*, 278.
150 Asahi Glass Co. Ltd., Jpn. Patent 05080205-A **1993**.
151 Hitachi Ltd., Eur. Patent Appl. No. 0533030 A2 **1992**.
152 J. Gilberts, A. H. A. Tinnemans, *J. Sol-Gel Sci. Technol.* **1998**, *11*, 153.
153 S. Wu, M. D. Soucek, *Polymeric Mater. Sci. Eng. Prepr.* American Chemical Society, **1997**, 76.
154 S. Wu, M. T. Sears, M. D. Soucek, *Polymer*, **1999**, *40*, 5676.
155 S. Wu, M. T. Sears, M. D. Soucek, *Prog. Org. Coat.* **1999**, *36*, 89.
156 S. Wu, M. D. Soucek, *RadTech. Proceedings*, **1998**, 719.
157 S. Wu, M. D. Soucek, *Polym. Prepr.* **1998**, *39*, 540.
158 S. Wu, M. D. Soucek, *J. Coat. Technol.* **1998**, *70*(887), 53.
159 A. H. Johnson, M. D. Soucek, *Polym. Adv. Tech.* **2005**, *16*(2–3), 257.
160 J. D. Basil, C. Lin, H. Kittanning, "Optically Transparent UV-Protective Coatings," United States Patent Number 4,799,963 **1989**.
161 C. Decker, S. Biry, K. Zahouily, *Polym. Degrad. Stab.* **1995**, *49* 111.
162 D. Deffar, G. Teng, M. D. Soucek, *Macromol. Mater. Eng.* **2001**, *286*, 204.
163 C. R. Wold, H. Ni, M. D. Soucek, *Chem. Mater.* **2001**, *13*(9), 3032.
164 G. Teng, J. R. Wegner, G. J. Hurtt, M. D. Soucek, *Prog. Org. Coat.* **2001**, *42*(1–2), 29.
165 D. Deffar, G. Teng, M. D. Soucek, *Surf. Coat. Int.: Part B* **2001**, *84* (B2), 147.
166 R. L. Ballard, J. P. Williams, J. M. Njus, B. R. Kiland, M. D. Soucek, *Eur. Polym. J.* **2001**, *37*, 381.
167 R. A. Sailer, M. D. Soucek, *Prog. Org. Coat.* **1998**, *33*(2), 117.
168 C. R. Wold, M. D. Soucek, *J. Coat. Tech.* **1998**, *70*(882), 43.
169 R. A. Sailer, M. D. Soucek, *Prog. Org. Coat.* **1998**, *33*, 36.
170 M. D. Soucek, S. J. Tuman, *J. Coat. Technol.* **1996**, *68*(854), 73.
171 S. Frings, C. F. Van Nostrum, R. Van der Linde, *J. Coat. Technol.* **2000**, *72*(901), 83.
172 K. H. Haas, S. Amberg-Schwab, K. Rose, G. Schottner, *Surf. Coat. Technol.* **1999**, *111*, 72.
173 K. H. Haas, S. Amberg-Schwab, K. Rose, *Thin Solid Films*, **1999**, *351*, 198.

Index

a

abalone nacre   284
abrasion resistance   435
absolute emission quantum yield   371
absorption   353, 354, 369, 420
absorption ability   339
absorption spectra   355, 358
absorption spectrum   355
acetic acid   340
acetylene   198
acidic polyelectrolytes   409
acid-to-alkoxide ratio   455
acrylates   164, 252
acrylic acid   139
acrylics   434
acrylonitrile   126, 199
active waveguides   391
actoxidation   443
additives   433
adenine   264
adhesion   433, 451
adhesives   161, 471
adsorption   112, 115, 118, 221
aerogels   16, 22, 175, 192, 248
aerosols   88
AFM   467
ageing   183
agents   112
agglomeration   25, 72, 99
aircraft coatings   456
Al_2O_3 (corundum)   57
alanine   287
albumin   289, 298
alginic acid   329
aliphatic amines   193
alizarin   289
alkali oxides   303
alkaline earth oxides   303
alkanediols   279

alkanethiols   101
alkoxyamine initiators   128
alkoxysilane   90, 362, 412, 436, 442
alkoxysilane groups   352
alkyds   434, 449
alkylamine   212
alkylaromatic isomerization   181
allergies   312
allografts   307
allyl trimethoxysilane   140
alumina   320
alumina-silicate   154
alumine   127
aluminophosphates   180
aluminosilicates   153, 180, 193
aluminum alkoxide   446
aluminum oxide   313
aluminum substrates   442
aluminum-tris-(2-butylate)   350
amelogenin   276
amino acid sequence   277
amino acids   262, 279, 296
aminopropyltriethoxysilane   304
γ-aminopropyltriethoxysilane   330
3-aminopropyltriethoxysilane (APTES)   341
aminopropyltrimethoxysilane   140
aminosilane   453
ammonium DL-tartrate   292
ammonium ions   101
amorphous growth   261
amphiphilic molecules   182, 196
amplifiers   376, 391
AMS   140
anion exchange capacity   166
antenna effect   376
antibodies   293, 441
anti-inflammation   334
anti-reflective coating   384
antireflective layers   350

Hybrid Materials. Synthesis, Characterization, and Applications. Edited by Guido Kickelbick
Copyright © 2007 Wiley-VCH Verlag GmbH & Co. KGaA, Weinheim
ISBN: 978-3-527-31299-3

apatite 257, 285, 286, 310, 324
apatite deposition 327, 329, 331
apatite formation 314
applications 39, 40, 253, 401
　– electrochemical 401
　– electronic 40, 401
　– optical 39
　– optoelectronic 40
APTMS 140
aragonite 257, 270, 287
arginine 285
L-arginine 279
art preservation 352
Arthobacter sp. 200
artificial organs 306
asparagines 285
aspartic acid 263, 287
atom transfer polymerizations 242
atom transfer radical 242
atom transfer radical polymerization (ATRP) 33, 124
aurum 53
autografts 307
autolysis 200
automobile industry 352
automotive area 168
auxochromic shifts 464
2,2′-azo(bis) isobutyramidine dihydrochloride (AIBA) 117

b

bacillus subtilis 200
band gap 385, 405
band structure 53
barium 371
barium sulfate 372
barium titanate 302, 404
barrier layer 452
barrier properties 168
basic building units BBU 181
$BaSO_4$ (barite) 57
batteries 429
battery applications 426
battery electrolytes 402
bending strength 317
benzene-1,3,5-tricarboxylate 215
1,4-benzenedicarboxylate (BDC) 213, 216
benzene-1,4-dicarboxylates 215
benzotrialkoxysilane 426
bidentate ligands 17, 31, 185
bifunctional precursors 191
bimodal pore structure 332
bioactive hybrid electrode 419
bioactive sol–gel glasses 441

bioactivity 89, 313, 321, 325, 327
biocatalysis 199
bioceramic 441
biocompatibility 199, 294, 297, 441
biodegradability 327
biodegradation 310, 322, 326
Bioglass® 286, 314
biohybrid materials 9
bio-hybrids 256
bioinert 310, 313
bioinspired hybrid materials 281
biological control 275
biological performance 294
bioluminescence 353
biomaterials 255, 286, 296, 312, 313, 317, 319, 320, 328
　– ceramics 319
　– evaluation 296
　– glasses 319
　– mechanical properties 255, 286, 317
　– metals 319
　– organic-inorganic hybrids 319
　– pins 312, 313, 320
　– polymers 319
　– screws 312, 313
　– surface modifications 328
　– tooth roots 320
　– wires 313, 320
biomedical applications 255
biomedical materials 307
biomimetics 2, 255
biomineral deposition vesicles 275
biomineralization 2, 255, 259, 269, 272, 280, 284
　– control 272
　– growth 259
　– nucleation 259
　– roles of the organic phase 280
biominerals 257, 270
　– functions 270
　– properties 270
biomolecular recognition 292
biomolecules 32
bioreactor, hybrid organ 321
bioreactors 306, 331
bioresorption 310
biosensors 292, 441
biosilica 270
biosynthesis 262
biotolerant 310
Bioverit® 314, 317
4,4′-biphenyldicarboxylate 216
4,4′-bipyridine 215

birefringence 251
bis(ethylenedithio)tetrathiafulvalene (ET) 407
bis(triethoxysilyl)ethane 209, 212
bis(triethoxysilyl)methane 209
1,4-bis(triethoxy-silyl)benzene 212
bivalves 287
blends 5, 20
blistering 454
block copolymers 33, 37, 74, 76, 96, 103, 134, 136, 212, 240, 256, 262, 282, 303, 337
– organic 262
blood clotting 310, 318, 329
blood coagulation 329
blood compatibility 329, 330
blood plasma 311, 315
blood vessels 268, 321
blood-compatible materials 317
blue spectral region 368
body fluid 310
bonated apatite 258
bone 2, 175, 256, 258, 266, 268, 272, 276, 283, 302, 304
– mechanical properties 283
bone cell proliferation 308
bone cement 290, 313
bone defects 326
bone formation 269, 281, 328
bone morphogenic proteins (BMP) 322
bone plates 312, 320
bone repair 297
bone replacement materials 310
bone stiffness 273
bone strength 273
bone substitutes 320
bone tissue 315
bone tissue generation 308
bone toughness 273
bone-replacement materials 303
bottom-up approach 36, 178, 294, 337
Bragg's equation 76, 157
branching chain growth 239
branching linear 239
brass 303
bridged bis(trialkoxysilyl) molecules 190
bridged organosilanes 210
bridged polysilsesquioxanes 245, 249
bridged silsesquioxanes 228, 248
Bronsted acids 234, 235
Bronsted bases 235
brush structure 328
brushite 257
building block 5, 8, 11, 12
– clusters 5
   inorganic 5, 8

building block approach 23, 24, 27, 29, 32, 33, 214
– *in situ* functionaliation 29
– inorganic building blocks 24
– macromolecules 33
– organic building blocks 32
– organic colloids 33
– organic particles 33
– post-synthetic modification 27
– small organic molecules 32
– surface functionalization 24
build-the-bottle-around-the-ship 194
buoyancy 270
butadiene 164
butyl acrylate 127, 139
n-butyl acrylate 118
butyl methacrylate 118
t-butyltriethoxysilane 236

c

C18-TMS 140
$CaCO_3$ (aragonite) 57
$CaCO_3$ (calcite) 57
cadmium sulfide 53, 127, 133
calcite 257, 270, 272
calcium carbonate 257, 271, 275, 287
calcium oxalate 257
calcium phosphates 257, 269, 284
calcium reservoir 283
calcium store 257
calculus 315
calorimetry 159
cancellous bone 283, 324
capacitance problem 415
capping agents 31, 99, 292
capping agents–passivators 101
capralactone 164
caprolactone polyols 461
caramer concept 437
carbamic acid 451
carbazole 363
carbohydrates 264, 276
carbon 290
carbon black 65
carbon ceramic composite electrode (CCE) 404, 411, 419
carbon fiber-reinforced plastics (CFRP) 301
carbon nanotube 153
carbon oxides 153
carbonates 269
carboxylic acids 31
casein 289
catalysis 78, 219, 274, 292
catalysts 162, 181, 183

catalytic activity   59, 132
catalytic cracking   181
catalytic electrode materials   402
cathode ray tubes   350
cathodoluminescence   353
cationic photoinitiators   462
CdS   133
CdSe   54
CdTe   71
cell adhesion   295
cell attachment   310
cell compatibility   328
cell differentiation   334
cell membrane receptors   296
cell proliferation   310, 321, 328
cells   199, 304
cellular regulation   275
cellulose   266, 287, 311
cephalopods   270, 287
Cerabone A-W®   314, 317
ceramer   436
ceramer coatings   446
ceramic character   451
ceramics   313
Ceravital®   314, 317
cermets   301
cetyl trimethyammonium bromide   101
characterization of hybrid materials   41
  – NMR   41
characterization   42–44, 54
  – atomic force microscopy (AFM)   44
  – electron microscopy   42
  – scanning electron microscopy (SEM)   43
  – small-angle X-ray scattering (SAXS)   45
  – thermal analysis techniques   45
  – transmission electron microscopy (TEM)   43
  – UV/vis   54
  – X-ray diffraction   44
  – X-ray photoelectron spectroscopy (XPS)   42
chemical resistance   167
chemical sensors   402
chemicaluminescence   353
chemiluminescence   421
chemisorption   293
chirality   216
chitin   266, 287, 311
α-chitin   276
β-chitin   276
chiton teeth   270
chitosan   308, 311

chitosan-based hybrids   334
chitosan–silicate hybrids   327
chitosan–silicate porous hybrids   333
chloroauric acid   101
chlorobenzyl trimethoxysilane   140
chlorophyll   200, 385
cholesterol   267
chondroitin sulfate   276, 311
chromates   453, 458
chromaticity coordinates   366
chromaticity diagram   366, 379
chromatography   222, 253
chromophores   338, 342, 365, 376, 395
class I hybrid   402, 411, 414, 425, 427
class I hybrid materials   4, 20, 342
class II hybrid   395, 402, 411, 414, 425
class II hybrid materials   4, 21, 342
clay   21, 78, 108, 121, 154, 192, 196, 289, 429
  – exfoliation   21
  – intercalation   21
  – modification   154
clay nanocomposites   168
  – applications   168
clays and layered silicates   153
Clean Air Act   443
clinical materials   307
clip coating   352
clots   329
clusters   7, 24, 25, 27, 99, 260, 342, 361, 429
  – ligand exchange   25
  – surface functionalization   24, 27
clusters of   31
CMS   140
CO_2 evolution   458
coagulation factors   318
co-assembling   178
coating   39, 161, 330, 455, 456
  – polyurethane/polysiloxane   455
  – self-priming   456
coatings   36, 161, 168, 306, 321, 338, 343, 351, 433, 434, 443, 450, 451, 459, 460, 461, 465, 466, 468, 470
  – alkyd   450
  – antireflective   470
  – antistatic   470
  – applications   470
  – categories   434
  – diffusion barriers   470
  – film morphology   468
  – inorganic/organic hybrid   433
  – models   465
  – organic   433

– packaging   470
– polyurea   451
– polyurethane   451
– polyurethane/polysiloxane   466
– processing   434
– radiation curable   459
– seed oils   443
– ultraviolet-curable   459
– UV-curable   461
cobalt octanoate   446
cobalticinium cations   194
coccolith   258
coccolith formation   281
coccolithophores   272
co-condensation   191
Co—Cr alloys   312
Cole-Cole plot   416
collagen   262, 266, 276, 285, 286, 290, 292
collagen fibers   283
collagen fibrils   301, 315
colloidal aggregates   246
colloidal crystals   34, 38, 91, 87, 103
colloidal domains   441
colloidal templates   132, 136
colloids   3, 14, 38, 50, 87, 106, 305, 313
– polymer   106
color coordinates   367
color coordinates, Hue   365
color purity   365, 368
combustion   168
Commission International de L'Éclairage (CIE)   365
compact bone   286
composite building units (CBU)   181, 214
composites   1, 49, 255
concanalvin A   293
condensation   229, 230
condensation kinetics   188
conducting particles   412
conducting polymer   199
conduction band   373
conductive particles   401
cone calorimetry   159, 166
cone cells   365
confinement effects   219
continuous phase   49
controlled release   97, 334
– of ions   99
coordination chemistry   213
coordination polymer   184
copolymer of ethylene and vinyl acetate (EVA)   81
copolymerization   12

copper acetate   289
copper chlorophthalocyanine   194
copper phthalocyanine   194, 361
coprecipitation   161
core-shell   121, 124, 130, 131, 139, 141
core-shell particles   34
core-shell-corona (CSC)   135
cork   175
cornea   266, 268
corrosion   352
corrosion inhibition   458
corrosion inhibitor   456
corrosion protection   343, 447, 452
corrosion resistance   435
corrosion–protection coatings   441
cortical bone   283, 301
co-surfactant   95, 139
Coulomb repulsion   374
Coulomb's law   277
Coulombic repulsion   282
coupling agents   155, 208, 462
CPTMS   140
crab cuticle   276
cracking   234
critical micelle concentration (CMC)   94
critical nucleus radius   260
crosslinked materials   12
crosslinkers   433, 452
crosslinking   5, 14, 26, 140, 243, 247, 352, 383, 443, 461
crosslinking density   207, 211
12-crown-4   421
crown ether   155
crystal growth   261
crystallite size   45
crystallization temperature   167
cubic arrangements   183
cubic packing   179
cubic pore geometry   212
curing temperature   461, 469
cuttlebone   292
cyanatopropyl triethoxysilane   140
cyborg   306
cycloaliphatic epoxide   459
cycloaliphatic polyesters   456
cylinders   282
cysteine   262
L-cysteine   294
cystine   262
cytochrome C   200
cytochrome P 450   195
cytocompatibility   328, 333
cytosine   264
cytotoxicity   333

d

2-(diethyl amino ethyl) methacrylate   127
Debye–Scherrer equation   45
Debye–Silver equation   64
N-(decadecyl styrene) trimethyl ammonium chloride   117
Decay curve   359
decoration   435
deflection prism   393
degradation   377
delaminated hybrids   156, 159
delamination   159, 162
denaturation   200
dendrimers   34, 90, 95, 97, 107, 110, 134, 137, 228, 248
– convergent synthesis   97
– divergent synthesis   97
dental   286
– ageing   286
– arch wires   312
– caries   286
– composite   240
– disease   286
– materials   40
– resin cements   302
dentin   276, 285, 286
dentures   312
deoxyribonucleic acid   264
deposition   29, 107
detection device   355
deterrent   257
devices   270
diamond dust   315
diatom biosilica   281
diatom shells   276
diatoms   175, 272, 275
diblock copolymers   74
dichroic materials   81
dichroism   56, 64
didodecyldimethylammonium bromide   101
dielectric constant   277
dielectric properties   302
Diels–Alder reaction   445
differential scanning calorimetry (DSC)   45
diffraction angle   157
diffuse layer   105
diffuse reflectance spectra   371
diffuse reflectances   371
diffusion   196
diffusion coefficient   406, 417
dihydridotetramethyl disiloxane   140
dimensional stability   167
dimethydihydroxysilane oligomers   304
N-dimethyl-N-[(ω-methacryloyl)-ethyl] alkyl ammonium chloride   117
dimethyl dimethoxysilane   140
dimethylacetamide   20
dimethylsiloxane   323
dimethylsulfoxide   234
diols   279
dip   36
direct writing   387
disaccharide sucrose   265
dispersants   433
dispersion   89, 92, 155, 166
disproportionation   181
di-urethanepropyltriethoxysilane   341
di-urethanepropyltriethoxysilane (d-UPTES(600))   341
di-urethanepropyltriethoxysilane (d-UtPTES(300))   341
2-(dimethyl amino ethyl) methacrylate   127
DMMS   140
DMTA   451
DNA   262, 264, 282, 292
dodecylmethacrylate   208
dominant wavelength   365
donor–acceptor (D–A) pairs   372
double layered hydroxide   164
driers   445
drug delivery   91, 96, 256, 290
drug delivery systems   334
drug release   256
drug–receptor complex   277
drying   250
drying oil   434, 443, 444
– auto-oxidation mechanism   444
DSC   451
d-spacing   157, 165
DTDS   140
dyes   251, 289, 338, 385
dye-sensitized solar cells   422
dynamic mechanical analysis (DMA)   168
dynamic mechanical spectroscopy   455

e

EDAX   466, 468
elastomers   77
electrical conductivity   403, 405, 413
electrical double layer   105
electrical insulators   401
electrical resistance   404
electrical susceptibility   393
electroactive materials   9
electrocatalysts   420
electrocatalytic application   415
electrochemical diffusion   417

electrochemical dye-sensitized solar cells   385
electrochemical sensors   419
electrochemically active films   414
electrochromic composite   413
electrochromic devices   402
electrochromic materials   9, 422
electrodes   414
   – chemically modified   414
electrogenerated chemiluminescence   421
electroluminescence   353, 359
electroluminescent   89
electroluminescent diodes   339
electrolytes   423
   – proton conducting   423
electron copping   403
electron hopping   402, 406, 411
electron microscopy   246
electron paramagnetic resonance (EPR)   374
electron self-exchange reaction   407
electron/hole pair   372
electronegativity   277
electronic nose   414
electro-optic properties   353
electro-osmotic drag   410
electrophoretic mobility   105
electrostatic potential   277
embedment   49
emission decay time   375
emission intensity   373
emission peak   375
emission quantum yield   368
emission quenching   377
emission spectra   358, 378
emission wavelength   357
emitting centers   339, 372
emitting lanthanide centers   344
emulsion polymerization   165
emulsions   88, 92, 162
enamel   270, 286
enamelins   276
encapsulation   91, 95, 113, 115, 135, 194, 198, 225, 251, 338, 377
endoskeleton   257
energy transfer processes   339
entrapment   338
entropy reduction   282
enzymes   199, 262, 419, 441
epoxides   434, 435, 459
epoxidized norbornyl linseed oil (ENLO)   466
epoxidized seed oil   459
epoxidized soybean   448
epoxy   166

epoxy primer   456
epoxy/silanol/curing/acrylic ESCA   460
EPR signal   375
ESI-MS   466
esterase   311
eterrent   257
ethylenediamine   203
EVA (ethylene-vinyl acetate copolymer)   78, 80
excitation power   373
excitation source   355
excitation spectra   358
excitation spectrum   355
excitation wavelength   371
excited state   353, 359, 380
   – singlet (S_1)   353
   – triplet (T_1)   353
excitons   372
exfoliated hybrids   156
exfoliation   121, 163
exoskeleton   257
expanded graphites   164
expoxy   167
external stimulus   407
external surface   203
extracellular matrix   267
extraction   189
extrusion process   160
eye lens   257

f

Faradaic current   417
fatty acid process   449
fatty acid triglycerides   443
fatty acids   267, 443, 448
faujasite X   195
Fe_2O_3 (hematite)   57
Fe_3O_4 (magnetite)   57
femoral bone tissues   313
femoral heads   320
femur heads   312
Fermi level   422
ferrihydrite   257, 270, 281
ferrites   302
ferritin   281, 282
ferroelectric domains   282
fiber-optic sensors   351
fibers   36, 49, 306
fibrillogenesis   268
fibrils   267, 285
fibrin   318
fibrinogen   298
fibroblastic cells   328, 329
fibronectin   296

fibrous tissue 311
fibrous tissue formation 310
filaments 268
fillers 50, 151, 433
films 161, 306
fire retardancy 167
fitania 462
flame retardance 80
flexural modulus 151
flocculation 93
fluorapatite 258
fluorescence 353, 354
fluoroimmunoassays 376
fluoromica 153
fluoropolymer 352, 410
foaming 433
foams 88, 235
food packaging 168
forced hydrolysis 99
formic acid 234, 341
four-point probe resistivity 418
fractal growth 470
fragmentation and transfer (RAFT) 124
free radical 461
free radical polymerization 89
freeze-drying 306, 332
Friedel–Crafts alkylation 181
D-fructose 265
FSM Materials 182
FSM, folded sheet materials 182
fuel cell 401, 471
fuel cell applications 410, 423
full width at half maximum (FWHM) 371
fullerenes 361
fumed silica 418, 422, 428
fungi 270

g

gallium arsenide 359, 385
gas adsorption 215
gas sensor 79, 420
gas storage 221
gastropods 287
gel polymer electrolytes 409
gelatin 54, 82, 308, 311, 326
gelatin–silicate hybrids 326
gelatin–silicate porous hybrids 332
gelation 3
gels 14, 239, 305, 325, 328, 332, 340
genetic regulation 275
geothite 270
Gibbs free energy 260
Gla (γ-carboxyglutamic acid) 283
Gla-containing proteins 276

glass 303
glass fibers 301
glass transition temperature 78, 93, 326, 408, 428, 455, 465, 469
glass-ceramics 303
D-glucose 265
glucose oxidase 419
glutamic acids 287
glutamine 285
3-glycidoxipropyltrimethoxysilane (GPTMS) 326, 350
glycidoxypropyl triethoxysilane 140
glycidoxypropyl-trialkoxysilane 20
glycidoxypropyltrimethox-ysilane 304
glycidyl ether 459
glycidyl propyl trimethoxysilane 436
glycine 267, 287
glycoproteins 269, 274, 276, 283
glycosaminoglycan disaccharide 265
glycosidic bond 264
glycosylated proteins 266
glycosylation 262
GLYMO 140
goethite 257
gold 25, 53, 60, 63, 80, 87, 127
GPTMS 332
Grätzel cells 385
grains 301
granite 287
graphite 164, 404
graphite oxide 164
graphite sulfuric acid 165
graphitic carbon 199
gravity device 257
Grignard chemistry 229
Grotthus mechanism 409, 425
growth 258
growth factors 321, 334
guanine 264
guest 192
gypsum 270

h

Hall mobility 362
HAPEX 317
hardness 443
HDI 456, 468
heat distortion temperature (HDT) 151, 167
Heck reaction 79
hectorite 122, 153, 166
α helix 262, 277
Henry equation 105
heterocoagulation 103
heterogeneous catalysis 181, 219, 220

– convergent approach   220
– sequential approach   220
heterophase nucleation   167
hexagonal arrangements   183
hexagonal packing   179
hexagonal pore geometry   212
hexamethyldisilazane   140
1,6-hexamethylene diisocyanate (HDI)   453
hexanediodimethacrylate   208
hierarchical assembly   268
hierarchical materials   36, 91
hierarchical order   2
hierarchical organization   269
hierarchical structures   282
high capacity adsorbents   249
high energy density electrode materials   428
high performance liquid chromatography, HPLC   222
high-density polyethylene (HDPE)   329
hip joint prosthesis   320
histopathology   297
HMDS   140
hexagonal molecular sieves   184
hollow metallic spheres   133
hollow silicate capsule   108
hollow titania shells   108
homeostasis   312
homopeptides   279
horse-radish peroxidase (HRP)   200
host   192
host–guest reactions   180
human hepatocellular carcinoma cells   332
human sponge bone   324
hyaluronic acid   311
hybrid biomaterials   255, 289
– artificial   289
hybrid ceramer coatings   442
hybrid gel electrolyte   422
hybrid materials   2, 337, 403, 411
– conductivity mechanisms   403
– electrochemically-active   411
– history   2
– optical applications   337
hybrid redox polyethers   405
hybridization   304
hydrocarbon permeability   167
hydrogel   135
hydrogen   101
hydrogen bond   251
hydrogen bonding   202, 231, 241, 245, 383
hydrogen hexachloro platinate IV   101
hydrogen peroxide   101, 330
hydrogen storage   218, 221

hydrogenation   132
hydrolases   311
hydrolysis   229, 230
hydroperoxides   444
hydrophobic effect   280
hydrosilation   141, 242, 411
hydrosilation reaction   17, 27
hydroxides   269
hydroxyapatite   262, 274, 283, 286, 290, 302, 308, 313, 315, 322
hydroxyethyl methacrylate (HEMA)   20
N-hydroxyl methyl acrylamide   139
α,ϑ-hydroxyl poly(ε-caprolactone) (PCL)   441
hydroxymethylcellulose   92

i

illuminance   370
immunoglobulin G   293
immunological response   294
immunology   262
impedance   416
impedance spectroscopy   415
implant   294, 301, 307
implant metals   312
impregnation   192
in situ modification   182
in vivo medical monitoring   199
in vivo testing   297
indigo   289
indium tin oxide electrode   363
indium-doped tin oxide   422
infiltration   22
initiator   92, 94, 114, 117, 119, 124, 162
integrated nonlinear optics   387
integrated optics (IO)   339
integrating sphere   369
intensity loss   56
inter system crossing   354
interactions   4, 10, 11, 21, 27, 60, 62, 67, 70, 112, 185, 186, 192, 212, 274, 277, 279, 282, 292, 293, 295, 310, 318
– antigen-antibody   292
– complexation   112
– coordination   67
– coordinative   112
– coordinative ionic   185
– covalent   11, 21, 192
– electrostatic   4, 27, 67, 70, 112, 279, 293
– hormone-receptor   293
– hydrogen bonding   4, 21, 27, 277, 279, 282
– hydrophobic forces   282
– interfacial   274

– ionic interactions 282
– noncovalent 186
– nucleic acid-DNA 293
– π–π 212
– physical 277
– protein–surface 295
– tissue–material 310, 318
– van der Waals 4, 60, 277, 282
intercalated hybrid system 159
intercalated hybrids 156
intercalation 121, 165
intercalation compounds 151
interface 10
interface area 59
 – of a single particle 59
 – per volume unit 59
interface stabilization 433
interleukin 322
internal conversion 354
internal surface 203
interpenetrating networks 5, 11, 22, 138
intractions 28
 – ionic interactions 28
inverse micelles 102
inverse opal structures 192
inward curving surface 178
ion conducting electrolyte 402, 416
ion exchange 187, 191, 192
ion transport 274
ion-pairing 408
ion–product (IP) 315
ionic conductivity 167, 408, 415
ionic crystals 282
ionic electron transport 403
ionophores 421
iron 158
iron oxide 116, 289
iron porphyrin 421
iron store 257
irradiance 370
isocyanate 451, 455
3-isocyanatopropyltriethoxysilyl 324
isocyanurates 453, 456, 468
isoelectric point (IEP) 105
isophthalic acid (IPA) 449
isopropoxide 464

k

$K_2Al_6Si_6O_{22} \cdot 2H_4O$ (muscovite) 57
kaolinite 153, 166
keratin sulfate 265, 276
Kerr effect 394
Kerr Pockels coefficients 394
kidney stones 269

kinetic growth 261
knee joints 312
Knudsen transport 177

l

ladder polymers 243
lamellae 282
lamellar additives 153
lamellar arrangements 183
lamellar materials 153
lamellar packing 179
lamellar structures 245
lanthanide chelates 377
lanthanide ions 342
lanthanide-doped hybrids 374
lanthanoid ions 185
laponite 122
Laporte's selection rule 376
large surface 218
laser dyes 381
laser efficiencies 338
laser emission 381
lasers 391
latex 34, 38, 89, 94, 104, 106, 122, 130, 135, 138, 434
layer-by-layer 28, 107
layer-by-layer deposition 69
layered double hydroxides (LDHs) 153, 166
layered inorganic materials 21
LC phase 212
leaching 200
LEGO approach 9
lepidocrocite 257, 270
leucine 263
lewis acids 234
ligand excited sates 378
ligand field interactions 376
light conversion efficiency 385
light conversion molecular devices (LCMDs) 376
light interference pigments 435
light scattering 56
light-emitting properties 353
limpet teeth 270
linolenic acid 444
linseed 443
linseed oil 447
lipid bilayer 266
lipids 266, 275
liposome 95
liquid crystal displays 81
liquid crystals 182, 212
liquid electrolytes 411
liquid sensor 420

lithium conducting electrolytes   426
lithium ion conductivity   407
lithium tetrahydroaluminate   101
lithium tetrahydroborate   101
lithium-conducting polymer electrolytes   418
lithium-ion batteries   426
lithium-polymer battery   403, 426
living polymerization   96
loading   205, 209, 212
longitudinal relaxation time   158
low κ dielectric formulations   248
luminance   365, 369, 370
luminescence   338, 353, 365, 370, 374
– quantifying   365
luminescence intensity   357
luminous efficiencies   361
luminous flux   370
luminous intensity   370
lustrin A   276
lustrin proteins   284
lyotropic phase   183, 282
lysine   263, 279
L-lysine   279
lysozyme   200, 311, 327

m

M41S   175, 179, 182, 185, 188, 205, 219
M41S material   175, 182, 191, 196, 201
macromonomers   117, 119
macropores   177, 306
macroscopic phase separation   188
madder   289
magadiite   153, 166
magnesium electrolyte   429
magnesium oxide   371
magnet   257
magnetic bacteria   276
magnetic domains   282
magnetic exchange   216
magnetite   257, 270, 281
magnetosomes   270
magnetotactic bacteria   270, 275
maleic anhydride   126, 129, 164
manganese oxide   403
marrow   283
marrow cells   328
mass spectroscopy   237
matrices   179
maturation   262
MCM material   208
MCM-41   203
MCM-41 material   199, 343
MCM-48 material   343
mechanical loading   297

mechanical stresses   297
medical applications   301
melt blending   162, 165
melt compounding   163
melt intercalation   160
melting temperature   167
membranes   9, 88, 266, 277, 280, 321, 322, 327, 343, 471
mercaptopropyl triethoxysilane   140
mercaptopropyl trimethoxysilane   140
mesopores   177
mesoporous channels   386
mesoporous materials   22, 38, 182, 191, 196, 211, 251
– integrated organic groups   211
mesoporous solids   195, 198, 201, 206
– co-condensation reactions   206
– doping with polymers   198
– doping with small molecules   195
– grafting reactions   201
mesostructured silicas   343
mesostructured materials   187
metal alkoxides   31, 90, 98, 435, 441, 446
metal catalysts   241, 445
metal chalcogenides   153
metal cluster   141
metal complexes   193, 199
metal oxide   25 80, 98, 102, 106, 108, 113, 135, 138, 153, 175, 435
– metallic and semiconductor particles   108
metal phosphates   153
metal phthalocyanines   194
metal salts   74, 90, 99, 132, 135, 141
metal–organic framework (MOF)   175, 177, 184, 190, 213
metal–oxo clusters   447, 450
metals   175
3-(methacryloxypropyl)trimethoxysilane   196
N-[(ω-methacryloyl)-decadecyl] trimethyl ammonium chloride   117
N-[(ω-methacryloyl)-ethyl] trimethyl ammonium chloride   117
methacroylpropyl-trialkoxysilane   20
3-methacryloxypropyltrimethoxysilane 208
γ-methacryloxypropyltrimethoxysilane   329
methacrylate groups   202
methacryloxypropyl trimethoxysilane   140, 236, 304, 325
methanol fuel cell applications   424
1-methyl-3-propylimidazolium iodide   422
methyl methacrylate   72, 126, 139, 198
methyl trimethoxysilane   140

methylacrylate 72
– surface-functionalization 72
methylene blue 194
methylmethacrylate poly(ethyleneoxide) 119
methyltrimethoxysilane 233, 247, 419
$Mg(OH)_2$ (brucite) 57
$MgCO_3$ (magnesite) 57
mica 154
micelles 94, 114, 134, 182, 266, 282
micrococcus luteus 200
microcomposites 156, 158, 160, 166
microelectrodes 419
microemulsion 92, 102, 138, 140, 342
microgel 134, 141
micro-optical elements 393
microphase separation 179, 196, 440
micropores 177
microporous materials 180, 191, 196, 209
– integrated organic groups 209
microporous solids 192, 198, 201, 205
– co-condensation-reactions 205
– doping with polymers 198
– doping with small molecules 192
– grafting reactions 201
microstructure 36
mineral phases 258
– growth 258
– nucleation 258
miniemulsion 92
mixed valent oxides 403
mixed-valent charge state 407
methyl methacrylate 127, 139
MMS 140
$MnFe_2O_4$ 127
modulus 167
moisture-curing process 453
molecular brush 306
molecular brush layers 329
molecular recognition 109, 276, 283
molecular sieves 180, 205
mollusc 270, 287
mollusc shells 274, 276
molybdenum sulfide 153
monochromator 355, 357
monoglyceride process 449
monolayer protected gold nanoclusters (MPCs) 406
monoliths 234, 246
monomer adsorption 115
monomodal 183
monomode waveguides 389
monosaccharide 264
montmorillonite (MMT) 121, 127, 151, 153, 158, 166, 429

monument preservation 352
morphologies 121
MoS_2 (molybdenite) 57
γ-MPS 140, 325
MPS 112, 139
MPTES 140
MPTMS 140
multicrystalline silicon solar cells 385
multidentate 28
multidentate ligands 17, 185
multifunctional materials 7
multijunction 385
multimode waveguides 389
multiphonon emission 338
multiplayer hybrid structures 29
multireversible photochromic systems 383
myoglobin 200

n

nacre 2, 276, 286, 287, 290
nacrein 276
NADH 415
Nafion 409, 423
nanobiotechnology 292
nanocomposites 3, 6 11, 13, 21, 34, 40, 49, 63, 151
– clay 151
– history 63
nano-dispersion 156
nanofiber 152
nanoparticles 27, 31, 32, 35, 37, 50, 54, 55, 58, 62, 63, 65, 68–71, 74, 75, 79, 80, 82, 90, 99, 100, 102, 106, 108, 115, 117–119, 124, 125, 131–137, 139, 140, 142, 152, 283, 292, 293, 342, 351, 361, 404, 406, 424
– absorption 58
– alumina 125
– band gap 54
– cadmium selenide 54, 70, 102
– cadmium sulfide 71, 125, 133, 136
– cadmium tellurile 71
– capping agent 32
– CdS 31, 79
– clay 125
– cobalt 75, 132
– copper 74, 137
– core-shell 108, 139
– gold 31, 68, 74, 80, 82, 99, 115, 119, 125, 132, 135, 137, 142, 293, 406
– history 63
– lead sulfide PbS 54, 62, 69, 82

- magnetic   118, 125
- metal sulfide   74
- metal   100, 124, 131, 134, 292, 361
- nickel   75, 132
- palladium   132, 135, 137
- platinum   62, 65, 75, 79, 137, 424
- polymer   106
- polyorganosiloxane   140
- refractive index   55
- rhodium   132
- scattering   58
- semiconductor   102, 124, 131, 292
- silica   118, 124, 125
- silver   74, 80, 117, 132, 137, 293
- SiO_2 65
- templates   37
- titania TiO_2 65, 351
- transmittance   58
- zinc oxide   74
- zirconia   404

nanoreactors   91, 102, 134
nanorods   387
nanoscale fillers   408
nanotube   152
2,6-naphthalenedicarboxylate   216
naphthenate   446
National Institute of Standards and Technology (NIST)   158, 169
natural polymers   64
neovasculation   322
Nernst-Einstein equation   417
network functionalizer   5
Ni—Ti alloys   312
nitrides   291
nitrogen sorption isotherms   218
nitroxide-mediated polymerization (NMP)   124
NMR spectroscopy   285
nonaqueous sol–gel process   230
noncentrosymmetric lattices   395
nonhydrolytic sol–gel process   340
nonlinear effects   342
nonlinear optical effects   251
nonlinear optical properties   216
nonlinear optics   393
nonradiative processes   354
nonradiative transitions   353
nuclear magnetic resonance (NMR)   158, 418
nuclease   311
nucleation   258, 284, 317
nucleic acids   264
number average particle diameters d_{na} 51
nylon   311

nylon 6,6   57
nylon coatings   435

O

O/I colloids   89
O/I particles   90
- *in situ* synthesis   90
- self-assembly   90
- simultaneous   90
- synthesis   90

n-octadecyl trimethoxysilane   140
octadecyltrimethylammonium   212
octylamine   101
odontoblasts   286
oil-in water microemulsion polymerization   95
- microemulsion   95
- oil-in-water   95

oil-in-water (o/w) systems   92
oil-in-water miniemulsion polymerization   95
OLED screens   471
OLEDs   376
oligoethers   193
oligomers   33
oligonucleotides   294
oligosaccharide   266, 296
oligosiloxame   18
oligosilsesquioxane   226
ophthalmic lens   352
optical amplifiers   391
optical applications   337, 339
- mechanical   343
- synthesis strategy   339

optical clarity   167
optical filters   80
optical information storage media   384
optical spectrum   55
optical transparency   50, 80, 338, 439, 441, 465
optical transparency properties   343
optically active hybrids   339
optically passive hybrids   339
optoelectronic applications   421
optoelectronics   292
oranically-modified clays   154
organic light emitting diodes (OLEDs)   359
organic polymers   196
organic solar cells   429
organic spacers   190
organic–inorganic hybrids   378
- emission spectra   378

organic/inorganic (O/I) particles   89
organically-modified clay   161, 164

organism 255
organoalkoxysilanes 200, 207, 324
organoalkoxysilane hybrids 324
organoclays 164
organogels 291
organolithium chemistry 229
organometallic catalysts 247
organometallic networks 242
organosilanes 188, 201, 442
organotitanate coupling agents 112
organotrialkoxysilanes 225, 233, 234, 239, 240, 247, 426
organotrichlorosilanes 229, 234, 237, 240, 241
organotriethoxysilanes 229
organotrihalosilanes 225
organotrimethoxysilanes 229, 235
organs 255, 301, 308
ORMOCER® 390, 392
ormosil 323, 327, 331
ormosil nanocomposite 252
osteoblast 283, 315
osteoblastic cells 328, 333
osteocalcin 276
osteoclast 315
osteoconductive 322
osteoconductive properties 297
osteocytes 283
osteoid 284
osteoinductive 321
osteonectin 276, 284
osteopontin 276
Ostwald ripening 95
outward curving surface 178
oxalates 215, 269
oxides 255, 269
oxidiazole 363

p

paint 2, 91, 289
palladium chloride 101
palmitic acid 267
partially stabilized zirconia (PSZ) 320
particles 27, 38, 50, 52, 53, 59, 60, 66, 67, 72, 87, 91, 92, 98, 99, 103, 104, 107, 108, 110, 111, 115, 130, 131, 176, 279, 290, 302, 411, 422
  – agglomeration 60
  – core-shell 91, 110, 130, 131, 290
  – core-shell structures 60
  – dense packing 38, 53
  – encapsulation 60
  – hollow 131
    inorganic 52, 98
  – interface area 59
  – interface free energy 60
  – metal 53, 91, 99, 103, 108
  – metal oxide 103
  – metal selenides 103
  – metal sulfides 103
  – mixing with polymers 67
  – optical absorption 53
  – oxides 104
  – physical properties 53
  – polymer 92
  – porous 176
  – semiconductors 53, 91, 108
  – silica 115, 279, 302, 411
  – size distribution 50
  – surface free energy 60
  – surface functionalization 27, 66, 72, 111
  – synthesis 92, 103
  – titania 107
  – tungsten oxide 422
passivation 447
passive optical properties 343
passive optical waveguides 389
pathogens 294
PbS (galena) 57
PDMS–Silica Porous Hybrids 331
pearl 287
PEM fuel cells 403
penicillin acylase 200
pentaerythirtol 450
peptidase 311
peptides 282
percolation 168
percolation threshold 412
periodically mesoporous organosilicas (PMOs) 38
periodically mesostructured organosilicas 209
permittivity 277
perovskite 362
peroxide initiator 128
peroxy free radicals 444
PFSA membrane 424
phase segregation 247
phase separation 9, 10, 11, 19, 22, 179, 205, 243, 245, 303
phase transformation 275
phenethyltrimethoxysilane 205
phenolics 434
phenyl trimethoxysilane 140
phenylalanine 263
phenylsilanetriol 232
phenyltrimethoxysilane 247, 351

phosphatase   311
phosphate reservoir   283
phosphate-buffered solution (PBS)   334
phosphates   269
phosphatidylcholine   267
phosphazene   408
phospholipid   95
phospholipid vesicles   275
phosphoproteins   284
phosphorescence   353, 354, 369
phosphoric acid   425
phosphorylation   262
phospomolybdate   429
photoactive optical properties   351
photobleaching   377, 384
photocatalysis   58, 195
photochemical stability   384
photochromic materials   384
    – photochemical stability   384
    – thermal stability   384
photochromic photovoltaic devices   381
photochromic process   382
photochromic responses   338
photochromism   381
photoelectric effect   384
photo-electrochemical cells   338
photoequilibrium   381
photoinitiator   465
photoluminescence   353, 372
photoluminescence measurements   356
photon   380
    – energy   353
photonic applications   376
photonic crystals   136
photonic hybrid materials   337
photonic materials   342
photonic properties   465
photonic response   365
photonics   393
photonics applications   251
photopolymerization   392
photoresist   392
photostability   338, 377
photosynthesis   385
photovoltaic cell   350, 384, 404
photovoltaics   384
phthalic anhydride (PA)   449, 450
pigments   91, 104, 289, 352, 433, 453
pillared clays   177
planar waveguides   387
plant silica   276
plasma   310
plastisols   434
PMMA   133, 422

PMS   140
point of zero charge (PZC)   104, 118
polarization effects   393
polarized light   55
poly(dimethyldiallylammonium) (PDDA)   155
poly(methyl methacrylate)   162
poly(NIPAM)   107
poly(N-isopropylacrylamide) (NIPAM)   136
poly(vinyl alcohol) (PVA)   155
poly(vinylpyrrolidone) (PVP)   155
poly(1,3-butadiene)   57
poly(2-vinyl pyridine)   51, 74
poly(4-vinyl pyridine)   51, 119
poly(acryl amide)   51
poly(acrylic acid)   51, 57
poly(acrylonitrile)   57, 199, 408
poly(allylamine hydrochloride) (PAH)   70, 108
poly(amic acid) (PAA)   441
poly(amide imide)   79
poly(amidoamine) (PAMAM)   68, 137
poly(aniline) (PANI)   51, 68, 71, 76, 79
poly(arylene ether ether ketone) (PEEK)   410
poly(arylene ether phosphene oxide)   440
poly(arylene ether)s   410
poly(benzyl methacrylate)   125
poly(chloroprene)   57
poly(diallyl dimethylammonium chloride) (PDADMAC)   108
poly(dimethylsiloxane) (PDMS)   57, 164, 310, 439, 441
poly(dimethylsiloxane) PDMS–silica hybrids   323
poly(dodecylmethacrylate)   208
poly(ether ketone)   440
poly(ethyl acrylate)   198
poly(ethylene glycol)   405, 421, 427, 428, 407, 408
poly(ethylene glycol) dimethyl ether   418
poly(ethylene)   51, 57, 60, 62, 65, 77, 81, 82
poly(ethylene oxide)   62, 67, 82, 206
    – PbS   62
poly(ethyleneimine)   70
poly(ethyleneoxide)   51
poly(glycolic acid) (PGA)   308
poly(imide)s   410
poly(lactic-coglycolic acid) (PLGA)   308
poly(L-lactic acid) (PLA)   308
poly(methyl acrylate)   57, 77
poly(methyl methacrylate) (PMMA)   51, 57, 72, 78, 82, 122, 155, 166, 408, 409
poly(methyl methacrylate-co-methacrylic acid)   133

poly(methylmethacrylate) (PMMA)   309, 313
poly(N-isopropyl acrylamide)   101
poly(N-methylpyrrole)   119
poly(N-vinyl pyrrolidone) (PVP)   51, 133
poly(N-vinylcarbazole)   76
poly(organosiloxane/vinylic) copolymer hybrids   137
poly(oxymethylene)   57
poly(oxypropylene)   57
poly(propylene)   51, 81
poly(p-xylylene)   75, 76
poly(pyrrole)   76
poly(silsesquioxanes)   211
poly(stryrene-co-styrylethyltrimethoxysilane)   208
poly(styrene)   51, 57, 198
poly(styrene sulfonate)   420
poly(styrene)-block-poly(4-vinylpyridine)   79
poly(styrene/methacrylic acid) copolymers   117
poly(styrene-block-2-vinylpyridine)   74
poly(styrene-block-ethylene oxide)   74
poly(tetrafluoroethylene)   57, 309
poly(tetramethylene oxide) (PTMO)   324, 439
poly(thiourethane)   82
poly(thylene oxide) (PEO)   124
poly(urethane)   71
poly(vinyl acetate)   57, 78
poly(vinyl alcohol)   51, 57, 78, 92, 166
poly(vinyl carbazoie)   51
poly(vinyl chloride)   57
poly(vinyl pyrrolidone)   101
poly(vinyl sulfate)   166
poly(vinylidene fluoride)   57, 408
poly[bis(methoxyethoxy)ethoxy-phosphazene] (MEEP)   429
polyacrylic acid (PAA)   108, 135
polyacrylonitrile   93
polyamide   166, 309
polyamide 6   155, 164, 167
polyaniline   118, 166, 198, 199, 413
polyanions   107
polybutylacrylate   131
polycarbonate   167
polycations   107
polycondensation   14, 140
polydiacetylene   199
polydimethylsiloxane   141
polyelectrolyte   29, 67, 90, 108
polyester coatings   449
polyesters   434, 448, 452
polyethylene   164, 199, 309, 311, 313, 329
polyethylene oxide   361

polyethylene oxide monomethylether mono methacrylate   117
polyethylenimine (PEI)   101, 108
polyhedral oligosilsesquioxane (POSS)   226, 233, 238, 240
polyhedral silsesquioxane   25, 29
polyimide   166
polyimide–silica materials   441
polyisocyanate   452
polylactic acid   311
polymerization   92
    – heterogeneous   92
polymer blends   302
polymer brushes   124
polymer electrolyte applications   407
polymer lamellar material nanocomposites   153
polymer melt-direct intercalation   160
polymer nanocomposites   75–78, 151
    – catalytic properties   78
    – electrical conductivity   75
    – magnetic properties   76
    – photoconductivity   76
    – properties   75
    – toughness   77
    – Young's modulus   77
polymer-clay nanocomposites (PCN)   160
    – synthesis   160
polymeric binder   433
polymeric stabilizers   101
polymerization   33, 89, 93–96, 111, 115, 124, 126, 127, 139, 161, 162, 325, 459
    – atom-transfer radical   127
    – controlled free radical   162
    – controlled radical   124
    – dispersion polymerizations   93
    – emulsion   94, 115
    – free radical   22, 89, 126, 139
    – free radical cationic   459
    – grafting   124
    – grafting   124
    – in presence of inorganic particles   111
    – in situ   161
    – ionic   22
    – living   96
    – miniemulsion   95
    – multifunctional initiators   26
    – nitroxide-mediated   126
    – oil-in-water   94, 95
    – precipitation   93
    – radical   325
    – reversible addition-fragmentation chain transfer (RAFT)   127

polymers   5, 19, 22, 32, 33, 49, 50, 196
  – conducting   196
  – organic   19, 50
  – solvent   33
polymetallosiloxane (PMS)   441
polymorphs   257, 259
polynuclear clusters   185
polyols   452, 459, 465
polyoxometalates   429
polypeptides   200, 262, 267, 285
polyphosphazenes   408, 410
polypropylene   164, 311
polypyrrole (PPy)   118, 166, 198, 404
polysaccharides   255, 266, 269, 275, 327, 329
polysiloxanes   18, 408, 428, 453
polysilsesquioxane gels   246
polysilsesquioxanes   225, 226, 228, 243, 252, 253
  – gelation   243
  – macromolecule-bridged   252
  – mechanical properties   253
  – synthesis   228
  – thermal stability   253
polysilsesquioxane–silica copolymers   247
polystyrene   110, 115, 119, 129, 141, 155, 159, 162, 164, 167
polystyrene latexes   107, 108, 119, 131, 133
polystyrene sulfonate (PSS)   108
polystyrene-b-poly(2-vinyl pyridine)-b-poly(ethylene oxide)   135
polystyrene-co-butylacrylate   122
polytetrafluorothylene   93
polythiophene (PT)   198, 414, 420
polyurea/polysiloxane coatings   454
polyureas   434
polyurethane   309, 311, 434, 465
polyurethane oligomer   461
polyvinyl alcohol   101
polyvinyl chlorides   434
polyvinylpyrrolidone   92
population inversion   380
porcelains   301
pore blocking   186, 188, 203, 212
pore loading   186
pore size   177, 181, 333
pore size distribution   191, 332
pore sizes   182, 332
  – tunable   182
pore walls   177, 180, 191
porogen   208, 306, 331
porosity   175, 283
porous ceramic matrix   421
porous glasses   192

porous hybrid materials   175, 192, 219
  – applications   219
  – classification   192
porous materials   21, 175, 177, 185, 187, 188, 196, 199, 201, 203, 308, 321, 331, 334, 338, 342, 412
  – biomedical applications   331
  – chemical modification   185
  – co-condensation reactions   188, 203
  – doping with polymeric species   196
  – grafting reactions   201
  – $in\ situ$ modification   188
  – incorporation of biomolecules   199
  – liquid-phase modification   187
  – post-synthesis modification   185
porous particles   176
post-synthetic ion exchange reactions   182
postsynthetic treatment   191
potassium bromide   371
precipitation   92, 245, 261, 273
processing   9
  – thin films   9
programmed assembly   282
programmed recognition   282
prohesion protocol   453
proliferated cells   332
proline   262, 267
propagating modes   388
  – conductivity   11
  – electric transverse   388
  – hydrophobicity   9
  – inorganic   8
  – long-term stability   21
  – magnetic transverse   388
  – mechanical   1, 9, 11, 39, 160, 167, 180, 196
  – optical   39, 180
  – optical transparency   9, 23
  – organic materials   8
  – refractive index   11, 23
  – rheological   160, 167
  – surface area   10
properties   1, 8–11, 21, 23, 39, 160, 167, 180, 196
propylamines   276
protective coating   248
protein adsorbtion   295–297
protein expression   283
protein folding   277, 283
proteins   199, 255, 262, 269, 275–277, 281, 283, 284, 290–293, 297, 304, 311, 330, 334
proteoglycans   268, 275, 276, 283
proton conductivity   401, 409, 424
proton exchange membrane   423

proton mobility 409
proton-conducting electrolytes 409
proton-conducting membrane 410
proton-exchange membrane (PEM) 401
pulsed excitation source 357
pulsed field gradient NMR 418
pulsed gradient spin-echo (PGSE) 418
purine 264
pyrazine 215
pyridimine 264
pyridine 421
pyrrole 117
PZC 118

q
quantum dots (QD) 251, 361, 435
quantum yield 340, 365, 368
o-quinone 415

r
radial-flow bioreactor (RFB) 332
radiance 368–370
radiant flux 370
radiant intensity 369, 370
radiation confinement 391
radiative processes 354
radiolaria 175
Raman effect 394
Raman spectroscopy 451
rare earth elements 374
rare earth ions 342
raspberry-like morphology 118
recombination mechanisms 372
recombination probability 373
redox films 402
redox melt 406
redox reactions 198
redox-active groups 411
redox-active ormocers 402
reducing agents 101
reflectance 370
refractive index 52, 54, 82, 343, 352, 370, 387, 391, 394, 435
t-resins 227
resin cements 302
reverse microemulsion 102
reversed-photochromism (rpm) 383
reversible photochromic systems 383
reversible radical addition 124
rgioselectivily 219
rheology 433
rhodium chloride 101
ripening 262
RNA 264
roentgenoluminescence 353
rubber 290

s
safflower 443
salts 255
saponite 153, 166
SAXS 448
Santa Barbara Amorphous 184
scaffolds 282, 308, 310, 319, 331, 332, 333
scratch resistance 9, 39, 352, 441, 443
sea urchins 275
second harmonic generation 394
second order non-linear hybrids 396
second order nonlinear optical properties 395
secondary building unit (SBU) 214
secondary structures 264
second-order nonlinear optical responses 338
selective hydrogenation 79
self-assembly 36, 103, 110, 179, 214, 282, 291
 – electrostatically driven 103
self-organization 36
SEM 468, 469
semiconducting oxides 404
semiconductors 362, 372, 435
semi-solid electrolytes 403
sensing 256, 292
sensitizers 404
sensors 9, 225, 247, 270, 338, 343, 413
serine 285
β sheet 262, 277
shells 256
ship-in-the-bottle-synthesis 182, 186, 192, 195
shrinkage 36, 459
silaffins 276
^{29}Si CP-MAS NMR 332
^{29}Si NMR 227, 231, 235, 247
simulated body fluid 324, 327, 333
^{29}Si-MAS NMR 209
silane coupling agents 20, 27, 112, 155, 202, 248, 302
silanes 61, 326
silanol 112, 325
silanol groups 112, 115, 124, 131, 201, 229, 233, 239, 305, 329, 339, 377, 383, 389, 439
silica 3, 18, 25, 27, 61, 65, 72, 76, 115, 118, 126, 166, 191, 199, 201, 203, 208, 248, 256, 258, 275, 279, 290, 291, 302, 322, 338, 383, 411, 414, 420, 424
 – silicon alkoxides 3

silica gel   222, 239, 243, 246, 316, 421
silica sol   427
silica surfaces   240
silica-based matrices   377
silicatein   276
silicatein α   276
silicates   411, 434
silicic acid   232, 279
silicification   279
    – bioinspired   279
silicon   359
silicon-29   231
silicon alkoxides   3, 452
silicon tetrachloride   229
silicone   329, 434
silicotungstic acid   425
silk fibrils   311
siloxane   226, 229, 237, 242, 337, 401, 408, 411, 462
siloxane rings   230, 237, 243
siloxane-based polymer   427
silsesquioxanes   18, 188, 225
silver   61, 63, 87
silver nitrate   101
silver tetraoxyl chlorate   101
silylation   200
simulated body fluid (SBF)   315, 324, 327
sintering   301
SiO_2 72
SiO_2 (quartz)   57
size distribution   183
size exclusion chromatography (SEC)   455
skeletal fixation   256
skeleton   257
skin   268
slate   287
small angle neutron scattering (SANS)   158
small angle X-ray scattering (SAXS)   158, 439
smart material   245
smectite clays   154
smectites   166
smoke emission   167
Smoluchowski approximation   105
sodium carbonate   101
sodium chloride granules   331
sodium citrate   101
sodium polyacrylate   101
sodium polyphosphate   101
sodium salicylate   371, 372
sodium styrene sulfonate   127
sodium tetrahydroborate   101
sodium tungsten bronze   403
soft tissue   314, 326
soft tissue replacement   321
soft tissue substitutes   320
soft/hard particles   131
sol   3, 14, 88, 305
solar cell   384
solar cell performance   387
solar energy conversion   338, 422
sol–gel   7, 9, 199, 208
    – trialkoxysilanes   9
sol–gel precursors   435, 447, 450, 462, 465
sol–gel process   3, 13–17, 19, 23, 33, 38, 66, 98, 130, 138, 211, 225, 229, 230, 231, 233, 234, 237, 305, 318, 323, 327, 337, 339, 377, 383, 401, 411, 424, 435, 437
    – acid catalyzed   14, 23, 230, 437
    – acidic   23
    – ageing   15
    – alcohol exchange   233
    – base catalyzed   14, 23, 230, 437
    – condensation   229, 230
    – gelation point   15
    – hydrolysis   229, 230
    – kinetics   14, 229, 237
    – nonhydrolytic   16
    – organic polymers   19
    – shrinkage   15, 23
    – solvent   15, 19
    – solvent effects   15, 19, 33
    – substituent effects   231
    – supercritical drying   16
    – tetraethoxysilane   66
    – tetraalkoxysilanes   16
    – transparency   23
    – trialkoxysilane   17
solid solutions   302, 303
solid-state dye-lasers   379
solubility product   273, 315
solubility product constant   259
solubility regulation   274
solution mixing   161
solvents   433
soybean   443, 447
soybean oil   450
spatial effects   281
spectral locus   366
spectral region   357
spheres   282
spherosilicates   25, 242
spicule formation   275
spicules   270
spin coating   36, 68, 352
spinel   301
spines   256
spin-orbit interaction   374

sponge 175, 270, 272, 291
sponge silica 276
spontaneous emission 380
stainless steel 312
standard phosphor 371
star polymers 242
starch 266
Stark effect 374, 379
stearic acid 267
stereochemical interactions 281
stereochemistry 277
Stern layer 105
stimulated emission 380
Stöber particles 27, 30, 142
storage-modulis 167
strawberry-like 121
strength 167
streptavidin 293, 294
stress-strain curve 151
structural engineering 35
structure-directing agents 180, 182, 193, 196, 199, 205
styrene 118, 126, 164, 165, 198, 252
Styrofoam® 246
sucrose 331
sucrose granules 331
sugar 331
sulfates 269, 371
sulfides 291
supercapacitors 429
superparamagnetism 76
supersaturation 259, 273, 275, 284, 315
superstructures 62
supramolecular arrangement 196
supramolecular assemblies 182
supramolecular chemistry 291
surface appearance 167
surface area 88
surface charge 28, 104
– isoelectric point 28
surface confined redox waves 402
surface coupling 112
surface energy 11, 155
surface modifications 328
surface functionalization 305
surfactant 20, 27, 33, 37, 94, 101, 114, 115, 139, 155, 165, 182, 199, 212, 251, 282, 291, 433, 462
surfactant catalyst 189
suspension 92, 162
Suzuki reaction 79
swelling properties 413
synthesis of hybrid materials 3, 5, 12, 17, 21, 37
– biological molecules 17
– biomolecules 17
– building block approach 12
– entrapment 5
– entrapped 5, 17
– in situ formation 13
– M41S-materials 21
– sol–gel 3
– sol–gel process 3, 17
– synthetic strategies 12
– template-directed synthesis 37
synthesis of nanocomposites 67, 71, 73–75
– concomitant formation of particles and polymers 74
– in-situ particle preparaction 67
– polymer melts 73
– polymerization 71
– simultaneous evaporation 75
synthesis of polymer nanocomposites 65, 66
– in situ 65
– polymer melt 66

t
talc 153
tall oil 450
teeth 256–258, 270
telechelic polymers 252
telecommunication windows 338
TEM 467
template 37, 38, 91, 109, 130, 134, 179, 182, 183, 192, 196, 201, 208, 251, 282, 291
– colloidal 109
– colloidal crystals 38
– nanoparticles 37
– removal 183
– supramolecular 182
– surfactants 37
template exchange 187
template removal 187, 189, 197, 205, 206
tendon 268
tennis ball 168
tensile modulus 151
tensile strength 151, 267
TEOS 139, 140, 188, 208, 331, 422, 442, 461
TEOS loading 458
TEOS oligomers 456, 466, 469
4,4′-terphenyldicarboxylate 216
tertiary structure 262, 264
tetraalkoxysilanes 18, 138, 188, 203, 207, 209, 230, 232, 245, 247, 304
tetraalkoxytitanes 304

tetraalkylammonium ions   193
tetraethoxyorthosilicate (TEOS)   340
tetraethoxysilane   131, 140, 188
tetraethyl orthosilicate (TEOS)   99, 310, 317, 323, 438, 439, 452
tetraisopropylorthotitanate   317, 318, 324
tetramethoxyorthosilane (TMOS)   352
tetramethoxysilane   108, 188, 239, 247
tetramethoxysilane, methyltrimethoxysilane   207
tetramethyl orthosilicate (TMOS)   419, 439
tetramethylammonium bromide   182
tetramethyoxymesoporphyrin   419
tetraoctylammonium bromide   101
tetraphenylphenylenediamine   363
tetrathiafulvalene (TTF)   407
thermal decomposition   73
thermal stability   167
thermochromic memory cell   384
thermogravimetric analysis (TGA)   45
thermoreversible gels   245
thermosetting   434
thermotropic liquid crystals   282
thin films   35, 40, 251, 350, 370, 433
third order non-linear hybrids   397
third order nonlinear properties   395
threonine   285
thrombi   318
thymine   264
Ti alloys   286
Ti6Al4V   312, 313
time-resolved emission spectra   359
TiO_2 81
TiO_2 (anatase)   57
TiO_2 (rutile)   57
tissue   255, 267, 268, 283, 285, 294, 302, 304, 305
tissue bonding   322
tissue engineering   331, 334
tissue engineering materials   307
tissue engineering scaffold   321
tissue ingrowth   322
tissue-bonding ability   306
tissue-compatibility   305
tissue-material bond formation   310
titania   106, 107, 130, 153, 291, 404, 420, 424, 464
titanium   313, 464
titanium (di-i-propoxide diacetyleacetonate)   438
titanium ethoxide   436
titanium isopropoxide   436, 450
titanium methacrylate triisopropoxide (TMT)   304, 330
titanium substrates   330
titanium tetraethoxysilane ($Ti(OEt)_4$)   352
titanium tetra-i-propoxide   438
TMMS   140
TMOS   188, 208
TMS   140
tobacco mosaic virus   292
tooth enamel   276
tooth pulp   286
tooth roots   312
tooth root implant   322
top-down   36
top-down approach   178
total ionic conductivity   415
toughness   9
toxicity   297
trabecular bone   283
transcribing process   291
transesterification catalyst   448
transference number   417
transition metal   185
transition temperature   167
transition-metal alkoxides   16
  – transition-metal alkoxides   16
translucence   50, 80
transmission electron microscopy (TEM)   156, 160
transmittance   56
transplants   307
triacylglycerols   266
trialkoxysilane   5, 28, 29, 138, 179, 207, 212, 232, 241, 205
trialkoxysilane groups   304, 414
trialkoxysilane-modified alkylamine   425
trialkylphosphine   362
triboluminescence   353
tricalcium phosphate   313, 322
trichlorosilane   29
triethoxysilane   140
triglycerides   449
trimethoxysilane groups   326
3-trimethoxysilyl propyl methacrylate   112
trimethoxysilane-modified ferrocene   411
trimethy methoxysilane   140
tri-n-octyl phosphine (TOP)   102
tri-n-octyl phosphine oxide (TOPO)   102
trioctylphosphine   101
tripepetide   266
triple helix   268
tripropylamine (TPA)   421
tristimulus   366
trypsin   200
tryptophan   263
tubules   286

tuneable lasers 376
tung oil 443
tungsten oxide 403
two-package (2K) coatings 452
Tyndall effect 80, 88
n-type dopants 404
tyrosine 285

u

ultraviolet patterning 392
UV curing 208
UV radiation 80
UV stabilizers 80
UV/vis spectra 77, 81
UV-protection 465

v

valence band 373
van der Waals 277, 282
vanadium oxide 429
vanadium pentoxide 403
vaterite 257
vegetable oils 443
vehicle mechanism 409
vermiculite 153, 166
vesicle membranes 284
vesicles 90, 95, 107, 275, 280
 – magnetic 107
vinyl acetate 139, 198, 434
4-vinyl pyridine 117, 118
vinyl triethoxysilane 436
vinyl trimethoxysilane 140
vinylalkoxysilane 324
vinylferrocene 411
vinylidene fluorides 434
vinylpyridine 20
vinyltrimethoxysilane (VTMS) 304, 325, 329
viruses 291
vitamin B12 200
vitamin E 200
vitamin E-TPGS 201
VMS 140
VOC 449, 452, 459
volatile organic compounds 449, 452, 459
volume-weighted average diameter 52
vulcanization 225

w

water 167
water-in-oil (w/o) systems 92
water-in-oil microemulsions 102
waveguide attenuation 389
waveguides 387
weddellite 257, 270
wood 175, 286, 287, 292
wood fibers 64

x

xerogel 16, 22, 247, 250, 305, 327, 383, 429
XPS data 458
X-ray crystallography 241
X-ray diffraction (XRD) 156, 160

y

yeast cells 292
Young's modulus 317
yttrium 132
yttrium oxide 320

z

zeolite 108, 153, 175, 177, 180, 187, 191, 196, 198, 201, 205, 213, 259, 415
zeolite NaY 205
zeolite Y 194, 198
zeolites with organic groups as lattice 209
zeozymes 195
zeta potential 105
zinc acetate 213, 447
zinc oxide 404
zinc oxide titania 81
zinc phosphate 447
zirconium 133
zirconium alkoxides 31
zirconium hydrophosphate 153
zirconium isopropoxide 439
zirconium n-propoxide 450
zirconium oxide 313
zirconium phosphate 424
zirconium tetra-n-propoxide 438
zirconyl chloride 425
ZnO 81
ZnO (zincite) 57
ZOL 209

Related Titles

Gómez-Romero, P., Sanchez, C. (Eds.)
Functional Hybrid Materials

2004
ISBN 3-527-30484-3

Schmid, G. (Ed.)
Nanoparticles
From Theory to Application

2004
ISBN 3-527-30507-6

Laeri, F., Schüth, F., Simon, U., Wark, M. (Eds.)
Host-Guest-Systems Based on Nanoporous Crystals

2003
ISBN 3-527-30501-7